EMERGING SEMICONDUCTOR TECHNOLOGY

A symposium
sponsored by
ASTM Committee F-1
on Electronics
San Jose, CA, 28–31 Jan. 1986

ASTM SPECIAL TECHNICAL PUBLICATION 960
Dinesh C. Gupta, Siliconix, Inc., and
Paul H. Langer, AT&T Bell Laboratories,
editors

ASTM Publication Code Number (PCN)
04-960000-46

 1916 Race Street, Philadelphia, PA 19103

Library of Congress Cataloging-in-Publication Data

Emerging semiconductor technology.

(ASTM special technical publication; 960)
Papers presented at the Fourth International Symposium on Semiconductor Processing.
"ASTM publication code number (PCN) 04-960000-46."
Includes bibliographies and index.
1. Semiconductor industry—Congresses. 2. Semiconductors—Design and construction—Congresses. I. Gupta, D.C. (Dinesh C.) II. Langer, Paul H. III. ASTM Committee F-1 on Electronics. IV. Symposium on Semiconductor Processing (4th: 1986: San Jose, Calif.) V. Series.
TK7871.85.E47 1987   621.3815'2   86-28761
ISBN 0-8031-0459-6

Copyright © by AMERICAN SOCIETY FOR TESTING AND MATERIALS 1987
Library of Congress Catalog Card Number: 86-28761

NOTE

The Society is not responsible, as a body,
for the statements and opinions
advanced in this publication.

Printed in Baltimore, MD
January 1987

Foreword

The Fourth International Symposium on Semiconductor Processing was held at San Jose, California on 28-31 January, 1986 under the chairmanship of Dinesh C. Gupta, Siliconix Incorporated. The Symposium was sponsored by ASTM Committee F-1 on Electronics and co-sponsored by National Bureau of Standards, Semiconductor Equipment and Materials Institute, Stanford University Center for Integrated Systems, and IEEE Components, Hybrids and Manufacturing Technology Society. The Technical Committee was headed by Paul H. Langer, AT&T Bell Laboratories and the Arrangements and Publicity Committee was headed by Carl A. Germano, Motorola Incorporated.

The Symposium was successful because of the efforts of many persons who participated in the Advisory Board, and various committees, namely, the Technical Committee, the Arrangements and Publicity Committee, Registration Committee and the Spouse Committee. These persons included: Winthrop A. Baylies, ADE Corporation, Kenneth E. Benson, AT&T Bell Laboratories, W. Murray Bullis, Siltec Corporation, Paul Davis, Semiconductor Equipment and Materials Institute, James R. Ehrstein, National Bureau of Standards, Bruce L. Gehman, Deposition Technology Incorporated, Kathleen Greene, ASTM, Gilbert A. Gruber, Siliconix Incorporated, Lou Ann Gruber, Vijay Gupta, John A. Imbalzano, E.I. DuPont DeNemours & Company, Sharon Kaufmann and Philip L. Lively, ASTM, George E. Moore, Sr., Semiconductor Equipment and Materials Institute, James D. Plummer, Stanford University Center for Integrated Systems, Robert I. Scace, National Bureau of Standards, William R. Schevey, Allied Chemical Corp., Donald G. Schimmel, AT&T Bell Laboratories, Robert B. Swaroop, Fairchild Semiconductor, Gail Wesling, and Paul Wesling, Tandem Computers Incorporated.

In addition, the guidance was provided by the Chairman and the officers of ASTM Committee F-1 on Electronics, its various subcommittees including the Executive subcommittee. The following persons presided the technical and workshop sessions: J. Albers and A. Baghdadi, National Bureau of Standards, K. E. Benson, AT&T Bell Laboratories, J.O. Borland, Applied Materials, Inc., R.H. Bruce, Xerox Palo Alto Research Center, M. Buehler, Jet Propulsion Laboratory, W. M. Bullis, Siltec Corporation, S. Cox, AT&T Technologies, M. Current, Applied Materials, Inc., J. R. Ehrstein, National Bureau of Standards, B. Fay, Micronix Corporation, T. Francis, Air Products & Chemicals, G. A. Gruber and D. C. Gupta,

Siliconix, Inc., T. I. Kamins, Hewlett-Packard Laboratories, G. Koch, Flexible Manufacturing Systems, P. H. Langer, AT&T Bell Laboratories, J. Matlock, SEH America, R. K. Pancholy, Gould AMI Semiconductor, M. Pawlik, GEC Research, D. Perloff, Prometrix, Inc., J. Plummer, Stanford University Center for Integrated Systems, D. Walters, Varian Associates, Inc., W. R. Schevey, Allied Chemical Corporation, L. Shive, Monsanto Company, E. R. Sirkin, Zoran Corporation, F. Voltmer, Intel Corp., and W. Weisenberger, Ion Implant Services.

We are indebted to Richard D. Skinner, President, Integrated Circuit Engineering Corporation who presented a dinner speech on "Semiconductor Industry - An Economic Review", Richard A. Blanchard, Vice President, Siliconix Incorporated, Pat Hill Hubbard, Vice President, American Electronics Association, Richard Reis, Assistant Director, Stanford University Center for Integrated Systems, Robert I. Scace, Deputy Director, Center for Electronics and Electrical Engineering, National Bureau of Standards, James E. Springgate, President, Monsanto Electronic Materials Company, and James A. Thomas, Vice President, ASTM for the keynote speeches on the various topics on the first day of the Symposium.

We are grateful to the members and guests of ASTM Committee F-1 and Standards Committees of SEMI who were called upon from time to time for special assignments during the two-year planning of the Symposium.

Over one hundred and fifty scientists participated all over the world in the review process for the papers published in this publication. Without their participation, this publication would not have been possible.

And finally, we acknowledge the hard work and efforts of the staff of publication, review, editorial and marketing departments of ASTM in bringing out this book.

A Note of Appreciation to Reviewers

The quality of the papers that appear in this publication reflects not only the obvious efforts of the authors but also the unheralded, though essential, work of the reviewers. On behalf of ASTM we acknowledge with appreciation their dedication to high professional standards and their sacrifice of time and effort.

ASTM Committee on Publications

Related ASTM Publications

Semiconductor Processing, STP 850 (1984), 04-850000-46

Silicon Processing, STP 804 (1983), 04-804000-46

Lifetime Factors in Silicon, STP 712 (1980), 04-712000-46

Laser-Induced Damage in Optical Materials: 1982, STP 847 (1984), 04-847000-46

Laser-Induced Damage in Optical Materials: 1981, STP 799 (1983), 04-799000-46

Laser-Induced Damage in Optical Materials: 1983, STP 911 (1985), 04-911000-46

Preface

The papers in this volume were presented at the Fourth International Symposium on Semiconductor Processing held in San Jose, California on 28-31 January, 1986. The Symposium was sponsored by ASTM Committee F-1 on Electronics, and co-sponsored by National Bureau of Standards, Semiconductor Equipment and Materials Institute, Stanford University Center for Integrated Systems, and IEEE Components, Hybrids and Manufacturing Technology Society. In addition to the technical presentations, the symposium included two well-attended workshops, impressions of which are provided in appendix I.

The symposium addressed new problems in semiconductor technology and day-to-day problems in semiconductor processing for the mid 80's which arise from the rapid increases in device complexity and performance, emergence of integrated systems on-a-chip, automated factories, and silicon foundries. In the face of these demands, the realization of acceptable yields and reliability requires greater manufacturing and process-control disciplines from starting materials to finished devices. The symposium theme was, again this year, chosen to be Quality Through Measurement and Control.

The symposium opened with the talks on Standards and Product Quality by James A. Thomas, ASTM, and Standards for the Semiconductor Industry from ASTM and SEMI by Robert I. Scace, National Bureau of Standards. These presentations were followed by two papers giving the overview of silicon technology and relating it to device requirements. The requirements of silicon materials were described by James E. Springgate, Monsanto Electronic Materials Company. The process and equipment considerations were discussed by Richard A. Blanchard, Siliconix Incorporated.

The opening general session included a presentation and a discussion on Graduate Education for the Electronics Industry. Pat Hill Hubbard, Vice President, American Electronics Association discussed various programs that the Foundation is involved in in order to make doctoral study and academic careers more attractive. She said, "The need to have an adequate supply of quality engineers and sufficient faculty to educate them is considered of paramount importance to the health of the high tech industry and to the economic health of the nation." Richard Reis, Assistant Director, Stanford University Center for Integrated Systems presented a graduate education mix from Stanford's point of view, noting the exceptions which make Stanford different from other schools in the nation.

The response to the symposium was extremely favorable once again. The involvement of industry, academia and government including the participation of foreign institutions confirmed a continued need for a regular forum to discuss technology topics in the context of measurement and control, a consistent theme which the Symposium established in 1982 involving the understanding and day-to-day control of the complex process technologies required for VLSI and other advanced device concepts.

The plans for the next symposium in 1988 in the series of symposia to be held at two-year intervals are underway. The problem areas and standardization needs identified in these symposia will provide the feedback to the research community and voluntary standards system essential for the future growth of the industry.

The cooperation and support of the ASTM staff in the formulation of this publication is appreciated. We are indebted to our industrial, government and university colleagues who contributed to the Symposium and the Proceedings.

Dinesh C. Gupta
San Jose, California.

Paul H. Langer
Allentown, Pennsylvania.

Contents

Introduction   1

KEYNOTE ADDRESS

Silicon and Semiconductors: Partners in the Late 1980's—
J. E. SPRINGGATE   7

STANDARDS FOR SEMICONDUCTOR INDUSTRY

ASTM and SEMI Standards for Semiconductor Industry—
R. I. SCACE   15

EPITAXIAL TECHNOLOGY

Low Temperature and Low Pressure Silicon Epitaxy by Plasma-Enhanced CVD—R. REIF   21

Thin Silicon Epitaxial Films Deposited at Low Temperatures—
H.-R. CHANG AND J. S. ROSCZAK   24

Thin Epitaxial Silicon by CVD—
S. M. FISHER, M. L. HAMMOND, AND N. P. SANDLER   33

Effects of Gettering of EPI Quality for CMOS Technology—
C.-C. D. WONG, J. O. BORLAND, AND S. HAHN   51

Silicon Epitaxial Growth on N+ Substrate for CMOS Products—
R. B. SWAROOP   65

Characterization of the In Situ HCL Etch for Epitaxial Silicon—
J. W. MEDERNACH AND V. A. WELLS   79

DIELECTRICS AND JUNCTION FORMATION TECHNIQUES

Doped Oxide Spin-On Source Diffusion—V. RAMAMURTHY   95

Effect of a Shallow Xenon Implantation on a Profile Measured by
Spreading Resistance—E. LORA-TAMAYO,
J. DU PORT DE PONTCHARRA, AND M. BRUEL   108

Measurements of Cross-Contamination Levels Produced By Ion
Implanters—L. A. LARSON AND B. J. KIRBY   119

Some Aspects of Productivity of a Low Pressure CVD Reactor—
S. MIDDLEMAN   129

Deposition and Properties of Ultra-Thin High Dielectric Constant
Insulators—S. ROBERTS, J. G. RYAN, AND D. W. MARTIN   137

The Electrical Properties of MOS Transistors Fabricated with Direct
Ion Beam Nitridation—H.-S. LEE   150

PLASMA TECHNOLOGY AND OTHER FABRICATION TECHNIQUES

RIE Damage and Its Control in Silicon Processing—S. J. FONASH
AND A. ROHATGI   163

The Bonding Structure and Compositional Analysis of Plasma
Enhanced and Low Pressure Chemical Vapor Deposited
Silicon Dielectric Films—S. V. NGUYEN, J. R. ABERNATHEY,
S. A. FRIDMANN, AND M. L. GIBSON   173

Monte Carlo Simulation of Plasma Etch Emission Endpoint—
E. J. BAWOLEK   190

Profile Control of Plasma-Etched Polysilicon Using Implant
Doping—T. ABRAHAM AND R. THERIAULT   204

The Effects of Plasma Processing of Dielectric Layers on Gallium
Arsenide Integrated Circuits—K. C. VANNER, J. R. COCKRILL,
AND J. A. TURNER   220

Quality Control and Optimization During Plasma Deposition of
a-Si:H—M. KUNST, A. WERNER, G. BECK, U. KUPPERS,
AND H. TRIBUTSCH   241

Effects of Deep UV Radiation on Photoresist in Al Etch—S. C. LEE
AND B. CHIN   250

Influence of X-Ray Exposure Conditions on Pattern Quality—
V. STAROV   257

Palladium Silicide Contact Process Development for VLSI—
R. N. SINGH   266

MATERIAL DEFECTS, OXYGEN AND CARBON IN SILICON

Characterization of Silicon Surface Defects by the Laser Scanning Technique—H. M. LIAW, J. W. ROSE, AND H. T. NGUYEN   281

Damage Aspects of Ingot-to-Wafer Processing—L. D. DYER   297

Hydrogen in Silicon and Generation of Haze on Silicon Surface in Aging—T. SHIRAIWA AND S. INENAGA   313

Identifying Gettered Impurities in Silicon by LIMA Analysis—
M. C. ARST   324

Effect of Bulk Defects in Silicon on SiO_2 Film Breakdown—
H. SUGA AND K. MURAI   336

Free Carrier Absorption and Interstitial Oxygen Measurements—
W. K. GLADDEN AND A. BAGHDADI   353

High Reliability Infrared Measurements of Oxygen and Carbon in Silicon—N. INOUE, T. ARAI, T. NOZAKI, K. ENDO, AND K. MIZUMA   365

YIELD ENHANCEMENT AND CONTAMINATION CONTROL ASPECTS

Nature of Process Induced Si-SiO_2 Defects and Their Interaction with Illumination—S. KAR AND M. TEWARI   381

A Strategy for Reducing Variability in a Production Semiconductor Fabrication Area Using the Generation of System Moments Method—E. C. MAASS   393

Computerized Yield Modeling—C. H. BECK   404

Particle and Material Control Automation System for VLSI Manufacturing—M. D. BRAIN   414

Semiconductor Yield Enhancement through Particulate Control—
N. D. CASPER AND B. W. SOREN   423

Particulate Control in VLSI Gases—J. M. DAVIDSON AND T. P. RUANE   436

DOPANT PROFILING TECHNIQUES AND IN-PROCESS MEASUREMENTS

Spreading Resistance Measurements—An Overview—J. R. EHRSTEIN   453

Some Aspects of Spreading Resistance Profile Analysis—J. ALBERS   480

Spreading Resistance: A Comparison of Sampling Volume Correction Factors in High Resolution Quantitative Spreading Resistance—M. PAWLIK   502

Comparison of Impurity Profiles Generated by Spreading Resistance Probe and Secondary Ion Mass Spectrometry—G. G. SWEENEY AND T. R. ALVAREZ   521

Monte Carlo Calculation of Primary Kinematic Knock-On in SIMS—
J. ALBERS   535

A Comparative Study of Carrier Concentration Profiling Techniques in Silicon: Spreading Resistance and Electrochemical CV—
M. PAWLIK, R. D. GROVES, R. A. KUBIAK, W. Y. LEONG, AND E. H. C. PARKER   558

Analysis of Boron Profiles as Determined by Secondary Ion Mass Spectrometry, Spreading Resistance, and Process Modeling—
G. W. BANKE, JR., K. VARAHRAMYAN, AND G. J. SLUSSER   573

Mapping Silicon Wafers by Spreading Resistance—R. G. MAZUR   586

Production Monitoring of 200mm Wafer Processing—
W. A. KEENAN, W. H. JOHNSON, AND A. K. SMITH   598

Applications of X-Ray Fluorescence Analysis to the Thin Layer on Silicon Wafers—T. SHIRAIWA, T. OCHIAI, M. SANO, Y. TADA, AND T. ARAI   615

Qualification of GaAs and AlGaAs by Optical and Surface Analysis Techniques—J. F. BLACK, J. M. BERAK, AND G. G. PETERSON   628

FAB EQUIPMENT: AUTOMATION AND RELIABILITY

Wafer FAB Automation, An Integral Part of the CAM Environment—C. A. FIORLETTA, R. LENNARD, AND J. G. HARPER   653

Computer Integrated Manufacturing: The Realities and Hidden Costs of Automation—M. S. LIGETI   662

Industry Considerations in Determining Equipment Reliability—J. C. GREINER   673

APPENDIXES

Appendix I—Workshop and Panel Discussions   683

Appendix II—Graduate Education for the Electronics Industry   691

Indexes   695

Introduction

This volume is organized into seven sections: (1) Standards for Semiconductor Industry; (2) Epitaxial Technology; (3) Dielectrics and Junction Formation Techniques; (4) Plasma Technology and Other Fabrication Techniques; (5) Material Defects, Oxygen and Carbon in Silicon; (6) Yield Enhancement and Contamination Control Aspects; (7) Dopant Profiling Techniques and In-Process Measurements; and (8) Fab Equipment: Automation and Reliability. The papers describe the emerging semiconductor processes used in device fabrication, from its simplest use to discrete circuits through the complex applications to very-large-scale-integrated (VLSI) circuits.

After the introduction, unedited excerpts of one keynote paper, "Silicon and Semiconductors: Partners in the Late 1980's" are given. The synopses on workshops and graduate education are presented in two appendixes, I & II at the end of the volume.

STANDARDS FOR SEMICONDUCTOR INDUSTRY

Quality of measurement is the cornerstone of ASTM's system for the development of voluntary consensus standards. From the inception of the test method, through balloting and interlaboratory testing, the volunteers assure a high level of precision. The standard test methods, nomenclature, and specifications developed by ASTM and SEMI for the semiconductor industry support many acceptance tests and online measurements. The history of this collaborative work and its expected future course is discussed in this paper. The standards activities of foreign organizations and the interactions among these groups are also discussed.

EPITAXIAL TECHNOLOGY

The epitaxial layer is a backbone of the device structure. Major emphasis in epitaxial technology is to lower the defect level and improve the dopant distribution within the layer. The papers in this section discuss various techniques to improve epilayer quality. R. Reif presents a low pressure CVD system to deposit epitaxial films both with and without plasma enhancement at temperatures as low as 650°C. Chang and Rosczak adopt a more conventional atmospheric CVD system at 825°C to deposit epitaxial films. Swaroop and Fisher, et al present methods to improve epitaxial quality with respect to as-grown defects and electrical parameters. Medernach and Wells study the vapor etch, and Wong et al present methods to improve the epilayer quality using intrinsic gettering techniques.

DIELECTRICS AND JUNCTION FORMATION TECHNIQUES

Deposition and properties of ultra-thin dielectric insulators are presented by S. Roberts and others. Various aspects of both, the conventional and implantation techniques for junction formation are discussed. These include doped oxide spin-on source diffusion, and measurement of cross-contamination levels produced during implantation. The ion beam nitridation and a CVD reactor (productivity model) are also given in this section.

PLASMA TECHNOLOGY AND OTHER FABRICATION TECHNIQUES

A wide variety of plasma technology issues are presented in six papers. The topics of these papers include: RIE damage, bonding structure and chemical analysis of PECVD and LPCVD dielectric films, plasma etch emission endpoint, profile control, quality control and optimization during plasma deposition. Also presented in this section are the effects of UV radiation on photoresist in Al etch and palladium silicide contact process.

MATERIAL DEFECTS, OXYGEN AND CARBON IN SILICON

Material defects may be classified in various catagories: bulk defects, surface defects, process-induced defects, deep levels, gettered impurities etc. Most of these catagories are discussed in the papers in this section. Liaw et al and Rose suggest the use of wafer scanners to screen the incoming wafers for defects. Dyer discusses many defects introduced in ingot-to-wafer processing which may lead to device degradation. Shiraiwa and Inenaga explain the haze on wafers. The haze may be due to silicon oxide nodules which grow on the silicon surface in the density of about 1000 to 10 000 per square centimeter. Arst describes a laser-induced mass analysis technique to identify impurities captured in defect structures. Suga and Murai discuss the effects of bulk defects on the intrinsic and extrinsic gettering in silicon.

YIELD ENHANCEMENT AND CONTAMINATION CONTROL ASPECTS

Device yields are dependent upon a number of factors. Processing defects, variability in fabrication, and contamination during device processing are just a few factors which can impact the device yields. These aspects are discussed in this section. Kar and Tewari attempt to identify the nature of defects induced by e-beam evaporation at the silicon-oxide interface. Maass describes the application of the Generation of Moments method and relates device parameters to processing variables. He shows that tightening the distributions of key device parameters results in an enhancement and prediction of yields. Beck explains that an overall circuit yield is the product of two independent factors, namely, the device physics limitations yield factor and the process defect loss yield factor. The papers on contamination control emphasize the following points: the need to provide a clean environment for fabrication including ultra clean SMIF boxes, dedicated robotic mechanisms and clean air equipment enclosures, particulate control on the wafers and in gases and chemicals.

DOPANT PROFILING TECHNIQUES AND IN-PROCESS MEASUREMENTS

A number of papers were presented on the dopant profiling techniques. These techniques included SIMS, Rutherford backscatter, spreading resistance, and capacitance-voltage. These papers are listed in this section. A workshop was also held on this topic. The synopsis of workshop is given in appendix I.

FAB EQUIPMENT: AUTOMATION AND RELIABILITY

Computer-aided Manufacturing [CAM] and Computerized Integrated Manufacturing [CIM] are discussed in detail in this section. The development of a flexible wafer fab automation system is described. The latter performs three basic functions, real-time inventory control, material distribution throughout the fab, and automated loading of cassette-to-cassette process equipments.

In as much as the automation and mechanization are essential to the future of our industry, so is the understanding of both, the raw capability of each component of a system and the capability of each component with human factors integrated. The paper by Greiner isolates and defines components of real factory time and formulates them in two distinct ways: with and without human interfacing. This paper is the result of a SEMI document, presently in preparation by the SEMI Standards Committee.

Dinesh C. Gupta
Symposium Chairman and Co-Editor

Keynote Address

SILICON AND SEMICONDUCTORS: PARTNERS IN THE LATE 1980'S

James E. Springgate

[The material presented below did not go through the review process and is presented for information only. It was prepared by Dinesh C. Gupta, Co-Editor from the recordings of the presentation and the excerpts provided by J.E.Springgate.]

Forecasts in Semiconductor business these days are about as reliable as a weather forecast. The semiconductor industry has historically been playing "Crack the Whip" with the electronic end equipment buyers handling the whip.

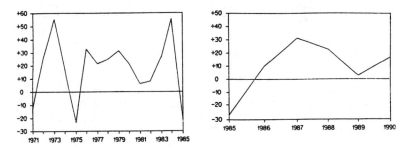

Fig.1 - U.S.Semiconductor Sales
[Percent annual-dollar change]

The challenge facing us is to achieve this growth profitably in the face of an escalating technology race and to successfully compete in a global market that is becoming dominated more by government and economic policy than by individual company initiative. According to device manufacturers, late 1980's will lead them to specific ICs, shrinking design rules, the inexorable move to larger diameter wafers, and worldwide competition. These needs have driven our planning the past few years and we, the silicon manufacturers are ready to meet them with application - specific wafer zone engineering, a technique used to tailor wafer

James E. Springgate is the President, Monsanto Electronic Materials Company, 755 Page Mill Road, Palo Alto, California.

characteristics to circuit requirements. Advanced development of larger diameter wafers, such as 200 and 250 mm is being actively pursued. Reduced feature sizes cannot be produced without improved wafer flatness. Over the last 15 years, wafer flatness has tremendously improved. The evolution is shown in Fig. 2.

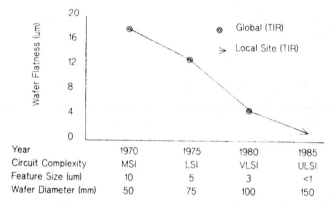

Fig. 2 - Wafer Flatness Evolution

A similar relationship exists across all wafer parameters, especially towards improved tolerances. The concept behind wafer zone engineering is to engineer the wafer zones, depending upon the requirements of a particular circuit, such as ULSI, VLSI, LSI or MSI circuit. Figure 3 depicts three different wafers. The top area is the circuit zone. Flatness in this zone is critical to the photographic depth of focus required for feature size resolution. Surface contamination, such as haze and defect artifacts must also be controlled in this zone. The defect-free denuded zone, which is just below the circuit active zone, is developed during thermal processing that accompanies the formation of the internal gettering zone. The latter zone getters process-induced metal contaminants from the circuit active region during IC fabrication. It apparently has not played a major role during LSI or MSI circuit fabrication, where back-surface mechanical damage gettering has been more prevalent. But internal gettering is a key to VLSI device fabrication, where low levels of surface contamination are essential.

As we move into finer line geometries and more dense circuitry in VLSI devices, channel length is reduced, junction depths are shallower and gate oxides are thinner. For this type of device, the backside cleanliness of the wafer becomes more critical and must be reduced in particle count. Monsanto has patented an

**LSI/MSI APPLICATION
SERIES WAFER
SCHEMATIC**

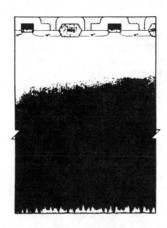

**VLSI APPLICATION
SERIES WAFER
SCHEMATIC**

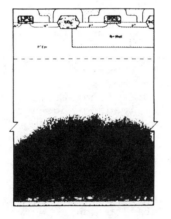

FIGURE 3 — Three typical customized wafers.

**ULSI APPLICATION
SERIES WAFER
SCHEMATIC**

enhanced gettering process that incorporates a thin polysilicon layer on the backsurface. This creates numerous external gettering sites which getter process-induced contaminants away from the circuit active region. Enhanced-gettered wafers tend to exhibit longer lasting gettering than mechanical backside gettering through several IC fabrication steps.

The schematic also includes an epitaxial layer, which is typically used in CMOS applications to reduce latch-up conditions in circuits. The substrate sinks the transient currents responsible for initiating latch-up. For epi-wafers, an oxide seal may be incorporated on the wafer backsurface to minimize auto doping.

The wafer used for ULSI circuit fabrication shows the continued evolution to reduced channel lengths, shallower junctions and thinner gate oxides. Trench structures further prevent latch-up conditions in circuits.

The multi-zone wafer products have been designed with specific mechanical, chemical, and structural characteristics necessary to develop the electrical characteristics that support the IC density and electronic performance goals of each level of IC integration. Mechanical characteristics are more important than ever for leading edge ICs.

Oxygen is a major impurity and carbon a subsidiary impurity influencing silicon's mechanical strength as well as the nucleation and growth of bulk oxygen precipitates for internal gettering. At the same time it is essential the bulk oxygen precipitate is controlled to minimize in-process warpage and maintain the desired wafer flatness. Crystallographic perfection near the circuit active region is important to reduce excess leakage current.

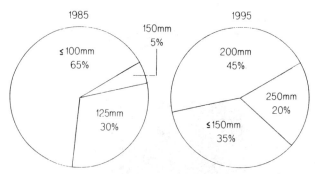

Figure 4 - The Silicon Wafer Market - Diameter Mix

The economic factors that created the need for the 150 mm diameter wafer will drive the industry to 200 mm diameter wafer by the end of this decade. However, the speed of this transition will depend to a large extent on the economic health of the industry. Figure 4 gives the silicon wafer market in terms of diameter mix - present and future.

CONCLUSION

We have seen dramatic changes in the silicon and semiconductor industries the past few years, and we anticipate significant changes in the next five years. 200 mm and 250 mm diameter wafers, wafers tailored to specific applications, wafers designed specifically for the emerging ULSI applications and advanced epitaxial wafers are among the more significant trends we foresee.

Standards for Semiconductor Industry

Robert I. Scace

ASTM AND SEMI STANDARDS FOR THE SEMICONDUCTOR INDUSTRY

REFERENCE: Scace, R. I., "ASTM and SEMI Standards for the Semicon-ductor Industry," <u>Emerging Semiconductor Technology, ASTM STP 960</u>, D. C. Gupta and P. H. Langer, Eds., American Society for Testing and materials, 1986.

ABSTRACT:
This article, based on an introductory talk at the symposium, points out the needs for standards in the semiconductor industry and briefly describes the activities of the organizations that develop them. The direct cooperation between the several organizations in the world active in semiconductor standards development is briefly described. Readers are invited to join these development activities.

KEY WORDS:
standards, standards applications, participation in standards development, standards development organizations, international cooperation in standards

Why standards in the semiconductor industry? After all, ASTM F-1, SEMI, and NBS are three of the sponsors of the symposium that resulted in this book. All three organizations care about standards. Standards help you, believe it or not. Did you ever have this kind of dialogue with a supplier or a customer?

> Buyer: "You can't measure what I want."
> Seller: "You can't either."

Standard test methods help prevent this sort of heartburn. When both parties to a transaction measure in the same way these problems don't occur so often. ASTM F-1 has developed well over 150 standard test methods for the semiconductor industry. They are in volumes 10.04 and 10.05 of the annual ASTM Book of Standards.

Robert I. Scace is Deputy Director, Center for Electronics and Electrical Engineering at the National Bureau of Standards, Building 220 Room B358, Gaithersburg, MD 20899.

Or have you been in this situation?

> Buyer: "I want it pure and I want it now."
> Seller: "Bug off."

Standard specifications can define what "pure" means in your situation so buyer and seller understand each other. You still have to fight over the definition of "now". SEMI and ASTM both have product specifications developed specifically for the semiconductor industry. ASTM's are mostly for metals such as for lead frames and bonding wire. SEMI's are for silicon, chemicals, gases, and a host of others. They are published in the annual Book of SEMI Standards.

ASTM F-1 has been developing semiconductor test methods for nearly 30 years starting with material properties of germanium and silicon. As the industry has grown, other topics have been addressed, and there are now quite a few subcommittees at work.

> F-1 SUBCOMMITTEES
>
> Semiconductor Physical Properties
> Electrical and Optical Measurements
> Microelectronics Imaging
> Processing Environments
> Gallium Arsenide
> Enclosures, Substrates, and Films
> Metallic Materials
> Encapsulants
> Interconnection Bonding
> Quality and Hardness Assurance

When SEMI began issuing standard silicon specifications about 10 years ago, ASTM F-1 took the opportunity to work with SEMI to develop the test methods needed to support those specifications. Every property defined in a specification needs a test method to measure it. SEMI has also grown to cover more topics:

> SEMI STANDARDS DIVISIONS
>
> Materials
> Chemicals
> Micro-Patterning
> Packaging
> Equipment Automation
> Safety

Each of these major subject areas is served by several committees of SEMI volunteers who now number in the thousands.

SEMI is an international organization serving an international industry. ASTM's standards, though developed in the U.S., are used

around the world. The same measurement and specification problems
that need to be resolved with domestic customers or suppliers also
have to be dealt with in other countries. ASTM F-1 has been working
closely for 18 years with the German DIN committee NMP-221 on materials for semiconductor technology

> DIN SUBCOMMITTEES
>
> Elemental Semiconductors
> Compound Semiconductors
> Process Chemicals

to assure that DIN and ASTM test methods use the same technical
approach to the measurement problem. Today, you can use either a DIN
or an ASTM method to measure, say, silicon resistivity, and get the
same answer because the test methods are technically equivalent.

All three organizations, ASTM F-1, SEMI, and DIN NMP-221, are also
cooperating with the French industry association SITELESC and with the
Japan Electronics Industry Development Association (JEIDA) on topics
of common interest. With JEIDA these include measurements of oxygen
and carbon in silicon and equipment communications.

> There have always been measurement problems
> in the semiconductor business.

We -- all of these organizations -- have not solved all the semiconductor world's problems. We can't measure double ordering very
well at all.

Some properties can only be measured destructively, like epitaxial
film resistivity in many cases. Other properties can't be measured
with useful accuracy

> Witchcraft and alchemy will always be with us.

or if a good measurement is possible, it may not be available at the
right time. As processing becomes more and more automated, measured
data are required more promptly. Real-time data would be ideal, but
furnaces and vacuum systems are often difficult places in which to
make useful measurements.

> Any measurement that is possible
> is not necessarily meaningful.

GOOD measurements are important. Test procedures that define GOOD
measurements are candidates for standards.

> There aren't enough standard tests, specifications, or reference materials.

A GOOD technique has to be developed and proved to work. The test method has to be written down and agreed to. Consensus takes time and a lot of hard thinking.

> Standards are never ready on time.

Changing technology drives the standards process. We need standard tests and specifications and they need to be kept in step with changing demands. The inevitable result of this chain of events is this:

> There will never be enough standards and the ones you get will always be late.

That's the bad news. The good news is that we can do better than we have done in the past. Maybe the standards process can never get ahead of changing technology, but it can follow a lot closer.

SEMI and ASTM can use as many enthusiastic volunteers as they can find. We can offer you work to fill your off hours without pay dealing with tough problems that get solved with frustrating slowness. We also offer good friends and colleagues, working toward a common goal, who are often helpful sources of technical ideas that will help you in your work.

We offer the first look at emerging standards in time to take advantage of them. There is nothing illegal about being better informed than your competitor. Have you thought about what value your competitor sees in standards work?

We offer the chance to bring changes to standards. Don't forget that standards are living, breathing, evolving documents. While anyone can propose changes to standards, that can be done more effectively from inside a committee than from outside.

ASTM and SEMI are the world leaders in standards for this industry. The only requirement of you is that you be willing to vote on issues. If you can think, write, or argue that's fine, but you must vote. Come join us. We'll all be glad you did.

Epitaxial Technology

Rafael Reif

LOW TEMPERATURE AND LOW PRESSURE SILICON EPITAXY BY PLASMA-ENHANCED CVD

REFERENCE: Reif, R., "Low Temperature and Low Pressure Silicon Epitaxy by Plasma-Enhanced CVD", *Emerging Semiconductor Technology, ASTM STP 960*, D.C. Gupta and P.H. Langer, Eds., American Society for Testing and Materials, 1986.

ABSTRACT: This paper reviews the most recent results obtained using a very low pressure, plasma enhanced chemical vapor deposition technique for low temperature (650-800°C) silicon epitaxy. Initial results on autodoping studies and on p-n junctions and MOS transistors fabricated in these films are briefly discussed.

KEYWORDS: silicon epitaxy, autodoping, channel mobility, buried layers

A low temperature procedure for the deposition of silicon epitaxial films is expected to become important in future bipolar and CMOS technologies to minimize autodoping and solid state diffusion effects. Several approaches have been proposed in the literature [1-3], and some more are discussed in this volume [4]. This article reviews the most recent results obtained using a very low pressure (10^{-2} Torr), plasma enhanced chemical vapor deposition process to grow silicon epitaxial films at temperatures as low as 650°C. The system and deposition procedure used in this work are described elsewhere [5,6]. The autodoping results will be mentioned first, followed by the results obtained from the fabrication of p-n junctions and MOS transistors.

AUTODOPING

The reincorporation of dopants into a growing film from the substrate or other surfaces in the reactor is normally referred to as autodoping. Generally, most of these unwanted dopants originate from heavily doped regions on the substrates. Autodoping directly above the heavily doped regions is referred to as vertical autodoping, and autodoping off to the sides of these regions is referred to as lateral autodoping. In order to study autodoping effects in epitaxial films deposited at temperatures ≤800°C, intrinsic epi layers were deposited on wafers containing heavily doped buried layers of arsenic

Dr. Rafael Reif is an Associate Professor of Electrical Engineering at the Massachusetts Institute of Technology, Cambridge, MA 02139

(10^{20}/cm^3), phosphorus (10^{20}/cm^3), and boron (10^{19}/cm^3) [7]. In all cases, the vertical autodoping profiles measured by SIMS were extremely abrupt, with the measurement probably limited by the resolution of the instrument. In addition, as expected, lateral profiles did not exhibit any lateral autodoping. The absence of autodoping is attributed not only to the low deposition temperature, but also to the low-temperature predeposition cleaning used to remove the native silicon dioxide, and to the very low operating pressure [7].

P-N JUNCTIONS

It is important to examine the performance of p-n diodes fabricated in these low-temperature epitaxial films because they are the basic building blocks of any IC technology. Diodes fabricated in these films exhibit the following features [8]: sharp turn-on characteristic, no sign of soft breakdown out to 10 volts reverse bias, and an ideality factor (forward bias regime) of 1.10.

MOS TRANSISTORS

PMOS transistors have also been fabricated in these low temperature epitaxial films [9]. Drain current versus drain voltage plots for these transistors exhibit classical square-law behavior in the saturation region, with an on-current to off-current ratio of 10^7. The threshold voltage variations measured across a two-inch wafer were less than 0.1 volt. Plots of low-field drain current versus gate voltage for these transistors exhibit the expected linear behavior. The hole channel mobility calculated from the slope of this line is 213 cm^2/V-sec, versus 216 cm^2/V-sec in an identically processed bulk control substrate. NMOS transistors fabricated in these films also exhibited excellent channel mobilities, with electron channel mobilities of 520 cm^2/V-sec versus 560 cm^2/V-sec in a bulk control substrate.

CONCLUSIONS

Silicon epitaxial films deposited at temperatures ≤800°C by a very low pressure, plasma enhanced chemical vapor deposition technique have been studied in terms of autodoping and electrical properties. Neither vertical nor lateral autodoping were found on epitaxial layers deposited on substrates containing heavily doped buried layers of arsenic, phosphorus, and boron. High quality diodes with no sign of soft breakdown and an ideality factor of 1.10 can be fabricated in these films. PMOS and NMOS FET's having hole and electron channel mobilities, respectively, nearly identical to those obtained in bulk silicon controls can also be fabricated in these films.

ACKNOWLEDGEMENT

This work was sponsored by the Semiconductor Research Corporation (Contract No. 83-01-033) and the NSF Presidential Young Investigator Award (Grant No. 8352399-ECS).

REFERENCES

[1] D. Brasen, S. Nakahara, and J.C. Bean, "Study of MBE growth of Ge_xSi_{1-x} on {111} vicinal surfaces of Si substrates", Journal of Applied Physics, Vol. 58, No. 5, 1 September 1985, pp. 1860-1863.

[2] P.C. Zalm and L.J. Beckers, "Ion beam epitaxy of silicon on Ge and Si at temperatures of 400 K", Applied Physics Letters, Vol. 41, No. 2, 15 July 1982, pp. 167-169.

[3] N.T. Quach and R. Reif, "Solid phase epitaxy of polycrystalline silicon films: effects of ion implantation damage", Applied Physics Letters, Vol. 45, No. 8, 15 October 1984, pp. 910-912.

[4] H-R. Chang and J.S. Rosczak, this volume.

[5] T.J. Donahue, W.R. Burger and R. Reif, "Low-temperature silicon epitaxy using low pressure chemical vapor deposition with and without plasma enhancement", Applied Physics Letters, Vol. 44, No. 3, 1 February 1984, pp. 346-348.

[6] T.J. Donahue and R. Reif, "Silicon epitaxy at 650-800°C using low-pressure chemical vapor deposition both with and without plasma enhancement", Journal of Applied Physics, Vol. 57, No. 8, 15 April 1985, pp. 2757-2765.

[7] T.J. Donahue and R. Reif, to be published.

[8] W.R. Burger and R. Reif, "Electrical characterization of epitaxial silicon deposited at low temperatures by plasma-enhanced chemical vapor deposition", IEEE Electron Device Letters, Vol. EDL-6, No. 12, December 1985, pp. 652-654.

[9] W.R. Burger and R. Reif, to be published.

Hsueh-Rong Chang and Joseph S. Rosczak

THIN SILICON EPITAXIAL FILMS DEPOSITED AT LOW TEMPERATURES

REFERENCES: Chang, H-R and Rosczak, J.S., "Thin Silicon Epitaxial Films Deposited at Low Temperatures", Emerging Semiconductor Technology, ASTM STP 960, D.C. Gupta and P.H. Langer, Eds., American Society for Testing and Materials, 1986.

ABSTRACT: Specular epitaxial silicon layers have been successfully deposited on 3-inch wafers at 825°C using an atmospheric pressure chemical vapor deposition process. No plasma or high temperature etching is involved in this process. Predeposition cleaning of the substrate surface is the key to achieve epitaxial growth at this low temperature. Quantitative characterization of the low-temperature epitaxy quality has been performed by X-ray diffraction, UV reflectance, Hall mobility measurement, and diode breakdown measurement. All test results demonstrated that an epitaxy comparable to that of high temperature epitaxy has been achieved. This low temperature epitaxy process greatly reduces the out-diffusion and autodoping, thus leading to significant improvement in device dimension control.

KEYWORDS: silicon epitaxy, CVD, low temperature, surface cleaning

The epitaxial growth of silicon is widely used in the fabrication of semiconductor devices. The essential importance of epitaxy lies in the fact that the dopant type, its concentration and distribution can

Dr. Chang is a research engineer and Mr. Rosczak is a specialist at General Electric Company, Research and Development Center, 1 River Road, Schenectady, N.Y. 12345.

be varied independently of the impurity type and concentration in the substrate. High quality silicon epitaxial layers with abrupt interfaces are the key to high frequency, high efficiency , and high reliability device operation.

The trend towards increasing IC chip sizes with finer geometry and higher packing density requires silicon epitaxial layers with low defect levels, minimum autodoping and solid-state out-diffusion, and improved uniformity in resistivity as well as thickness.

To date, CVD epitaxy is the most commonly used technique for silicon epitaxial growth due to its enormous advantages over the other techniques in terms of high film quality and throughput. The conventional CVD epitaxy process is a thermally driven process. Epitaxial silicon is grown in a hydrogen atmosphere from silane or chlorosilanes in the range of 1000-1250°C. This high temperature process gives undesirable solid-state out-diffusion and autodoping of dopants from the substrates (1,2), which in turn, degrade device performance. If deposition temperature goes below 1000°C, a poor quailty epitaxial layer, or even a polycrystalline layer, would typically result.

More recently, effort has been placed on developing low temperature epitaxial processes to minimize auotdoping and solid-state diffusion effects for future bipolar and CMOS technologies (2,3). Several low temperature processes have been proposed in the literature such as molecular beam epitaxy (4) ion cluster beam deposition (5), ion beam epitaxy (6), and plasma-enhanced chemical vapor deposition (PECVD) (7-10). Only PECVD appears to be compatible with manufacturing requirements in terms of high throughput and easy operation and maintenance. However, the epitaxy quality using these processes is the major concern for widespread use in the semiconductor industry.

In the present study, specular epitaxial silicon films have been successfully deposited at 825°C on 3-inch silicon substrates using an atmospheric pressure CVD process. This is the lowest deposition temperature reported for silicon vapor phase epitaxy without involving plasma in the process. Surface cleanliness is the key to producing epitaxial layers with such a low temperature process. Researchers reporting low temperature epitaxial growth have all used an in situ clean prior to epitaxial growth. Typical in situ cleaning techniques used include heating (11), etching at high temperature in hydrogen chloride (12), or sputtering cleaning (10). The pre-epitaxial cleaning process removes native oxide and surface contamination and provides a good interface for epitaxial growth.

High temperature cleaning treatments reduce the advantages of lower temperature processing by increasing both solid-state out-diffusion and autodoping effects. In situ sputter cleaning requires additional expensive equipment. In this paper, the growth of device quality epitaxial films at temperatures of 875°C or lower can be achieved using only wet chemical cleaning prior to loading the wafers into the reactor.

EXPERIMENTAL

All the epitaxial layers examined in this study were grown in a vertical-type CVD reactor Model AMV 1200 made by Applied Materials, Incorporated. The cross section of the reactor chamber is shown in Fig. 1. The susceptor had a 30.48 cm diameter and was made of high purity graphite coated with a layer of silicon carbide. The gases were introduced from a 9.5 mm (O.D.) injection tube through the center of the susceptor into the reactor chamber. During operation, the susceptor rotated continuously and a constant hydrogen carrier gas flow of 50 liters/min. was used. Before loading the wafers for epitaxial growth, the susceptor was etched with hydrogen chloride at 1200°C, then coated with a fresh layer of undoped silicon having an impurity concentration less than 1×10^{14} cm^{-3}. Only one substrate was used per experiment. The temperature of silicon wafers was measured with an optical pyrometer corrected for emissivity and wall transmission losses.

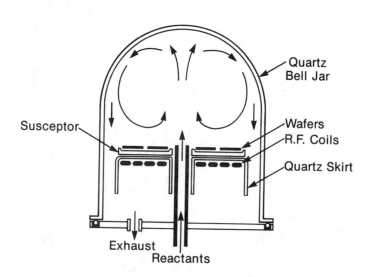

Fig. 1--Silicon Epitaxial Reactor

Depositions were made at temperatures ranging from 825°C to 875°C using silane as the silicon source. The thickness of the deposited films ranging from 0.75 µm to 2.5 µm were measured by the spreading resistance technique using an instrument made by Solid State Measurements, Incorporated. Both p⁻ and n⁻ type, 3-inch diameter silicon substrates were used with <100> orientation.

Quantitative characterization of the low-temperature epitaxial quality has been performed by X-ray diffraction, UV reflectance, Hall mobility measurements, and p-n junction diode breakdown measurements.

RESULTS AND DISCUSSION

The effects of predeposition surface cleaning procedures on the thin film structure and morphology is summarized in Table 1. For films deposited at 825°C on wafers which went through a dilute HF acid clean followed by a brief rinse in deionized water prior to loading into the reactor, a specular epitaxial layer was obtained even when the use of high temperature (1150°C) cleaning steps were omitted. Fig. 2 shows an X-ray diffraction photograph of the low-temperature epitaxial film where the spot pattern of a single crystal structure can be seen. If the HF acid clean was omitted, poly-crystalline films were obtained on wafers without a high temperature (1150°C) predeposition cleaning treatment. This is due to the native oxide on the substrate surface which was sufficiently thick (20Å) to mask the crystal structure of the underlying substrate. Fig. 3 shows the Nomarski optical micrographs of films grown at 875°C with and without an in situ clean (high temperature heating). Without removing the native oxide, polycrystalline silicon films were formed at such low deposition temperature. This demonstrates the significance of surface preparation in low temperature epitaxial processes.

TABLE 1--Effect of Predeposition Cleaning Treatment On Thin Film* Structure and Morphology

Predeposition Cleaning	Layer Crystallinity Type
1. No Wet Chemical Clean Preheat at 925°C for 10 Min.	Polycrystalline
2. Wet Chemical Clean Preheat at 925°C for 10 Min.	Epitaxy (Specular)
3. No Wet Chemical Clean Preheat at 1150°C for 10 Min.	Epitaxy (Specular)

*Deposition Temperature 825°C, Deposition Time 120 min. Growth Rate 62.5A/Min.

28 EMERGING SEMICONDUCTOR TECHNOLOGY

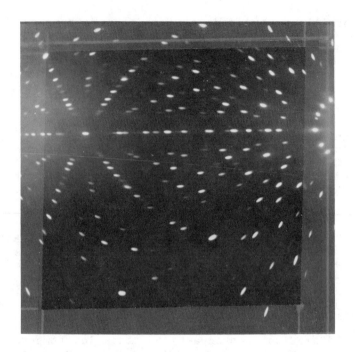

Fig. 2--X-ray diffraction photograph of a 0.75 μm silicon epitaxial layer deposited at 825°C.

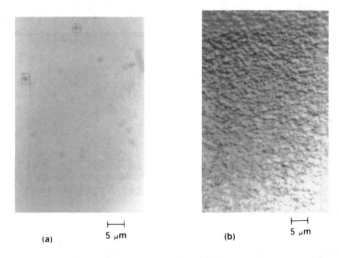

Fig. 3--Nomarski optical micrographs of films grown at 875°C (a) with prebaking cycle (1150°C) film - film is smooth and monocrystalline, defects are marked with squares (b) without prebaking cycle - film is rough and polycrystalline.

In the present study, it was found that the low-temperature process could produce epitaxial films with varying degrees of cloudiness by visual observation. Similar results were reported by Donahue et al. (10) using a plasma enhanced CVD process. The appearance of haze is a visual measure of the crystalline imperfection and surface roughness of the silicon film and could translate into low device fabrication yields. The UV reflectance method is a rapid, non-destructive technique for haze determination (13). In this work, the reflectance at 280 and 400 nm were measured relative to bulk silicon and the difference ΔR_{280} - ΔR_{400} is the haze rating. As shown in Fig. 4, epitaxial films grown at 875°C have identical reflectance as bulk silicon wafers. This indicates that low temperature epitaxial films do show good surface quality. For epitaxial films deposited at temperatures below 875°C, surface roughness was noticeable.

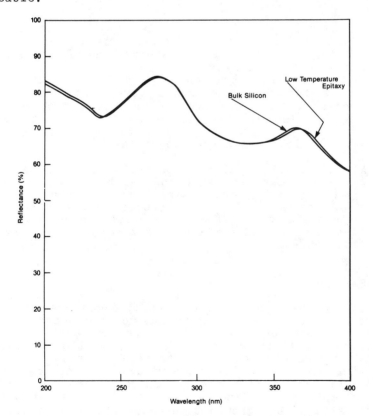

Fig. 4--The UV reflectance of bulk and low-temperature epitaxial silicon surfaces in the wavelength range of 200-400 nm. Aluminum mirror was used as the reference.

The electron mobility of the epitaxial films grown at 875°C was measured by the Van der Pauw method. For a film with a phosphorus concentration of 9×10^{16} cm^{-3}, the electron mobility was determined to be 795 cm^2/V-sec, which is similar to the value of 825 cm^2/V-sec in the bulk silicon. This indicates that the growth of good quality epitaxial films is achievable at such a low temperature. No data on mobility measurement of epitaxial films grown at temperatures below 875°C are available at present.

P$^+$-n junction diodes were fabricated on the epitaxial layers grown at 875°C with boron implantation. The junction profile is shown in Fig. 5. To a good approximation, the breakdown voltages of p-n diodes with a cylindrical junction are given by (14)

$$BV \simeq 3.47 \times 10^{13} \, N_D^{-3/4} \, V$$

where N_D is the carrier concentration in the epitaxial layer. Fig. 6 shows the distribution of breakdown voltages of the diodes fabricated on an epitaxial layer with a carrier concentration of 4×10^{16} cm^{-3}. Only two out of 146 diodes measured gave soft breakdown. The majority of the diodes have breakdown voltages of 11.75 volts which is very close to the theoretically calculated 12.3 volts. These results imply low defect levels in the low-temperature epitaxial layers.

All the test results demonstrate that an epitaxy quality comparable to that of a high temperature epitaxy has been achieved using temperature as low as 875°C. This low temperature process greatly reduces the solid-state out-diffusion and autodoping effects, thus leading to significant improvement in device dimension control.

SUMMARY

It has been shown that silicon epitaxial growth can be achieved at 825°C in an atmospheric pressure CVD system without using in situ high temperature pre-deposition cleaning steps. The predeposition cleaning of the substrate surface is extremely important for the epitaxial growth at low temperatures. Wet chemical cleaning prior to film deposition could produce device quality epitaxial films at 875°C. Crystal quality deteriorates as the growth temperature is decreased.

ACKNOWLEDGEMENT

The authors acknowledge with appreciation the help of Marx, Barbara in the preparation of this manuscript.

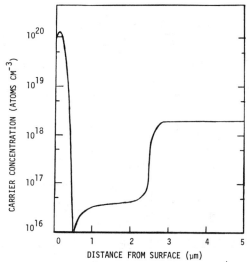

Fig. 5--Dopant distribution in p^+-n junction diode.

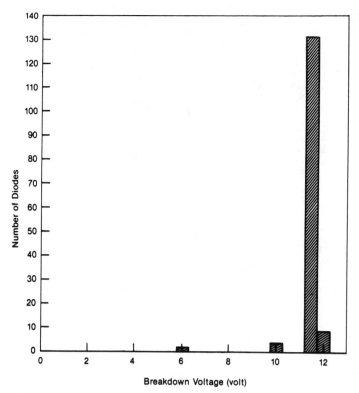

Fig. 6--Histogram of breakdown voltage of p^+-n diodes fabricated in an epitaxial layer grown at 875°C.

REFERENCES

(1) Srinivasan, G.R., *Journal of Electrochemical Society*, Vol 125, pp. 146 (1978).
(2) Reif, R. and Dutton, R.W., ibid., Vol. 128, pp. 909 (1981).
(3) Reif, R., *Proceedings of the Ninth International Conference on Chemical Vapor Deposition,* pp. 359, ECS Softbound Symposium, 1984.
(4) Ota, Y., *Journal of Applied Physics,* Vol. 51, pp. 1102 (1980).
(5) Takagi, T., Yamada, I., and Sasaki, A., *Thin Solid Films*, Vol. 39, pp. 207 (1976).
(6) Zalm, P.C. and Beckers, L.J., *Applied Physics Letters*, Vol. 41, pp. 167 (1982).
(7) Townsend, W.G. and Uddin, M.E., *Journal of Solid-State Electronics*, Vol. 16, pp. 39 (1973).
(8) Suzuki, S., Okuda, H., and Itoh, T., *Japanese Journal of Applied Physics*, Vol. 19, pp. 647 (1979).
(9) Suzuki, S. and Itoh, T., *Journal of Applied Physics,* Vol. 54, pp. 1466 (1983).
(10) Donahue, T.J., Burger, W.R., and Reif, R., *Applied Physics Letters,* Vol. 44, pp. 346 (1984).
(11) McFee, J.H., Schwartz, R.G., Archer, V.D., and Finnegan, S.N., *Journal of Electrochemical Society*, Vol. 130, pp. 214 (1983).
(12) Richman, D., and Arlett, R.H., ibid., Vol. 116, pp. 872 (1969).
(13) Cullen, G.W., Abrahams, M.S., Corboy, J.F., Duffy, M.T., Ham, W.E., Jastrzebski, L., Smith, R.T., Blumenfeld, M., Harbeke, G., and Lagowski, J., *Journal of Crystal Growth,* Vol. 56, pp. 281 (1982).
(14) Ghandhi, S.K., *Semiconductor Power Devices*, John Wiley & Sons, New York, 1977.

Fisher, Stephen M., Hammond, Martin L., and Sandler, Nathan P.

THIN EPITAXIAL SILICON BY CVD

REFERENCE: Fisher, S.M., Hammond, M.L., and Sandler, N.P. "Thin Epitaxial Silicon by CVD", *Emerging Semiconductor Technology, ASTM STP 960*, D.C. Gupta and P.H. Langer, Eds., American Society for Testing and Materials, 1986.

ABSTRACT: Device applications for thin silicon epitaxy require sharp transition widths and uniform layers on wafers up to 200 mm diameter in production quantity. Uniform silicon epitaxy nominally 1 um thick deposited with transition widths of 0.2 - 0.3 μm over heavily doped buried layers using reduced pressure CVD technology are discussed. Both boron and arsenic buried layers can be accommodated in the 100 - 200 torr range. Thin selective epitaxy, with and without uniform polycrystalline silicon overgrowth on the oxide, is improved by reduced pressure deposition. Epitaxial lateral overgrowth is another technology based on thin epitaxy that is improved by reduced pressure CVD. Silicon-on-Sapphire requires very high growth rates at atmospheric pressure for very short times to achieve acceptable crystal quality. Some results for these thin epitaxy processes are reviewed.

KEYWORDS: silicon epitaxy, vertical reactor, transition width, autodoping, reduced pressure, silicon-on-sapphire.

INTRODUCTION

The silicon epitaxy process permits the growth of lightly-doped (or oppositely-doped) single crystal silicon on top of heavily doped single crystal silicon. Many different configurations are possible. Epitaxy layer resistivity can be graded through the layer; multiple layers of different properties can be deposited; deposition can be area-selective, and epitaxial silicon can be grown on insulating substrates.

Mr. Fisher is the Applications Laboratory Manager at Gemini Research, 49026 Milmont Dr., Fremont, CA 95438. Dr. Hammond is Vice-President of Tetron, a subsidiary of Gemini Research, and Mr. Sandler is a Senior Applications Engineer at Gemini Research.

Epitaxial silicon layer thicknesses are being driven to lower values as device feature sizes shrink and integrated circuit density increases. Today, most IC's require epitaxial layer thicknesses of 3-10 µm. By 1990, many devices will require epitaxy layers in the 0.5-2 µm range.

Commercial epitaxial silicon deposition requires control of many parameters, [1] for example:

o Thickness and resistivity uniformity,

o Dopant transition width and profile,

o Buried layer pattern shift and distortion,

o Crystal quality,

o Surface quality,

o Wafer flatness,

o Metallic contamination.

In addition, these parameters must be controlled in production quantities on ever increasing wafer sizes.

Epitaxial silicon has been prepared by a wide range of technologies, including: molecular beams, ion beams, evaporation, sputtering, closed tube vapor transport, open tube hot wall vapor deposition, plasma-enhanced vapor deposition, and regrowth from the melt or by recrystallization. To date, epitaxial silicon by open tube cold wall chemical vapor deposition (CVD) is the preferred method for devices manufactured in commercially significant quantities.

Two principal epitaxy reactor geometries are in commercial use: the radiantly heated cylinder reactor and the induction heated vertical reactor. Both systems are capable of epitaxial deposition at reduced pressure and both systems meet the present day requirements for thickness and resistivity uniformity, as well as surface and crystal quality [2].

Both reactor geometries can also meet the needs of thin epitaxy in the thickness range of 1-2 µm using existing technology. Many of the thin epitaxy device requirements can be met for thicknesses below 1 µm, especially as the existing technology is extended to lower temperatures, lower pressures, and lower growth rates. As epitaxy thicknesses approach 0.5 µm, CVD technology will be pushed to its limits. Other technologies will compete in the 0.5 µm thickness range, but the productivity, uniformity, and proven quality of the CVD process will be difficult to surpass.

DEVICE/MATERIALS REQUIREMENTS

Most silicon devices using epitaxy require sharp transitions from a heavily doped buried layer pattern to a lightly doped epitaxy layer. The doping in the epitaxy layer is the sum of three different contributions: 1) out-diffusion from the buried layer, 2) dopant transport from the substrate in the vapor phase (autodoping), and 3) dopant added to the deposition chemistry.

Assuming autodoping were reduced to zero, out-diffusion would still make a noticeable contribution to the transition width in thin, lightly doped epitaxy layers. Therefore, the minimum transition width is controlled by dopant out-diffusion. Antimony should provide the sharpest transition, followed by arsenic, boron, and phosphorous.

The amount of autodoping is a function of pre-epitaxy processing, dopant chemistry, deposition pressures and temperatures, and reactor gas dynamics. Any process that reduces the dopant surface concentration and increases the rate of departure of the dopant from the silicon surface or the reactor process volume decreases autodoping. The higher the intentional dopant concentration in the epitaxy layer, the less the relative effect of autodoping. Therefore, thin epitaxial layers with resistivities in the 0.2-0.5 ohm-cm range are easier to control than layers with resistivities above 0.5 ohm-cm.

COLD WALL CVD TECHNOLOGY

Commercial silicon epitaxy is based on deposition from silane and chloro-silanes, which are readily available in high purity form. The nominal ranges for growth rate and temperature are noted in Table 1. Other silicon-halogen chemistries have been investigated but no overwhelming advantage has been found.

TABLE--1

Typical Epitaxial Silicon
Growth Conditions In Hydrogen

Chemical Deposition	Normal Growth Rate	Temperature Range	Allowed Oxidizer Level
$SiCl_4$	0.4-1.5 μm/min.	1150-1250°C	5-10 ppm
$SiHCl_3$	0.4-3.0 μm/min.	1100-1200°C	5-10 ppm
SiH_2Cl_2	0.4-2.0 μm/min.	1050-1150°C	<5 ppm
SiH_4	0.1-0.3 μm/min.	950-1050°C	<2 ppm

Generally, more thermodynamically stable silicon sources require higher deposition temperatures for the same growth rate. Lower deposition temperatures require lower growth rates to achieve the same crystal quality [3]. Most epitaxial silicon deposition takes place above 1000°C; however, device quality epitaxy has been deposited as low as 825-875°C in commercial systems [4,5].

When epitaxial silicon is deposited below atmospheric pressure, the rate of departure of autodoping atoms from the reactor is increased, generally leading to a lower autodoping contribution [6]. Reduced pressure processing also helps control pattern shift and distortion [6,7] and may permit better crystal quality at lower temperatures than achieved at atmospheric pressure by removing hydrogen and chlorine from the surfaces and exposing more surface lattice sites for epitaxial growth [8].

PRESENT COMMERCIAL CAPABILITY

SiH_2Cl_2 and SiH_4 are normally used for lower temperature epitaxial growth. SiH_4 offers the lowest growth temperature but the resulting film quality is extremely sensitive to the presence of oxidizer in the process volume. SiH_4 epitaxy usually requires growth rates below 0.3 μm/min. SiH_2Cl_2, deposited at reduced pressure, offers a wider range of growth rates and is somewhat less sensitive to oxidizers, especially for deposition temperatures above 1060°C. $SiCl_4$ has also been used, with good results, for lower temperature deposition (ca 1050°C) [9,10].

The transition width data reviewed here illustrate the capability of present processing to provide epitaxial layers in the 1-2 μm thickness range, using SiH_2Cl_2 and $SiCl_4$.

Because of its low vapor pressure and low diffusion coefficient, antimony should offer the least amount of out-diffusion and autodoping. Fig. 1 is a spreading resistance probe (SRP) profile, corrected for resistivity, for a 0.7 ohm-cm arsenic-doped epitaxy layer deposited over a 0.0021 ohm-cm antimony-doped buried layer. The layer was deposited with SiH_2Cl_2 at 1110°C and atmospheric pressure in a vertical reactor system.

The transition width in Fig. 1, defined here as the vertical distance required to increase the resistivity by two orders of magnitude, is approximately 0.28 μm.

Arsenic and boron have higher vapor pressures than antimony. Reduced pressure deposition is known to decrease arsenic autodoping, [6]; but reduced pressure deposition has been reported to increase boron autodoping for pressures below 100 Torr in a radiantly heated cylinder reactor. [11] In contrast, the data presented in Figs. 2-6 shows that boron autodoping is decreased in an induction heated vertical reactor by reduced pressure deposition in the 100-200 torr pressure range. [12]

Fig. 2 is a spreading resistance probe profile, corrected for resistivity, for an N/N+ structure (As/As) deposited at reduced pressure in an induction heated vertical reactor. Under the conditions noted, a transition width of 0.22 μm was required for a two order of magnitude change in resistivity. Fig. 3 illustrates similar data for an N/P+ structure (As/B). Transition widths for these structures, as defined here, were obtained for various pressures and plotted against pressure in Fig. 4.

The trend for both N/N+ and N/P+ is for the transition width to decrease with decreasing pressure to a minimum value which is probably determined by solid state diffusion from the heavily-doped substrate into the epitaxy layer.

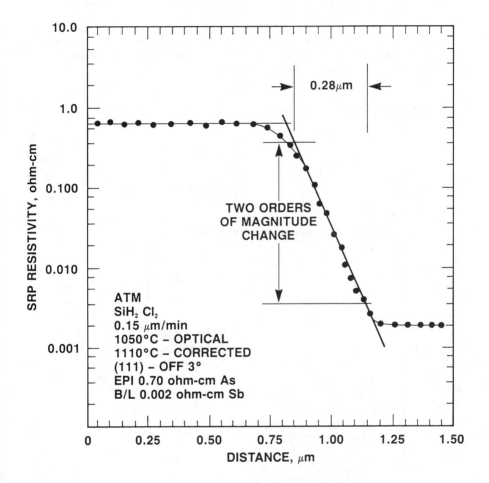

FIG.1 -- Vertical Distance Required to Change the Resistivity Two Orders of Magnitude, N/N+ Buried Layer (B/L) Structure (As/Sb) (Courtesy of J.C. Allison, National Semiconductor Corporation).

38   EMERGING SEMICONDUCTOR TECHNOLOGY

The transition zone of the N/P+ to N-type epitaxy. (Fig. 3) is complicated by N/P compensation; therefore, the distance required to go from 0.005 ohm-cm P+ to N-type epitaxy is substantially greater than the 0.35 μm transition width as defined here. In most IC structures, the P+ buried layer is used for isolation, and subsequent P-dopant diffusions will convert the silicon above the P+ buried layer to P+, eliminating the compensation peak in Figure 3. Autodoping effects can be complex and each device structure must be individually considered.

With dual buried layer structures (boron and arsenic), the concern is that autodoping from the P+ regions may cause a compensation or resistivity peak above the N+ regions.

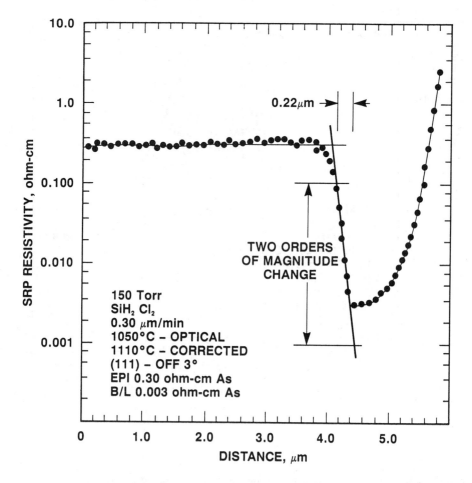

FIG.2 -- Vertical Distance Required to Change the Resistivity Two Orders of Magnitude, N/N+ Structure. (As/As).

Fig. 5 contains schematic beveled and stained sections through device structures with both N+ and P+ buried layer regions prepared using arsenic and boron. As the schematic insets illustrate, the appearance of beveled sections can indicate whether or not significant autodoping has occurred. In the case of significant P-type autodoping, the P-dopant flares out into the N-type epitaxy, narrowing the N+ pattern and producing a compensated resistivity peak in the resistivity profile taken through the N+ buried layer. For negligible autodoping, there is no P-type flare-out, and the SRP profile turns smoothly from the epitaxy layer into the N+ buried layer region.

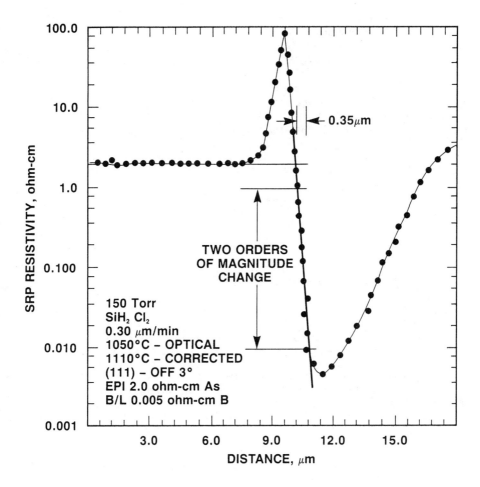

FIG. 3 -- Vertical Distance Required to Change the Resistivity Two Orders of Magnitude, N/P+ Structure (As/B).

Data from the device structure illustrated in Fig. 5 are provided in Fig. 6. The resistivity profile was taken 30 um from the P+ region using a 55 um probe spacing. For the photomicrograph, the P-regions were stained to delineate the P-N junction. Figure 6 clearly demonstrates that P+ autodoping is not significant at 150 torr in this device structure. There is no flare-out on the bevel/stain section and the transition width of the resistivity profile is 0.24 um, essentially the same as the 0.22 um value given in Fig. 2, which was obtained under similar deposition conditions.

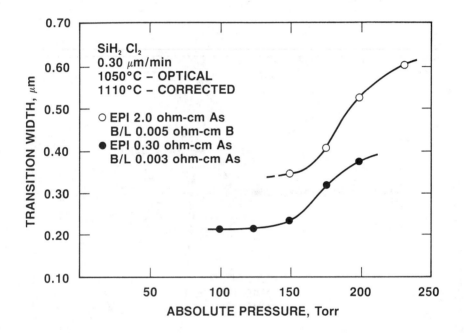

FIG. 4 -- Transition Width for Two Orders of Magnitude Resistivity Change versus Pressure, o: N/P+ (As/B) and o: N/N+ (As/As) Structures.

The reduced pressure transition width data presented in Figs. 1-6 are typical for the vertical reactor operating at atmospheric and reduced pressure.

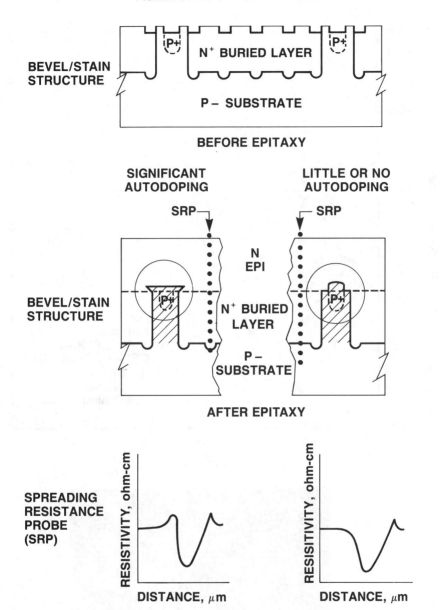

FIG. 5 -- Schematic Diagrams of Device Structures Illustrating the Effect of Autodoping.

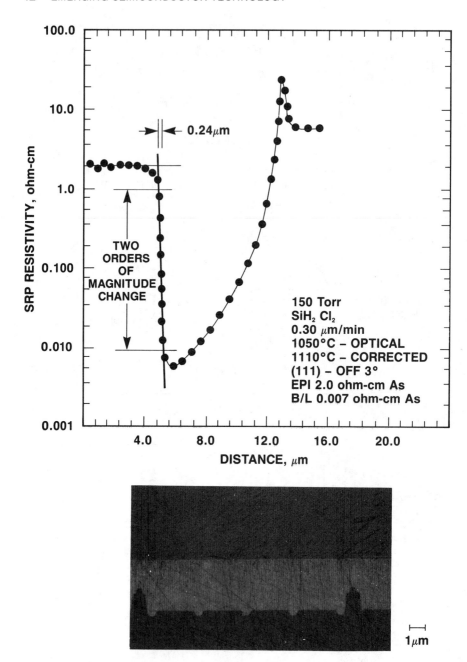

FIG. 6 -- Resistivity Profile and Bevel/Stain Photograph for Device Structure Shown in Fig. 5.

Ackerman and Ebert [13], using a cylinder reactor geometry, demonstrated the benefit of a pre-epitaxy bake for arsenic autodoping. With both pre-bake and deposition at 1150 C, flat resistivity zones of 0.4 - 0.7 μm thick were reproducibly deposited over arsenic buried layers.

Working with a vertical reactor at atmospheric pressure, Silvestri et al [9] demonstrated transition widths, as defined here, over arsenic buried layers of approximately 0.15 μm. Such sharp transitions were achieved using an 1150°C bake followed by a 1050°C deposit, as described by Srinavasan [14,15]. Device quality epitaxy was reported in the 0.5-0.9 μm thickness range using this high/low temperature technique.

It is clear from Figs. 1-6 here and the transition widths reported by others [9,10,13-15] that commercial epitaxial reactors can provide transition widths of 0.2-0.3 μm over heavily doped antimony and arsenic buried layers and that reactors operating in the 100-200 torr range can provide transition widths of 0.2-0.3 μm over buried layers prepared with arsenic and boron. Such transition widths are adequate for many devices requiring 1-2 μm epitaxy layers. In fact, high performance transistor structures with 0.5-0.9 μm epitaxy using the Srinivasan high/low temperature technique have been reported, [9].

NEW PROCESS CAPABILITY

The thin epitaxy capability presented above was limited to depositions above 1050°C. Under these conditions, an H2 bake removes native oxide and the addition of HCl helps reduce the silicon surface to permit high quality epitaxial deposition at 0.3-0.8 μm/min.

SiH4 offers the capability of depositing at even lower temperatures (<1030°C), but generally requires lower growth rates (<0.3 μm/min.). One process limitation for low temperature deposition is removal of the native oxide, which becomes more difficult as the deposition temperature is reduced. [16]

In this section, lower epitaxial deposition temperatures, selective deposition, epitaxial lateral overgrowth, and silicon-on-sapphire are briefly reviewed.

Lower Temperature Deposition

Conventional epitaxy reactors have been used to satisfactorily deposit epitaxial silicon in the 850-1000°C range [4,5]. Such processing normally requires a higher temperature pre-deposition in-situ H2 bake or HCl etch to clean the silicon surface. A liquid HF etch just prior to deposition has also given satisfactory results [4].

Photo-dissociation of silicon compounds has also been employed for low temperature (800-900°C) epitaxial deposition. Transition widths of 0.2-0.3 μm are reported for nominally 850°C deposition. [17-19] It would be expected that SRP profiles for epitaxial silicon deposited in the 850-1000°C range would have substantially sharper transitions than those of Figs. 1-6 which represent 1080 C deposition. It is possible that the low temperature resistivity transition profiles reviewed here [17-19] have been measured differently than those of Figs. 1-6.

Lowering temperatures and pressures even further, Myerson et al [20] demonstrated very sharp transitions and good quality 0.5 μm epitaxy at 750°C using silane at 2 x 10 exp(-3) torr in a load-locked hot wall reactor.

Plasma technology and low pressures have permitted epitaxial silicon deposition in the 650-850°C range [21] by CVD. Under such conditions, transition widths of approximately 0.1 μm over boron and antimony with good quality epitaxy have been reported [21, 22].

Epitaxial silicon processes operating below 1000°C have not been fully evaluated for commercial device production, but the trend is clear. With transition widths of approximately 0.1 μm, epitaxy thicknesses as low as 0.5 μm can be accommodated using CVD technology. Such sharp transitions may require a deeper understanding of SRP techniques to accurately reflect the transition profiles.

Selective Epitaxy

Selective epitaxial silicon deposition is another area where thin layers and sharp transitions are important. Structures with and without polycrystalline silicon-on-oxide overgrowth can be deposited and each has significant device potential [2,5,12].

Borland and Drowley [5] recently reviewed a wide range of device isolation concepts and reported on selective epitaxial deposition without polycrystalline silicon overgrowth on the oxide. The epitaxial deposition took place in a radiantly heated cylinder reactor. Epitaxial layer thicknesses were 2 um, deposited in the 826-1100°C range at nominally 25 torr using HCl with SiH_2Cl_2 to improve selectivity. Transition widths of about 0.1 um were reported for a 950°C bake, 25 torr, 826°C deposition. Drowley has achieved similar selective epitaxy results in an induction heated vertical reactor operating at 100 torr. [23]

Selective epitaxy with polycrystalline silicon overgrowth on the oxide is another promising device structure using thin epitaxy. In this case, epitaxial silicon is selectively grown simultaneously with a uniform deposit of polycrystalline silicon on the oxide. When this polycrystalline layer is patterned, it provides an oxide-isolated contact region adjacent to the single crystal device region. Such structures with 0.2 μm epitaxy and 0.2 μm polycrystalline layer have been demonstrated in a vertical reactor using SiH_4 at 175 torr [12].

Epitaxial Lateral Overgrowth

Selective epitaxy with lateral single crystal overgrowth onto the oxide is yet another application of thin epitaxy by CVD. In this technique, epitaxial silicon is selectively grown in openings etched in a patterned oxide. Growth is continued above the thickness of the oxide and conditions are arranged to permit single crystal growth laterally over the oxide until the growing films coalese [23]. Using this technique, Bradbury and Kamins [24] have achieved some remarkably low defect density films working at atmospheric pressure. Selectivity is one key to successful epitaxial lateral overgrowth and selectivity is significantly improved by reduced pressure processing.

Silicon-On-Sapphire

Silicon-on-sapphire (SOS) is one more example of thin epitaxy by CVD. SOS crystal quality is best achieved at 900-925 C using SiH_4 at approximately 2 μm/min apparent growth rate. Film thicknesses of nominally 0.6 μm are desirable; therefore, deposit times are only about 20 seconds.

With SOS, film thicknesses can be very precisely measured because of the large difference in optical properties at the silicon-sapphire interface and the lack of out-diffusion. [25] Therefore, SOS deposition offers a useful method of characterizing epitaxy reactor gas dynamics for short deposit times.

Gas flow through a present-day epitaxy reactor is of the mixed flow type; i.e., fresh process gas mixes with process gas already in the system before reacting at the heated surface. For an ideal, cold wall, mixed flow reactor, the concentration of process gas will increase exponentially at the beginning of the deposit step to its steady state value and, after the process gas valves have been closed, the concentration will fall exponentially back to zero. The time constant for these exponential changes will be the reactor volume divided by the total flow rate, corrected to standard temperature and pressure.

Figure 7 is a plot of apparent growth rate for SOS (total thickness divided by time deposit valves are open) vs deposit time in a vertical reactor [26]. The apparent growth rate is highest at very short deposit times and trends to a steady state value for deposit times greater than 30 seconds. With a process volume of 280 liters, a flow rate of 496 std. liters/min, and a gas temperature correction factor of 2.3 (300°K to 700°K), the reactor time constant is about 15 seconds.

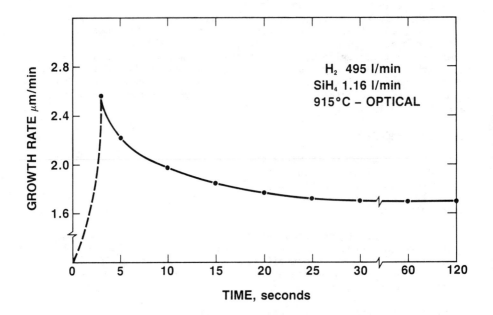

FIG. 7 -- Silicon-on-Sapphire Apparent Growth Rate versus Deposition Time.

The fact that the data points of Fig. 7 can be connected by a smooth curve illustrates that this vertical reactor is a well behaved, mixed gas flow system and it further illustrates that the process conditions are repeatable even for deposit times as short as 0.1 time constants. Fig. 7 also shows that constant growth rate conditions can be achieved in less than two reactor time constants. As epitaxy layer thicknesses diminish, this ability to reproducibly grow films in short deposit times will be very useful.

SUMMARY

Commercial epitaxy reactors, of either the vertical or cylindrical geometry, are capable of meeting the sharp resistivity transition profiles required for thin silicon epitaxy in the 1-2 um thickness range. By operating in the 100-200 torr pressure range, both arsenic and boron buried layers can be accommodated. Commercial reactors have demonstrated the necessary uniformity, film quality, pattern shift/distortion control, and productivity on wafers up to 150 mm diameter. In addition, the capability to process wafers up to 200 mm diameter has also been reported [27].

The capability to create unusual structures such as selective epitaxy, with and without polycrystalline silicon overgrowth on the oxide, epitaxial lateral overgrowth, and silicon-on-sapphire gives CVD technology wide application in future device technology.

Extending commercial epitaxy reactor capability to lower temperatures, lower pressures, and lower growth rates will provide sharper resistivity transition profiles and will permit commercial systems to satisfactorily meet epitaxy device needs in the 0.8-1 μm range.

The use of plasma technology, very low pressure, photo-dissociation, and new chemistry may permit device quality epitaxy with transition widths of less than 0.1 μm or less, as will be required for 0.5 μm thick layers.

Many technologies will compete with CVD in the 0.5-1 μm thickness range but the commercially proven performance of CVD gives it a substantial advantage.

REFERENCES

1. Hammond, M.L., "Silicon Epitaxy", Solid State Technology, Vol. 21, No. 11, Nov. 1978, pp. 68-75.

2. Liaw, H.H., Rose, J., Fejes, P.L. "Epitaxial Silicon for Bipolar Integrated Circuits", Solid State Technology, Vol. 28, No. 5, May 1985, pp.135-143.

3. Bloem, J., "High Chemical Vapor Deposition Rates of Epitaxial Silicon Layers", Journal of Crystal Growth, Vol. 18, No. 1, Jan. 1973, pp 70-76

4. Fok, T.Y., Wright, G.L., Atkinson, C.J., "Sub-micron Silicon Epitaxial Films Deposited at Low Temperature", Abs. 498 RNP in Abstracts of Recent Newspapers. Journal of the Electrochemical Society, Vol. 130, No. 11, Nov. 1983, p. 441C.

5. Borland, J.O. and Drowley, C.I., "Advanced Dielectric Isolation Through Selective Epitaxial Growth Techniques", Solid State Technology, Vol. 28, No. 8, August 1985, pp. 141-148.

6. Herring, R.B., "Advances in Reduced Pressure Silicon Epitaxy", Solid State Technology, Vol. 22, No. 11, Nov. 1977, pp. 75-80.

7. Lawrence, L.H., McDiarmid, J., Hammond, M.L., "Reduced Pressure Epitaxy in an Induction-Heated Vertical Reactor", Proceedings, Ninth International Conference on Chemical Vapor Deposition, McD. Robinson, C.H.J. van der Brekel, G.W. Cullen, J.M. Blocker, Jr., Eds., The Electrochemical Society, Pennington, NJ. 1984, p. 454.

8. Duchemin, M.J., Bonnett, M.M., Koelich, M.F., Journal of the Electrochemical Society, Vol. 125, No. 4, April 1978, pp. 637-644.

9. Silvestri, V.J., Srinivasen, G.R., and Ginsberg, B., "Submicron Epitaxial Films", Journal of the Electrochemical Society, Vol. 131, No. 4, April 1984, p. 877.

10. Srinivasan, G.R., "Modeling and Applications of Silicon Epitaxy in Silicon Processing", Silicon Processing, ASTM STP 804, D.C. Gupta, Ed., American Society for Testing and Materials, 1983, pp.151-173.

11. Graef, M.W.M., Leunissen, B.J.H., and de Moor, H.H.C., "Antimony, Arsenic, Phosphorous, and Boron Autodoping in Silicon Epitaxy", Journal of the Electrochemical Society, Vol. 132, No. 12, Dec. 1985.

12. Fisher, S.M., Hammond, M.L., and Sandler, N.P., "Reduced Pressure Epitaxy in an Induction Heated Vertical Reactor", to be published in Solid State Technology, Vol. 29, No. 1, Jan. 1986.

13. Ackermann, G.K., and Ebert, E., "Autodoping Phenomenon in Epitaxial Silicon", Journal of the Electrochemical Society, Vol. 130, No. 9, Sept. 1983, pp. 1910-1915.

14. Srinivasan, G.R., U.S. Patent No. 4, 153, 486, "Silicon Tetrachloride Epitaxial Process for Producing Very Sharp Autodoping Profiles and Very Low Defect Densities on Substrates with High Concentration Buried Layers Utilizing a Pre-heating in Hydrogen", May 8, 1979.

15. Srinivasan, G.P., "Silicon Epitaxy for High Performance Integrated Circuits", Solid State Technology, Vol. 24, No. 11, Nov. 1981, pp 101-110.

16. Srinivasan, G.R., and Meyerson, B.S., "Current Status of Reduced Temperature Silicon Epitaxy by Chemical Vapor Deposition", Abstract 265 in Extended Abstracts, The Electrochemical Society, Pennington, NJ., Vol. 85-2, Oct. 1985, p. 400.

17. Frieser, R.G., "Low Temperature Silicon Epitaxy" Journal of the Electrochemical Society, Vo. 115, No. 4, April 1968, pp. 401-405.

18. Kumagawa, M. Sunami, H., Terasaki, T. and Nishizawa, J., "Epitaxial Growth with Light Irradiation", Japan Journal of Applied Physics, Vol. 7, No. 11, Nov. 1968, pp. 1332-1341.

19. Yamazaki, T., Ito, T., and Ishikawa, H., "Disilane Photo-Epitaxy for VLSI", 1984 Symposium on VSLI Technology, Sept. 10-12, 1984, IEEE and Japan Society of Applied Physics, IEEE Catalogue #84 CH 2061-0, pp 56-77.

20. Meyerson, B.S., Gannin, E., and Smith, D.A., "Low Temperature Silicon Epitaxy by Hot Wall Ultra High Vacuum/Low Pressure Chemical Vapor Deposition Techniques", Abstract 266 in Extended Abstracts, The Electrochemical Society, Pennington, NJ., Vo. 85-2, Oct. 985, p. 401.

21. Burger, W.R., and Reif, R., "Device Performance in Epitaxial Silicon Deposited at Low Temperature by Plasma-Enhanced Chemical Vapor Deposition", Abstract 267, Ibid. p. 403.

22. Donahue, T.J., and Rief, R., "PECVD of Silicon Epitaxial Layers", Semiconductor International, August 1985, pp. 142-146.

23. Drowley, C.I., Private Communication. Also see reference [12].

24. Bradbury, D.R., and Kamins, T.I., "Device Isolation in Lateral CVD Epitaxial Silicon-on-Insulators", Abstract 523 in Extended Abstracts, The Electrochemical Society, Pennington, NJ., Oct. 1984, pp 767-768.

25. Cullen, G.W., Duffy, M.T., Jastrzebski, L., and Lagowski, J. "The Characterization of CVD Single-Crystal Silicon on Insulators: Heteroepitaxy and Epitaxial Lateral Overgrowth", Journal of Crystal Growth, Vol. 65, No. 12, Dec. 1983, pp. 415-438.

26. Sandler, N.P., "Silicon-on-Sapphire in an Induction Heated Vertical Reactor", IEEE SOS/SOI Technology Workshop, Oct. 1-3, 1985, Park City, Utah. (Unpublished data).

27. "200 mm Silicon Epitaxy", Information Release, Gemini Research, Fremont, CA 94538, July 28, 1985.

C.-C. Daniel Wong, John O. Borland, Sookap Hahn

EFFECTS OF GETTERING ON EPI QUALITY FOR CMOS TECHNOLOGY

REFERENCE: Wong, C.-C. D., Borland, J. O., and Hahn, S., "Effects of Gettering on Epi Quality for CMOS Technology," Emerging Semiconductor Technology, ASTM STP 960, D. C. Gupta and P. H. Langer, Eds., American Society for Testing and Materials, 1986.

ABSTRACT: The quality of silicon epitaxial layer deposited on n+ substrates has been improved by applying a pre-epitaxial intrinsic gettering process to the substrates. The epilayer defect density was decreased, while the junction breakdown voltage and the product yield of 16K CMOS Static RAM were both increased. The bulk microdefects generated by the IG process were observed and believed to be the key to the quality improvement of epitaxial layer. The better quality of epitaxial layer deposited on heavily phosphorus doped substrates, compared to those deposited on heavily antimony doped substrates, was attributed to the microdefects formed close to the epi/substrate interface. Although the epi/substrate transition width of n/n+(P) is wider than that of n/n+(Sb), no significant difference in current gain of parasitic bipolar transistors has been observed.

KEYWORDS: silicon epitaxy, intrinsic gettering, latch-up, junction breakdown voltage

Dr. Wong is senior process development engineer at Integrated Device Technology, Inc., 3236 Scott Blvd., Santa Clara, CA 95051; Mr. Borland is senior staff engineer at Applied Materials, Inc., 3050 Bowers Ave., Santa Clara, CA 95051; Dr. Hahn is analytical lab manager at Siltec Corp., 423 National Ave., Mountain View, CA 94043.

INTRODUCTION

In recent years, CMOS technology has attracted a great deal of attention. Its inherent low operational power gives rise to higher device packing density on a single chip without suffering from high power dissipation as in the cases of bipolar and NMOS technologies. With the continued push for CMOS technology with minimal design rule, latch-up becomes a key issue [1-3]. There are several ways to minimize and prevent latch-up in CMOS technology ranging from using an epitaxial wafer to complex device design and processing modification involving retrograde well formation and use of dummy collectors. A very simple and effective technique to eliminate latch-up reported in the literature has been the use of lightly doped epilayers grown over heavily doped substrates forming either an n/n+ or p/p+ epitaxial structure [4,5]. The choice of using an epitaxial structure to prevent latch-up is extremely attractive to IC manufacturers, since it requires only minimal change in the existing CMOS device fabrication process thereby eliminating costly process development efforts.

Besides latch-up prevention, the use of epitaxial structures has also been observed to improve other device parameters and yield [6,7]. This usually applies to the use of p/p+ epitaxial structure but not n/n+ epitaxial structure due to the lack of microdefects formed in the bulk of n+ substrates [8,9]. A special gettering technique is usually required for n/n+ epitaxial structure to improve the epilayer quality. However, the use of n/n+ epitaxial structure is attractive for certain CMOS device applications, such as alpha particle immunity for Static RAM using the p-well in n-substrate approach [10].

In this paper, the improvement of epilayer quality on n/n+ given by a modified three-step pre-epitaxial intrinsic gettering technique will be discussed. The influence of doping species in n+ substrates on epilayer quality will also be reported. Product yield enhancement of 16K CMOS Static RAM on n/n+ wafers by the pre-epitaxial intrinsic gettering technique will be demonstrated. Brief discussion on latch-up suppression by n/n+ structure is also included.

EXPERIMENTAL METHOD

Low resistivity (100) oriented, 100 mm diameter n-type wafers were used for this investigation. Dopants were antimony and phosphorus. Resistivity was in the range of 0.01 - 0.02 ohm-cm. For these heavily doped wafers, no direct measurement of oxygen content was possible by the IR method. However, in the growing process used to produce these crystals, oxygen content of lightly phosphorus doped materials ranged from 13 to 18 ppma (using the conversion factor based upon the ASTM F121-81). High resistivity (n-) wafers (10 - 12 ohm-cm) were also added to our tests as a control group. Interstitial oxygen content of these wafers typically ranged from 16 to 18 ppma and the carbon concentrations were less than 0.5 ppma.

Prior to epilayer deposition, some of the wafers were subjected to a modified high-low-medium three-step pre-epitaxial intrinsic gettering process. It started with an oxidation cycle at 950°C in dry oxygen for 45 minutes and then a denudation cycle at 1150°C in nitrogen for 2 hours. An extended nucleation cycle was then carried out at 700°C in nitrogen for 16 hours. Finally, a 24-hour anneal was done at 950°C in nitrogen to induce oxygen pricipitation and microdefect formation inside the bulk of the wafers.

N-type arsenic doped epilayers, 10 to 15 ohm-cm, were grown in an AMC-7810 radiantly heated barrel reactor. SiHCl3 was used as the source and deposition occurred at 1100°C with a growth rate of 0.75 μm/min. Epilayers grown on lightly phosphorus doped substrates and heavily antimony doped substrates were 15 μm thick, while those on heavily phosphorus doped substrates were 18 μm in thickness. In order to avoid auto-doping problems, the heavily phosphorus doped substrates were oxide backside sealed prior to epilayer deposition.

After epitaxial layer deposition, wafers were processed through IDT's CEMOS 1 process line. CEMOS 1 is a twin-tub, double poly, single metal CMOS process with minimum poly gate length of 2.5 μm. The gate oxide thickness is 500 Å. This process has successfully manufactured high performance, low power 16K CMOS Static RAM. The majority of the wafers went through the complete CEMOS 1 process for final product yield analysis and latch up characterization. Test wafers, subjected only to the front end of the CEMOS 1 process up to gate oxidation without receiving any masking step or ion implantation, were also included for material defect characterization. The results of the material defect characterization had been reported by Hahn, et al [11].

RESULTS AND DISCUSSIONS

Experimental results from fully processed lots are summarized and discussed in this section. Some of the results from test wafer lots are also included for review and reference.

Doping Profiles

In this study, the epilayer quality was characterized in terms of the material defect and the final product yield. In order to compensate for the difference in diffusivity for different dopants in n+ substrates, the epilayer thickness was tailored to CEMOS 1 process so that the final p-well doping profile was the same for various substrates. As a result, the device parameters, such as threshold voltages of n and p channel MOSFET transistors, remained unaffected by the dopant used in n+ substrates. Typical p-well doping profiles of various n/n+ epilayer structures are shown in Figure 1. The high diffusivity of phosphorus in n+ substrates resulted in strong outdiffusion as wafers subjected to the CEMOS 1 process. Therefore, n/n+(Sb) epiwafers exhibited sharper transition width than n/n+(P) epiwafers. Because the as-deposited epilayer thickness had already been tailored to the CEMOS 1 process, the final p-well doping profile was not affected by the strong outdiffusion of phosphorus.

It has also been noted that the pre-epitaxial intrinsic gettering process resulted in a wider transition width. This effect is suggested to be related to the enhanced diffusion of dopant due to the increase of point defects in n+ substrates by pre-epitaxial intrinsic gettering process.

Defect Characterization

(i) <u>Bulk defect analysis</u>: Oxygen precipitation and microdefect formation in the bulk of n-type substrates was found to be strongly suppressed by high doping concentration of P and Sb throughout the CEMOS 1 process (see Figure 2). However, for lightly phosphorus doped wafers, significant amount of precipitates was observed as shown in Figure 3.

Even though the behavior of heavily phosphorus doped substrates is similar to that of heavily antimony doped substrates, there were two noticeable differences. In the case of IG n/n+(Sb) epilayer structure, a rather well

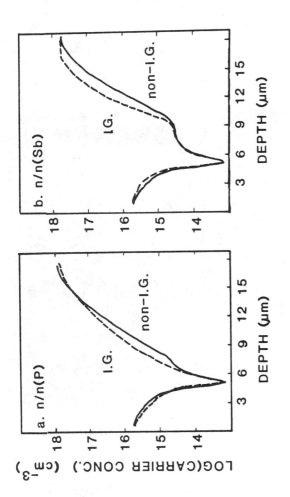

Figure 1 -- Typical doping profile of non-gettered and IG n/n+ epitaxial structures after full CEMOS 1 process; (a) for phosphorus doped substrates, and (b) for antimony doped substrates.

defined denuded zone with some noticeable amounts of microdefects inside the bulk (as shown in Figure 4) was observed. However, in the case of IG n/n+(P) epilayer structure, no noticeable etch features related to oxygen precipitation were found. This phenomenon has been confirmed by our recent synchrotron radiation section topograph studies [12]. Finally, in the case of n/n+(P) epilayer structure, a rather pronounced defect layer close to the epi/substrate interface was observed while no such layer was present in n/n+(Sb) epilayer structure.

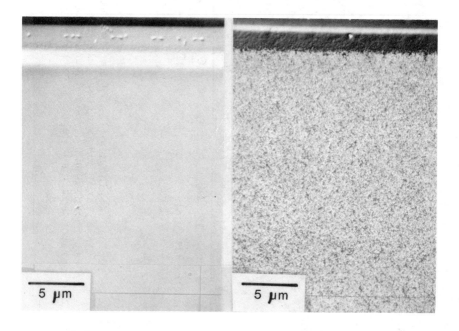

a. n/n(P)　　　　　　　　　b. n/n(Sb)

Figure 2 -- Cross section optical photomicrographs for (a) non-gettered n/n+(p) and (b) non-gettered n/n+(Sb) epi structures after full CEMOS 1 process (vertical scale is 5 times horizontal scale).

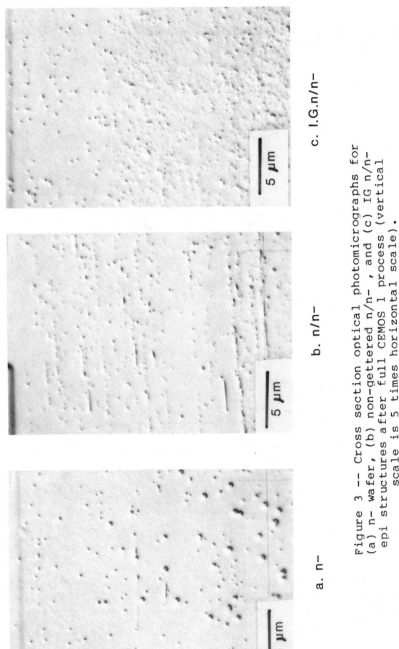

Figure 3 -- Cross section optical photomicrographs for (a) n- wafer, (b) non-gettered n/n-, and (c) IG n/n- epi structures after full CEMOS 1 process (vertical scale is 5 times horizontal scale).

58 EMERGING SEMICONDUCTOR TECHNOLOGY

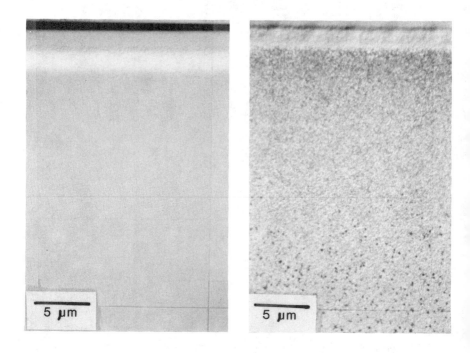

a. I.G.n/n(P)   b. I.G.n/n(Sb)

Figure 4 -- Cross section optical photomicrographs for (a) IG n/n+(p) and (b) IG n/n+(Sb) epi structures after full CEMOS 1 process (vertical scale is 5 times horizontal scale).

(ii) <u>Surface defect analysis</u>: The surface defect analysis was carried out on the test wafer lot after the simulated CEMOS 1 process had been completed up to gate oxidation. The thermally grown oxide on the surface of the wafers was removed before the wafers were etched in Wright solution for 3 minutes to delineate the defects. The surface defect density was then determined by the 9-point count method [13]. The surface defect density of various n/n-, n/n+ epi wafers and n- substrates is summarized in Table 1.

TABLE 1--Summary of surface defect density

Surface defect density (#/cm2)

Substrate	Etch pit	Stacking Fault
n-	0	0
n/n-	314	105
IG n/n-	0	523
n/n+(P)	698	87
IG n/n+(P)	314	0
n/n+(Sb)	1343	140
IG n/n+(Sb)	419	52

Two types of defect, etch pit and stacking fault, were observed on the surface of epi wafers; whereas the n- substrates were found to be free of any crystallographic defects. In general, the density of etch pit was higher than that of stacking fault on epi wafers. For the case of IG n/n- epi structure, the density of stacking fault was higher than that of etch pit. This can be attributed to the extended low temperature cycle of the pre-epitaxial intrinsic gettering process. Extended nucleation time has been reported [14] to be detrimental to the surface quality of n- CZ wafers due to the residual oxygen precipitates in the "denuded" zone. The IG process used in this study was optimized for the n+ substrates only. The extended nucleation time was required to create oxygen precipitates and microdefects in the bulk of n+ substrates.

The effect of pre-epitaxial intrinsic gettering on n/n+ epi structure was clearly demonstrated by the significant reduction in the density of etch pit and stacking fault. Both the etch pits and the stacking faults were effectively gettered by the oxygen precipitates and microdefects formed in the pre-epitaxial IG n+ substrates.

The epilayers on heavily phosphorus doped n+ substrates generally exhibited lower defect density than those on heavily antimony doped substrates. The defect layer close to the epi/substrate interface observed in n/n+(P) is believed to be closely related to the better quality of epilayer grown on these substrates. The combination of pre-epitaxial intrinsic gettering technique and the use of phosphorus as the dopant for n+ substrates gave rise to the best epilayer quality in this study.

Junction Breakdown Voltage And Product Yield Analysis

The breakdown voltage of n+ to p-well junction diode with interdigited finger structure was measured on all wafers to electrically characterize the effectiveness of the pre-epitaxial intrinsic gettering process. The summary of breakdown voltage for various n/n-, n/n+ and n- substrates is given in Table 2. The breakdown voltage reported has been defined as the reverse biasing voltage when the diode leakage current is equal to 1 nA. The pre-epitaxial intrinsic gettering process is found to improve the junction breakdown voltage of n/n+ epilayer structure effectively. As for the IG n/n- structure, since the pre-epitaxial intrinsic gettering process had not been optimized for the n- substrates, the junction breakdown voltage was found to be degraded. It has been noted that the junction breakdown voltage tracks the density of surface stacking fault very well; the higher the density of surface stacking fault, the lower the junction breakdown voltage.

TABLE 2--Summary of junction breakdown voltage

Substrate	Junction Breakdown Voltage (volts)
n-	17.5
n/n-	16.1
IG n/n-	13.4
n/n+(P)	14.4
IG n/n+(P)	17.5
n/n+(Sb)	14.2
IG n/n+(Sb)	14.8

Final product yield improvement is the ultimate goal in applying the pre-epitaxial intrinsic gettering technique. The normalized yields of 16K CMOS static RAM fabricated on various n/n-, n/n+ epi wafers and n- CZ wafers are summarized in Table 3. All the yields have been normalized to the die yield of the control n- CZ substrate group. The product yield on n/n+ epilayer structure, both n/n+(P) and n/n+(Sb), has improved 25% by using the pre-epitaxial intrinsic gettering technique. The product yield on n/n+(P) is also higher than that on n/n+(Sb). The yield enhancement of n/n+ epilayer structure, either by pre-epitaxial intrinsic gettering on n+ substrates or by simply using heavily phosphorus doped substrates, is consistent with the epilayer quality improvement observed from the results of the test wafer lot.

TABLE 3--Summary of normalized product yield

Substrate	Normalized Product Yield
n-	1
n/n-	0.76
IG n/n-	0.29
n/n+(P)	0.71
IG n/n+(P)	0.92
n/n+(Sb)	0.69
IG n/n+(Sb)	0.83

Latch-up Analysis

The latch-up resistance of the n/n+ epilayer structure was characterized by measuring the current gain of the vertical npn and lateral pnp parasitic bipolar transistors and the triggering current injecting from the output node of a CMOS inverter. The values of current gains and triggering current are summarized in Table 4. The p+ to p-well spacing for the lateral current gain measurement was 8 μm. For triggering current measurement, the inverter layout was 7 μm for p+ to p-well spacing, and 4 μm for n+ to p-well spacing.

TABLE 4--Summary of parasitic bipolar current gain and triggering current

Substrate	Bipolar Current Gain		Triggering Current (mA)
	Vertical	Lateral	
n-	293	0.87	0.9
n/n-	313	0.80	1.0
IG n/n-	311	0.67	0.9
n/n+(P)	349	1.41	no latch-up
IG n/n+(P)	340	1.31	no latch-up
n/n+(Sb)	354	1.15	no latch-up
IG n/n+(Sb)	350	1.24	no latch-up

The lateral current gain was consistently higher for n/n+ structure than for n/n- and n- substrates. This was due to the electric field generated at the epi/substrate interface, where high-low doping concentration gradient existed (see Figure 1). The holes injecting from the emitter (p+) into the base region (n-) were repelled back towards the surface and collected by the collector contact. Therefore, the lateral current gain was higher in n/n+ than in n/n-. Since the epilayer thickness had been tailored to the CEMOS 1 process to compensate for the outdiffusion of dopant and the p-well doping profile is the same for all substrates, no change in vertical current gain caused by n+ substrates was observed.

Although the product of vertical current gain and lateral current gain had increased by using n/n+ epilayer structure, no latch-up was detected in our inverter due to the tremendous reduction in substrate resistance. In contrast, the latch-up triggering current was about 1 mA for n- or n/n- substrates.

CONCLUSIONS

The quality of epitaxial layer deposited on heavily doped n+ substrates has been shown to be greatly improved by applying the modified three-step pre-epitaxial intrinsic gettering process to the n+ substrates. The microdefects formed in the bulk of n+ substrates by pre-epitaxial IG process exhibited effective getterability during the epitaxial layer deposition and the full twin-tub CMOS process. Product yield enhancement of 16K CMOS Static RAM on n/n+ epilayer structure has been achieved through the reduction of defect density in the epilayer and the increase of n+/p-well junction breakdown voltage.

Moreover, the use of phosphorus as the dopant for n+ substrates has been shown to further improve the epilayer quality and hence the product yield. This improvement is believed to be related to the layer of microdefects observed at the epi/substrate interface for the n/n+(P) epilayer structure.

ACKNOWLEDGEMENT

The authors are indebted to the manufacturing group in IDT for their support in making this work possible.

REFERENCE

1. "High Resistivity Epi May Solve MOS Problems," Semiconductor International, Editorial, April 1980, pp. 71-75.

2. Borland, J. O. and Singh, R. S., "Improved p-well CMOS Latch-up Immunity and Device Performance Through Intrinsic Gettering Techniques," in VLSI Science and Technology/1985, ECS PV 85-5, W. M. Bullis and S. Broydo, Ed., The Electrochemical Society, 1985, pp. 77-87.

3. Choi, K. H., Borland, J. O., and Hanh, S., "Improved n-well CMOS Latch-up Immunity Through The Optimization of Epilayer Thickness and Resistivity," in VLSI Science and Technology/1985, ECS PV 85-5, W. M. Bullis and S. Broydo, Ed., The Electrochemical Society, 1985, pp. 88-95.

4. Takacs, D., Harter, J., Jacobs, E. P., Werner, C., Schwabe, U., Winnerl, J., and Lange, E., "Comparison of Latch-up in p- and n-well CMOS Circuits," in 1983 International Electron Devices Meeting Technical Digest, The Institute of Electrical and Electronics Engineers, Inc., pp. 159-163.

5. Huang, C. C., Hartranft, M. D., Pu, N. F., Yue, C., Rahn, C., Schrankler, J., Kirchner, G. D., Hampton, F. L., and Hendrickson, T. E., "Characterization of CMOS Latch-up," in 1982 International Electron Devices Meeting Technical Digest, The Institute of Electrical and Electronics Engineers, Inc., pp. 454-457.

6. Borland, J. O., Kuo, M., Shibley, J., Roberts, B., Schindler, R., and Dalrymple, T., "Influence of Epi-substrate Point Deffect Properties on Geter Enhanced Silicon Epitaxial Processing for Advance CMOS and Bipolar Technologies," in VLSI Science and Technology/1984, ECS PV 84-7, K. E. Bean and G. A. Rozgonyi, Ed., The Electrochemical Society, 1984, pp. 93-106.

7. Fejes, P. L., d'Aragona, F. S., and Rose, J. W., "Electrical Characterization of Epitaxial Wafers for Use in CMOS," in VLSI Science and Technology/1985, ECS PV 85-5, W. M. Bullis and S. Broydo, Ed., The Electrochemical Society, 1985, pp. 118-127.

8. Pearce, C. W., "Effects of Heavy Doping on The Nucleation and Growth of Bulk Stacking Faults in Silicon," in *Impurity Diffusion and Gettering in Silicon*, Materials Research Society Symposia Proceedings, volume 36, R. B. Fair, C. W. Pearce, and J. Washburn, Ed., Materials Research Society, 1985, pp. 231-238.

9. Pearce, C. W. and Rozgonyi, G. A., "Intrinsic Gettering in Heavily Doped Si Substrates For Epitaxial Devices," in *VLSI Science and Technology/1982*, ECS PV 82-7, C. J. Dell'Oca and W. M. Bullis, Ed., The Electrochemical Society, 1982, pp. 53-59.

10. Minato O., Masuhara T., Sasaki T., Nakamura, H., Sakai, Y., Yasui, T., and Uchibori, K., "2K x 8 Bit Hi-CMOS Static RAM's," *IEEE Journal of Solid-state Circuits*, Vol. SC-15, No.4, August 1980, pp. 656-660.

11. Hahn, S., Borland, J. O., and Wong, C.-C. D., "Effects of n-type Substrate on Epilayer Quality For Twin Tub CMOS Technology," in *VLSI Science and Technology/1985*, ECS PV85-5, W. M. Bullis and S. Broydo, Ed., The Electrochemical Society, 1985, pp. 96-105.

12. Tuomi, T., Hahn, S., Wong, C.-C. D., and Borland, J. O., to be presented at Materials Research Society 1986 Spring Meeting.

13. Cheng, D. and Hahn, S., in *Defects in Silicon*, ECS PV 83-9, W. M. Bullis and L. C. Kimerling, Ed., The Electrochemical Society, 1983, pp. 453-462.

14. Wong, C.-C. D., Malwah, M., and Pollock L., "Nucleation Time Effects on Intrinsic Gettering," in *Impurity Diffusion and Gettering in Silicon*, Materials Research Society Symposia Proceedings, volume 36, R. B. Fair, C. W. Pearce, and J. Washburn, Ed., Materials Research Society, 1985, pp. 239-244.

Robert B. Swaroop

SILICON EPITAXIAL GROWTH ON N+ SUBSTRATE FOR
CMOS PRODUCTS

> REFERENCE: Swaroop, R. B., "Epitaxial Growth on N+ Substrate for CMOS Products," <u>Emerging Semiconductor Technology</u> <u>ASTM STP 960</u>, Dinesh C. Gupta and P. H. Langer, Eds., American Society for Testing and Materials, 1986.
>
> ABSTRACT: An N-type epitaxial layers were grown over low resisitivity N-type substrate with and without pre-epi internal gettering (IG) cycle. Epitaxy was grown using either Silicon tetrachloride (SIL) or Silicon dichloride (DCS) at 1200° and 1100°C respectively. The epitaxial structures were characterized for as-grown microdefects and electrical characteristics (minority carrier lifetimes, breakdown voltage and C-t holdtime). The results indicate that epitaxial layers grown at a low growth rate and high temperature produced a minimum density of microdefects. These defects may further be reduced during subsequent thermal cycles especially with pre-epi IG-cycle. Consequently, electrical characteristics of epi layer were also improved.
>
> KEYWORDS: Epitaxial growth, microdefects, internal gettering.

A n/n+ or p/p+ epitaxial silicon substrate offers two main advantages in the manufacturing of high density CMOS products such as 256K RAMS. First the epi structure serves as an excellent ground plane for damping substrate noise/cross talk or latch-up. Secondly, in conjunction with diffused guard rings or isolation trenches it forms an almost impenetrable barrier to injected minority carriers.

The quality of epitaxial layer over silicon substrate is an important factor to produce a reduced leakage current

---

Dr. Robert Swaroop is a Manager at the Fairchild Semiconductors, 545 Whisman Road, Mountain View, CA 94039.

Two groups of epitaxial thickness of 15 $\pm$ 1 and 2.8 $\pm$ 0.2 microns were grown using phosphorous dopant (N-type), having resisitivity of 3 $\pm$ 0.5 ohm-cm. The spreading resistance profiles (SRP) were taken to examine the transition width of dopant (dopant concentration gradient) from the substrate to the epitaxial layer. Epitaxial structures were subjected to a thermal simulation of a CMOS process in which initial oxidation was performed at 1050°C for 4 hours in a dry-wet-dry O_2 environment. This thermal cycle contributed toward the growth of oxide nuclei to precipitate-defect-clusters (PDC) which are required for efficient gettering during subsequent process steps. The highest and lowest temperatures of the process were 1200° and 550°C respectively.

The wafers were evaluated for microdefects on the surface as well as in the bulk of epi and substrate. The latter was achieved by angle-polishing a cleaved section of the wafer. The microdefects were delineated by using a preferential etch using Wright etchant. The defect density was determined by counting at the selected nine points across the wafer using Normarski differential interference contrast microscope at 200X.

Minority carrier generation lifetimes were measured since this parameter influences the holdtime for RAM devices and also affects the source-drain leakage current. The minority carrier lifetimes were measured using a modified Zerbst Method (6,7). These measurements were done on a minimum of five capacitors across the wafer. The oxide breakdown was also measured to investigate the integrity of MOS gate oxide. The measurements were done on MOS capacitors having aluminum gate over 750 A° of gate oxide. To investigate the effect of intrinsic gettering on various types of epi layers, C-t holdtimes were measured on p-channel gate oxide capacitors.

TABLE 1--Epitaxy Deposition Parameter

Silicon Source	Temperature* (°C)	Growth Rate (micron/min)	Designation
$SiCl_4$ (SIL)	1200	0.8	SIL (80)
$SiCl_4$ (SIL)	1200	0.5	SIL (50)
SiH_2Cl_2 (DCS)	1100	0.48	DCS (48)
SiH_2Cl_2 (DCS)	1100	0.18	DCS (18)

(*) T $\pm$ 5°C

along with a low minority carrier concentration in the degenerated substrate (1). The interdependence of epitaxial growth defects on the substrate's surface defects in conjunction with epitaxial processing parameters plays a major role in the quality of epi layer. Further intrinsic (or internal) gettering (IG) to have oxygen precipitation in the bulk of the substrate after or prior to epitaxial growth reduces microdefects concentration in the epi layer and thus increases generation lifetime (2-5).

In the present work we investigated N-type epi layer over heavily doped N-type substrate. The quality of the epi layers was evaluated by growing epi layers from two different silicon sources and at two temperatures. We studied the type and concentration of the microdefects produced during epi growth and after thermal simulation of a CMOS process. The relationship between electrical characteristics (minority carrier lifetimes, breakdown voltage and C-t holdtime) and the microdefects was examined.

EXPERIMENTAL METHOD

The N^+ silicon wafers with doping concentration 10^{18} antimony atoms per cc were 100 mm in diameter and of (100) orientation. The resistivity range was between 0.008 and 0.025 ohm-cm. The interstitial oxygen range for these wafers was estimated to be between 13 and 17 ppma (ASTM F121-81)*. Since the oxygen concentration in heavily doped (especially Sb-doped) can not be measured by Fourier Transform Infrared (FTIR) spectroscopy because of free carrier absorption, the oxygen concentration was controlled by the wafer supplier using crystal growth conditions known to produce the lightly doped crystals for similar oxygen range. Furthermore, due to antimony trioxide evaporation during crystal growing, the estimated oxygen range may be slightly lower than that indicated above.

One-half of the experimental wafers were processed through pre-epi nucleation thermal cycle of 24 h at 750°C in nitrogen. Some of the epitaxial layers were grown using SIL at 1200°C and at a growth rate of 0.50 and 0.80 microns per min. The other type were grown using DSC at 1100°C and a growth rate of 0.18 and 0.48 microns per min. (Table 1). For both, SIL and DCS processes, an in-situ HCl etching at 1200°C was used to remove approximately 800 to 1600 A° of silicon surface prior to deposition. All epitaxial layers were grown at one atmospheric pressure using a barrel type radiant heat reactor.

---

* provided by the silicon wafer vendor

RESULTS

The results from this investigation are presented in three sections: epi-substrate interface, process-induced defects and electrical characteristics. We should emphasize that the results presented here are based on data obtained from silicon wafers with a given thermal history and specific test cycles and thus may not be generalized.

Table 2--Epi-Substrate Interface

Deposition Process	Interface Transition Width[a] (um)	
	After Epitaxy	After CMOS Simulation
SIL (80)	1.4	2.5
SIL (50)	1.6 ± 0.2	2.8 ± 0.2
DCS (48)	1.2	2.3
DCS (18)	0.35 ± 0.1	(0.8 ± 0.1)[b]

(a) Average of three readings
(b) Modified CMOS process

Epi-Substrate Interface

Table 1 lists the deposition parameters of epitaxy which was grown at two temperatures and various growth rates. For simplification we have designated $SiCl_4$ (0.80 um/min; 1200°C) as SIL (80), $SiCl_4$ (0.50 um/min; 1200°C) as SIL (50), SiH_2Cl_2 (0.48 um/min; 1100°C) as DCS (48) and SiH_2Cl_2 (0.18 um/min; 1100°C) as DCS (18).

The transition width due to outdiffusion of the substrate dopant (Sb) at the epi-substrate interface was calculated from Spreading Resistance Profiles (SRP); two of these are shown in figure 1. All other data is given in table 2. The transition width ranged between 1.2 and 1.6 um for 15 um thick epi while for 2.8 um thick epi it was 0.35 um. These transition widths extended further into epi after the wafers have been processed through CMOS simulation. However the usable epi depth was almost 10-12 um for 15 um epi and 2 um for 2.8 um thick epi. It is apparent from the data given in table 2 that the transition width did not show strong dependence on growth rate of epitaxy. However there may be an indication that outdiffusion of Sb at 1100°C is smaller than that at 1200°C and this was expected.

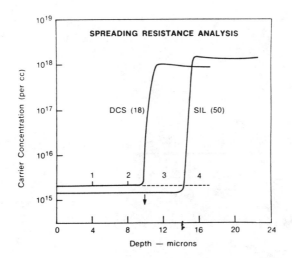

FIGURE 1: Spreading Resistance Profiles of As-Deposited Epi Layers.

Process-Induced Defects

The defects on the as-grown epi surface were examined before and after preferential etching. The epi surface under high intensity light appeared clean and free of haze. However after preferential etching we could observe the surface defects under the microscope as shown in figure 2 (a). These microdefects after initial oxidation as shown in figure 2(b), mainly appeared as "hillocks" with some OISF (Oxidation-Induced Stacking Faults). After completing the CMOS simulation process, we mainly observed OISF with some enlarged hillocks or punched out dislocation loops, as shown in figure 2(c).

FIGURE 2: Epi Surface Defects
(All at 200X)

(a) Microdefects in As-grown Epi

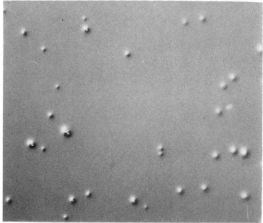

(b) "Hillocks" after Epi-Oxidation

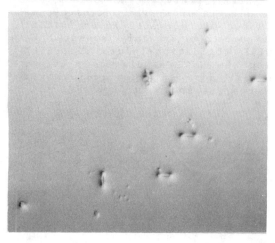

(c) Stacking Faults and hillocks after CMOS simulation process

FIGURE 3: Angle Polished Section of Epi-Substrate Structures

(All at 200X)

(a) Standard Wafer

(b) SIL (50) IG-Wafer

(c) DCS (48) IG-Wafer

TABLE 3: Surface Microdefects* at Various CMOS Process Steps

Deposition Process	After Epi		After Initial-OX		After Complete Process	
	STD.	IG	STD.	IG	STD.	IG
SIL (80)	17	16	17	3	3	0.8
SIL (50)	9	7	8	2	1.8	0.4
DCS (48)	115	108	110	25	11	1
DCS (18)	94	80	85	15	8	0.7

(*) In 10^3 per cm^2; average of five wafers at each step and 9 points - count for each wafer.

As shown in Table 3, the density of microdefects was approximately the same after epi and initial oxidation (initial-Ox) for standard wafers (without IG thermal cycle); however this density was reduced by approximately one order after completing the CMOS simulation process. The wafers with IG-thermal cycle always produced smaller density of microdefects by approximately one to two orders than standard wafers. The nature of the surface defects was the same whether epi was grown using SIL or DSC.

The angle-polished section of IG-wafers as shown in figures 3(b) and 3(c) show PDC after CMOS process. We observed consistently a denuded zone just below the epi layer mostly in the case of SIL deposition process. We believe that the denuded zone was formed during epitaxial growth at 1200°C. (Figure 3(b))

Electrical Characteristics

Table 4 summarizes the data from minority carrier generation lifetime measurements. The results are given for SIL (50) and DCS (48) deposition processes. The lifetimes of standard wafers after epi and CMOS simulation remained approximately the same or slightly decreased. However IG-wafers after epi indicated a slight increase (2 to 3 times) in lifetimes while this increase after CMOS simulation was 10 to 20 times. The lifetime was consistently higher for SIL (50) wafers than from DCS (48) wafers.

TABLE 4: Minority Carrier Generation Lifetimes*

Deposition Process	Standard Wafers		IG Wafers	
	After Epi	After CMOS	After Epi	After CMOS
SIL (50)	2.5 ± 0.6	2.2 ± 0.8	8.5 ± 2	180 ± 40
DCS (48)	1 ± 0.3	0.85 ± 0.4	2.5 ± 1.5	30 ± 25

(*) in μsec

TABLE 5: Gate Oxide Integrity

Deposition Process	Percentage BV Above 55V*		Percentage BV Above 70V*	
	STD.	IG	STD.	IG
SIL (50)	81	100	74	86
DCS (48)	83	92	71	79

(*) at 10 nA

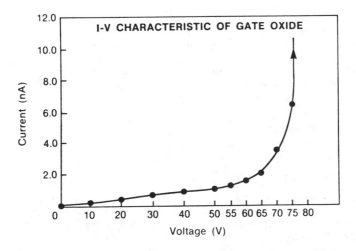

FIGURE 4: A Typical I-V curve from Breakdown Voltage Measurements

The data from breakdown voltage (BV) measurements on MOS capacitor having gate oxide thickness of 750 A° is given in Table 5. The data is obtained from a minimum of twenty devices for each case. The breakdown voltage was defined as the potential required for 10 nA leakage current. Table 5 also compares the percentages of good devices on standard wafers with those on IG-wafers. The results indicated that SIL (50) wafers were consistently better than DCS (48) even though percentages of good devices are quite comparable for both the cases: standard-or IG-wafers. Figure 4 shows a typical I-V curve from breakdown voltage measurements.

TABLE 6: Normalized* C-T Holdtimes for P-Channel Devices

Deposition Process	Standard Wafers	IG Wafers
SIL (50)	1.3	6.4
DCS (48)	1	9.5

(*) Normalized against DCS (48) and data was obtained from a minimum of twenty devices for each case.

The C-t holdtimes were measured on the p-channel SRAM devices made on standard-and IG-wafers. The results indicate a definite advantage by internal gettering thermal cycle because holdtime is improved at least 5 and 9.5 times over the standard wafers for SIL (50) and DCS (48) epitaxial layers respectively.

DISCUSSIONS

By comparing the microdefects and electrical characteristic results for epitaxial layers listed in Table 1, we may conclude that SIL (50) has given slightly better quality epi than others. It appears that the silicon source (SiH_2Cl_2 or $SiCl_4$) may not be an important factor but deposition temperature and growth rate are controlling factors for the quality of epitaxial layer. The quality appears to be best for SIL (50) deposition process because of the low growth rate (0.5 um/min) at high temperature (1200°C). Even though DCS (18) deposition process has the lowest growth rate (0.18 um/min) than others; however it did not produce a minimum density of microdefects. As observed, a combined effect of low growth rate and high temperature produced a minimum number of as-grown microdefects. The dopant (Sb) outdiffusion during epi deposition did cause a dopant gradient across the epi-substrate interface but not to the extent as observed by Secco d'Aragona et. al. (8). However, the magnitudes of outdiffusion observed in this study are comparable to that observed by Srinivasan (9). Therefore we believe that outdiffusion is process dependent and not simply controlled by the fundamentals of diffusion process.

The microdefects on epi surface are revealed only by preferential etching and their reduction with subsequent CMOS thermal cycles is observed. These "S" pits type of microdefects are converted to "hillock" type of defects after initial oxidation and their density is approximately the same as those of "S" pits in as-grown epi surface. A reduction in surface defects for IG-wafers is observed after initial oxidation and after the CMOS simulation process. Such a reduction in standard wafers is also observed but not to the extent as in IG-wafers simply because gettering due to oxygen precipitation is missing in these wafers. These observations are quite similar to those observed by Secco d'Aragona et. al. (8) who identified metallic impurities to be the cause of the "hillock" formation. The causes for formation of "S" pits in the epi surface is not well understood. However, the density of "S" pits depends on the deposition process. It may be that subsequent to epi deposition cooling rate and environment are affecting the kinetics of these microdefects. More work is needed to understand the mechanism of microdefects formation.

We believe that one pre-epi IG-thermal cycle of 24h at 750°C is sufficient in bringing an effective internal gettering process (10,11) rather than 72h multiple cycles (12). However, the time and temperature of one cycle to multiple cycles as explored by Dyson et. al. (12) depends on the thermal history of silicon crystal during growth, pre-anneal of wafers at the vendor, the interstitial oxygen concentration of the wafer and the process thermal cycles at IC manufacturer.

The minority carrier generation lifetimes, breakdown voltage and C-t holdtime results point mainly two major conclusions: SIL (50) deposition process produced a good quality epi-substrate structure and wafers with IG-cycle indicated reduced microdefects and improved electrical characteristics. These results agree well with the previous investigations (8,12,13). However C-t holdtimes data (Table 6) for IG-wafers do not follow the trend of results observed from the microdefects, minority carrier lifetimes and oxide breakdwon voltage given in tables 3, 4 and 5. Minority carrier lifetimes data for IG-wafers indicated best lifetime for SIL (50) epi-substrate structure while C-t holdtime data indicated best results for DCS (48) epi-substrate structure. We can explain this incompatibilty in the results as follows.

The formation of PDC in the substrate right below the epi layer (see figure 3(c)) provides adequate recombination centers to produce a reduced carrier lifetime below the device region (14) and thus controlling or reducing the parasitic vertical npn transistor gain. Such an effect would thus increase the latch-up immunity. On the other hand if a denuded zone is formed underneath the epi (see figure 3(b)), the mean free path of the carriers is increased to the extent that carriers can latch-up the adjacent cell. In such a case C-t holdtime may not be the best as observed from the data given in table 6. Essentially, we need an improved carrier lifetime near the device region and not in the bulk of silicon substrate. We observed similar results for the DRAM device yield on the basis of refresh cycle (15). An optimum combination of epi thickness with or without denuded zone in a degenerated substrate (produced by PDC) would then be required for desired latch-up immunity for a CMOS structure as already substantiated by Borland and Singh (14).

CONCLUSIONS

We conclude from this investigation that the quality of an epi layer depends on the temperature and growth rate of epitaxy. As-grown epitaxy has surface microdefects. These may effectively be reduced during subsequent thermal cycles with a pre-epi internal gettering thermal cycle. Further, minority carrier lifetimes, breakdown voltage and C-t holdtime measurements indicated a tremendous improvement with an internal gettered epi-substrate structure.

ACKNOWLEDGEMENTS

The author wishes to thank J. Buncayao and J. Miller of Fairchild for assistance in some of the experimental work and Dr. Namsoo Kim of AMI for assisting in part of the electrical characterization work. Thanks are also due to Fairchild's management for supporting the work.

REFERENCES

[1] Yaney, D.S. and Pearce, C.W., "The Use of Thin Epitaxial Silicon Layers for CMOS VLSI", Proc. IEDM, 1981, p.236.

[2] Tan, T.Y., Gardner, E.E. and Tice, W.K., "Intrinsic Gettering by Oxide Induced Dislocation in Cz-Si", Appl. Phys. Lett. 30, 1977, p.175.

[3] Swaroop, R.B., "Oxygen Control and Intrinsic Gettering in Cz Silicon" Defects in Silicon eds. Bullis, W.M. and Kimerling, L.C. Electrochem. Socy. PV 83-9, 1983, p.180.

[4] Dyson, W., Hellig, L., Moody, J., and Rossi, J., "Gettering of P+ (100) Si Substrates for Epitaxial Growth", ibid, p.246.

[5] Borland, J.O. and Deacon, T., "Advanced CMOS Epitaxial Processing for Latch-Up Hardening and Improved Epi Layer Quality" Solid State Tech. 27 (8), 1984, p.123.

[6] Zerbst, M. and Agnew, "Relaxationseffekte an Halbleiter-Islator-Grenzflaechen" Z. Phys., 22, 1966, p.30

[7] Heiman, F.P., "On the Determination of Minority Carrier Lifetime from the Transient Response of an MOS Capacitor" IEEE Trans Electron Devices ED-14, 1967, p.781

[8] Secco d'Aragona, F., Rose, J.W., and Fejes, P.L., "Outdiffusion, Defects and Gettering Behavior of Epitaxial N/N+ and P/P+ Wafers used for CMOS Technology" VLSI Science and Technology eds. Bullis, W.M. and Broydo, S., Electrochem Socy. PV 85-5, 1985, p.106.

[9] Srinivason, G.R., "Modeling and Applications of Silicon Epitaxy" Silicon Processing ASTM STP 804 Gupta, D.C. ed. American Society for Testing and Materials, 1983, p.151.

[10] de Kock, A.J. and Van de Wijgert, W.M. "The Influence of Thermal Point-Defects on the Precipitation of Oxygen in Dislocation Free Silicon Crystal", Appl. Phys. Lett. 38, 1981, p.888

[11] Pearce, C.W., Rozgonyi, G.A., "Intrinsic Gettering in Heavily Doped Si Substrate for Epitaxial Devices" VLSI Science and Technology eds. Dell'Oca, C.J. and Bullis, W.M., Electrochem Socy., 1982, PV 82-7, p.53.

[12] Dyson, W., O'Grady, S., Rossi, J.A., Hellig, L.G., and Moody, J., "N+ and P+ Substrate Effects on Epitaxial Silicon Properties" VLSI Science and Technology, eds. Bean, K.E. and Rozganyi, G.A., Electrochem Socy. pv84-7, 1984, p.107.

REFERENCES-cont

[13] Fejes, P.L., Secco d'Aragona, F. and Rose, J.W., "Electrical Characterization of Epitaxial Wafers for Use in CMOS" <u>VLSI Science and Technology</u> eds. Bullis, W.M. and Broydo, S., Electrochem Socy. PV 85-5, 1985, p.118
[14] Borland, J.O. and Singh, R.S., "Improved P-Well CMOS Latch-Up Immunity and Device Performance through Intrinsic Gettering Techniques" ibid p.77.
[15] Swaroop, R.B., "Effects of Oxygen on Process-Induced Defects and Gettering in CZ-Silicon" <u>Semiconductor Processing ASTM STP 850</u> Gupta, Dinesh C. ed. American Society for Testing and Materials, 1984, p.230.

John W. Medernach and Victor A. Wells

CHARACTERIZATION OF THE **IN SITU** HCL ETCH FOR EPITAXIAL SILICON*

REFERENCE: Medernach, J. W., Wells, V. A., "Characterization of the **in situ** HCl Etch for Epitaxial Silicon," Emerging Semiconductor Technology, ASTM 960, D. C. Gupta and P. H. Langer, Eds. American Society for Testing Materials, 1986.

ABSTRACT: An experimental design strategy was used to characterize the **in situ** HCl etch used in the epitaxial silicon deposition and the selective epitaxial silicon processes. The dependence of the etch rate, haze and etch anisotropy is determined as a function of the HCl flow and etch temperature. Results show that temperatures less than 1025°C are the only limitation for the **in situ** HCl etch used in the epitaxial silicon process. Patterned oxide substrates which are used in a selective epi process show different etch rate and haze characteristics than unpatterned substrates. Etch anisotropy on the oxide patterned substrates limits the operational region for the selective epi process.

KEYWORDS: Epitaxial Silicon, High Temperature Vapor Etching, **in situ** HCl Etch, Barrel Reactor, Haze, Etch Anisotropy, Etch Rate

The use of high temperature vapor phase etching techniques for silicon etching has been extensively reported [1-8]. Although many different gases were studied, the majority of etchants are the halogens or hydrogen halides [1-3]. Some of the non-halides studied include H_2S [4], H_2O [5], O_2 [6], H_2 [7,9], and SF_6 [10]. Etch temperature greater than 850°C were used in these studies. The use of an **in situ** $HCl-H_2$ etch is commonly employed prior to the deposition of epitaxial silicon [11,12]. The primary reason for the use of the **in situ** HCl etch is the elimination of surface damage or contamination which can act as a source to generate stacking faults, hillocks, spikes, or other structural imperfections in the epitaxial silicon layer. The etch must produce a

---

Dr. Medernach is a Member of the Technical Staff at Sandia National Laboratories, P. O. 5800, Albuquerque, NM 87185; and Dr. Wells is the Manager of MOS Development at Sperry Univac, 1500 Tower View Dr., Eagan, MN 55122.
*This work performed at Sandia National Laboratories was supported by the U. S. Department of Energy under contract number De-AC04-76DP00789.

high quality surface with a minimum of imperfections. Aside from the HCl-H_2 system other gases such as HBr [8] and SF_6 [10] have been used successfully as etchants in the production of epitaxial silicon layers. Cost and undesirable by-products were the major disadvantages of these gases. Recently there have been discussions on replacing HCl by H_2, because of submicroscopic precipitates [5] and heavy metal contamination from the HCl etch. These H_2 anneals are reported to result in a large reduction of defects during the epitaxial silicon process [9]. Investigators [1-3,13] who have studied the silicon etch rate using the H_2- HCl chemistry report a strong dependence on the partial pressure of the HCl and a weaker or no dependence on etch temperature.

The use of the HCl-H_2 etch as a preferential etchant is part of some new emerging technologies which use selective epitaxial silicon deposition [14]. Specific etch parameters must be established for preferential etching with the HCl-H_2 system. Either an oxide or nitride mask is used for pattern definition in the selective epi process. Reports [15,16] indicate that preferential etching of silicon occurs with the HCl-H_2 system in the presence of an oxide or nitride mask. An etch temperature of 1250°C was used, and all crystallograghic orientations showed some degree of preferential etching. A disadvantage of using an oxide mask is the occurrence of a second possible reaction at high temperatures [5]. A mass spectrometric study [17] of the HCl/H_2 - SiO_2/Si system shows that above 1200°C the vapor pressure of SiO becomes significant, which contributes to enhanced etching of the silicon.

The purpose of this study was twofold. First, to characterize the **in situ** HCl etch used in epi deposition, and second to determine the effect of using a high temperature **in situ** HCl etch in a selective epi process. This study was conducted using a statistical experimental design strategy [18]. No effort was made in the experimental design to incorporate a thermodynamic study of the **in situ** etch process. Our goal was to determine the response surfaces of etch rate, haze, and etch anisotropy as a function of the HCl flow and etch temperature parameter field.

EXPERIMENTAL

Statistical Design

A statistical design procedure is applied to minimize the total number of experiments required to characterize a process. This reduces the total number of trials by allowing more than one variable to change during successive trials. Recently, Kuhne and coworkers [19] used a statistical design to optimize a SiH_4-HCl-H_2 silicon epitaxial deposition process. This allows the investigator to obtain the maximum amount of information from a minimum number of experiments. Details on experimental organization and statisical analysis techniques are found in a number of sources [18,20]. The responses obtained from this type of experiment can be used to approximate the true, unknown, response surface with a low order polynomial.

In our case, we considered the models that are a subset of the full quadratic model:

$$Y = A_0 + A_1X_1 + A_2X_2 + A_{12}X_1X_2 + A_{11}X_1^2 + A_{22}X_2^2 \qquad (1)$$

The estimated response Y is a function of the values of the process variables X_1 and X_2. The A's are unknown coefficients which are estimated by least square regression. Providing the true response surface is smooth over the parameter field being studied, the estimated response surface then will likely be a good approximation of the true response surface. Duplicate experiments are used to evaluate the "goodness" of the approximation, and the analysis of the deviations between the estimated and observed responses at the experimental points also is used for checking the fit of the model. The use of a contour plot will reduce the difficulty in the interpretation of the somewhat abstract estimated response surface. Contour plots give detailed information concerning the behavior of the etch rate, haze or etch anisotropy as a function of the HCl flow and etch temperature. The design strategy used is a two-factor interaction with star experiment. This required trials at each corner of the square the center point and the mid-points. The variables in this study are HCl flow designated as (F) with units of standard liters per minute (slm), and etch temperature designated as (T) in °C. The responses are the silicon etch rate designated as ER_{Si} for unpatterned silicon, or ER_p for patterned oxide silicon both in units of microns per minute (μm/min). The other responses are haze and etch anisotropy designated as (H) and (S), respectively, both are dimensionless.

Process

The two types of substrates used in this study are described in Table 1. Substrate A had a patterned oxide and represented material which can be used in a selective epi process.

Table 1--Substrate Specifications

Parameter	Substrate A (oxide)	Substrate B (non-oxide)
Orientation	<100>	<100>
Diameter	100 mm	100mm
Thickness	625μm	625μm
Resistivity	3-4Ω-cm	0.02Ω-cm
Dopant	Phosphorus	Antimony
Oxygen Content	13-18 ppma	13-18 ppma*
Type	n	n+
Epitaxial Layer		
Structure	--------	n/n+
Dopant	--------	Phosphorus
Resistivity	--------	4.0+/-0.5Ω-cm
Thickness	--------	7.5+/-0.5μm
Silicon Source Gas	--------	Dichlorosilane

* Grown to a 13-18ppma specification (ASTM F121-83).

A thermal oxide was grown at 1100°C in dry O_2 to a thickness of 500nm (5000Å). A rectangular array pattern was defined in the oxide using photolithography. An AMT Model 7900RP epitaxial reactor, a barrel type reactor with radiant heating employing a SiC coated susceptor capable of holding 18 wafers, was used for the **in situ** etch runs. All etch runs were performed as occurred in normal production processing but without any silicon deposition. No special precautions were taken for the etch runs, so the normal scatter of experimental data would be obtained. Three wafers of each substrate type were used; other locations on the susceptor were filled with blank wafers. The blanks were unpatterned silicon wafers used to fill the remaining susceptor positions. All **in situ** HCl etch runs were performed at 625 Torr, this is atmospheric pressure in Albuquerque, NM. The etch temperature was varied from 900°C to 1200°C, while the HCl flow was varied from 1.82slm to 7.3slm, and the H_2 flow was a constant 154slm. The etch time was held at a constant 2.0 minutes, except for substrates which showed little or no etching after 2.0 minutes. In these cases a separate etch run was made with an increased etch time of 5.0 minutes. The etch depth, any lateral etch, and etch profiles were determined at three locations per wafer on the oxide patterned substrates using SEM inspection of fracture cross sections and photomicrograph measurements. The etch depth on the unpatterned substrates was determined by the difference of pre-etch and post-etch epitaxial thickness measurement using a Nicolet DX-ECO FTIR system. Five thickness measurements at different locations were made on each wafer. These were averaged for the etch depth determination. The measurement of haze was performed using a Tencor Surfscan. This is a relative measurement where the larger number indicates the presence of more haze. The analysis of all experimental **in situ** HCl etch rate and other data was performed using a VAX computer with a VMS operating system. Data manipulation was performed using the SAS (Statistical Analysis System) software package. The majority of the atmospheric etch rate data was used to develop the estimated etch rate and other response surfaces. This required that a number a extra etch runs were made in order to test the validity of these response surface equations.

RESULTS AND DISCUSSIONS

The silicon etch rates and haze formation for the patterned and unpatterned substrates were found to be different. This difference is attributed to the presence of the oxide mask on the patterned substrates, which contributes to enhanced etching from the formation of SiO at high temperatures and other possible secondary reactions including the formation of H_2O at lower temperatures. The estimated response surface for the silicon etch rate ER_{Si} for non-oxide, B substrates, is given as a function of HCl flow and etch temperature by equation 2.

$$[ER_{Si}]^{1/2} = -0.16367F + 0.0002FT \qquad (2)$$

The standard error of estimate is a measure of the model's prediction ability or scatter and is also frequently referred to as the root mean square error (RMS). We found this value in terms of ER_{Si} to be 0.0622 using 25 degrees of freedom. This indicates there is some scatter in estimated etch rate ER_{Si} has a stronger dependence on etch temperature

the experimental data. The cross product term of equation 2 indicates than on flow (partial pressure of HCl). A lack of an intercept indicates that at zero or low flows the etch rate is zero. A contour plot giving the estimated ER_{Si} for the entire parameter field was calculated using equation 2 and is shown in Figure 1.

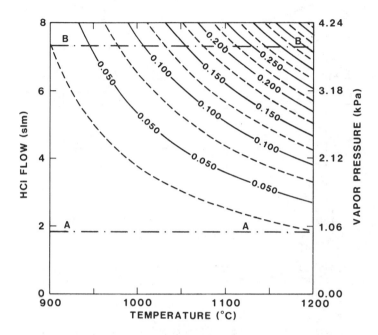

Fig. 1--Estimated etch rate ER_{Si} given as a function HCl flow and etch temperature. This contour plot is for unpatterned silicon and the etch rate contours are given in units of μm/min.

The region of Figure 1 outside the points labeled A and B, 0.0 to 1.82slm and 7.3 to 8.0slm, respectively, are extrapolated from equation 2. The lower region of the plot, less than 2.0slm, indicates that little or no etching occurs. Transformation from equation 2 to etch rate values requires substituting the desired F and T values, solving then squaring. The etch rate contours shown in Figure 1 indicate a stronger etch temperature dependence than previously reported [2,3,13,21]. Based on a review of previous work, all of which was performed using horizontal reactors and induction heating, we concluded that the etch rate dependence might be related to the reactor design and gas flow dynamics. Further study is needed to verify this finding.

Figure 2 shows the contour plot for the silicon etch rate ER_p for the patterned oxide, A-type substrate of Table 1, representing a silicon substrate in a possible selective epi process. This contour plot was calculated using equation 3, which is a polynomial for the estimated etch rate response surface of a patterned oxide substrate. We found the RMS error to be 0.0716 with 19 degrees of freedom. Again, the lack of an intercept in Figure 2 indicates that no etching occurs at low

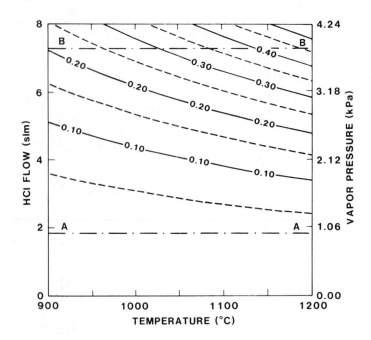

Fig. 2--Estimated etch rate ER_p contour plot for the patterned oxide substrate. The etch rate contours are given in units of μm/min.

or zero HCl flow. In this case, the cross product term of equation 3 is not as dominant as in equation 2.

$$[ER_p]^{1/2} = 0.03263F + 0.0001FT \qquad (3)$$

In contrast to Figure 1, the patterned substrate etch rate ER_p shows a stronger HCl flow dependence and a much weaker etch temperature dependence. We observe higher etch rates across the entire parameter field. The stronger dependence on the HCl flow, or partial pressure of HCl, agrees with most previous work and recent studies [14]. These higher etch rates result from the presence of the oxide mask [5,16]. This enhancement in etch rate we attribute to the formation of SiO [5] and H_2O in the etching process [17]. This observed etch rate enhancement agrees with previous work [5,16].

Haze formation on the unpatterned silicon wafers is independent of HCl flow and only dependent on the etch temperature. This is illustrated in Figure 3 which indicates regions with and without haze. The formation of haze is always observed at temperatures less than 1025°C, while the no haze region is at a temperature greater than 1050°C. The area between 1025°C and 1050°C is a transition region which may be dependent on the surface mobility of a species, or reaction kinetics. Equilibrium calculations [17] indicate the major species in

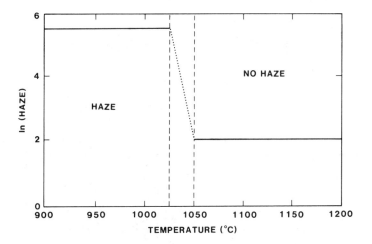

Fig. 3--Estimated haze plotted as a function of etch temperature. The ln(haze) is given for unpatterned silicon.

the 1025°C to 1050°C range are HCl, $SiCl_2$ and $SiCl_3H$. Whether these species contribute to haze formation will not be known until further study is completed. The unpatterned silicon substrates show a large operational zone. This includes all temperatures from 1050°C to 1200°C and all HCl flows between 1.82slm and 7.27slm. This leaves the process engineer with the decision of what depth of silicon should be removed from the substrate surface. Ritcher and coworkers [22] report that approximately 0.5μm is an optimum thickness to remove without generating excessive surface defects.

Equation 4 is the polynomial response surface for haze formed on the patterned oxide substrate given as a function of HCl flow and etch temperature.

$$\ln(H) = -31.48 + 0.0822T - 1.663F - 4.18(10^{-5})T^2 + 0.180F^2 \quad (4)$$

Figure 4 shows haze as having a complex dependence on temperature and HCl flow. We observed haze at all temperatures less than 1000°C and at all HCl flows. The patterned oxide wafer shows an entirely different response for haze than the bare silicon wafer. A broader transition between the haze and no haze areas was observed for the oxide patterned substrates. This wider transition region showing a complex dependence on the HCl flow and etch temperature is attributed to the presence of the oxide mask. The best results for the patterned oxide substrates were obtained between the temperatures of 1050°C and 1200°C at HCl flows from 3.35slm to 7.24slm. A RMS error of 0.472 was found with 20 degrees of freedom.

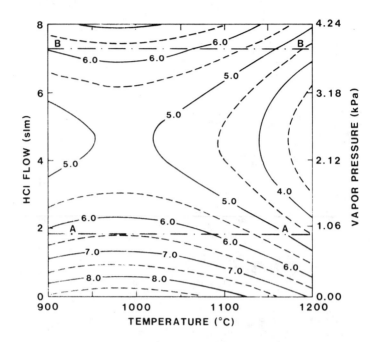

Fig. 4—Estimated haze contour plot for patterned oxide substrates. The ln(haze) is given as a function of HCl flow and Etch temperature. Haze is dimensionless quantity, and the contour lines are given in values of ln(haze).

Anisotropy is a measure of the lateral and vertical etch uniformity. We define the anisotropy $A = V_l/V_v$, where V_l is the lateral etch rate and V_v the vertical etch rate. The lateral etch is also defined as $V_l = V_l' + V_l''$, where V_l' is the standard lateral etch rate and V_l'' is the silicon-oxide encroachment etch rate. When both the lateral and vertical etch rates are equal, the etch is referred to as isotropic, which also suggests that the encroachment etch rate is zero.

We observed in this study that under certain etch conditions, enhanced lateral etching occurred with the patterned oxide substrates. Variations in the anisotropy ranged from 2 to 30 across the parameter field. The polynomial response surface of the anisotropy for the parameter field is given by equation 5.

$$\ln(A+1) = 18.34 - 0.0456T + 2.76(10^{-5})T^2 + 8.54(10^{-4})TF - 8.51(10^{-7})T^2F \quad (5)$$

which shows the anisotropy to be strongly dependent on etch temperature. A contour plot calculated using equation 5 is shown in Figure 5.

The quantity $\ln(A+1)$ is used in the response equation 5 to obtain smoother contours and eliminate the problem of dealing with quantities. The contour lines are given in values of $\ln(A+1)$ of the natural log less

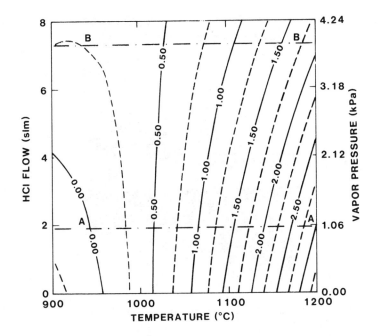

Fig. 5--Estimated anisotropy dependence on etch temperature. Contour plot of etch anisotropy ln(A+1) as a function of HCl flow and etch temperature for patterned oxide substrates. The contour lines are given in values of ln(A+1).

than 1.0. The enhanced lateral etching was observed at temperatures greater than 1100°C and only at the Si-SiO$_2$ interface. The cause is the formation SiO, which has a high vapor pressure at temperatures greater than 1100°C [5]. Lin [17] reports that the vapor pressure of SiO is significant above 1200°C, however, we see appreciable lateral etching at temperatures less than 1200°C. Our results agree with other investigators [14] demonstrating the strong dependence of the etch anisotropy on etch temperature; however, the SEM cross sections indicate severe anisotropy at temperatures less than 1200°C. We illustrate this in Figure 6 which shows SEM cross sections obtained by backscattering indicating various degrees of lateral etching for different etching conditions.

One observes severe lateral etch at low HCl flows and 1200°C, Figure 6a, but as the flow increases, the amount of the lateral etch is not nearly as great. Likewise, a decrease in etch temperature accomplishes the same result. Our results indicate that at low HCl flows and high temperatures SiO appears to be a controlling factor, then as the HCl flow increases from 1.82slm to 7.27slm the dominance of the SiO is reduced by the presence of the HCl. A continued temperature reduction eliminates the enhanced lateral etching and also changes the etch profile as shown in Figure 7. Figure 7a shows that the onset of the SiO lateral etch dominance initiates at 1100°C; note the slight

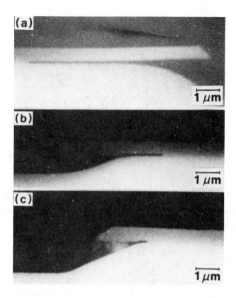

Fig. 6--Lateral etching observed on patterned oxide substrates.
(a) Etch temperature 1200°C; HCl flow 1.82 slm.
(b) Etch temperature 1200°C; HCl flow 6.34 slm.
(c) Etch temperature 1170°C; HCl flow 5.25 slm.

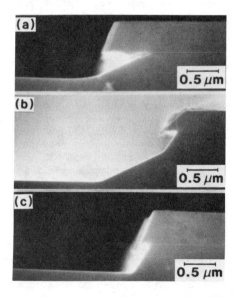

Fig. 7--SEM cross sections of oxide patterned substrates. Cross sections are for different etch conditions.
(a) Etch temperature 1100°C; HCl flow 4.72 slm.
(b) Etch temperature 1050°C; HCl flow 4.72 slm.
(c) Etch temperature 1000°C; HCl Flow 7.27 slm.

lateral etch. Based on our observations with the 1200°C samples, the amount of lateral undercut was larger at low HCl flows and smaller at higher HCl flows. This supports the etch anisotropy dependence on etch temperature illustrated in Figure 5. Decreasing the etch temperature to 1050°C shows no enhanced lateral etching, and a further decrease to 1000°C also show no enhanced etching.

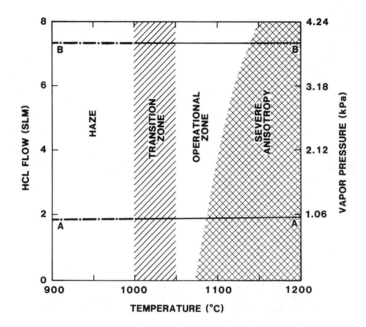

Fig. 8--A representation of operational zone for selective epi. The operational zone is given as a function of HCl flow and etch temperature. The haze, transitional, operational and severe anisotropy zones are illustrated.

An in situ HCl etch operational zone for a selective epi process is more restrictive than for an epitaxial silicon process. All temperatures less than 1050°C are eliminated because of haze formation, while the high temperature region is restricted by the etch anisotropy. This is illustrated in Figure 8 which shows the operational etch zone for a selective epi process. This can be used as a guide for the process engineer who is using a selective epi technology must decide whether it is feasible to use a HCl in situ etch. This operational zone is bounded on the left by the transition zone and on the right by a region of severe anisotropy. Etch profiles similar to Figure 7b and 7c are typically observed in this region. Etch rate also must be considered when selecting an operational region for the selective epi process.

SUMMARY/CONCLUSIONS

We have demonstrated that statistical design methods can be applied to process characterization, giving a complete picture of an entire parameter field with a limited number of experiments. A large operational zone for the HCl etch used in the epi process was determined. More study is required to understand the unexpected etch rate dependence of the unpatterned substrates. We also have demonstrated that a high temperature **in situ** etch can be applied to a selective epi process provided that necessary precautions are taken. The use of an oxide mask in the selective epi process does place restrictions on operational parameters. We also have shown that the presence of the oxide mask enhances the **in situ** etch rate and haze formation. The **in situ** HCl etching with the patterned oxide substrates at temperatures greater than 1100°C is detrimental due to the formation of SiO and H_2O. The formation of SiO contributes to severe etch anisotropy. This enhanced lateral etch is attributed to the formation of SiO at the Si-SiO_2 interface. The operational zone for the **in situ** HCl etch used in an epitaxial silicon deposition is large and does not impose any major limitations on the epi process; whereas, the operational zone for a selective epi process has major limitations of haze and etch anisotropy. Operational parameters should be selected with care.

ACKNOWLEDGEMENTS

The authors wish to thank the following individuals, K. M. Baumgardner, H. Praefcke, and M. Lynn for performing the HCl etchs, L. Witherspoon for sample preparation; E. Thomas for discussions and suggestions related to the statistical design, and P. Ho and B. Brieland for their discussions and review of the manuscript.

LITERATURE REFERENCES

[1]   G. A. Lang & T. Stavish, "Chemical Processing of Silicon with Anhydrous Hydrogen Chloride", RCA Review, Vol. 24, 1963, pp. 488-498.

[2]   W. H. Shepherd, "Vapor-Phase Deposition and Etching of Silicon", J. Electrochem. Soc., Vol. 112, No. 10, 1965, pp. 988-993.

[3]   L. V. Gregor, P. Balk & J. Campagana, "Vapor-Phase Polishing of Silicon with H_2-HBr Gas Mixtures", IBM J. Res. Dev., Vol. 9, July 1965, pp. 327-332.

[4]   A. Reisman & M. Berkenblit, "The Etching and Polishing Behavior of Ge and Si with HI", J. Electrochem. Soc., Vol. 112, No. 8, 1965, pp. 812-816.

[5] T. L. Chu, G. A. Gruber & R. A. Stickler, "In situ Etching of Silicon Substrates Prior to Epitaxial Growth" J. Electrochem. Soc., Vol. 113, No. 2, 1966, pp. 156-158.

[6] F. W. Smith & G. Ghidini, "Reaction of Oxygen with Si(111) and (100); Critical Conditions for the Growth of SiO_2", J. Electrochem. Soc., Vol. 129, No. 6, June 1982, pp. 1300-1306.

[7] J. G. Gaultier, M. J. Katz & G. A. Wolff, "Gas Etching and its Effect on Semiconductor Surfaces", Z. Krist., Vol. 114, 1960, pp. 9-22.

[8] W. Kern & C. A. Deckert, "Chemical Etching", Thin Film Processes, J. L. Vossen & W. Kern Eds., Academic Press, New York, 1978 pp. 401-496.

[9] H. R. Chang, "Defect Control For Silicon Epitaxial Processes Using Silane Dichlorosilane, And Silicon Tetrachloride", Defects in Silicon, 83-9, W. M. Bullis & L. C. Kimmerly Eds., The Electrochemical Soc. Inc., Pennington NJ, 1983, pp. 549-557.

[10] L. J. Stinson, J. A. Howard & R. C. Neville, "Sulfur Hexafluoride Etching Effects in Silicon", J. Electrochem. Soc., Vol. 123, No. 4, 1976, pp. 551-553.

[11] P. Rai-Choudhury, "Substrate Surface Preparation and its Effect on Epitaxial Silicon", J. Electrochem. Soc., Vol. 118, No. 7, 1971, pp. 1183.

[12] P. M. Petroff, L. E. Katz & A. Savage, "Stacking Fault Nucleation and Growth Mechanism in CVD Silicon Epitaxial Films", Semiconductor Silicon '77, The Electrochemical Society, Pennington NJ, 1977, pp. 761-772.

[13] F. Ritcher, Th. Morgenstern & R. Sperling, "On Batch Homogeneity in Horizontal CVD Reactors (II) HCl Gas-Phase Etching", Crystal Res. & Technol., Vol. 18, No. 5, 1983, pp. 641-650.

[14] M. Druminski & R. Gessner, "Selective Etching and Epitaxial Refilling of Silicon Wells in the System $SiH_4/HCl/H_2$", J. Crystal Growth, Vol. 31, 1975, pp. 312-316.

[15] K. Sugawara, Y. Nakazawa & Y. Sugita, "Observation of Facets on Silicon Surface Formed by Hydrochloride Acid Selective Vapor Etching", Electrochem. Technol., Vol. 6, 1968, pp. 295-296.

[16] K. Sugawara, "Facets Formed by Hydrogen Chloride Vapor Etching on Silicon Surfaces Through Windows in SiO_2 and Si_3N_4 Masks", J. Electrochem. Soc., Vol. 118, No. 1, 1971, pp. 110-114.

[17] S. Lin, "Mass Spectrometric Studies on High Temperature Reaction Between Hydrogen Chloride and Silica/Silicon", J. Electrochem. Soc., Vol. 123, No. 4, 1976, pp. 512-514.

[18] D. Doehlert, Experimental Strategies, Edgework, Inc., Redmond WA 98052.
[19] H. Kuhne, Th. Morgenstern & Ch. Kuhne, "Temperature Profile and Homogeneity of the Growth Rate of CVD Silicon from $SiH_4-HCl-H_2$ Mixtures", Crystal Res. & Technol., Vol. 17, No. 9, 1982, pp. 1105-1116.

[20] W. J. Diamond, Practical Experimental Design For Engineers And Scientists, Lifetime Learning Publications, Belmont CA, 1981.

[21] F. Richter, G. Weidner, H. Borchardt, E. Bugiel, M. Kittler, K. Schmalz, M. Weidner & H. Raush, "The Influence of Gas-phase Etch Removal, Depostion Temperature and Substrate Properties on the Crystallographic Perfection of Silicon Layers", Crystal Res. & Technol., Vol. 18, No. 12, 1983, pp. 1521-1531.

[22] Th. J. M. Kuijer, L. L. Giling & J. Bloem, "Gas Phase Etching of Silicon with HCl", J. Crystal Growth, Vol. 22, 1974, pp. 29-33.

Dielectrics and Junction Formation Techniques

Ramamurthy, V.

DOPED OXIDE SPIN-ON SOURCE DIFFUSION

REFERENCE: Ramamurthy, Vallivazhai., "Doped-oxide Spin-on Source Diffusion," Emerging semiconductor Technology, ASTM STP 960, D.C. Gupta and P.H. Langer, Eds., American Society for Testing and Materials, 1986

ABSTRACT: The advantages offered by doped oxide spin-on source diffusion are briefly out-lined from thermodynamic as well as techno-economic considerations. Diffusion characteristics are presented for spin-on sources of gallium, gallium with boron, gallium with arsenic, boron and phosphorus, keeping in view their potential applications. Sheet resistivity and doping profile measurements show that the dopant distributions which essentially exhibit a time varying surface concentration, cannot be explained by any of the existing models for solid-state diffusion available in the literature. With three specific examples it is shown how this technique can be employed for the fabrication of different types of semi-conductor devices. The investigations also show that the results of diffusion are independent of gas-flow rates as this type of diffusion dispenses with gas-solid interactions.

KEYWORDS: Silicon, diffusion, sheet-resistivity, spreading resistance, doping profile, junction-depth

There is an ever increasing interest in the use of doped oxide spin-on diffusion source for junction formation in silicon because of a number of advantages inherent to this technique [1]-[3]. While diffusion in sealed quartz ampoules has been successful for both gallium and arsenic, they are cumbersome, involve high rate of quartz consumption and high surface concentration is still very difficult with ion-implantation and CVD techniques. Further, ion-implantation facility requires high capital costs and CVD technique involving arsenic is extremely

Dr. V. Ramamurthy is Senior Engineer at Bharat Heavy Electricals Limited, Corporate Research and Development Division, Vikasnagar, Hyderabad-500 593, India.

hazardous. Diffusion with spin-on sources promises to solve many of these problems. It is relatively inexpensive, easily automatable, non-toxic and has capacity for high throughputs. Therefore this technique is very attractive from techno-economic considerations for introducing controlled amounts of dopant atoms into host silicon lattice required in the production of electron devices and integrated circuits.

When compared to the conventional diffusion process, kinetics of diffusion from doped oxide sources of this type differ in the following aspects:

1. There is a serious constraint on the availability of the diffusing species for the host lattice, not infinite as in the conventional pre-deposition process.

2. Surface concentration is not solubility limited, in the sense, even at higher temperatures (which might be dictated by other considerations) we can realise surface concentration values which are lower than the solubility limit at that temperature.

3. The doped oxide is deposited at low temperatures. Therefore the diffusion kinetics are primarily determined by the rate kinetics at the oxide-silicon interface. The diffusion characteristics during the actual high temperature annealing are insensitive to gas flow dynamics, as no gas-solid interactions are involved.

Because of these thermodynamic considerations, spin-on source diffusion from doped oxides offers greater flexibility in choosing diffusion schedules and device configurations, compared to the conventional predep-drive in diffusion techniques.

In this paper results of work done on five important spin-on dopants for formation of p-n junctions in silicon are presented. Diffusion characteristics are given for the diffusion of gallium, boron, and phosphorus. Diffusions involving simultaneous diffusion of gallium with boron and gallium with arsenic from a single source are also discussed. In discussing the implications of these results, specific examples are cited with results and other potential applications are suggested.

EXPERIMENTAL METHOD

Silicon wafers from Wacker Chemitronic were polished to mirror finish using a combination of mechanical and chemical polishing for all investigations. For solar cells and high power devices, the samples were etch polished prior to diffusion. After the usual degreasing procedures, the samples were invariably made hydrophylic by boiling in a mixture of hydrogen peroxide and concentrated sulphuric acid for 15 minutes, to ensure good adhesion of the spin-on dopant. A fifteen minute rinse in 16-18 Meg.

ohm-cm de-ionised water was followed by a bake at 150°C for at least 15 minutes.

The spin-on formulations are essentially doped silicates dissolved in an organic solvent. The dopant is spun on the samples using a standard photoresist spinner. While a spin time of 15-20 seconds was adequate for most applications, choice of spin speed was normally dictated by the viscosity of the dopant as well as the diffusion schedule. The samples were prebaked for 15 minutes at 150-200°C to densify the spun film and drive out excess solvent. Diffusions at high temperature were carried out according to a predetermined time temperature schedule. The ambient was normally nitrogen/argon while in some specific cases controlled amounts of oxygen were introduced. The gas flow rates were varied over a wide range from a few ml/min. to a few litres/min. to examine its effects on results. After diffusion the doped silicates were deglazed in hydroflouric acid and rinsed in de-ionised water. Samples were characterised for sheet-resistivity and junction depth by four point probe and angle lapping respectively. Some samples were analysed for dopant profile by spreading resistance probe model ASR 100C/2 from Solid State Measurements Inc.

RESULTS AND DISCUSSION

Gallium

Open tube diffusion of gallium in silicon at 1200° and 1250°C was investigated for achieving deep diffusions and reasonable surface concentrations required for power devices like thyristors. Polished n-type samples with resistivity 300-350 ohm-cm were used for this purpose. Spin-on dopant supplied by Emulsitone was used.

Experiments carried out at 1250°C for 16 hours yielded a junction depth of 35 µm. Fig.1. shows the doping profile of gallium in silicon obtained from spreading resistance probe measurements. From Fig.2, it is clear that sheet-resistivity varies considerably with time. These data cannot be explained by the conventional models of diffusion which use erfc and gaussian distribution functions, because of the different boundary conditions that exist for the diffusion from a thin film of doped oxide. Further, we could not fit this data to any of the existing models for diffusion from doped oxide sources [4]-[5]. The junction depth was however found to vary linearly with square-root of time of diffusion.

Gallium With Arsenic

The gallium silica film with arsenic is a liquid dopant source that can be used to realise simultaneous diffusion of arsenic and gallium into silicon to varying depths. This is possible because, at a tiven temperature, gallium diffuses much faster than arsenic, into n-type silicon substrate.

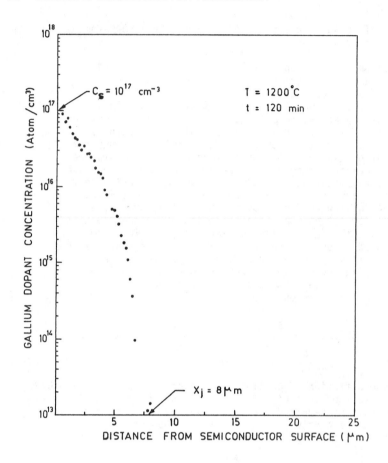

Fig. 1--Doping profile of gallium in silicon

Variation of junction depth and sheet-resistivity of arsenic as a function of diffusion temperature is shown in Fig. 3. and variation of gallium junction depth with background resistivity of n-type silicon substrate is given in Table 1. In all cases, diffusion was carried out at 1200°C for one hour duration.

TABLE 1--Gallium with Arsenic

Substrate resistivity (ohm-cm)	30	50	100	300	1000	5000
Gallium junction depth* (µm)	1.6	2.1	3.6	4.0	5.5	6.4

*for diffusion at 1200°C for 1 hour.

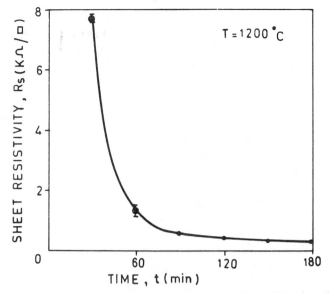

Fig. 2--Sheet-resistivity variation with time for gallium diffusion in silicon at 1200°C

An npn structure realised with a single step diffusion is shown in Fig. 4. This, coupled with the fact that silicon dioxide can selectively mask arsenic diffusion allowing gallium to diffuse through, makes this an attractive choice for the fabrication of power transistors with a single step diffusion using n/n^+ epi-substrates. Apart from process simplicity, this procedure helps in the preservation of carrier lifetime considerably.

Gallium With Boron

This is a diffusion souurce containing both gallium and boron that can be used to achieve high surface concentration of boron and deep gallium junction in one diffusion. This could solve the twin problems of ampoule diffusion and back junction removal, which make the conventional diffusion technique very cumbersome and uneconomical for the fabrication of high power diodes.

The dopant profile of boron/gallium in 300-350 ohm-cm n-type silicon for 2 hours diffusion at 1200°C measured by spreading resistance probe is shown in Fig. 5. It is clear from the figure that high surface concentration and deep junction are both realisable in a single diffusion step. In subsequent trials involving diffusion at 1250°C for 24 hours, surface concentration better than 10^{19} cm^{-3} and junction depths of the order of 55 μm were achieved. From these results one can infer that it is indeed possible to make high power diodes using this technique.

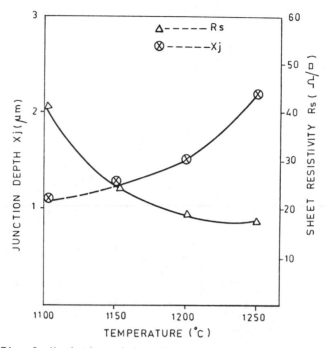

Fig. 3--Variation of junction depth and sheet-resistivity of arsenic with temperature for one hour diffusion in silicon

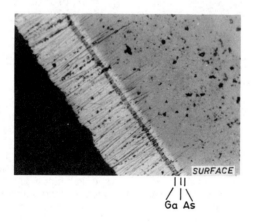

Fig. 4--Angle lapped silicon specimen showing arsenic (white) and gallium (black) stains (angle:5°, magnification:100x)

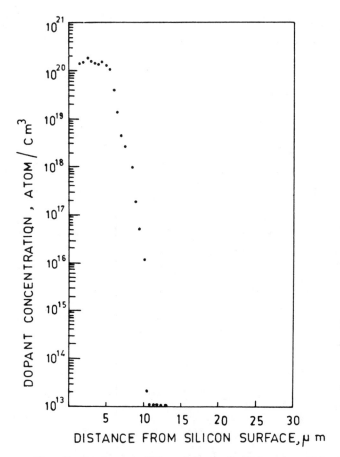

Fig. 5--Doping profile of boron/gallium in silicon for two hours diffusion at 1200°C

Boron

Studies were conducted using spin-on dopants of boron, mainly keeping in view its usefulness as a back junction for BSF solar cells and p-i-n diodes.

Table 2 shows sheet-resistivity and junction depth data as a function of temperature for one hour diffusion of boron into 300-350 ohm-cm n-type silicon. In Table 3 Trumbore's solid solubility data for boron in silicon [6] is compared with the surface concentration data computed from the results of Table 2 assuming gaussian distribution for boron. The values obtained from spreading resistance probe are also shown alongside for 1050° and 1200°C. Though the surface concentration values are in good agreement, the doping profile of boron in silicon shown in Fig. 6 clearly shows that the doping profile cannot indeed be characterised as gaussian.

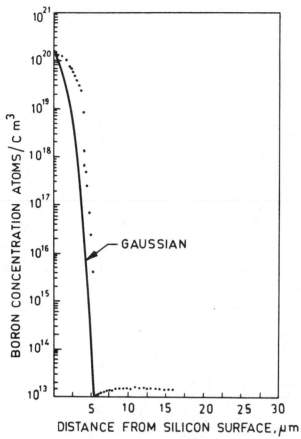

Fig. 6--Doping profile of boron in silicon for one hour diffusion at 1200°C. The gaussian distribution (solid line) is shown for comparison

TABLE 2--Diffusion Data for Boron

Temperature T (°C)	Sheet-Resistivity RS	Junction Depth X_j (μm)
1050	108.9	1.9
1100	30.0	2.6
1150	3.3	4.5
1200	2.9	5.0

TABLE 3--Boron surface Concentration (C_s)

Temperature T (°C)	C_s from Table 2 (cm^{-3})	Trumbore's solid Solubility data (cm^{-3})
1050	2×10^{19} (4×10^{19})a	8×10^{20}
1100	5.5×10^{19}	4×10^{20}
1150	3×10^{20}	4.5×10^{20}
1200	3.5×10^{20} (1.5×10^{20})	5×10^{20}

[a]values in parenthesis indicate data obtained from spreading resistance probe measurements.

Phosphorus

Phosphorus diffusion using spin-on dopant was studied for its application in the fabrication of solar cells and thyristors. Dopant film designated as "phosphoro silica film 2×10^{20} and 10 ohm-cm p-type wafers" were used in these investigations.

Sheet-resistivity variation with time for diffusion at 1000°C is shown in Fig. 7 for short diffusion times. Fig. 8 shows a typical dopant profile obtained using spreading resistance probe. Deep diffusions have also been carried out to realise junction depth of the order of 18 μm and surface concentrations better than 5×10^{19} cm^{-3}.

It is clear from the results that we can vary sheet-resistivity at will to match the grid pattern for metallisation used in solar cell fabrication.

APPLICATIONS

In this section three specific examples are given to demonstrate the application of spin-on diffusion technique for fabrication of semiconductor devices.

p-i-n Diodes

High power p-i-n diodes are used for switching microwave power in phased array radars. In the conventional processing, very high resistivity (>1000 ohm-cm), n-type silicon is chosen as starting material and subjected to two high temperature diffusion steps and two high temperature oxidation steps.

In the present work both the p$^+$-n and n-n$^+$ junctions were simultaneously formed using spin-on dopants of boron and phosphorus supplied by Transene Inc. Multi-layer metallisation was done through a metal mask and the diodes were subsequently me etched. Exposed junctions were passivated using a spin-on oxide film.

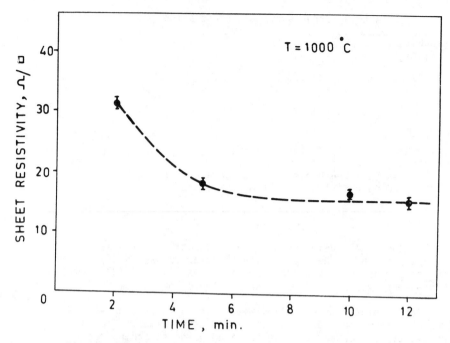

Fig. 7--Sheet resistivity variation with time for phophorus diffusion in silicon at 1000°C.

The diodes exhibited a forward voltage drop of 600 mV and breakdown voltage better than 1100 V. The depletion capacitance was 1.0 pF and the minority carrier lifetime as measured by the standard double-pulse technique was 8 μs. The diodes showed an rf resistance of 1.5 ohms. The insertion loss and isolation [7] of the diodes measured in the series configuration were 1.5 dB and 25 dB respectively at 3 GHz.

Because of four fold reduction in high temperature processing, a starting material resistivity of 350 ohm-cm was adequate to realise the desired results.

High-power Thyristor

Fabrication of high power thyristors involves a deep diffusion of gallium into n-type silicon to create a p-n-p structure and a final diffusion of phosphorus to form a deep n-junction-retaining at the same time a high surface concentration of phosphorus.

An attempt was made to replace the final phosphorus diffusion, which is conventionally carried out using $POCl_3$ source, by spin-on source diffusion of phosphorus. Starting with a 50 ohm-cm, n-type silicon wafer of 25 mm dia., conventional processing was carried out to create the desired p-n-p structure. Spin-on diffusion was carried out to form a 20 μm deep junction. The surface concentration was estimated to be better than 5×10^{19} cm^{-3}. As can be seen from Table. 4, the results are encouraging.

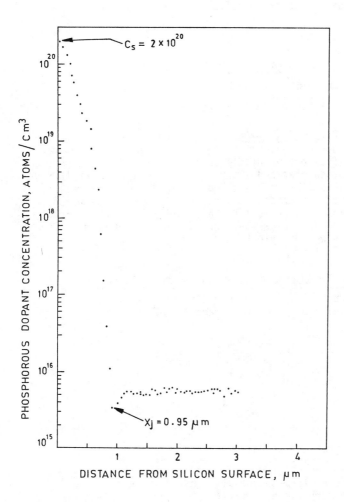

Fig. 8--A typical doping profile of phosphorus in silicon for diffusion at 1000°C for 10 mnts.

TABLE 4--Thyristor Characteristics

1.	Forward Voltage drop	:	1.53V
2.	Minimum gate Triggering Current	:	175 mA
3.	Minimum gate Triggering Voltage	:	0.95 V
4.	Forward Blocking Voltage	:	1000 V
5.	Reverse Blocking Voltage	:	1000 V
6.	Device Turn-off Time	:	120 μs

Solar Cells

N^+-p solar cells were fabricated on CZ grown single crystal silicon of 1.0 ohm-cm resistivity using spin-on source of phosphorus for diffusion. The diffusion was carried out at 900°C on 2 inch dia. samples and after Ti-Pd-Ag metallisation through a metal mask, samples were scribed into 2x4 cm² solar cells. for purposes of comparison n^+-p solar cells were made using $POCl_3$ diffusion. Lighted I-V characteristics were obtained at 28°C using ORIEL solarsimulator under AMI conditions The results are shown in Table 5.

TABLE 5--Lighted I-V Characteristics of Solar Cells*

Cell Type	Open ckt. Voltage	Short-Circuit Current density	Fill Factor	Active Area Efficiency
	V_{oc}(mV)	J_{sc}(mA/cm²)	FF	η(%)
$POCl_3$ diffused[a]	590	26.5	0.76	13
Spin-on diffused[b]	570	24.5	0.77	11.9

*without anti-reflection coating

It is evident from the results that spin-on diffused cells stand a very close comparison to the base line cells.

SUMMARY AND CONCLUSIONS

The potential advantages offered by doped oxide spin-on source diffusion technique are briefly outlined from thermodynamic as well as technoeconomic considerations. Diffusion characteristics are investigated for spin-on sources of gallium, gallium with boron, gallium with arsenic, boron and phosphorus, keeping in view their potential applications. From sheet-resistivity and doping profile measurements, it has been shown that the dopant distributions which essentially exhibit a time varying surface concentration, cannot be explained by any of the existing models of diffusion available in the literature. With three specific examples it is demonstrated how this processing technology can be useful in the fabrication of semiconductor devices. Our investigations have also shown that the results are insensitive to gas flow rates as gas solid interactions are not involved in this type of diffusion.

To briefly state the disadvantages, the dopants usually have a limited shelf life (3-6 months) and specific storage temperature (4-10°C). Wile fluctuations in the ambient humidity and temperature at the spinning area affect the results considerably so also the nature of the silicon surface over which the dopant is spun-on.

ACKNOWLEDGEMENT

The author acknowledges substantial contributions made by Dr. J. Subrahmanyam, Dr. P.R. Vaya and (Mrs) P. Tiku. The author is thankful to Dr. K.L. Jasuja for his help in measurements with spreading resistance probe. Thanks are also due to Mr. V. Radhamohan for his active assistance in device processing. Acknowledgment is also due to Dr. P.R. Krishnamurthy who was largely responsible for initiating this work at this centre.

REFERENCES

[1] Justice, H.B., Harnish, D.F., Jones, H.F., "Diffusion Processing of Arsenic Spin-on Diffusion Sources," Solid State Technology, Vol.21, No.7, July 1978, pp.39-42.
[2] Beyer, K.D., "A New Paint-on Diffusion Source," Journal of Electrochemical Society, Vol.123, No.10, October 1976, pp. 1556-1560.
[3] Flowers, D.L. and Schi-yi Wu, "Diffusion in Silicon from a Spin-on Heavily Phosphorus Doped Oxide Source," Journal of Electrochemical Society, Vol.129, No.10, October 1982, pp.2299-2302.
[4] Barry, M.L. and Olofsen, P., "Doped Oxides as Diffusion Sources", Journal of Electrochemical Society, Vol.116, No.6, June 1969, pp. 854-860.
[5] Ghoshtagore, R.N., "Model of Doped Oxide Source Duffusion in Silicon", Solid State Electronics, Vol.17, No.10, October 1974, pp.1065-1073.
[6] Trumbore, F.A., Bell Systems Technical Journal, Vol.39, 1960, pp. 205.
[7] Chaturvedi, P.K., Ramamurthy, V. and Kakati, D., "Thermal Design of Microwave p-i-n diodes," Solid State Electronics, Vol.22, No.2, February 1979, pp. 129-134.

E. Lora-Tamayo, J. Du Port de Pontcharra, and M. Bruel

EFFECT OF A SHALLOW XENON IMPLANTATION ON A PROFILE MEASURED BY SPREADING RESISTANCE

REFERENCE: Lora-Tamayo, E., Pontcharra, J., and Bruel, M., "Effect of a shallow Xenon implantation on a profile measured by spreading resistance", Emerging Semiconductor Technology, ASTM STP 960, D.C. Gupta and P.H. Langer, Eds., American Society for Testing and Materials, 1986.

ABSTRACT: Shallow xenon ion implantations were performed into silicon wafers previously doped by BF_2^+ implantation and subsequent annealing. The major extended changes in the spreading resistance profile of the P-type doping, arising from the shallow xenon implantations, are presented here. A possible explanation of this effect on the electrical profile, based on "carrier spilling" described and analyzed by S.M. Hu (5) is presented.

KEYWORDS: Amorphization, Xenon ionic implantation, spreading resistance profile

The spreading resistance (SR) technique is nowadays a useful tool widely used to obtain data on uniformity, reproducibility and absolute values of the electrical activity of doped layers or bulk materials (1).

Also SR measurements on bevelled samples associated with mathematical processing provide the in-depht profile of the electrical activity in a very quick and smart way.

Dr. Lora-Tamayo is research scientist at the National Center of Microelectronics-CSIC, UAB Bellaterra (Barcelona-Spain); Dr. Pontcharra and Dr. Bruel are research scientists at LETI-IRDI, CEA-CENC, 85, 38041 Grenoble Cedex (France)

This profile is quite complementary to the atomic one obtained by using physico-chemical methods such as SIMS or RBS.

The ability of the SR technique to measure in-depth resistivity profiles led us to use it to characterice layers amorphized by ion implantation (defects and amorphization generated by ions give rise to an increase of silicon resistivity).

This kind of experiment is related to the technology of shallow junction which is one of the present challenges in microelectronics.

In order to avoid the need for high annealing temperatures with the undesirable effect of profile diffusion, a solid phase epitaxy (SPE) mechanism of amorphized layer is prefered (2). This process provides good electrical properties even at low annealing temperatures in the range 500-800°C.

Thus ion implantation step for shallow junctions applications must give rise to an amorphized upper layer, where the carrier profile will be created by thermal annealing. This is very easy for heavy ions such as arsenic, but becomes rather difficult for light ions such as boron.

An alternative for such a case is to use, in addition to boron, heavier ions such as silicon, germanium, xenon, etc... so that the electrical characteristics of the junction are not altered. This means that the thickness of the amorphized layer has to be compared to the junction depth in order to avoid residual defects in the space charge region.

Therefore the SR seems to be an elegant method for this comparison. When the amorphizing implantation is done after the shallow junction has been made, an unexpected effect appears, which will be described in this paper.

DESCRIPTION OF THE EFFECT (Fig. 1)

The original material is N-type bulk silicon boron-doped to form a shallow P^+/N junction in the 200 to 300 nm range.

A xenon implantation is subsequently made. The ion energy is chosen, so that the mean penetration depth be much lower than the junction depth in order to keep the disordered region small compared to the P-type profile.

110 EMERGING SEMICONDUCTOR TECHNOLOGY

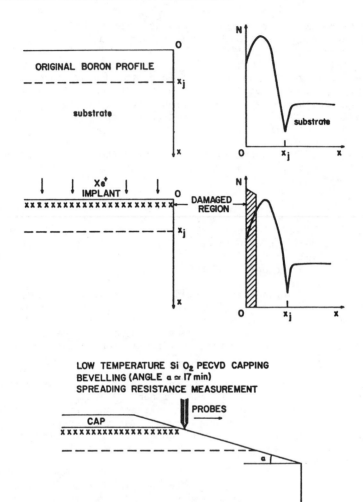

FIG. 1- Diagram of the situation and procedure

With this kind of experiments the expected results were:

- a strong modification of the upper part of the profile (corresponding to the damaged region caused by the Xe implantation).

- a minor alteration of the remaining part of the boron profile.

However the experiments led to the conclusion that

the Xe^+ ion implantation gives rise to a modification of the entire measured profile, even in the region where no defects were created.

Moreover for high xenon doses, above the amorphization dose (calculated with the Morehead and Crowder model, (3) , an unexpected resistivity increase appears in the remaining part of the profile, a flat resistivity region with the same value as the original bulk one is observed, as if the original profile were no longer there.

EXPERIMENTAL

We used P and N-type silicon wafers, 100 oriented and 1 to 10 $\Omega \cdot cm$ resistivity. The primary junction was made using BF_2^+ implantation followed by a 600°C, 30 min thermal annealing in nitrogen ambient. The wafers were encapsulated by 100 nm of $PECVD-SiO_2$.

The standard conditions of the BF_2^+ implantation were 50 KeV-10^{15} cm^{-2}; higher energies were also used without any change in the results. Monoatomic B^+ ions

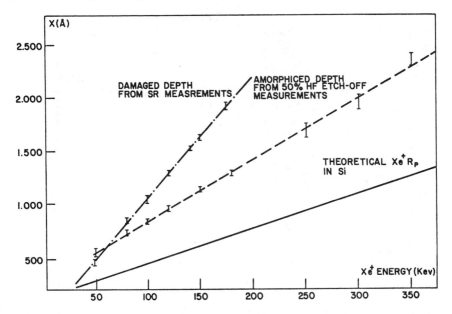

FIG. 2- Theoretical R_p and experimental damaged and amorphized depth in function of Xe^+ implant energy.

instead of BF_2^+ were also used to study the possible influence of fluorine on the observed effect: the conclusion was that there is no effect due to the fluorine. The type of the substrate N or P has no influence in the described effect. Of course in the case of substrate P we cannot talk of PN junction.

Once the capping SiO_2 was eliminated, the Xe implantation was done at different energies and doses.

To avoid the round edge bevel effect, for the SR measurements, the wafers were covered with a new layer of 500 nm PECVD SiO_2. The bevel angle was 17', the step 0.5 to 1 μ, the probe load 15 gr and the spacing between probes was 70 μ.

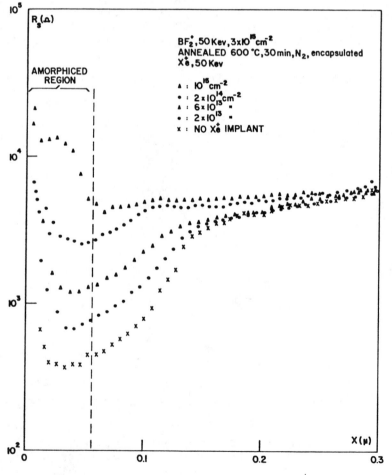

FIG. 3- Experimental SR profiles in shallow P^+P implanted with Xe^+ at difference doses.

The thickness of the amorphized layer created by the Xe bombardment was independently measured (Fig. 2) by etch-off of the amorphized Silicon in HF (50% dilute) and Talystep* measurement (4). In some cases wafers were annealed in an N_2 ambient at 600-700°C to study the recovering of the original profile.

RESULTS AND DISCUSSION

A first set of experiments was conducted with a P sustrate, keeping the Xe ion energy constant, the xenon ion dose varying from 2.10^{13} cm^{-2} to 10^{15} cm^{-2} (Fig. 3). In this figure the dashed line representants the limit of the amorphized region created by Xe bombardment (this depth was measured by HF etch-off). At low and medium doses no clear defference or discontinuity in the shape of the profile appears between the damaged region and the remaining part up to the "junction", but the whole profile is affected as if the carrier density were smaller.

For high incident xenon ion doses such as 10^{15} cm^{-2} (well above the threshold for amorphization) an increase in the resistivity is clearly visible in the damaged region (which is not surprising) but the remaining part of the profile has flattened as if electrical carriers due to boron had disappeared.

A second set of experiments (Fig. 4) was conducted to evaluate the effect of the ratio between the depth of boron profile and the depth of the damaged zone. Now the BF_2^+ implant on a P silicon sustrate was made at an energy of 100 KeV and a dose of 3×10^{15} cm^{-3}. The xenon dose was kept constant (10^{15} cm^{-2}) and the ion energy was chosen within the 40-120 KeV range. The dotted line represents the limit of the amorphization zone as measured by etch-off.

The "junction" is much deeper (6000Å) than in the former set of experiments because of the higher BF_2^+ energy. One can see some chanalization effects, but this has no importance for our discussion. This could explain the difference between the two sets of experiments for the case of Xe implantation at a dose of 10^{15} cm^{-2} : for the shallower junction this implantation leads to a disappearance of the profile, whereas for the deeper junction the alteration is minor.

The ratio between junction depth and thickness of the damaged zone therefore seems to play an important role. This is confirmed by the evolution of the observed

* Talystep is a trade mark from the Rank Organization.

profiles corresponding to increasing Xe implanted. As can be seen in Fig. 4, a Xenon implantation 120 KeV and 10^{15} cm^{-2} shows a similar behaviour (increased resistivity part) than the 10^{15} cm^{-2}, 50 KeV of the former set of experiments.

The effect of a shallow disordered region in the measured SR profile therefore appears to be very important and unexpected.

TABLE 1- Effect in two different BF_2^+ profiles

"Junction" depth, Å	Xe$^+$ Energy, KeV	Amorphized thickess, Å	Limit of profile disappearance cm^{-2}
3000	50	500	10^{15}
6000	120	1000	10^{15}

To show this "long range effect" a dashed line in Fig. 4 represents the extreme limit of the zone where SR measurements show some dispersion, assuming that this dispersion could be attributed to defects created by xenon bombardment. The results corresponding to this evaluation are reported as a function of energy in Fig. 2. This makes the comparison possible between etch-off measurements and these results. Measurements arising from SR dispersion give thickness about twice as great as those of etch-off. Nevertheless these values (like those which can be deduced from theoretical evaluations) remain much lower than the profile depth. It could therefore be said that the shallow damaged zone has a long range effect on the SR measured profile.

Other experiments were conducted using primary profiles made by monoatomic boron implantation and subsequent annealing. The results are quite similar to those presented herein. The type of junction P$^+$/N or P$^+$/P has no influence.

SIMS measurements before and after the Xe$^+$ implantation show no variation in the atomic profiles of boron (and also fluorine).

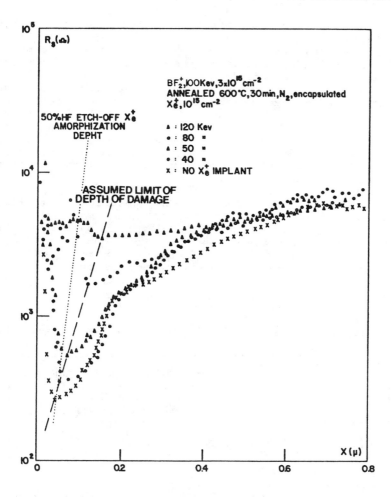

FIG. 4- Experimental SR profiles in P^+P junctions implanted with Xe^+ at different energies.

Other experiments were made by measuring SR of Xenon implanted in P and N-type silicon wafers. The SRP shows high resistance up to a depth similar to that observed in previously reported experiments. Although this is the expected amorphization depth, a RBS analysis could be done to check whether there is some Xe chanalization effect.

The explanation of the described effect could be

based on the "carrier spilling" phenomenon described and simulated by S.M. Hu(5). Carrier spilling could exist between the undamaged boron profile and the damaged one. In this shallow zone the majority of carriers are trapped in the created defects and the weak carrier concentration may encourage a carrier spilling mechanism from the non damaged zone similar to the one described by S.M. Hu in high-low junctions. The spilled carriers decrease the total charge of the non damaged zone, and due to this, the SR increases. The carriers in the damaged zone do not have any electrical manifestation because they are trapped and/or the mobility in this amorphized zone is very low.

Fig. 5 shows a diagram of the described situation for a P^+/N junction. According to this, when the Xe^+ implant energy rises, the depth X_A is greater, the "volume" of the damaged zone that can trap carriers also becomes greater, and the charge transferred from the non damaged region to the damaged one can increase. The SR of the non damaged zone will therefore be increased.

At constant energy, the higher the Xe^+ dose, the

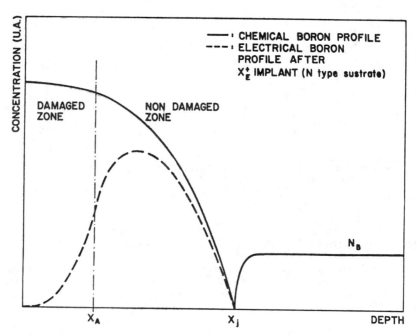

FIG. 5- Diagram of a possible model

greater the defects in the damaged zone, so that the capacity to trap carriers coming from the non damaged zone also increases. When the carrier level in this zone reaches the substrate value, the latter behaves as an infinite carrier source and the SR does not increase

The thermal anneal at 600-700°C produces a recrystallization of the damaged zone which eliminates defects. Trapped carriers are released and the

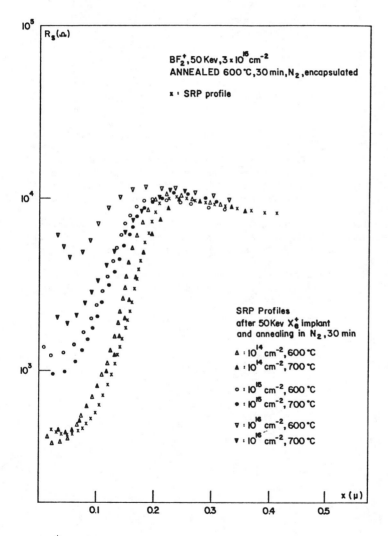

FIG. 6- BF_2^+ profil recovered after annealing

profile coincides with the original one. Fig. 6 shows the "aparition" of the original profile after this thermal treatment. More work may be needed to substantiate the hypothesis aforementioned. This work is continued and will be published in the future.

ACKNOWLEDGEMENTS

The authors wish to thank the whole team of Ionic Implantation of the LETI for their help.

REFERENCES

(1) Spreading Resistance Symposium. NBS Special Publication 400-10, June 1974.
(2) J.F. Gibbons, Process and device simulation for MOS-VLSI circuits. NATO ASI Series, 1983 Martinus Nijhoff Publishers.
(3) F.F. Morehead, B.L. Crowder, "Model for the formation of amorphous Si by ion bombardement, Radiation Eff. 6, 27, (1970).
(4) M.Y. Tsai, B.G. Steetman, "Recristallization of implanted amorphous silicon layers. I. Electrical properties of silicon implanted with BF_2^+", J. Appl. Phys., 50 (1), Jan 1979, pp 183-187.
(5) S.M. Hu, "Between carrier distributions and dopant atomic distribution in beveled silicon subtrates", J. Appl. Phys., 53 (3), March 1982, pp 1499 - 1510.

Lawrence A. Larson and Bradford J. Kirby

MEASUREMENTS OF CROSS-CONTAMINATION LEVELS PRODUCED
BY ION IMPLANTERS

REFERENCE: Larson, L.A. and Kirby, B.J., "Measurements of cross-contamination levels produced by Ion Implanters," Emerging Semiconductor Technology, ASTM STP 960, D.C. Gupta and P.H. Langer, Eds., American Society for Testing and Materials, 1986.

ABSTRACT: The contamination of implants by other species previously used in the same machine has been shown to be of significant concern for high-dose implants. This paper reports the results of a survey of the contamination levels extant in high-volume production implanters. Two classes of machines are represented. The first includes machines where wafer holders have been dedicated to a particular species. The second is machines where no precautions regarding cross-contamination have been taken. The machines have different production histories, especially for different species, and it is felt that the range of contamination levels reported is typical of what might be seen.

KEYWORDS: Ion implantation, contamination, Rutherford Backscattering Spectroscopy, Auger Electron Spectroscopy

INTRODUCTION

With the increasing use of high dose ion implantation in semiconductor processing a number of significant problems are coming to the fore. One of the most serious is the contamination of the implanted dopant with other elements, either other dopant species [1-3] or materials present as part of the implanter fixturing [1,4,5]. The extent of this problem has not been generally recognized in production areas. Present implanter designs incorporate measures to

Drs. Larson and Kirby are Engineering Section Head and Member Technical Staff respectively at National Semiconductor Corp., M/S CT035, 2900 Semiconductor Dr., P.O. Box 58090, Santa Clara CA 95052-8090

reduce the possibility of contamination from fixturing. These include wafer clamps designed to minimize resputtering, and low sputter rate, non-metallic coatings for areas exposed to the beam [6]. Despite these efforts, cross-contamination remains at significant levels in production machines.

Our previous work focused on the process effects of dopant cross-contamination, the most serious being the enhanced diffusion of phosphorus in the presence of high doping levels of arsenic and antimony. These diffusion effects have been explained by Fair and Meyer [7]. An earlier work by Larson [2] demonstrated them in a production simulation, where as little as 0.03% phosphorus contamination in a bipolar antimony collector produced an observable tail on the carrier concentration profile. Further work defined a table of tolerance limits for dopant cross-contamination, Table 1 [3]. A full matrix of the possible combinations of implant species and contaminants was run in silicon and modeled using SUPREM3 [8]. For this test the contamination was simulated by low energy implants of each species: see reference [3] for details. The tolerance level was defined as that causing a 5% shift in junction depth (X_j) or sheet resistance (R_s).

TABLE 1 -- Tolerance limits for dopant cross-contamination

Implant Species	CONTAMINANTS			
	B	P	As	Sb
B	*	1.5% (R_s)	2% (R_s)	1.5% (R_s)
P	7% (R_s)	*	6% (R_s)	8% (R_s)
As	0.2% (X_j)	0.01% (X_j)	*	4% (R_s)
Sb	0.05% (X_j)	0.03% (X_j)	2% (R_s)	*

These results fall into two classes. First, when the contaminant diffuses at a rate roughly equal to or less than that of the implant species, there is a small percentage change in the sheet resistance of the layer. As a matter of practice, this effect will be included in a functioning process as an uncontrolled variable with little result on the yield. The second class is more serious. When the contaminant diffuses faster than the implant species there is a variation in junction depth. This class is considerably more sensitive to contamination, as evidenced in Table 1.

It should be noted that the tolerances of Table 1 were determined for a simulated bipolar collector anneal. The results--particularly for junction depth--can be influenced considerably by the choice of anneal. For instance, boron contamination can be removed by the use of an oxidizing anneal, a common process, with the tolerance level increased by as much as two orders of magnitude. On the other hand, newer processes require junction depths an order of magnitude less than those tested. For these, the sensitivity to cross-contamination is higher.

The present work documents the levels of cross-contamination experienced in production equipment. These are initial results of efforts to develop routine monitors for engineering use.

METHODS

Ion implants were run on test wafer grade silicon wafers, typically 3 to 20 Ohm-cm resistivity. These wafers were then analyzed for surface contamination. The implants were done on various production ion implanters. Some of the implanters have been used for multispecies work with no precautions taken against sputter contamination. In others, wafer holders are dedicated to particular species implants to eliminate them as a source of cross-contamination. The results show maximum dopant contamination levels (as a percentage of the test implant dose) determined by several surface analytical methods. Data taken for our in-house monitoring was analyzed by RBS (Rutherford Backscattering Spectroscopy) and AES (Auger Electron Spectroscopy); some profiles were obtained by using SIMS (Secondary Ion Mass Spectrometry). The following sections discuss these methods in more detail.

RBS: Rutherford backscattering spectrometry using 2.275 MeV alpha particles was used on all samples. This method has been well covered by Chu et. al. [9]. In short, RBS is the observation of backscattered alpha particles from the near surface of the sample. The energy loss of the alpha particles is a well known function of both the mass of the scattering atom and its depth within the substrate.

Due to this double functionality, the RBS signal from the substrate will obliterate the signals from any low level contaminants of an atomic number near to or less than that of the substrate. In order to measure phosphorus and lower atomic number contaminants, we used a carbon substrate. Photoresists are an obvious possibility, but all resists tried contained other interfering species (generally chlorine). We chose a colloidal graphite suspension in iso-propanol (DAG), manually applied to the wafer surface. This system reduced the silicon background to an acceptable level, although experimental difficulties sometimes left a noticeable signal. For elements of higher atomic number than silicon various noise mechanisms decrease the sensitivity. Table 2 lists sensitivity levels of different species estimated from the data of this experiment.

The contaminant levels were determined by reference to the test implant species signal using the constant energy approximation equation [9].

$$C = \#/S \qquad (1)$$

Here C is the surface concentration (atoms/cm^2), # is the total number of counts contributed by the measured species, and S is the coulomb cross section for scattering of an alpha particle from the target nucleus. (For calculating ratios, the latter can be approximated by the square of the target's atomic number.)

TABLE 2 -- Minimum detectable concentrations ($\times 10^{14}/cm^2$) of various species using RBS analysis.

B	N/A
F	20
Na	10
P	5
Fe	1
As	0.8
Sb	0.1
W	0.1

Because the alpha particle will loose energy to electronic scattering processes in the substrate, it is an approximation to use a single value for the cross section. But at the implant energy used in this experiment, a more precise treatment would reduce the measured concentration of the test implant species by less than one percent. Further, our results support the hypothesis that the contaminant atoms are within a few hundred angstroms of the surface, so the effect on their measured concentration is even smaller. The combined effect is well below the resolution of our results.

A sample scattering spectrum is shown in Figure 1. Visible are the signals from the carbon substrate (the large feature below channel #150), from the test implanted argon, and from two contaminating species, arsenic and antimony. For the latter three, the approximate

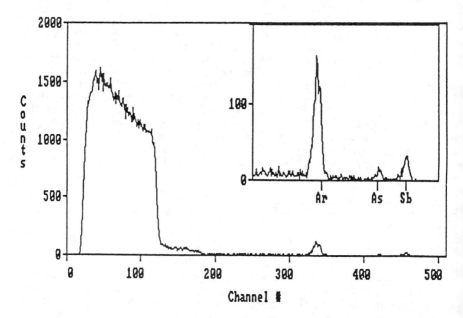

FIGURE 1 -- RBS energy spectrum. 1 $\times 10^{16}$ Argon/cm^2 implant in DAG

count totals are 1200, 75, and 225, respectively. Using the atomic number approximation, one calculates contamination levels of 1.9% (As) and 2.3% (Sb) of the test argon dose.

AES: The sensitivity of RBS to light elements in not sufficient for the actual contamination levels experienced. For this reason, Auger spectroscopy was used for detecting boron and phosphorus contamination. Historically, Auger sensitivities have been considered to be on the order of one atomic percent. This is due to the rather large secondary electron background and the various combinations of electron beam and detector noise that may occur. We have shown that considerably better sensitivities can be achieved for the case of ion implants in silicon by using background subtraction and integrating the resulting EN(E) Auger peak. The details of this work are reported separately [10].

For calibration, 1×10^{16} /cm^2 implants of the four standard dopant species (B, P, As, Sb) were profiled. The EN(E) difference spectrum at the depth of the peak of the phosphorus implant profile is shown in Figure 2. The phosphorus Auger peak is observed at 118 eV. Signal/noise characteristics of this spectrum were evaluated with different choices of available smoothing routines and integration endpoints. The optimal parameters gave a sensitivity of 0.04 atomic percent or 2.0×10^{19} P/cm^3. For the boron calibration implant we achieved a sensitivity of 0.6 atomic percent or 3×10^{20} B/cm^3.

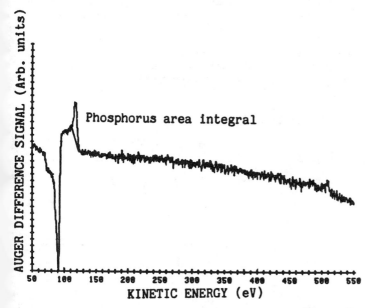

FIGURE 2 -- Auger difference spectrum. 1×10^{16} P/cm^2 in silicon

A significant assumption was made for the Auger data reported in the following tables, specifically that the concentration profile of the contaminants follows that determined by SIMS. SIMS profiles typically show a very high surface level and decrease logarithmically to 10^{-3} of the surface level by a depth of 100 nm. To convert Auger measurements to equivalent dose, several SIMS curves were integrated and referenced to the concentrations displayed by the curves at a depth of about 5 nm. This yielded a conversion factor of 2.6×10^{-6} cm (+/- 20%). An Auger measurement of 5×10^{19} P/cm^3 at 5 nm depth is converted to an equivalent dose of 1.3×10^{14} P/cm^2, which in turn would be reported as 1.3% of the standard 1×10^{16}/cm^2 test implant. Using this conversion, the minimum detectable level of phosphorus contamination is 0.5% and the minimum detectable level of boron contamination is 7.8%.

AES uses surface removal by ion sputtering for depth profiling. We found that the spatial separation of the contaminant from the implanted species was sufficient to allow us to distinguish the two, even when they were the same element. We made no similar attempt with RBS.

RESULTS

The results of several surveys testing implanters for cross-contamination levels are reported in the following three tables. The first two columns of the tables give the type of implanter and/or wafer holder and the implant test species. The remaining columns show the contamination levels observed. These are reported as percentages of the test dose. For most of the tests this dose is 1×10^{16}/cm^2 at 80 KeV. Included with each result is a letter indicating the method used to obtain it: A for Auger, R for RBS, or S for SIMS. Blank entries indicate where no measurements were made. "---" indicates levels below the sensitivity of the indicated method.

TABLE 3 -- Contamination levels in non-dedicated implanters

Equipment	Test Species	Sb		As		P		B	
Eaton NV10(1)	Ar	---	R	5.8%	R	(~ 3%	R)	---	A
Eaton NV10(2)	Ar	2.9%	R	5.9%	R	---	A	0.9%	A
	Sb			0.17%	S	1.8%	S		
	Sb[a]			4.5%	R				
Varian 80-10 [a]	Ar	2.0%	R	---	R				
Varian 120-10	Ar	---	R	3.5%	R	0.1%	A	---	A
Varian 200-1000	As	---	R			0.15%	S	---	A
Varian CF3000	As	---	R			---	A	0.2%	A
AIT III-X [a]	Ar	1.5%	R	---	R	(+ to 8% Ta R)			

[a] Unpublished results, M.I. Current, Xerox PARC

Table 3 reports results for implanters in which it is known that the wafer holders have been used for more than one of the standard dopant species. These include medium dose implanters at National Semiconductor and several machines at other sites.

TABLE 4 -- Contamination levels in Eaton NV10 implanters with non-dedicated disks

Equipment	Test Species	Contaminant Sb		As		P		B	
3" Full Ring	Ar	---	R	---	R	---	R		
3" Two Point	Ar	0.3%	R	2.7%	R	---	R		
4" Full Ring	Ar	0.1%	R	0.8%	R	---	R		
4" Full Ring	Ar	0.2%	R	1.2%	R	---	A	---	A
4" Two Point	Ar	1.8%	R	5.4%	R	---	A	---	A
4" Two Point	Ar	2.0%	R	0.9%	R	---	R		
5" Two Point	Ar	2.6%	R	1.5%	R	---	R		

Table 4 reports results for implants run on Eaton NV10 series machines at National Semiconductor, using wafer holder disks that are not limited to one species. These may be considered somewhat equivalent to the non-dedicated machines of Table 3, with the difference that their historical use for some species may be much lower. The first column of the table now lists the type of disk.

TABLE 5 -- Contamination levels in Implanters with dedicated wafer holders.

Equipment	Test Species	Contaminant Sb		As		P		B	
: 4" Full Ring	B	---	R	---	R	---	A	1.6%	A
: 4" Full Ring	Ar	0.1%	R	1.9%	R	0.8%	A	---	A
	P	---	R	0.7%	R	0.7%	A	0.2%	A
s: 4" Full Ring	Ar	---	R	5.5%	R	---	A	---	A
	As	---	R			---	A	---	A
b: 4" Two Point	Ar	1.3%	R	1.0%	R	---	A	---	A
	Sb			(0.3%	R)	---	A	---	A
b: 5" Two Point	Ar	2.2%	R	1.6%	R	---	R		
c: AIT III-X	Ar	--- (+0.1% Ta	R R)	---	R	---	A	---	A

Table 5 reports results for implants run on machines at National Semiconductor and for which the wafer holders are dedicated to one species, though some may have a history that includes use with other species. (The machines themselves are used for all species.) Except for the AIT III-X, all entries are for Eaton NV10 implanters and the type of wafer disk is indicated. The first column also includes the dedicated species. One might expect this table to show the maximum levels for the dedicated species and minimal amounts for all others.

DISCUSSION

Although this data presents no clear-cut conclusions on the relationship between implanter and wafer holder usage and resulting contamination levels, a number of implications can be seen. The operating assumption that we have been using in our work to control contamination is that the primary mechanism is the sputtering and/or evaporation of previously implanted atoms from the wafer holder onto the wafer. We feel the data supports that assumption. These atoms may also be subject to drive-in by collisional processes with the implanted ions [3]. This will increase the probability of contamination of the diffused layer during further thermal treatments.

Contamination levels:

The highest contamination levels observed were between 1.3% and 2.9% for antimony, between 5.5% and 5.9% for arsenic, between 1.5% and 1.8% for phosphorus, and 3.8% for boron. This is for argon as the test implant species. With the sputtering/evaporation hypothesis, significant differences would be expected for different test species. What data there is for other test species supports this expectation, but there is too little from which to draw any clear trends. It is noted that for the NV10 machines, the two different wafer holder designs (full ring clamp and two point clamp) give contamination levels of comparable magnitude, although one has much more sputtering surface above the plane of the wafer.

Memory Effect:

A previous publication [3] has estimated that it would take an argon dose on the order of 2×10^{17} Ar/cm^2 at 50keV in order to "sputter clean" an aluminum surface in the implanter. As mentioned earlier, some of the NV10 disks that are now species dedicated had been used for other species to some extent in the past. For example, the one listed in table 5 as dedicated to phosphorus is known to have originally been used for all species. But since being dedicated it is known to have received a total phosphorus dose of at least 2×10^{18}/cm^2. Although one might argue that the contamination is from other parts of the machine, this is not supported by the results for those dedicated disks that show no cross-contamination. It is more likely that subsequent implants drive in previously implanted species, either by thermal diffusion or knock-on.

Tolerance levels:

Based on the calculated tolerance levels these results are encouraging. Most of the measured levels in Table 1 are for the class of slow diffusers which is characterized by a shift in sheet resistance but little effect on the junction depth. With the exception of some of the maximum levels all the measurements fall within the tolerance limits. Due to the nature of this type of contamination and the long-lived memory effect, it is probable that these sheet resistance changes are incorporated into existing processes as a minor, but uncontrolled, process variable.

A similar situation exists for boron contamination of arsenic and antimony. The work that produced Table 1 showed this to be nearly as disruptive as phosphorus contamination but the silicon that was tested in parallel with those calculations showed very little sensitivity. This was due to the effect of the anneal. With an oxidizing anneal, boron preferentially segregates to the surface and is incorporated into the oxide, greatly increasing the tolerance. The sensitivity listed in Table 1 requires a careful inert anneal.

This leaves phosphorus contamination of arsenic and antimony implants as the remaining area of significant concern. The tolerance limits given in Table 1 (0.01% in an arsenic implant, 0.03% in an antimony implant) are relatively tight. But they have been proven necessary in actual bipolar processes. Our results suggest that wafer holder dedication can achieve this goal, if it is instituted for new holders. But it is also clear from our work that neither RBS nor AES has the sensitivity necessary to confirm the required levels. We conclude that any process that is sensitive to such phosphorus levels should be monitored with high resolution SIMS.

SUMMARY

1) The maximum contamination levels observed were 2.9% for antimony, 5.9% for arsenic, 1.8% for phosphorus, and 3.8% for boron.

2) Wafer holder dedication considerably lowers the amount of contamination but the memory effect may be very long lived.

3) The observed levels are generally within the tolerance limits set by Table 1, except for:

4) Phosphorus contamination of arsenic and antimony implants is the remaining area of significant concern. Better monitoring methods are necessary to test this to the accuracy needed.

REFERENCES:

[1] Current, M.I., "Contamination of Ion Implants by Sputtered Films", Bulletin APS 28(9) (1983), p.1372

[2] Larson, L.A., "Phosphorus in Antimony: A Case Study in Implant Cross-contamination", Advance Applications in Ion Implantation, SPIE Proceedings 530 (1985), pp.50-54

[3] Larson, L.A. and Current M.I., "Metallic Impurities and Dopant Cross-contamination Effects in Ion Implanted Surfaces", in Ion Beam Processes in Advanced Electronic Materials and Device Technology MRS Proceedings Vol.45 (1985) 381

[4] Barraclough K.G. and Ward P.J.,"Iron Contamination in Silicon Processing" in Defects in Silicon, Electrochemical Society Proceedings 83-9, Bullis W.M. and Kimmerling L.C. Eds., ECS(1984), pp.388-395

[5] Haas, E.W., Glawischnig, H., Lichti, G., and Blair A., Journal of Electronic Materials 7(1978), 525

[6] Ryding, G.,"Evolution and performance of the NOVA NV-10 Predep^{TM} Implanter" in Ion Implantation Techniques, Eds. Ryssel, H. and Glawischnig, H. , Springer(1982) 319-342

[7] Fair, R.B. and Meyer, W.G., "Modeling Anomalous Junction Formation in Silicon by the Codiffusion of Implanted Arsenic with Phosphorus" in Silicon Processing, ASTM STP 804, Gupta D.C. Ed., American Society for Testing and Materials (1983), pp. 290-305

[8] Ho, C.P., Hansen, S.E., Fahey P.M., "SUPREMIII - A program for Integrated Circuit Process Modeling and Simulation", Integrated Circuits Laboratory Tech. Rept. SEL84-001, Stanford University(1984)

[9] Chu, W-K, Mayer, J.W. and Nicolet M.A., Backscattering Spectrometry, Academic Press N.Y. (1978)

[10] Larson, L.A. , "High Sensitivity Auger Spectroscopy" in Materials Characterization, MRS Spring Meeting Proceedings Vol.69 (1986)

Stanley Middleman

SOME ASPECTS OF PRODUCTIVITY OF A LOW PRESSURE CVD REACTOR

REFERENCE: Middleman, S., "Some Aspects of Productivity of a Low Pressure CVD Reactor," *Emerging Semiconductor Technology, ASTM STP 960,* D. C. Gupta and P. H. Langer, Eds., American Society for Testing and Materials, 1986.

ABSTRACT: A mathematical model of a CVD reactor is presented, from which performance criteria such as deposition uniformity, and reactor productivity, may be obtained in terms of design and operating parameters. An example is presented in which the interwafer spacing, and the number of wafers to be processed as a single batch, are determined to maximize productivity while maintaining a constraint on uniformity.

KEYWORDS: reactor design, CVD, uniformity, productivity

The methods of traditional chemical reactor analysis [1] are now being brought over from the field of chemical engineering and finding application to the design of reactors for chemical vapor deposition of thin solid films [2]. Until the *input* to these methods, especially the relevant sets of reactions, rate equations, kinetic parameters, and gas dynamics is on a more secure basis, the designer will not be able to use *output* from these methods with any great security. Nevertheless, the exercise of producing mathematical models of reactor performance is more than a sterile activity. In many cases of industrial importance there is a sufficient, if incomplete, understanding of the physics to allow the generation of useful qualitative guidelines for reactor design and operation. Further, these mathematical models can provide insight into the design of critical experiments, and also suggest methods of correlating experimental data in such a way as to guide extrapolation of the data outside the range of the selected laboratory variables.

The rate of reaction (deposition) that can be achieved, and its dependence upon design and operating conditions, is often of primary concern to the engineer responsible for production. Traditionally, once the reactor operates at the appropriate rate, additional considerations enter. One of these is the *uniformity* of deposition, both wafer-to-wafer, as well as across the surface of each individual wafer. Often, good engineering design represents a compromise among conflicting demands for high productivity, uniformity of product, and economical utilization of equipment and resources. The goal of this paper is to illustrate some of the methods of reactor analysis which can be used in support of CVD reactor design and operation. Particularly simple *input* assumptions are made so that the output variables can be obtained by analytical methods. More realistic physical assumptions can be handled by the general approach illustrated here, but this usually requires extensive numerical computation, and it is then not so easy to *illustrate* the use of these modelling techniques.

Dr. Middleman is Professor of Chemical Engineering at the University of California, San Diego, B-010, La Jolla, CA 92093.

A MODEL CVD REACTOR

Let us suppose that we have a reactor designed along the lines suggested in Figure 1. Wafers, stacked in holders, sit inside a chamber which provides control of pressure and temperature. The feed gas is introduced in such a way that the combination of the feed distribution system, and the momentum of the gas flow, provides effective mixing of the feed in the region *external* to the stacked wafers. In order to prevent exposure of individual wafers to a directed gas flow, which might vary depending on the proximity to a feed port, the wafer holder is shielded by a perforated "cage." The cage does not provide any significant diffusion resistance; it only insures that each wafer is exposed, just beyond its circumferential boundary, to a gas composition that is uniform throughout the entire reactor.

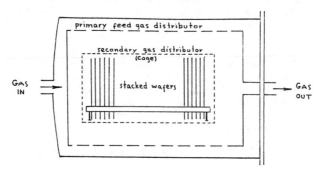

1. Stacked wafers in a CVD reactor

Our goal now is to examine a model for the *production rate* of wafers having some fixed film thickness deposited on them, subject to a constraint that the *radial uniformity* of the deposited film must satisfy some design criterion. Note that the issue of wafer-to-wafer uniformity does not arise *by definition* of the problem: we *assert* that each wafer is subjected to an identical environment for deposition.

THE MATHEMATICAL MODEL

The mathematical model, consistent with the physical ideas expressed above, may be developed now. With reference to Figure 2, we divide the reactor into two zones. One is the gas space external to the cage(s), which is regarded as well-mixed, and within which no reactions are assumed to occur. The other zone is the wafer surface(s) upon which reaction occurs, and the gas space between wafers. This latter gas space is assumed to be stationary, *i.e.*, unmixed. In this model it is assumed that reactants must diffuse through this second zone in order to reach active sites on the wafer surface(s), whereupon a heterogeneous reaction occurs.

As a specific example, and one that permits an analytical solution for the performance of the reactor, we take the case of a silicon layer grown from a feed gas of silane, highly diluted in helium. We assume that both zones are at constant and identical pressure and temperature. The model begins with a mass balance on the reactant:

$$(FC)_{in} - (FC)_{out} = 2n\pi R^2 \eta \, \mathcal{R}\, (\overline{C}) \tag{1}$$

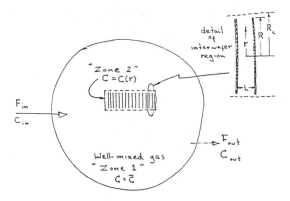

2. A two-zone model of a CVD reactor

The volumetric flowrate F and the molar concentration C will be evaluated at the temperature and pressure within the reactor. The rate of reaction $\mathcal{R}$ (per unit of reacting surface) is a function of the molar concentration on the reacting surface. (In Eq. (1) we assume reaction occurs on both sides of n wafers.) Because of a finite diffusion resistance the surface concentration may vary across the radius of a wafer. Hence the average rate of reaction on a single side of a wafer is obtained from an integral over the wafer surface:

$$\pi R^2 \langle \mathcal{R} \rangle = \int_0^R 2\pi r \, \mathcal{R}\,(C[r])\, dr \qquad (2)$$

We define an effectiveness factor η so that

$$\langle \mathcal{R} \rangle = \eta \, \mathcal{R}(\bar{C}) \qquad (3)$$

where $\bar{C}$ represents the average concentration of reactant in the well-mixed zone of the reactor. Hence, the right-hand side of Eq. (1) represents the average rate of reaction on *one* side of a wafer, multiplied by the number of sides of wafers in the reactor, 2n.

The assumption of a well-mixed reactor (for zone one) implies that the outlet concentration is identical to the mean concentration within the reactor, or

$$C_{out} = \bar{C} = C \qquad (4)$$

(We now use the notation C, without subscript or overbar.)

For dilute reactant we may ignore any volume change due to reaction, so that

$$F_{in} = F_{out} = F \qquad (5)$$

Thus Eq. (1) becomes

$$F(C_{in} - C) = 2n\pi R^2 \eta \, \mathcal{R}(C) \qquad (6)$$

In order to obtain η it is necessary to know the radial concentration distribution $C(r)$ in the interwafer region. This follows from a solution to the diffusion equation for the interwafer region, taking into account the kinetics of reaction on the wafer surface. This problem has been solved for a first order reaction, *i.e.*,

$$\mathscr{R}(C) = k\,C \tag{7}$$

and the final result of concern here is (see Appendix)

$$\eta = \frac{2\,I_1(\Phi)}{\Phi\,I_0(\Phi)} \tag{8}$$

where

$$\Phi^2 = \frac{2R^2 k}{DL} \tag{9}$$

Figure 3 shows the function $\eta(\Phi)$. I_0 and I_1 are Bessel functions.

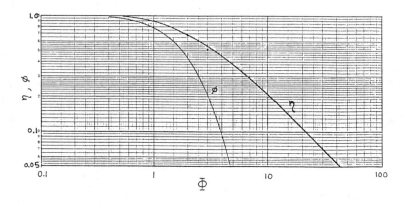

3. The functions η and ϕ

This result assumes that the "cage" radius is identical to the wafer radius R. For a cage radius $R_c > R$ one can show that

$$\eta = \frac{2\,I_1(\Phi)}{2[I_0(\Phi) - \Phi I_1(\Phi)\,\ln(R/R_c)]} \tag{10}$$

We may solve for the mean reactor concentration C and write the solution in the form

$$\frac{C}{C_{in}} = \left[1 + \frac{2\pi R^2 k}{F}\,n\,\eta(\Phi)\right]^{-1} = (1 + \eta\kappa)^{-1} \tag{11}$$

In Eq. (11) we have introduced a dimensionless rate parameter

$$\kappa = \frac{2\pi R^2 nk}{F} \tag{12}$$

The ratio C/C_{in} is a measure of the *depletion* of reactant arising from reaction within the interwafer zone. Reaction does not occur at the feed concentration, because of this depletion effect. However, finite diffusion resistance in the stagnant interwafer zone reduces the depletion somewhat. This latter effect is accounted for by the effectiveness factor η.

Another result that follows from the analysis of diffusion in the interwafer region is the radial profile of deposition. (For a first order reaction, this is identical to the radial concentration profile, $C(r)$.) From this, a nonuniformity index may be defined and determined. A number of measures of the uniformity of film deposition can be calculated. Middleman (3) presents two of these: a root mean square variation, and a center-to-edge difference normalized to the mean. For the purpose of this presentation we use a simple *ratio* of the film thickness at the center to the edge. Under the assumptions invoked above we find this ratio, ϕ, to be given by

$$\phi = 1/I_o(\Phi) \tag{13}$$

Figure 3 shows the function $\phi(\Phi)$.

We may now put some of these results together and evaluate the performance of this particular reactor as a function of various design and operating parameters, subject to constraints on uniformity.

REACTOR PERFORMANCE

The primary performance variable is the rate at which solid film is deposited, usually measured as area of coverage (to a specified film thickness H) per unit of time. For a given value of H, this is equivalent to the volumetric rate of deposition divided by H. Thus we define the *reactor productivity* $\mathscr{P}$ as

$$\mathscr{P} = \frac{2n\pi R^2}{H} \eta \mathscr{R}(C) \left[\frac{M}{\rho_s} \right] \tag{14}$$

The ratio of silicon molecular weight M to silicon solid phase density ρ_s is used to convert the molar rate of reaction (deposition) $\mathscr{R}$, per unit of area, to the volumetric rate per area. From Eqs. (7) and (11) we obtain

$$\mathscr{P} = k\, C_{in} (1 + \kappa\eta)^{-1} \left[\frac{2\pi M}{\rho_s H} \right] nR^2 \eta \tag{15}$$

Equation 15 is a model for reactor productivity in terms of reactor design and operating variables ($C_{in}, \kappa, n, R, \eta$). The operating temperature does not appear explicitly, but rather implicitly through the rate coefficient k, and further through the dependence of η on k (note Eq. (9)).

The maximum rate of production occurs if all of the feed material is instantly converted to solid film, *i.e.*,

$$\mathscr{P}_{max} = \frac{M}{\rho_s H} FC_{in} \tag{16}$$

In other words, at $\mathscr{P}_{max}$, growth is limited only by the feed rate. We may rewrite Eq. (15) as

$$\frac{\mathscr{P}}{\mathscr{P}_{max}} = \frac{\kappa\eta}{1 + \kappa\eta} \tag{17}$$

In this form we see that, under a given set of feed conditions (a fixed value of $\mathscr{P}_{max}$), the degree of approach to maximum utilization of the feed is governed by both κ and η. A criterion for efficient performance, then, is seen to be

$$\kappa\eta \gg 1 \tag{18}$$

This means that for a given value of $\mathscr{P}_{max}$, once $\kappa\eta$ becomes large, additional design or operating changes that increase $\kappa\eta$ further do not yield significant improvement.

Let us examine some possible strategies for improving the productivity of a reactor. For example, suppose we increase the number of wafers in a reactor, at a fixed value of $\mathscr{P}_{max}$. We assume, first, that this is done by adding additional wafer holders, perhaps necessitating the use of a larger reactor. Specifically, however, we assume that the wafer spacing is unchanged by the addition of more wafers. Then, since Φ is unchanged, there is no change in η, but κ increases with increasing n. Our model will show that the reactor productivity can be increased by treating more wafers in a batch, but with diminishing degree of improvement as n increases.

As a specific example, consider a reactor designed to grow a silicon layer upon six inch (diameter) wafers. We wish to recommend an interwafer spacing, and the number of wafers to be processed in the reactor. Assume that the conditions for the feed are

$F = 1.0$ liter/min at STP (298°K and 1 atm)
5% silane in helium (by volume)
Reactor temperature at 650°C
Reactor pressure at 1.0 torr

Assume that the constraint on uniformity, ϕ, is $\phi \geq 0.9$. From Fig. 3 this gives $\Phi = 0.6$. An estimate for k is obtained in the form [2]

$$k = 1.25(10^9) \exp(-18,500/T) \frac{\text{mol}}{m^2 - s - atm} \tag{19}$$

At $T = 923°K$ we find $k = 18.7$ cm/s. (In addition to converting the rate expression of (2) from meter to centimeter units, it is also necessary to use a factor $R_G T$, where $R_G = 82.05$ is the Gas Constant, and T is absolute temperature, in order to obtain consistent units.) A diffusion coefficient may be estimated using

$$D_{SiH_4} = 0.62 \left(\frac{T}{300}\right)^{1.7} \left(\frac{760}{p}\right) \tag{20}$$

where $T = °K$ and $p =$ torr. At the design conditions we find $D_{SiH_4} = 3180$ cm²/s.

For a six inch wafer, $R = 7.6$ cm. From Eq. (9) we find

$$L = \frac{2R^2 k}{D \Phi^2} = 1.9 \text{ cm} .$$

Any interwafer spacing less than this value will have a deleterious effect on the uniformity of deposition. Note that, at this spacing, the effectiveness factor η (see Fig. 3) is nearly unity.

So long as we stay above this value of the interwafer spacing we may add wafers to the reactor, and improve the productivity. A practical limit to improvement is given by Eq. (18). Since $\eta \approx 1$, we select some value for κ, say $\kappa = 10$. (At this value, we are within 10% of the maximum productivity $\mathscr{P}_{max}$, that can be achieved under the specified feed conditions.) From Eq. (12) we find

$$n = 10 \frac{F}{2\pi R^2 k} \tag{21}$$

The feed rate $F = 1$ liter/min at STP corresponds to 4.5 10⁴ cm³/s at reactor conditions. Thus we find a value of $n = 66$. While to some degree this number is the result of the specific choices for F, R, T, and p used in this example, it is still clear that this value of n is higher than the number of wafers usually processed in a single batch. (At this wafer spacing we would have about one meter of stacked wafers.)

This result illustrates an important point that must be kept in mind in any modelling exercise. The model used here deals with only one aspect of the design problem: the interaction of kinetics and mass transfer. Other equally important considerations such as mechanical design of the reactor may dictate the character of the final design and operation. While this precaution is important to recognize, it is equally important to understand the conclusion one would draw from this exercise: a consideration of kinetic and diffusion phenomena leads to a reactor design model with which one may determine how close a particular design is to its limiting productivity. One may choose to base the final design on other more critical limitations (mechanical, thermal, safety), but at least one should know how far the real reactor is from ideal performance. In this way, the incentive for process improvement can be placed in its proper perspective.

REFERENCES

[1] Smith, J. M., _Chemical Engineering Kinetics,_ 3rd Ed., McGraw-Hill, NY, 1981; Hill, C. G., Jr., _An Introduction to Chemical Engineering Kinetics and Reactor Design,_ Wiley, N.Y., 1977.
[2] Jensen, K. F. and D. Graves, "Modeling and Analysis of Low Pressure CVD Reactors," J. Electrochem. Soc., _130,_ 1950 (1983).

APPENDIX

Diffusion in the interwafer region controls the rate of arrival of a reacting species onto the wafer surface. In the simplest case of a single first-order surface reaction, the radial distribution of the reacting species is found from a solution of the diffusion equation, which takes the form

$$0 = \frac{D}{r} \frac{d}{dr} \left(r \frac{dC}{dr} \right) + kC \qquad \text{(A-1)}$$

Equation A-1 implies a constant diffusivity. Diffusion in the direction normal to the wafer surface is assumed to be much faster than radial diffusion; hence axial concentration gradients do not appear in Eq. A-1. Most CVD reactions result in an increase in the number of moles of gas, and as a consequence a radial convection term should be included in Eq. A-1. It can, however, be neglected for the common situation of reactants diluted with a large excess of inert carrier gas. We imply that assumption here. Boundary conditions on C correspond to symmetry at the axis:

$$\frac{\partial C}{\partial r} = 0 \quad \text{at} \quad r = 0 \qquad \text{(A-2)}$$

and constant concentration of reactant in the gas phase external to the wafer stack:

$$C = \overline{C} \quad \text{at} \quad . \, r = R \qquad \text{(A-3)}$$

The solution to this system of equations is easily found to be

$$C(s) = \overline{C} I_o (\Phi s) / I_o (\Phi) \qquad \text{(A-4)}$$

where I_o is a modified Bessel function of the first kind, and where a dimensionless radial coordinate is

$$s = r/R \qquad \text{(A-5)}$$

and

$$\Phi = (2R^2 k / LD)^{1/2} \qquad \text{(A-6)}$$

is a Thiele modulus.

When Eqs. 2, 3, and 7 of the main body of this paper are used, with Eq. A-4 for $C(r)$, the effectiveness factor, η, may be calculated, and we find Eq. 8:

$$\eta = \frac{2I_1(\Phi)}{\Phi I_o(\Phi)} \qquad (8)$$

Eq. A-4 also permits us to calculate the ratio of reaction rates at the center and edge of a wafer, and we find the result presented above:

$$\phi \equiv \frac{\mathscr{R}(0)}{\mathscr{R}(1)} = \frac{1}{I_o(\Phi)} \qquad (13)$$

NOMENCLATURE

- C molar concentration (moles/cm^3)
- D diffusion coefficient (cm^2/s)
- F volumetric flowrate (cm^3/s)
- H final solid film thickness (cm)
- k first order rate constant (cm/s)
- L interwafer spacing (cm)
- M molecular weight of silicon
- n number of wafers
- $\mathscr{P}$ reactor productivity (cm^2/s)
- p reactor pressure (torr)
- R wafer radius
- $\mathscr{R}$ molar reaction rate (moles/cm^2-s)
- T absolute temperature (°k)

GREEK

- η effectiveness factor
- κ dimensionless rate parameter
- ρ_s solid density (gm/cm^3)
- ϕ uniformity index
- Φ Thiele modulus

Stanley Roberts, James G. Ryan, and Dale W. Martin

DEPOSITION AND PROPERTIES OF ULTRA-THIN HIGH DIELECTRIC CONSTANT INSULATORS

REFERENCE: Roberts, S., Ryan, J. G., Martin, D. W., "Deposition and Properties of Ultra-Thin High Dielectric Constant Insulators", Emerging Semiconductor Technology, ASTM STP 960, D. C. Gupta and P. H. Langer, Eds., American Society for Testing and Materials, 1986.

ABSTRACT: Exploratory studies have been carried out with reactively sputtered thin films of several transition metal oxides and co-sputtered mixtures with SiO_2. Good insulation behavior is observed following post-deposition oxidation anneals. Best overall electrical properties are observed with mixtures of HfO_2 and SiO_2, in combination with additional layers of thermal SiO_2 and CVD Si_3N_4. Significant reduction in dielectric constant is observed with all the transition metal oxides with sub-30 nm films. High capacitance with stacks containing thermal SiO_2, mixtures of HfO_2 and SiO_2, and CVD Si_3N_4 may be due to a charge storage mechanism.

KEYWORDS: thin films, dielectrics, insulators, sputtering

INTRODUCTION

Current silicon integrated circuit (SIC) applications have introduced more demanding capacitor requirements, resulting in a renewed interest in the application of high dielectric constant insulators(1). Additional applications include coupling capacitors(2) and gate insulators for metal gated thin film transistors (TFTs)(3,4). Most of these studies have focussed upon the use of anodized Ta_2O_5(5). Alternately processed Ta_2O_5 films, such as CVD(6,7), low temperature thermal oxidation of deposited Ta films on silicon substrates(8,9,10), and RF sputter deposition from a Ta_2O_5 target(3), have also been investigated. Studies of other high dielectric constant insulators have included HfO_2(11), ZrO_2(12,13), TiO_2(14), and rare earth oxides(15).

Dr. Roberts is an advisory engineer, Mr. Ryan is a staff engineer, and Mr. Martin is a senior associate engineer at IBM Corporation, General Technology Division, Essex Junction, Vermont 05452.

This study was intended to provide selective evaluation of some physical and electrical properties of reactively sputtered insulators on silicon. The physical properties include optical index of refraction and microstructural analysis. The electrical properties include standard DC biased capacitance, leakage, and breakdown. Derived properties, to be defined later, include the capacitance equivalence of the insulator to that observed with pure SiO_2, and a charge strage index, which is a useful parameter for the comparison of various insulators for the memory storage node of one device field effect transistor (FETs) random access memories (RAMs)(16). The greater the index value, the more effective is the storage insulator. The microstructural analyses were carried out, since previous studies with reactively sputtered Ta_2O_5(17) have suggested a relationship between electrical leakage and certain microstructural features that appear to be related to distributed subgrain microcracks.

EXPERIMENTAL

In this study, the method of film preparation was reactive sputtering from metal targets. The oxides investigated include HfO_2, Ta_2O_5, HfO_2-SiO_2 mixtures, and Ta_2O_5-SiO_2 mixtures. The films were deposited on bare Si <100> substrates, precleaned with hot peroxide solutions, followed by a buffered Hf dip and DI water rinse. Additional substrates included cleaned and thermally oxidized Si (3 to 6 nm of SiO_2), and carbon substrates used for Rutherford Backscatter Analysis (RBS). The sputtered films on the Si and SiO_2 substrates were annealed at 800°C in dry O_2 for 30 minutes, followed by a 1000°C, N_2, 30 minute anneal. The time at temperature in O_2 was chosen to limit the amount of SiO_2 grown at the sputtered insulator-substrate interface. Some of the thermal SiO_2-sputtered insulator layered films were coated with a 4 to 8 nm Low Pressure Chemical Vapor Deposited (LPCVD) Si_3N_4 film(18). These were given a wet O_2 oxidation at 800°C following the Si_3N_4 deposition.

The films were prepared in sputtering systems using planar magnetron metal cathodes for DC sputtering, and a silicon cathode for RF sputtering. Oxide mixtures were made by co-sputtering. Neither substrate bias nor applied substrate heating was used for the depositions. RBS of as-deposited films indicated the only significant impurity present within the hafnium target was zirconium (approximately 5 Wt.%). The zirconium impurity did not affect the insulation properties of the film.

The deposition systems were typically pumped down to less than 6.7×10^{-4} Pa. The sputtering targets were conditioned using the oxygen containing sputtering gas for several minutes prior to opening the shutter for deposition. No attempt was made to remove the oxide formed on the target from previous runs.

The HfO_2 and Ta_2O_5 were typically deposited at 1 and 1.2 nm/second, respectively, using an 80% Ar-20% O_2 sputtering gas mixture. To co-sputter the metal oxide-SiO_2 mixture films, the metal oxides were deposited at the previously mentioned rates, while the incident RF power to the Si target was varied to achieve the desired metal oxide: SiO_2 ratio. A typical SiO_2 deposition rate for the Hf_4SiO_{10} mixture was approximately 0.3 nm/second.

A series of experiments for the three-layer stacks (thermal SiO_2-HfO_2 + SiO_2-CVD Si_3N_4) examined the influence of a number of processing parameters including thickness variations of each of the three layers, post-deposition thermal processing, and the influence of low energy, low fluence arsenic implants. The film thicknesses were determined ellipsometrically by sequential measurement of each layer with test wafers(19). Two layers were measured using a dual layer program, from which growth of the SiO_2 layer, with oxidation, could be monitored.

Films were also physically characterized (with Si substrates) using Transmission Electron Microscopy (TEM) and Transmission Electron Diffraction (TED). The TEM samples were prepared by cutting out 3 mm diameter disks and etching through the backsides with a solution of HF and NHO_3. The etching solution stopped at the metal oxide. The metal oxide films were then examined using a Phillips 400T operating at 120 KV.

The samples were electrically characterized by high frequency (1 MHz) biased capacitance measurements, yielding C_{max}, V_{fb}, and ΔV_{fb}. The frontside dot electrodes were either in-situ phosphorus doped LPCVD polysilicon or evaporated aluminum. The polysilicon electrodes were defined as follows. An evaporated Al film (over the polysilicon) was defined lithographically with a photoresist dot mask plus wet chemical subtractive etching of the Al. The Al dots then provided the mask for subsequent wet etch (7 parts HNO_3/4parts HF/1 part H_2O at room temperature for a maximum of 15 to 20 seconds) of the polysilicon film, or plasma etched with CF_4 plus O_2(20). The Al was retained for low resistance probe contact during electrical measurements. Backside contact was obtained with deposited Al over the bare Si following oxide and nitride removal. Backside sintering and frontside electrode anneal was carried out in forming gas at 400°C. Some limited charge storage experiments utilized pulse biasing in the range of 10 seconds to 1 millisecond (± 7.5 to 10V)(22). Read measurements of ΔV_{fb} were carried out manually with a 30 second delay.

DC conduction measurements utilized a ramp rate of 0.5 V/minute with a 30 second hold between 0.1V ramp intervals. Repeat ramping was used where current saturation at 0.8 V_{bd} was not observed. The DC breakdown measurements used a ramp rate (in accumulation) of approximately 0.1V/second up to a 1 ma breakdown criteria (J = 55 ma/cm² for 1.52 mm dots).

Low energy arsenic implantation in the range of 4 to 10 KeV was carried out with $^+As^{75}$, using 2 ohm-cm, <100>, N-type Si substrates that were backside gettered to reduce substrate leakage. The profile of the implanted arsenic, diffused into the substrate, was measured using a Pulse C-V technique(23), with correlation of the dopant profile attempted by Secondary Ion Mass Spectroscopy (SIMS) analysis and LSS modelling(24). The arsenic doped shallow junctions comprise the lower electrode for transfer of the stored charge (under the storage node) to the transfer gate FET(16). Two processes for carrying out the implantations are described and discussed within the **RESULTS** section.

RESULTS

The deposition rates for HfO_2 and SiO_2 films as a function of the power applied to the cathode are shown in Fig. 1. The indices of refraction (λ = 632.8 nm, angle of incidence = 70°) for the as-deposited HfO_2 (measured on a bare Si substrate) varied from n

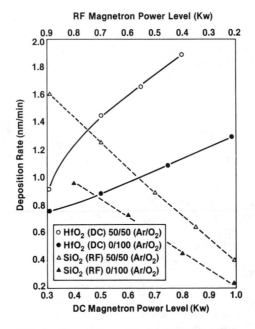

FIG--1. Deposition rates for HfO_2 and SiO_2

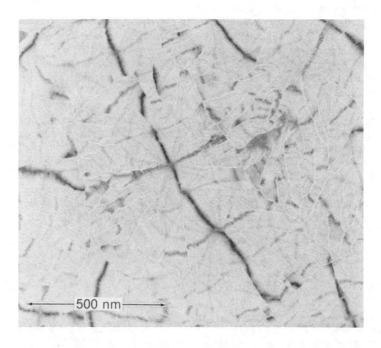

FIG 2a.--Reactively sputtered RF Ta_2O_5 film (31 nm) over Si substrate

FIG. 2b--Reactively sputtered "Hf_4SiO_{10}" 10nm, as deposited (150Kx)

= 1.90 to 2.05, without any correlation to deposition parameters. Similar results were observed for the Ta_2O_5(17). Some densification with annealing in N_2 is observed for the HfO_2, yielding a range of indices between 2.00 and 2.15. When oxidizing above 800°C, significant growth of thermal SiO_2 occurs beneath the HfO_2, as determined from both two-layer ellipsometry measurements and Auger analysis. The accuracy of the two-layer calculations were confirmed by etch removal of the oxidation-annealed HfO_2 in hot H_3PO_4 (170°C) and measuring the remaining SiO_2 by an ellipsometer. The growth rate for the underlying SiO_2 was approximately 0.12 nm/minute. Therefore, thermal processing of the three-layer films utilized 800°C, 30 minute oxidations in dry O_2 for the HfO_2 layer and 800°, wet O_2 for the Si_3N_4 layer.

The as-deposited Ta_2O_5 was amorphous and recrystallized above 600°C, with a net average film shrinkage of the order of 15%. This may have resulted in considerable film stress as indicated by an extensive microcrack configuration, as shown in Fig. 2a. The influence of film thickness upon microstructure and electrical properties has been reported elsewhere(17,25). The HfO_2 was shown by TED to be crystalline, as deposited, and exhibits densification with annealing. TEM indicates a very fine-grained, featureless microstructure, as-deposited (as shown in Fig. 2b), which does not change with annealing. The addition of SiO_2 does not change the microstructure.

For the electrical measurements, it is appropriate to define the calculated parameters, derived as follows from the maximum capacitance (at accumulation) and DC breakdown measurements. The T_{ox} EQUIV. is a representation of the area defined capacitance for an equvalent thickness of SiO_2 Then:

$$T_{ox} \text{ EQUIV(cm)} = 1/C_{accum.\text{(per unit area)}} \quad (1)$$

$$T_{ox} \text{ EQUIV} = E_o \times K_{SiO_2} \times A/C_{max} \quad (2)$$

With:

E_O = Free Space Permittivity

K_{SiO_2} = Relative Dielectric Constant of SiO_2 (3.9)

A = Capacitor Electrode Area (cm²)

C_{max} = Measured Capacitance in Accumulation (Insulation Capacitance -Farads)

We apply the basic definition for capacitance. Then:

$$Q_{charge\ stored} = C_{meas.insl.cap.(unit\ area)} \times V_{(voltage\ across\ insulator)} \qquad (3)$$

The maximum charge stored would be limited by the breakdown voltage (V_{bd}). Therefore (substituting from equations (2) and (3)):

$$Q_{max} = E_O \times 1/T_{ox}EQUIV(cm) \times V_{bd}(volts) \qquad (4)$$

(Charge Storage Index) (Coul/cm²)

The capacitor measurements were carried out for films oxidized at 600°C for two hours in dry O_2 and N_2 annealed at 1000°C for 30 minutes. The results are shown in

TABLE I: EXPLORATORY STUDIES WITH ALTERNATE HIGH CAPACITANCE INSULATORS

DEPOSITION PROCESS	INSULATOR FILM COMPOSITION	OXIDATION (1) ANNEAL	CAPACITANCE (2) T_{ox} EQUIV. (nm)	RELATIVE DIELEC. CONSTANT	DC LEAK AT 5V 1st PASS (na per cm²)	5th PASS (na per cm²)	V_{BD} MEAN (VOLTS)	CHARGE STORAGE INDEX (μCoul./cm²)
DC 60/40 Ar/O_2	Ta_2O_5	28	6.1	18	5	2	10.4	5.9
DC 100% O_2		10	5.5	7	5	2	6.2	3.9
DC+RF 50/50 Ar/O_2	Ta_8SiO_{22}	26	8.5	12	100	20	4.2	1.7
DC 100% O_2	HfO_2	12	6.2	7.5	0.3	0.28	9.5	5.3
DC+RF 50/50	Hf_4SiO_{10}	13	5.9	8.6	0.2	0.27	10.2	6.0

(1) 600°C: O_2, 120' + 1000°C: N_2, 30'
(2) Aluminum capacitor dot electrodes

Table I. The Hf_4SiO_{10} exhibited a higher relative dielectric constant than HfO_2, whereas, Ta_2O_5 exhibited a higher dielectric constant than the Ta_2O_5-SiO_2 mixture. The comparison of properties between Ta_2O_5, HfO_2, and Hf_4SiO_{10} is shown as a function of film thickness in Fig. 3. It is apparent that Ta_2O_5 is most favorable at 25 nm in thickness, while the HfO_2 or Hf_4SiO_{10} is useful at a thickness of 10 to 15 nm. The reason for the selection of the Hf_4SiO_{10}. composition (from deposition rate calibrations) can be seen in Fig. 4, which summarizes the results of capacitance and breakdown measurements as a function of deposited insulator composition. Growth of SiO_2 occurs for insulator compositions Hf_xSiO_{2x+2}, where $x < 2$ or $x > 4$. These results appear to favor the Hf_4SiO_{10} composition. TED studies indicated no compound formation, such as $HfSiO_4$ Only simple binary mixtures were observed. The two oxidation resistant compositions evident in Fig. 4 still require an outer layer of Si_3N_4.

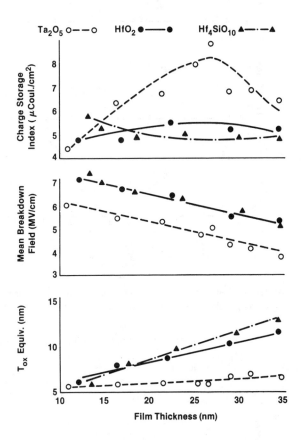

FIG. 3--Capacitive charge storage for three high dielectric insulators (Al cap dot electrodes)

The average capacitor yield, based upon the charge storage criteria, would appear to favor the Ta_2O_5 films when processed with Al electrodes only, but poor yield, based upon extensive low breakdown distributions, was observed with polysilicon electrodes (see also Ref. 17). Subsequent studies focussed upon the three layer configuration, with some results indicated in Fig. 4 and Table II.

Within the stack, the HfO_2 is superior to the Ta_2O_5 Replacing the initial thermal SiO_2 layer with nitrided oxide showed no increase in either breakdown strength or charge storage index. Additionally, removing the transition metal oxide layer does not improve the charge storage properties. The slightly lower breakdown values observed with the polysilicon electrodes, as compared to those observed with the Al electrodes, may be attributed to morphological factors contributing to enhanced Fowler-Nordheim (F-N) emission(26).

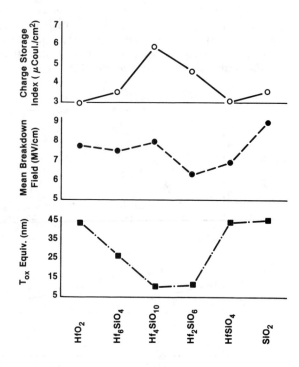

FIG. 4--Influence of assumed HfO$_2$ · SiO$_2$ composition upon the dielectric properties (poly Si electrodes)

TABLE II: COMPARISON OF INSULATOR STACKS WITH HfO$_2$ vs. Ta$_2$O$_5$

	INSULATOR STACK(1)						
FIRST THERMAL LAYER (nm)	SPUTTERED LAYER (nm)	CVD Si$_3$N$_4$ LAYER (nm)	ANNEALED OPTICAL THICK. (nm)	T$_{ox}$ EQUIV. (nm)	V$_{BD}$ MEAN (volts)	CHARGE STORAGE INDEX (Coul.) / cm^2	
SiO$_2$ (1.5)	HfO$_2$ (7.0)	6.0	15.0	7.8	9.1	4.0	
SiO$_2$ (4.0)	HfO$_2$ (5.0)	6.0	16.3	8.0	10.6	4.6	
SiO$_x$N$_y$ (4.0)	HfO$_2$ (5.0)	6.0	15.5	7.9	9.9	4.3	
SiO$_2$ (4.0)	Ta$_2$O$_5$ (15)	6.0	24.2	11.9	12.1	3.5	
SiO$_x$N$_y$ (4.0)	Ta$_2$O$_5$ (10)	7.5	21.0	11.4	12.6	3.8	
SiO$_x$N$_y$ (4.0)	none	6.0	10.0	7.5	8.2	3.8	

(1) Layers 2 and 3, oxidized 800°C: 30', O$_2$ followed by deposition of 300 nm of ISD phosphorus doped polysilicon, etch defined to electrodes

A design for the formation of a memory storage node requires a high N-type dopant concentration (10^{17} to 10^{19} (cm^{-3})) of arsenic) at the silicon-dielectric interface, along with a shallow junction depth (0.1 to 0.2 μm)(16). The junction must be self-aligned to the storage node electrode and removed from the adjacent zone containing the FET charge transfer device. Two implant processes were used. The first was a very low energy implant into the insulator stack with none of the forward straggle implanted into the Si substrate. The polysilicon film and electrode definition followed after which, the implanted insulator stack adjacent to the polysilicon electrode was removed by wet or dry etching. The second approach was to implant into a screen oxide utilizing the forward implant straggle to develop the Si substrate junction profile, etch remove the screen oxide, and deposit the insulator stack plus polysilicon gate film. Etching for storage line definition also removed the adjacent stack and surface of the adjacent Si substrate containing the implanted arsenic. The transfer gate oxidation process for both approaches provided the arsenic drive-in requirements.

Implant profiles, obtained by Pulse C-V, as a function of arsenic implant dose indicated that surface concentration and depth profile appear to be influenced more strongly by dopant fluence than the implant energy. Fig. 5 shows examples of the correlation between implant dose and arsenic surface concentration for both implant process, along with comparisons to LSS modelling and corroborative SIMS analysis. The electrical data appears to favor the second implant process (lower T_{ox} equivalent and higher charge storage index). Equivalent arsenic surface concentrations are observed for both.

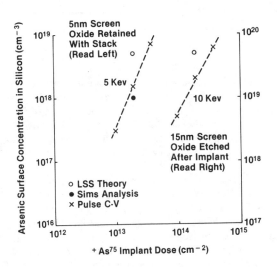

FIG. 5.--Formation of "N" skin with $^+As^{75}$ implant into screen oxide (SiO_2)

DISCUSSION

Some electrical properties comparison can be made with recent data reported for CVD ZrO_2(27):

TABLE III - ELECTRICAL PROPERTIES FOR Sp.HfO$_2$ vs. CVD ZrO$_2$ (27)

FILM	PROCESSING	ELECTRODE	FILM THICK. (nm)	T$_{0x}$ EQUIV. (nm)	DC LEAKAGE CURRENT [nano-a/cm^2]	V$_{bd}$ [volts]	E$_{bd}$ [MV/cm]	CHARGE STORAGE INDEX [uCoul./cm^2]
CVD ZrO$_2$	800°C O$_2$ + HCl	Al	30	7.5	10.0 [1.5MV/cm]	12.0	4.0	5.2
CVD ZrO$_2$	800°C	PolySi	35.0	7.5	10.0 [1.5MV/cm]	10.0	2.9	4.3
Reac.Sp. HfO$_2$	800°C/O$_2$	Al	35.0	11.5	0.15 [4.0MV/cm]	19.0	5.4	5.7
Reac.Sp. HfO$_2$	800°C/O$_2$	Al	12.0	6.2	0.30 [4.2MV/cm]	9.5	7.9	5.3
Reac.Sp. HfO$_2$	800°C/O$_2$	PolySi	12.0	5.9	0.50 [4.0MV/cm]	9.3	7.8	5.9

The HfO$_2$ appears to provide a performance advantage over the ZrO$_2$ with regard to leakage, breakdown, and charge storage. Since they both have very similar physio-chemical properties, the difference may be due to the method of film preparation, especially since other reported CVD films(6,7,12,13) appear to show high leakage and low breadown strength, as compared to the reactively sputtered and annealed HfO$_2$ based films (this study) and reactively sputtered and annealed Ta$_2$O$_5$ films(17).

The multi-layer films show interesting charge storage behavior with regard to film thicknesses and dielectric constant of the middle layer (Tables I and II). This would imply that the overall capacitance is not governed by the series capacitance of the individual layers, but is due to additional charge injected into the center layer, perhaps by a mechanism similar to that observed with MNOS-type structures(22). Since the thermal oxide thicknesses are within the tunneling range, this presents a plausible mechanism. The reasons for lower comparative capacitance exhibited by the Ta$_2$O$_5$ may be a result of greater conductance than the HfO$_2$, so that injected charge could leak out more rapidly within the Ta$_2$O$_5$. Comparative I-V measurements (not shown) have definitely indicated lower leakage for comparable fields for the HfO$_2$. The Hf$_4$SiO$_{10}$. may be more effective regarding charge storage than the pure HfO$_2$ due to the additional trapping sites provided by the added SiO$_2$. Both show comparable leakage behavior. Based upon the studies reported in Reference 22, it can be speculated that replacement of the Si$_3$N$_4$ layer with CVD Al$_2$O$_3$ would increase the charge stored and effective capacitance. Some evidence of the charge storage mechanism can be observed with the charge window shown in Fig. 6., whereby a 0.5V read window is observed, with 10V-millisecond write-erase pulsing. Because of the large read delay (30 seconds) following the write-erase pulsing, the real time charge window with short (millisecond range) read delays would most likely be larger.

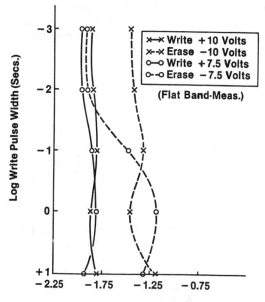

Read Voltage (Volts-30 Sec. Delay)

CONCLUSIONS

An evaluation of one, two, and three-layer thin dielectric films, which includes one-layer prepared by reactive sputtering from metal targets, has been carried out for FET memory node charge storage applications. The criteria for comparison of several high dielectric constant materials in single-layer and multi-layer combinations with thermal SiO_2 and CVD Si_3N_4 include: the capacitive equivalence to pure SiO_2 (T_{OX} equivalent), dielectric breakdown, DC leakage, and a calculated index for maximum charge storage capacity that includes the T_{OX} equivalent and DC breakdown.

The best storage performance for single-layer films is shown by Ta_2O_5, which yields T_{OX} equivalence in the range of 5 to 6 nm and charge storage index capacity in the range of 8 to 9 micro-coul./cm². Reduced values (in the range of 5 to 5.5 micro-coul./cm²) are shown by single films of HfO_2 and Hf_4SiO_{10}.. To fulfill high temperature device processing requirements, three layer films that include thermal SiO_2 as the first layer and CVD Si_3N_4 as the third layer are required. For these films, the Hf_4SiO_{10}. and HfO_2 provides superior capacitance performance to the use (as the middle layer) of Ta_2O_5, and yield storage properties in the range of 6 nm for the T_{OX} equivalent, and 4 - 7 micro-coul./cm² for the charge storage index.

ACKNOWLEDGEMENTS

The authors would like to thank J. Schaefer for film preparation, J. O'Connor for assistance in wafer processing, C. Koburger, III, and C. Schaefer for assistance in carrying out the electrical measurements, L. Nesbit for TEM and TED support, P. Pan for Auger support, and G. Slusser for SIMS support. Additionally, we would like to thank P. C. Velasquez and R. Geffken for program support, and J. Wursthorn for a critical reading of the manuscript.

REFERENCES

[1] K. Ohta, "A Stacked High Capacitor RAM", ISCC 80, pp. 66-69.
[2] M. E. Elta, "Tantalum Oxide Capacitors for GaAs Monolithic Integrated Circuits", IEEE Electron Device Letters, Vol EDL-3, No. 5, May 1982, pp. 127-128.
[3] S. Seki, T. Unagami, B. Tsujyama, "P Channel TFTs Using Magnetron Sputtered Ta_2O_5 Films as Gate Insulators", IEEE Electron Device Letters, Vol EDL-3, No. 6, June 1984, pp. 197-198.
[4] T. Kolfassand, E. Lueder, "High Voltage Thin Film Transistors with Ta_2O_5 as the Gate Oxide", Thin Solid Films, Vol. 61, 1979, pp. 259-264.
[5] L. I. Maisell and R. Glang, Eds., "Dielectric Properties of Thin Films", Handbook of Thin Films, McGraw Hill, N.Y., pp. 16-28 to 16-31, 1970.
[6] W. H. Kansenberger and R. N. Tauber, Selected Properties of Pyrolytic Ta_2O_5 Films", Journal of the Electrochemical Society, Vol. 120, No. 7, July 1973, pp. 927-931.
[7] E. Kaplan, M. Balog, D. Frohman-Bentchkowsky, "Chemical Vapor Deposition of Ta_2O_5 Films for MIS Device", Journal of the Electrochemical Society, Vol. 123, No. 10, Oct. 1976, pp. 1570-1573.
[8] A. G. Revesz, "Film Substrate Interaction in Si/Ta and Si/Ta_2O_5 Structures, Journal of the Electrochemical Society, Vol. 123, No. 10, Oct. 1976, pp. 1514-1519.
[9] D. J. Smith, L. Young, "Optical and Electrical Properties of Thermal Ta_2O_5 Films on Silicon", IEEE Transactions on Electron Devices, Vol. ED-28, No. 1, Jan. 1981, pp. 22-26.
[10] G. S. Oehrlein, A. Reisman, "Electrical Properties of Amorphous Ta_2O_5 Films on Silicon", Journal of Applied Physics, Vol. 54, No. 11, Nov. 1983, pp. 6502-6508.
[11] F. T. J. Smith, "Structure and Electrical Properties of Sputtered Films of Hafnium and Hafnium Compounds", Journal of Applied Physics, Vol. 41, No. 11, Sept. 1970, pp. 4227-4231.
[12] M. Belog, M. Scheiber, "The Chemical Vapor Deposition and Characterization of ZrO_2 Films from Organometallic Compounds", Thin Solid Films, Vol. 47, 1977, pp. 109-120.
[13] R. T. Taubner, A. C. Dumbri, R. E. Caffrey, "Preparation and Properties of Pyrolytic Zirconium Dioxide Films", Journal of the Electrochemical Society, Vol. 118, No. 5, May 1971, pp. 747-754.
[14] D. R. Harbinson, H. L. Taylor, in "Thin Film Dielectrics", F. Vantry, Ed., the Electrochemical Society, Princeton, N.J., 1969.
[15] T. Mahalingham, M. Radhakrishnan, C. Balasubramanian, "Dielectric Behavior of Lanthanum Oxide Thin Film Capacitors", Thin Solid Films, Vol. 78, Apr. 1981, pp. 229-233.
[16] E. Adler, "A High Performance High Density 256K DRAM Utilizing 1X Projection Lithography", Paper No. 13.3, IEDM, Washington, D.C., Dec. 5, 1983.

[17] S. Roberts, J. Ryan, L. Nesbit, "Selective Studies of Crystalline Ta_2O_5 Films, Journal of the Electrochemical Society, Vol. 133, No. 7, July 1986, pp. 1405-1410.

[18] R. B. Herring, "Advances in Reduced Pressure Epitaxy", Solid State Technology, Nove. 1977, pp. 75-79.

[19] F. Reizman, W. Van Gelder, "Optical Thickness Measurements of SiO_2-Si_3N_4 Films on Silicon", Solid State Electronics, Vol. 10, 1967, pp. 625-631.

[20] A. C. Adams, C. D. Capio, "Edge Profiles in the Plasma Etching of Polycrystalline Silicon", Journal of the Electrochemical Society, Vol. 128, No. 2, Feb. 1981, pp. 366-370.

[21] S. M. Sze, "MIS Diode and Charge Coupled Devices", Chapt. 7, p. 362 in Physics of Semiconductor Devices, John Wiley & Sons, N.Y., 1981.

[22] S. Zirinsky, E. A. Irene, "Selective Studies of Chemical Vapor Deposited Aluminum Nitride-Silicon Nitride Mixture Films", Journal of the Electrochemical Society, Vol. 125, No. 2, Feb. 1978, pp. 305-314.

[23] R. O. Denning, W. A. Keenan, "Techniques for Implant Profile Monitors", Solid State Technology, Vol. 28, 1985, pp. 163-170.

[24] J. Lindhard, M. Scharff, H. E. Schiott, "Range Concepts and Heavy Ion Ranges", Mat. Fys. Medd. Dan. Vid. Selk., Vol. 33, 1963, pp. 1-35.

[25] S. Seki, T. Unagami, B. Tsujiyama, "Electrical Characteristics of the R.F. Magnetron Sputtered Tantalum Pentoxide-Silicon Interface", Journal of the Electrochemical Society, Vol. 131, No. 11, Nov. 1984, pp. 2621-2624.

[26] R. M. Anderson, D. R. Kerr, "Evidence for Surface Asperity Mechanism of Conductivity in Oxide Grown on Polycrystalline Silicon", Journal of Applied Physics, Vol. 48, No. 11, Nov. 1977, pp. 4834-4836.

[27] J. Shappir, A. Anis, I. Pinsky, "Investigation of MOS Capacitors with Thin ZrO_2 Layers and Various Gate Materials for Advanced DRAM Applications", IEEE Transactions on Electron Devices, Vol. ED-33, No. 4, Apr. 1986, pp. 442-448.

Han-Sheng Lee

THE ELECTRICAL PROPERTIES OF MOS TRANSISTORS FABRICATED WITH DIRECT ION BEAM NITRIDATION

REFERENCE : Lee, H.S., "The Electrical Properties of MOS Transistors Fabricated With Direct Ion Beam Nitridation," Emerging Semiconductor Technology, ASTM STP 960, D.C. Gupta and P.H. Langer, Eds., American Society for Testing and Materials, 1986.

ABSTRACT : Direct ion beam nitridation was used to form a thin masking layer in the field oxide growth step. The resulting field oxide encroachment per device edge was reduced from 1.0 μm found in LOCOS devices to 0.33 μm. However, the transistors fabricated with the ion beam nitridation step displayed poor device characteristics. We found that with the growth of 50 nm of sacrificial oxide at 950°C and then subsequent removal of the oxide before the gate oxide growth, the adverse side effects, such as; low electron field effect mobility, high junction leakage current and low gate oxide breakdown strength can be drastically improved.

KEYWORDS : mobility, ion beam damage, surface roughness, MOS transistor.

The use of the local oxidation of silicon (LOCOS) process to form electrical isolation between devices has been the industry

Dr. Han-Sheng Lee is a senior staff research engineer at Electronics Department, General Motors Research Labs., 30500 Mound Rd., Warren, Michigan 48090-9055.

standard for years [1]. One of the problems associated with this process is the tapering of the isolation oxide into the active device region and consumption of the areas reserved for the active devices. The length of lateral encroachment is usually one to one and one-half times the isolation oxide thickness. The ratio of the areas consumed by the lateral encroachment to the active device areas will increase even more when device sizes are down-scaled. This problem will set a limit on the device packing density. Different isolation techniques have been developed by researchers to reduce or eliminate the lateral encroachment [2-6]. Usually, the reduction of the lateral encroachment is achieved by increasing the fabrication complexity. Among those techniques, seal-interface local oxidation (SILO) involves the least fabrication steps. With a nitridized thin film as the bottom masking layer in the SILO process, Hui et al [6] has obtained a lateral encroachment of 0.25 µm per device edge. On the control samples where the LOCOS technique was used, the lateral encroachment was 0.85 µm per edge. The field oxide thickness of their devices was 0.55 µm. However, when compared with the LOCOS step fabricated transistors, the devices involving the SILO process showed (1) lower slope in the I_{ds} (drain-to-source current) vs the mask channel width, W, plot, (2) higher n^+p junction leakage current and (3) inferior gate oxide quality.

In this report we will present the electrical properties of transistors with a thin ion beam nitridized (IBN) film as the mask in the field oxidation step. Our masking structure uses fewer process steps than the one fabricated by Hui et al. In their structure [6] the mask consists of three layers, as compared to one layer in our case. In the fabricated devices we also observed the same drawbacks. We found that by growing an oxide and removing this oxide before the gate insulator is grown, we can lower the three side effects that were observed in the SILO process. We refer to this intermediate oxide growth and removal as the sacrificial oxide (SO) treatment. By performing the sacrificial oxide treatment, one can benefit from the small lateral encroachment with minimum drawbacks. Since the sacrificial oxide treatment is a maskless procedure, the combination of the IBN/SO treatment is quite attractive.

EXPERIMENTS

The silicon wafers used in these experiments were p-type, <100>

oriented, 5 ohm-cm and 76 mm diameter. The oxidation masking films were formed by nitrogen bombardment at 2 keV, 1 mA/cm^2 for 2 minutes. It corresponds to a dosage of 7.5 x 10^{17} cm^{-2}. This ion beam nitridation process will form a 10 nm thick nitride layer at the surface. After patterning the device areas and chemical cleaning, field oxide was grown at 950°C in wet oxygen ambient to a thickness of 0.69 µm. Control samples prepared with the LOCOS isolation technique were also oxidized with the IBN wafers. After the field oxide growth, all wafers were etched in hot phosphoric acid to remove the residual nitride. Then some of the IBN wafers had 50 nm of sacrificial oxide grown at temperatures ranging from 900°C to 1000°C. After removing the sacrificial oxide, all wafers received channel implantation. Boron was implanted at 80 keV to 1.5 x 10^{12} cm^{-2} and then at 20 keV to 1.5 x 10^{12} cm^{-2}. After chemical cleaning, the gate oxide was grown at 900°C with the first 10 min in dry oxygen ambient followed by 4.5 min wet oxidation and then 10 min in dry oxygen. The wafers were annealed in the same furnace tube for 30 min with nitrogen flowing. The resulting gate oxide thickness was 37 nm.

The finished MOSFETs had silicon gates and arsenic implanted source and drain junctions. Arsenic was implanted at 80 keV to a dosage of 5 x 10^{15} cm^{-2}. The junction drive-in and implantation annealing was done at 900°C for 3.5 h in nitrogen ambient. On the finished chips we had transistors with fixed channel lengths but different channel widths and also transistors with fixed channel widths but different channel lengths. These devices were used to determine the electrical parameters, such as the effective channel length and the effective channel width, of the devices [7].

RESULTS

A convenient way to show the lateral encroachment of the devices is to plot the measured conductance factor, k, versus the mask channel width W. The conductance factor is defined as $k = W_{eff} \mu_n C_o / L_{eff}$, where W_{eff} and L_{eff} are the effective channel width and channel length of the transistor, respectively, μ_n is the electron field effect mobility and C_o is the gate oxide capacitance per unit area. The measurements in Fig.1 were done on transistors with the same channel length, 10 µm, and with devices operated at the linear region of channel conductance. Since the conduction factor is linearly proportional to the effective channel width, the non-zero intercept at the horizontal axis is the result of lateral encroachment from both sides of the devices. Assuming symmetry, the lateral encroachment per device edge is half the value of the intercept. The lateral encroachment per device edge are 1.0, 0.39 and 0.33 μm for the LOCOS devices and IBN samples without and with the SO treatment. The lateral encroachment in the sacrificial oxide treated sample is smaller because the edges of the encroached portion were slightly etched during the removal of sacrificial oxide. From the slope of each curve, we can calculate the electron field effect mobility of each device. These values are 617, 640 and 650 cm^2/V·s for the IBN samples with 50 nm sacrificial oxide grown at 900°C, 950°C and 1000°C, respectively. The electron field effect mobility of the control LOCOS sample was 635 cm^2/V·s as opposed to 573 cm^2/V·s for the IBN sample without having the SO treatment.

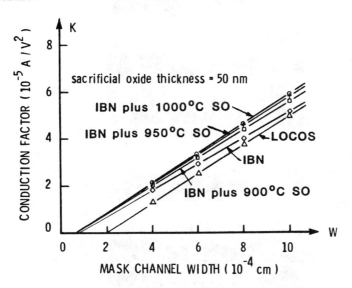

Fig. 1 - The conduction factor vs the mask channel width of the transistors with different masking methods to grow the field oxide. For those devices which had the sacrificial oxide (SO) treatment the oxide growth temperature is labeled.

Figures 2(a)-2(c) are SEM pictures of the fabricated transistors viewed along the channel width direction. These devices had the gate mask dimension of 10 μm in channel length and 4 μm in channel width. Measured channel widths are 2.0, 3.3 and 3.5 μm for the LOCOS and IBN samples without and with the SO treatment, respectively. These values are in good agreement with the electrically measured results shown in Fig. 1.

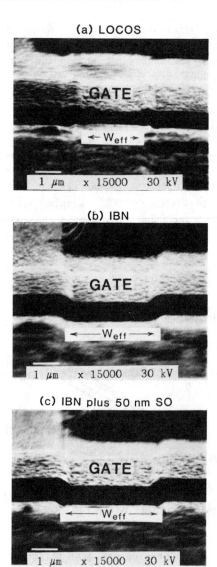

Fig. 2 - The SEM pictures of the transistors viewing along the channel width direction. The mask dimension of the width is 4 μm.

Figure 3 shows the gate oxide quality of the MOS capacitors which had different treatments. The diameter of the silicon-gate test MOS capacitors is 508 μm. The fabrication steps and the substrate used to grow the gate oxide of the capacitors are identical to those used in the transistor fabrication. In the measurements, the gates were positively biased with respective to the substrate and illuminated with light to ensure no deep depletion in the substrate. The electric field strength of the gate oxide improved as the temperature used in growing the sacrificial oxide is increased. We also found that the electric field strength of the gate oxide can be increased with samples annealed at 1100°C in nitrogen ambient without using the sacrificial oxide treatment.

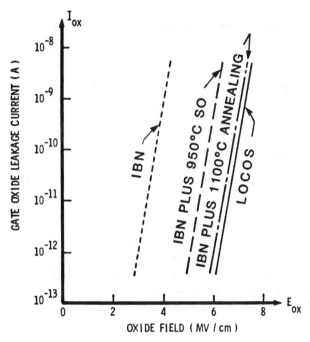

Fig. 3 - The gate oxide leakage current vs the electric field across the oxide. The diameter of the gate is 508 μm.

Measured results on n^+p junction leakage current are shown in Fig. 4. The test diodes had dimensions of 250 x 250 μm^2 with dopant distribution similar to the drain-to-substrate junction of the transistors and were on the same chip with the transistors. Reduction of the leakage current on samples which had SO treatment was observed.

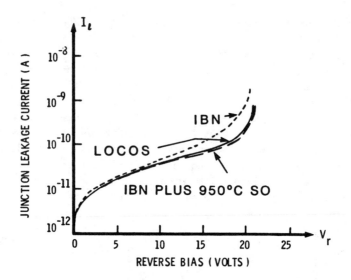

Fig. 4 - The n^+p junction leakage current vs the applied reverse bias. The area of the junction is 250 x 250 μm^2.

DISCUSSIONS

Experimentally, it was observed by Carim et al [8] that 1100°C annealing can smooth the Si/SiO_2 interface and increase the breakdown strength of the oxide. If the morphology of the silicon surface is the dominant factor in determining the gate oxide breakdown strength [9], the results of Fig.3 seem to imply that the surface of the IBN samples is not smooth and the sacrificial oxide treatment can smooth the silicon surface.

The maximum electron field effect mobilities of different gate sizes at different temperatures are shown in Fig.5. Results show the mobility is insensitive to the gate size. Therefore, the edge effect is not a major factor affecting μ_n. We also found that the curve corresponding to the IBN plus 950°C, 50 nm sacrificial oxide treatment (the upper curve) is very close to the results from the devices which used the LOCOS process. This implies that there exists a constant lowering factor in the IBN samples when compared with the LOCOS samples. The factor is independent of the temperature. One possible mechanism which can cause the mobility lowering and temperature insensitivity is the scattering caused by the surface roughness [10]. The scale of roughness could be a few atomic layers because the inversion layer thickness is thinner than 5 nm in our samples.

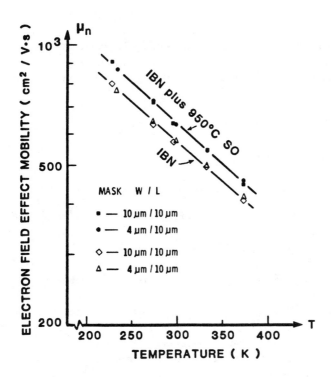

Fig. 5 - Measured maximum electron field effect mobility of IBN transistors with and without SO treatment at different temperatures.

The LOCOS samples used in this report did not receive SO treatment after removing the pad oxide. There could be some defects generated during the field oxide growth [11]. This could be the reason that the LOCOS samples had slightly lower μ_n than the IBN samples which had SO treatment.

The SO treatment had been used for the LOCOS samples before the growth of gate oxide to improve the oxide integrity [12]. However, as was suggested here, the reasons and the functions of using SO treatment for LOCOS and IBN samples are not the same. The SO treatment generally has been used after the LOCOS step to eliminate the gate oxide thinning problem by consuming the nitride patches formed at the substrate but under the masked areas during the field oxide growth [12-15]. However, we do not observe the gate oxide thinning problem in the IBN samples as evidenced by the 1100°C annealing experiment. This is because the local oxide thinning problem is not affected by annealing the samples in nitrogen ambient. The increase of the oxide breakdown strength after the annealing is attributed to the improvement in substrate smoothness. In conclusion, we believe the SO treatment on the IBN samples serve to anneal and smooth the damaged substrate caused by the ion beam bombardment.

REFERENCES

[1] Appels, J.A., Kooi, E., Paffen, M.M., Schatorje, J.J.H. and Verkuylen, W.H., "Local Oxidation of Silicon and Its Application in Semiconductor-Device Technology," Philips Research Reports, Vol. 25, 1970, pp.118-132.

[2] Oldham, W.G., "Isolation Technology for Scaled MOS VLSI," 1982 International Electron Devices Meeting Technical Digest, pp.216-219.

[3] Chiu, K.Y., Moll, J.L., Cham, K.M., Lin, J., Lage, C., Angelos, S. and Tillman, R.L., "The Sloped-Wall SWAMI- a Defect-Free Zero Bird's Beak Local Oxidation Process for Scaled VLSI Technology," IEEE Transactions on Electron Devices, Vol. ED-30, 1983, pp.1506-1511.

[4] Bondur, J.A. and Pogge, H.B., "Reactive Ion Etching Method for Producing Deep Dielectric Isolation in Silicon," U.S. Patent #4,139,442 (1979).

[5] Hui, J., Chiu, T.Y., Wong, S. and Oldham, W.G., "Sealed-Interface Local Oxidation Technology," IEEE Transactions on Electron Devices, Vol. ED-29, 1982, pp.554-561.

[6] Hui, J., Chiu, T.Y., Wong, S. and Oldham, W.G., "Electrical Properties of MOS Devices Made with SILO Technology," 1982 International Electron Devices Meeting Technical Digest, pp.220-223.

[7] Chern, J., Chang, P., Motta, R.F. and Godinho, N., "A New Method to Determine MOSFET Channel Length," IEEE Electron Device Letters, Vol. EDL-1, 1980, pp.170-173.

[8] Carim, A.H. and Bhattacharyya, A., "Si/SiO_2 Interface Roughness: Structure Observations and Electrical Consequences," Applied Physics Letters, Vol. 46, 1985, pp.872-874.

[9] Lee, H.S. and Marin, S.P., "Electrode Shape Effects on Oxide Conduction in Films Thermally Grown from Polycrystalline Silicon," Journal of Applied Physics, Vol. 51, 1980, pp.3746-3750.

[10] Cheng, Y.C. and Sullivan, E.A., "On The Role of Scattering by Surface Roughness in Silicon Inversion Layers," Surface Science, Vol. 34, 1973, pp.717-731.

[11] Irene, E.A., "Residual Stress in Silicon Nitride Films," Journal of Electronic Materials, Vol. 5, 1976, pp.287-298.

[12] Shankoff, T.A., Sheng, T.T., Haszko, S.E., Marcus, R.B. and Smith, T.E., "Bird's Beak Configuration and Elimination of Gate Oxide Thinning Produced During Selective Oxidation," Journal of Electrochemical Society, Vol. 127, 1980, pp.216-222.

[13] Kooi, E., Van Lierop, J.G. and Appels, J. A., "Formation of Silicon Nitride at a $Si-SiO_2$ Interface During Local Oxidation of Silicon and During Heat-Treatment of Oxidized Silicon in NH_3 Gas," Journal of Electrochemical Society, Vol. 123, 1976, pp.1117-1120.

[14] Nakajima, O., Shiono, N., Muramoto and Hashimoto, C., "Defects in a Gate Oxide Grown after the LOCOS Process," Japanese Journal of Applied Physics, Vol. 18, 1979, pp.943-951.

[15] Itsumi, M. and Kiyosumi, F., "Identification and Elimination of Gate Oxide Defect Origin Produced During Selective Field Oxidation," Journal of Electrochemical Society, Vol. 129, 1982, pp.800-806.

Plasma Technology and Other Fabrication Techniques

Stephen J. Fonash and Ajeet Rohatgi

RIE DAMAGE AND ITS CONTROL IN SILICON PROCESSING

REFERENCE: Fonash, S. J. and Rohatgi, A., "RIE Damage and Its Control in Silicon Processing," Emerging Semiconductor Technology, ASTM STP 960, D. C. Gupta and P. H. Langer, Eds., American Society for Testing and Materials, 1986.

ABSTRACT: Reactive ion etching (RIE) has become a necessary tool in much of silicon device manufacturing. When employing this important etching technique, one must be aware that its use can produce damage and contamination. The types of damage and contamination that can result from RIE are discussed for several etching chemistries. Control of RIE damage can be achieved by removing it by wet chemical etching or by furnace annealing. Control of RIE damage can also be achieved by hydrogen passivation or by the use of rapid thermal anneals. These latter techniques are of considerable interest since they are compatible with micron and submicron device geometries.

KEYWORDS: reactive ion etching, damage, contamination, damage control

It has been amply demonstrated that reactive ion etching (RIE) of semiconductors and insulators can result in damaged and contaminated surface regions [1-7]. The presence of these damaged and contaminated regions can affect subsequent contact formation, diffusion, silicidation, oxidation, etc. [1-7]. In many cases in current technology these obvious problems resulting from RIE are obviated by the use of subsequent wet etching and the use of high temperature processing in fabrication steps after the RIE exposure. This fortuitous situation will become more difficult to achieve, and hence RIE damage effects will demand more attention, as device geometries become smaller and processing temperatures are reduced. Consequently, RIE damage and contamination effects will demand more attention as Si technology continues to evolve.

In this report we survey the damage and contamination produced in Si by several different RIE etching chemistries and show how the

Dr. Fonash is Alumni Professor of Engineering Sciences at The Pennsylvania State University, University Park, PA 16802 and Dr. Rohatgi is Associate Professor of Electrical Engineering at Georgia Institute of Technology, Atlanta, GA 30332.

RIE-exposure effects observed fit into an RIE damage/contamination classification scheme first presented in reference [7]. We then examine the use of two different techniques for the control of RIE damage (viz, hydrogen passivation and rapid thermal annealing) which, unlike wet etching and furnace anneals, should be compatible with the drive to low temperature processing and submicron geometry.

RIE DAMAGE AND CONTAMINATION

Classification Scheme

The RIE damage/contamination classification scheme of reference [7] is reproduced here in Fig. 1 to aid in our discussion of RIE. As may be seen, this model classifies the different types of damage and contamination which have been observed to result from RIE etching. The model also shows their arrangement in an etched material. Three layers are shown to be possible, in general, as a result of exposure to RIE. Although it is difficult to convey in the figure, these layers are certainly not sharply defined; rather, they gradually blend

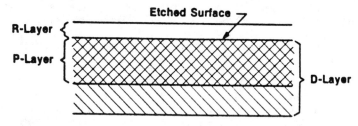

Fig. 1. Arrangement of damage/contamination in RIE etched Si. (From ref. 7)

into one another. The upper-most layer has been termed the R-layer (for residue layer). In the case of silicon it would be principally non-Si bearing layer. An R-layer may or may not be present after reactive ion etching; its existence depends on whether or not surface film layers are involved in the etching chemistry [7].

The next layer encountered as one moves toward the bulk of the etched sample has been termed the P-layer (for permeated layer). In the case of Si, this is a silicon layer containing the etching species or subunits of the etching species (i.e., Cl from CCl_4 RIE); it may also contain impurities introduced during etching. The etching species permeate the P-layer during etching due to various combinations of implantation [8,9], channeling-enhanced implantation, recoil implantation [10], and enhanced diffusion driven by non-equilibrium point-defect densities [9, 10]. In some situations the P-layer may also owe its existence, to some extent, to redeposition. For the ion kinetic energies encountered in RIE (from tens of eV up to < 1 keV), these P-layers may vary from tens of Angstroms to several hundred Angstroms in thickness.

The third layer produced by RIE has been termed the D-layer (for intrinsic bonding damage layer) in the model of Fig. 1. In this layer the silicon bonding has been damaged in the etched material. In the case where the etched material is a compound semiconductor, the stoichiometry may also be disturbed. The layer contains dislocations, vacancies, and interstitials. This D-layer may extend further into the material than the P-layer as seen in Fig. 1 or the two layers may coincide.

Damage/Contamination from Different RIE Chemistries

We now survey, and compare, the damage and contamination effects resulting from silicon exposure to four different RIE chemistries. The four different RIE etching chemistries to be considered are CCl_4 RIE, $CClF_3/H_2$ RIE, CF_4 RIE, and CF_4/H_2 RIE. This particular choice of etching chemistries will allow us to examine situations where R-layers are important as well as situations when they are not. It will also allow us the opportunity to determine the effects, on damage and contamination, of "doping" the plasma with hydrogen. Finally, this variety of etching chemistries will allow us to stress the ubiquity of the problem of silicon lattice damage (the D-layer) due to RIE exposure.

CCl_4 RIE: For the etching parameters used for the CCl_4 RIE data reported here [5, 6] (blanket etching of bare wafers at a total pressure of 20-27 m torr, RF power of 600 W, and a power density of ~0.5 W/cm^2) CCl_4 RIE of silicon does not result in a substantial

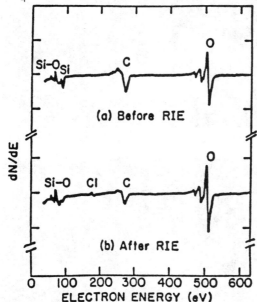

Fig. 2. Auger electron spectroscopy data for (a) a Si sample subjected only to an HF wet etch and (b) a Si sample subjected to the CCl_4 RIE. These data are mass scans obtained from the surface of the samples. (From ref. 5)

R-layer on the exposed Si. Figure 2 shows Auger electron spectroscopy (AES) data which substantiate this. Figure 2 is for Cz silicon that has not (before RIE, Fig. 2) and has (after RIE, Fig. 2) been exposed to this RIE. As may be noted from the figure, silicon is observed in the surface scan of CCl_4 RIE-exposed samples and the surface carbon signal is essentially the same for the etched and control samples indicating that there is no substantial residue layer present. However, it is noted that CCl_4 RIE does produce a P-layer. An indication of this is seen in the Cl signal present for the CCl_4 RIE- exposed sample of Fig. 2. Secondary ion mass spectroscopy (SIMS) profiling studies of this incorporation of Cl have shown that it resides in a surface layer ~25 Å in thickness [6]. An increase of the Si yield from SIMS and an enhancement of carbon and oxygen is found in this region also [6].

The lattice damage present after a CCl_4 RIE exposure of Si is easily discerned using reflected high energy electron diffraction (RHEED). Figure 3 presents RHEED data for a control (Fig. 3a) and for a sample subjected to CCl_4 RIE (Fig. 3b). As noted on the figure, the control sample was given a 5 sec. 5% HF etch and a subsequent deionized water rinse; the CCl_4 RIE-etched sample was not. The vertical streaks and Kikuchi lines present in Fig. 3(a) are seen to be completely absent in Fig. 3(b) indicative of damage, contamination, or both. Closer examination reveals that lattice damage is definitely

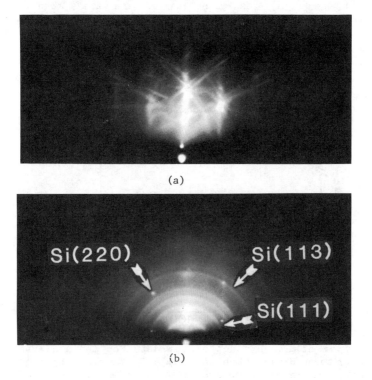

Fig. 3. Rheed data for (a) control and (b) CCl_4 RIE etched Si.

present since the ring pattern representing diffraction from {111} and {311} planes in a polycrystalline Si layer is seen to be observed for the CCl_4 exposed Si. Moreover, there are additional rings present in Fig. 3(b) whose origin cannot be attributed to polycrystalline Si. A check of x-ray powder diffraction data indicates that some of these additional rings can be attributed to the presence of SiC. However, there are other rings whose origin is not that clear; phases that one might readily suspect to be the sources of such unidentified rings (e.g., Si_3N_4, SiO_2, diamond, and graphite) have been checked and ruled out [5]. The comparison provided by Figs. 3(a) and 3(b) indicates that CCl_4 RIE, for etching parameters use in this study, results in a damaged Si layer (a D-layer), at least a portion of which is poly- crystalline silicon. In addition, the RHEED data shows there is a second phase layer present (which we assign to be part of the P-layer) consisting of Si, SiC, and some other unknown phase.

$CClF_3/H_2$ RIE: RIE-exposure of silicon to $CClF_3/H_2$ RIE for the etching parameters used in our work, is an example of etching which produces a non-silicon (residue or R-layer) surface region [11]. Figure 4 shows AES data for a $CClF_3/H_2$ etched sample (blanket etching of oxidized wafers at a total gas flow rate of 200 sccm, a pressure of 15 mtorr, RF power of 1000 W, and power density of 0.25 W/cm^2) obtained from a surface scan and after a 20 sec. sputtering (Ar sputter rate ~ 50 Å/ min.). The control data of Fig. 2 may be used also as the control for these data of Fig. 4. The AES data of Fig. 4, in contrast to that seen for CCl_4-etched Si, show that the surface of the CCl_4-etched Si, show that the surface of the $CClF_3/H_2$ etched Si is covered with a carbonaceous residue layer; fluorine and chlorine are seen to be present in this reside layer. Since the Ar

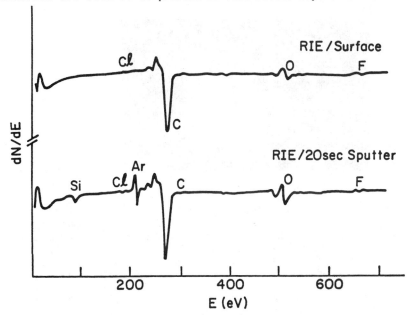

Fig. 4. AES scans for a Si surface and for Si after a 20 sec. Ar sputter etch for $CClF_3/H_2$ RIE exposed samples.

sputter rate is known, we are able to show, with deeper AES profiling than is shown here [11], that the residue layer is ≳35 Å for this sample. We note that we have shown in a more extensive report of our AES profiling data for $CClF_3/H_2$ etched Si that, as the sputtering proceeds, the silicon AES signal grows and there is a shape change in the carbon signature which is very suggestive of the formation of SiC [11]. Apparently SiC is also produced by $CClF_3/H_2$ RIE; this SiC may be in a very restricted region about the R-layer/silicon interface [11].

There is also a permeated (P-layer) region produced by this $CClF_3/H_2$ RIE. As we report elsewhere [11], hydrogen, for example, is found to be present above the background level as far down as 500 Å in $CClF_3/H_2$ exposed samples. In addition, the D-layer is present as the RHEED data of Fig. 5 clearly show. As may be seen, the RHEED pattern for the RIE-exposed sample displays a ring pattern characteristic of polycrystalline silicon as well as a background haze suggestive of amorphous material. Rings representing diffraction from {111}, {220}, and {311} Si planes are clearly seen and labeled in Fig. 5 [11]. As may be recalled, we have observed this polycrystalline Si ring RHEED pattern previously for CCl_4 reactive ion etched Si (see Fig. 3) [12]. The origins of this polycrystalline silicon may lie in redeposition or in regrowth of some of the damaged Si or in a combination of both. As we have noted elsewhere [11], further evidence for the presence of silicon lattice damage after $CClF_3/H_2$ RIE comes from Rutherford backscattering spectroscopy (RBS). This technique of characterizing a surface shows that there are more displaced Si near-surface atoms after the RIE exposure. In fact, RBS shows that the number of displaced Si atoms at a surface for the RIE samples is almost double that of a control [11].

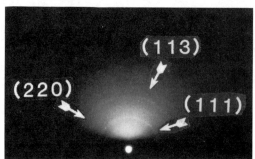

Fig. 5. RHEED pattern for $CClF_3/H_2$ exposed (50% over etched) Si.

CF₄ RIE: Just as was found to be the case for CCl_4 RIE of silicon, CF_4 RIE is not found to produce a significant residue layer on a silicon surface. This may be verified from Fig. 6. Using the control of Fig. 2 as the control for this figure and noting that the upper two traces of Fig. 6 are for CF_4 RIE, it can be seen that silicon (and oxygen) is clearly present at the surface for the etching parameters used (blanket etching of wafers using a gas flow rate of 25 sccm, a pressure of 20 mtorr, an RF power of 580 W. and an rf power density of 0.48 W/cm²). Contamination and permeation of the silicon by carbon and fluorine is indicated, however, by the data of Fig. 6 [12].

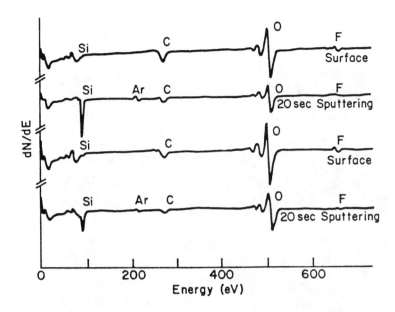

Fig. 6. AES scans (surface and after 20 sec. Ar sputtering for CF_4 RIE (upper two traces) and for CF_4/H_2 RIE (lower two traces). The Ar sputtering rate was 40 Å/min. (From ref. 12)

the silicon by carbon and fluorine is indicated, however, by the data of Fig. 6 [12].

We now turn to RHEED for a structural assessment of CF_4 RIE damage. Figure 7 presents the RHEED pattern for a Si surface after CF_4 RIE. The corresponding control is again provided by Fig. 3(a). It can be seen from this pattern that damaged Si clearly exists after CF_4 RIE and that the rings present indicate that at least a portion of the damaged Si must be polycrystalline as we observed for CCl_4 and $CClF_3/H_2$ RIE. The halo patterns and haze also present in Fig. 7b may be attributed to a possible residue layer, to an amorphized portion of the damaged Si or to both. However, on the basis of the AES data of Fig. 6, it may be deduced that these features are due to the presence of amorphized Si.

CF_4/H_2 RIE: By comparing the damage and contamination effects of CF_4/H_2 RIE with those of CF_4 RIE, we can assess the damage/contamination repercussions of doping the CF_4 plasma with hydrogen. First, we note from Fig. 6 that, for the CF_4/H_2 RIE parameters used here, there is no significant difference between contamination produced by CF_4 RIE and that produced by CF_4/H_2 RIE except that the data for the CF_4/H_2 exposed sample indicate a stronger oxygen signal. These data for this CF_4/H_2 RIE (blanket etching of wafers using a total gas flow rate of 25 sccm (12.5/12.5), a pressure of 20 m torr, total rf power of 580 W, and an rf power density of 0.48 W/cm^2) do not show the presence of the carbonaceous R-layer seen for some other CF_4/H_2 etching parameters [12].

Figure 7b shows the effects on the Si surface of using CF_4/H_2 RIE. As may be seen, there are hardly any rings when hydrogen is incorporated into the plasma; the dominant features are the halos

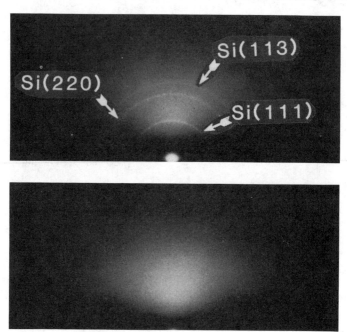

Fig. 7. RHEED patterns for (a) CF_4 RIE exposed Si and (b) CF_4/H_2 RIE exposed Si.

and haze. Contrary to the case for CCl_4, $CClF_3/H_2$, and CF_4 RIE, and RHEED data of Fig. 7b indicate that the Si surface is almost completely amorphized by the CF_4/H_2 RIE for these etching parameters. We note in passing that a comparison of the damage effects due to hydrogen using Figs. 7a and 7b is especially valid since the cathode voltages were chosen to be essentially the same in both cases (~700V) [12].

CONTROL OF RIE DAMAGE/CONTAIMINATION

In our survey and comparison of four different RIE etching chemistries we have noted that RIE exposure always produces silicon lattice damage. Generally, permeation of the silicon by various etching species is also present to varying degrees. In addition, residue layers may or may not be present depending on the etching chemistry and parameters.

As we have noted, wet etching can be used to remove the damage and contamination, and furnace anneals can be used to anneal out lattice damage. However, as we also noted, these approaches are not compatible with submicron geometries and low temperature processing.

Other approaches to dealing with RIE damage/contamination effects include hydrogen passivation and rapid thermal annealing (RTA). We have explored these in detail elsewhere [5, 6, 13].

Hydrogen passivation RIE damage is achieved by introducing atomic hydrogen into the RIE-damaged Si where it bonds with Si defects rendering them electrically inactive [13]. The atomic hydrogen may be introduced, for example, by implanting hydrogen ions using a low energy hydrogen ion beam [13]. Obviously, this approach does not remove damage and contamination but rather it electrically passivate the damaged near-surface region. Schottky barrier diodes and ohmic contacts made on these passivated surfaces display the same behavior as devices made on controls. This passivation is thermally stable at temperatures below 400-500°C, at temperatures > 450°C the passivating hydrogen will out-diffuse from Si. A drawback of this approach is that the introduction of hydrogen itself can further damage and contaminate a very thin surface layer. This necessitates the use of a mild HF etch after the introduction of the hydrogen to remove this thin surface layer [6, 13]. In general, such a mild HF etch will not remove the RIE damage and contamination itself [6].

We have also demonstrated successfully that RTA can be used to control RIE damage and contamination [5]. Unlike the hydrogen passivation approach, RTA does not passivate damage/contamination but rather it removes the damage by regrowth during the heat pulse. Also, the use of RTA does not necessitate any wet clean-up etch after that anneal. This technique for RIE damage/contaminaton control is very attractive since only the damaged and contaminated near-surface Si region is significantly heated, when the RTA is properly implemented. RTA cannot be effectively used if an R-layer is present. However, it does effectively regrow the D-layer and remove the P-layer, if the impurities are volatile, provided there is no R-layer.

SUMMARY

Reactive ion etching of silicon produces three distinct types of damage/contamination. These may be labeled as R-layers (principally non-silicon layer), P-layers (silicon permeated by impurities), and D-layers (lattice damaged Si). The R-layer may or may not be present depending on the RIE chemistry and etching parameters. The P-layer is always present to some extent (from tens of angstroms to hundreds of angstroms or more in depth) and the D-layer is also always present (from tens of angstroms to hundreds of angstroms in depth).

Wet clean-up procedures or wet clean-up procedures coupled with high temperature furnace annealing are required to remove RIE damage/ contamination effects. Currently these RIE damage/contaminaton control steps are often integrated into subsequent wafer processing steps. This fortuitous situation may not be possible as device geometries continue to shrink and wafer processing temperatures continue to be lowered. Alternative steps for the control of RIE damage and contamination must be developed. Two interesting alternative approaches are damage passivation by hydrogen and the use of rapid thermal annealing.

REFERENCES

[1] Ephrath, L. M. and DiMaria, D. J., Solid State Technology, 24 (4), 183, April 1981.

[2] Ashok, S. Chow, T. P., and Baliga, B. J., Applied Physics Letter, 42 (8), 587 (1983).

[3] Pang, S., Solid State Technology, 27 (4), 249 (1984).

[4] Climent, A. and Fonash, S. J., Journal of Applied Physics, 56 (4), 1063 (1984).

[5] Fonash, S. J., Singh, R., Rohatgi, A, Rai-Choudhury, P., Caplan, P. J., and Poindexter, E. H., Journal of Applied Physics, 58 (2), 862 (1985).

[6] Mu, X. C., Fonash, S. J., Yang, B. Y., Vedam, K., Rohatgi, A, and Rieger, J., Journal of Applied Physics, 58 (11), 4282 (1985).

[7] Fonash, S. J., Solid State Technology, 28 (4), 201 (1985).

[8] Fonash, S. J., Solid State Technology, 28 (1), 150 (1985).

[9] Oehrlein, G. S., Tromp, R. M, Lee, Y. H., and Petrillo, E. J., Applied Physics Letters, Vol. 45, 420 (1984).

[10] Mizutani, T. Dale, C. J., Chu, W. K., and Mayer, T. M., Nuclear Instruments and Methods in Physics Research, (1985).

[11] Mu, X. C., Fonash, S.J., Oehrlein, G. S., Chakravarti, S. N., Parks, C., and Keller, J., Journal of Applied Physics, 59, 2958 (1986).

[12] Mu, X. C., Fonash, S. J., Rohatgi, A., and Rieger, J., 48, 1147 (1986).

[13] Wang, J.-S., Fonash, S. J., and Ashok, S., IEEE Electron Device Letters, EDL-4, 432 (1983).

S. V. Nguyen, J. R. Abernathey, S. A. Fridmann and M. L. Gibson

THE BONDING STRUCTURE AND COMPOSITIONAL ANALYSIS OF PLASMA ENHANCED AND LOW PRESSURE CHEMICAL VAPOR DEPOSITED SILICON DIELECTRIC FILMS

REFERENCE: Nguyen, S.V., Abernathey, J.R., Fridmann, S.A. and Gibson, M.L., "The Bonding Structure and Compositional Analysis of Plasma Enhanced and Low Pressure Chemical Vapor Deposition of Silicon Dielectric Films", Emerging Semiconductor Technology, ASTM STP 960, D.C. Gupta and P.H. Langer, Eds., American Society for Testing and Materials, 1986.

ABSTRACT: Silicon dielectric films such as silicon nitride, oxide and oxynitride films deposited by Plasma Enhanced and Low Pressure Chemical Vapor Deposition (PECVD and LPCVD) processes were analyzed and compared using Fourier Transform Infrared (FTIR), X-ray Photoelectron, Auger, Electron Spin Resonance Spectroscopies and Nuclear Reaction Analysis for hydrogen. The plasma deposited films exhibit a more random structure with less long-range order and contain more hydrogen as compared to those of LPCVD films. However, marked similarities were observed in the bonding and its variation in films deposited by both processes. Analysis data indicates that the silicon oxynitride films (RI=1.75-1.78), deposited by both processes, may be the most stable oxynitride where mixed silicon oxynitride tetrahedral N_2-Si-O_2 bonding structures are most abundant.

KEYWORDS: Silicon oxynitride, silicon dielectrics, plasma deposition, low pressure deposition, bonding and structure.

Introduction

Silicon dioxide and silicon nitride have long played an important role in the manufacturing of integrated circuits. Recently, silicon oxynitride, a material with properties intermediate between silicon nitride and silicon dioxide, has become increasingly important in various microelectronic applications. The properties of these thin

---

M. L. Gibson is a laboratory specialist, and S. V. Nguyen, J. R. Abernathey, and S. A. Fridmann are engineers/scientists at IBM General Technology Division, Essex Junction, VT 05452.

dielectric films depend on the manner of deposition, which is typically either LPCVD or PECVD. Regardless of the deposition technique, considerable variation in film properties can result from variation in deposition conditions such as pressure, temperature, reactant gas types and flow rates, and in the case of PECVD, applied power and RF frequency as well.

LPCVD and PECVD silicon oxynitride films are well known for their outstanding properties such as chemical stability, masking ability, radiation hardness and low mechanical stress [1-3]. However, the bonding structure and compositional variations of deposited oxynitride films have not been analyzed in detail. In this paper, we present a systematic study of the bonding structure and composition of Si-N-O films prepared by both PECVD and LPCVD techniques. These films were analyzed and compared using various analytical techniques including FTIR, X-ray Photoelectron Spectroscopy (XPS), Auger Spectroscopy, Nuclear Reaction Analysis (NRA) for Hydrogen, and Electron Spin Resonance (ESR) at liquid helium temperature.

Experimental Procedure

Silicon nitride, silicon oxynitride and silicon oxide films were deposited on p-type silicon substrates or thermally grown oxide surfaces using plasma enhanced and LPCVD processing.

For PECVD processing, films were deposited in a high frequency (13.56 MHz) parallel plate, capacitively coupled plasma reactor with a ring-type gas injection system. The process parameters for good uniformity ($3\sigma=10\%$) silicon nitride, oxide and oxynitride deposition are given in Table I.

Table 1--Plasma Deposition Conditions for Silicon Dielectrics

*Parallel Plate Reactor, 13.56 MHz
Reactor radius: 30cm; Electrode Spacing: 2 cm; Wafer size: 10 cm

	Silicon Nitride	Silicon Oxide	Silicon Oxynitride
Power Density (w/cm^2)	0.12	0.06	0.12
Pressure (mTorr)	1200	1500	1200
Temperature (°C)	300	300	300
Gas types and flow rates (sccm)	SiH_4*(1500) NH_3 (90) He (800)	SiH_4*(1000) N_2O (1000)	SiH_4*(1500) NH_3(90), He(800) N_2O(8-18)
Refractive index	~1.92-1.95	~1.47	1.65 - 1.85 (Depending on N_2O flow rate)
Deposition Rate (nm/min)	7	33	

*SiH_4 = 1.9% SiH_4 in He

For LPCVD processing, silicon dielectric films were deposited in a hot-wall tubular type LPCVD reactor around 800°C. The reactant gases were SiH_2Cl_2, NH_3 and N_2O for nitride and oxynitride films. Silane and oxygen were used for oxide deposition. Typical deposition conditions for silicon nitride, oxide and oxynitride films are shown in Table 2.

Table 2--Deposition Condition of Silicon Dielectric by Low Pressure Process

Tube diameter: 16.5 cm; *Barrell Type, hot-wall, Tubular system; Wafer size: 10cm

Process Conditions	Type of Films		
	Silicon Nitride	Oxide	Oxynitride
Pressure (Torr)	0.3	0.3	0.3
Temperature (°C)	800	800	800
Gas Types and Flow rates (sccm)	$SiCl_2H_2$ (25) NH_3 (75)	SiH_4 (60) O_2 (40)	$SiCl_2H_2$ (25) NH_3 (8 - 60) N_2O (15 - 67)
Refractive index	2.00	1.46	1.65 - 1.95
Deposition Rate (nm/min)	4	8	2 - 3.5

Film thickness and refractive index were measured using a He-Ne laser ellipsometer (wavelength = 632.8 nm) and a step profilometer (Tallystep). FTIR studies were performed with 200-500 nm films deposited on bare silicon substrate. In most cases, the background absorption of the substrate was subtracted from the spectra to obtain bulk film spectra. The maximum resolution of the spectrometer was 1 cm^{-1}.

Auger compositional depth profiles were performed with 10-100 nm thick films on a silicon (or thermally grown oxide) substrate surface. The films were sputtered with a 1Kev Ar^+ ion to determine the composition as a function of depth. The detection limit of the Auger technique is about 1 atomic percent (a/o).

The XPS technique was used to analyze the film surface composition as deposited. Surface analyses were performed using a Hewlett-Packard HP 5950B photoemission spectrometer. Scans were made both in survey and high resolution modes: survey to detect the possible presence of unexpected (contaminant) elements on the surface, and high resolution to quantify the known elements and help establish the bonding environments.

ESR analysis for unpaired electrons (dangling bonds) was performed with 1-2 μm films deposited on bare silicon substrates. The measurements were taken at both room temperature and liquid helium temperature (~8-9°K). Double integrals were computed for all observed resonances and compared with standardized CVD Si_3N_4 material with known spin density. The reproducibility of the integral is about ± 20% and within this limit, integrals can be compared from one sample to another. For plasma films on bare silicon substrates, the deposited films were removed by wet etching in buffered HF solution.

The substrates were then measured for any spin resonance that may arise from plasma radiation. The detection limit of this ESR system is about 10E11 spin cm^{-3}.

Hydrogen depth profiles of various films (20-100 nm thick) were analyzed using the nuclear reaction:

$$^{15}N + {}^{1}H \longrightarrow {}^{12}C + {}^{4}He + 4.43 \text{ Mev Gamma Ray}$$

A detailed description of this Nuclear Reaction Analysis (NRA) technique has been given previously [4].

Results and Discussions

FTIR Analysis: For spectrum of LPCVD silicon nitride (RI=2.00), oxynitride (RI=1.76) and silicon oxide (RI=1.47) films. We observed the presence of N-H (~3350 cm^{-1}), Si-N (~850 cm^{-1}), Si-O (~1130, 800 and 480 cm^{-1}) and Si-Si (~600 cm^{-1}) bonds. We also observed a very small Si-H (2150 cm^{-1}) band at higher resolution in all LPCVD silicon nitride, oxide and oxynitride films. For silicon nitride films, the presence of a sharp Si-O band (1130 cm^{-1}) is due to interstitial oxygen in the substrate or native oxide. High resolution FTIR spectra of LPCVD films in the 600 - 1800 cm^{-1} region were also taken to monitor the variation of Si-N and Si-O bonding. Figure 1 shows the typical Si-O (~1130 and 800 cm^{-1}) band in silicon dioxide (RI = 1.46) and Si-N (~850 cm^{-1}) band in silicon nitride (RI = 2.0). For silicon

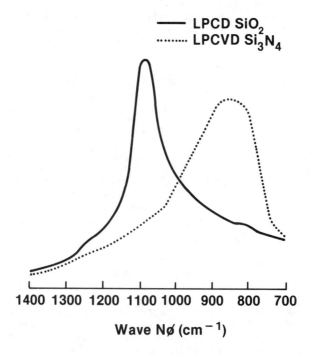

Fig. 1-- FTIR Spectra of LPCVD Si$_3$N$_4$ and Si

oxynitride films with refractive indices ranging from 1.92 to 1.61, we observed a broad band (800 - 1200 cm^{-1}) which is a combination of several overlapped bands as shown in Fig. 2. As the film refractive index is decreased, i.e., higher oxygen content in the film, the maximum of this broad band is shifted toward the shorter wavelength Si-O band (~1130 cm^{-1}). Because of the large overlap between small bands in oxynitride films, it is impossible to determine the exact location of each peak. However, these small bands can be qualitatively assigned as Si-O (~800 - 825, 930 - 1000 cm^{-1}) and Si-N (~840 - 880 cm^{-1}) in various N-Si-O bonding environments. The Si-O and Si-N vibrational band shift in oxynitride films, relative to Si-O and Si-N bands of stoichiometric oxide and nitride films, suggests that the bonding structures of oxynitride films are SiN_xO_y (x + y = 4) tetrahedral. The peak shifts result from differences in the total adjacent electronegativity near the silicon center as oxygen replaces nitrogen in a silicon nitride tetrahedral structure (or vice versa for silicon dioxide). This phenomenon is discussed in random network and bonding model calculation of amorphous silicon nitride and oxide by Knolle and others [5 - 7]. Since the mixed tetrahedral structure of amorphous silicon oxynitride includes five possible structures (Table 3), the overall shift of the broad band to shorter wavelengths as oxygen concentration increased correlates with increasing SiO_2 bonding charac-

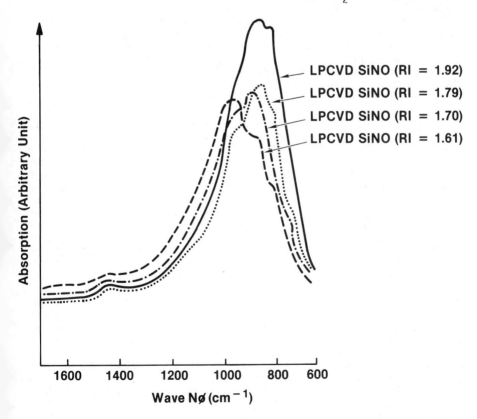

Fig. 2--FTIR Spectra of LPCVD SiNO at Various Refractive Index (t ≃ 300 nm)

Table 3--Mixed Tetrahedral Structures
of Silicon Oxynitride

Structure	Correspond bands wave number (cm^{-1})	Comment [based on references (5-7, 19)]
N_3-Si-N	~ 850	Silicon Nitride
N_3-Si-O	}	More Nitride Character
N_2-Si-O_2	Between 850-1130	Probably most stable structure
N-Si-O_3	}	More Oxide character
O-Si-O_3	~ 1130, 800	Silicon Dioxide

teristic as previously reported [5]. For PECVD silicon nitride, oxide and oxynitride films, the FTIR spectra are somewhat simpler. We observed the presence of N-H (3350 cm^{-1}, Si-H (2200 cm^{-1}), Si-N (850-900 cm^{-1}) and Si-O (1150, 800 and 450 cm^{-1}). No O-H bonding was observed in either silicon nitride or oxynitride films. For nitride films, only a small amount of Si-O bonding was found. The FTIR absorption spectra of PECVD silicon oxynitride is similar to those of LPCVD silicon oxynitride. However, the absorption bands of PECVD films are wider and the Si-O and Si-N bands become less distinguishable compared to those of LPCVD films. This is because PECVD films have a more random bonding structure as well as containing more hydrogen.

Auger Analysis: Representative Auger depth profiles of LPCVD silicon nitride and PECVD silicon oxynitride films are shown in Figs. 3 and 4. The depth profiles show the concentration of silicon, nitrogen and oxygen relatively uniform throughout the depth of the film. For all silicon nitride, oxide and oxynitride films, increased oxygen concentrations were observed at surfaces and at interfaces as compared to the bulk structure. This can be due to surface oxidation of deposited films, native oxide on silicon substrates, as well as slight oxygen contamination during the initial deposition process. For all plasma deposited films, slightly Si-rich interfaces were observed in the first 10-30 nm of deposited films. This is the result of the initial transient phenomena normally encountered in plasma deposition processes [2,8]. For LPCVD films, the variation of nitrogen and oxygen concentration in silicon oxynitride system with film's refractive index is nearly linear in the observed 1.60-1.95 range, Fig. 5. The oxygen and nitrogen concentration increased and decreased respectively as film refractive index decreased. For PECVD films, similar relations were observed in the oxynitride system although the variation does not appear to be linear. This is probably due to large amounts of hydrogen in the plasma films that alter the normal compositional analysis. The atomic compositions of plasma deposited films are shown in Table 4. It should be noted that the PECVD silicon nitride films have a substantial amount of excess silicon which is normally observed in PECVD films, and the silicon oxide films contain a small amount of nitrogen which comes from the N_2O reactant gas. Figure 6 shows the bulk silicon LVV Auger spectra of LPCVD silicon

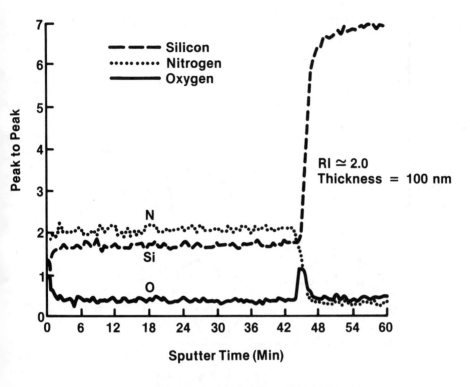

Fig. 3--Auger Depth Profile of LPCVD Silicon Nitride (RI≃2.00)

Table 4--Auger Analysis of PECVD Silicon Nitride, Oxide, and Oxynitride Films

Refractive Index	Average Atomic Percent Composition		
	Silicon	Nitrogen	Oxygen
1.95 (Silicon Nitride)	71	29	0
1.85	43	54	3
1.75	31	52	17
1.65	38	26	36
1.47 (Silicon Dioxide)	35	3.0	62.0

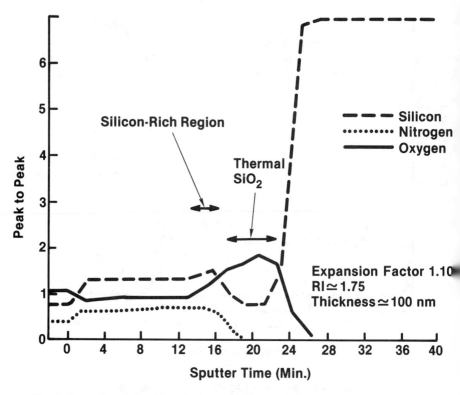

Fig. 4--Auger Depth Profiles of Plasma Deposited Silicon Oxynitride (RI = 1.75)

nitride, oxide and oxynitride (RI=1.75) films. The absence of 92 eV peak in the spectra indicates little or no excess silicon in the films to form Si-Si bonds. The main signal of the Si LVV spectra shifts to low energy monotonically with increasing oxygen content from ~88.5 eV for silicon nitride to ~76 eV for silicon dioxide. For silicon oxynitride films, a slight broadening and only one continuous LVV Si peak was observed. Once again, this indicates that the LPCVD silicon oxynitride has a continuous phase composition of SiN_xO_y with x+y=4, not a physical mixture of SiO_4 and SiN_4 tetrahedra. This result is similar to data reported on silicon oxynitride films formed by ion implantation of oxygen and nitrogen [9].

For PECVD films, the variation of the Si LVV peaks of silicon nitride, oxide and oxynitride systems, is similar to those of LPCVD films except that the peaks are slightly wider. This also indicates a more random bonding structure of PECVD films. Furthermore, we also observed excess silicon in silicon nitride films by the presence of an Si peak around 92 eV, and small amounts of nitrogen in silicon oxide films indicated by the presence of a small shoulder peak of Si-N around 88 ev. These results correspond well with Auger compositional analysis data of PECVD films shown in Table 4. Overall, the Auger bonding data of Si LVV peaks indicate that silicon oxynitride films

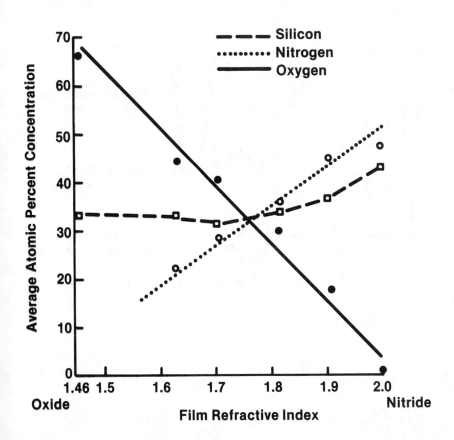

Fig. 5--Variation of Si, N and O in LPCVD Silicon Nitride, Oxide and Oxynitride

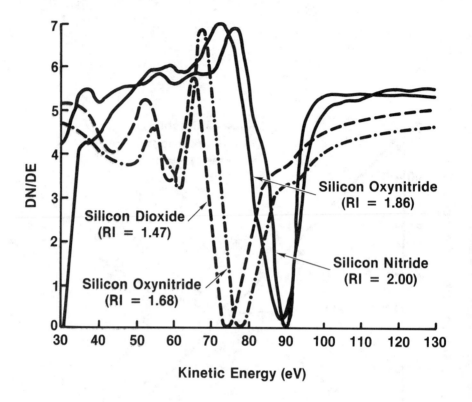

Fig. 6--Si LVV Peak of LPCVD Silicon Nitride, Oxide and Oxynitride

deposited by PECVD and LPCVD processes using the described conditions contain mostly homogeneous phases of SiN_xO_y (x+y=4) as shown in Table 3.

X-ray Photoelectron Analysis (XPS or ESCA): Table 5 lists the surface composition results, namely surface atom ratio, of Silicon 2p binding energy (B.E.), the full peak width at half maximum (FWHM) and relative atomic composition for the silicon peak. The binding energies are corrected for charging shifts by assigning the surface carbon signal to 285.0 eV. (A low level of surface carbon contaminant provides a reasonably reliable reference signal). Figure 7 shows the representative ESCA spectra of LPCVD and PECVD films. It is apparent that only subtle differences occur in binding energies and peak widths of the Si 2p signal from film to film, except for the notable uniqueness of the silicon oxide film. The binding energy for oxidized silicon is about 1 eV higher than that for the nitride and oxynitride films. The peak width is also somewhat narrower, suggesting that in the nitride and oxynitride films, there is a small component of oxidized silicon broadening those peaks.

There is a noticeable lack of difference in the binding energies for nitride vs. oxynitride films, and for LPCVD vs. PECVD films. The reason is that in all such films there is a substantial amount of

Table 5--ESCA Surface Composition of PECVD and LPCVD Silicon Nitride, Oxide and Oxynitride Films

	Surface Composition Data (ESCA)					
FILM TYPE	FILM RI	ATOM RATIOS Si	N	O	B.E. (eV)	FWHM (eV)
LPCVD Silicon Oxynitride	1.63	4.0	1.6	4.7	103.0	2.0
LPCVD Silicon Oxynitride	1.77	4.0	2.3	3.7	102.7	2.0
LPCVD Silicon Oxynitride	1.87	4.0	2.9	2.7	102.5	2.1
LPCVD Silicon Nitride	1.99	4.0	2.8	2.7	102.6	2.1
PECVD Silicon Nitride	2.02	4.0	2.9	2.1	102.4	2.1
PECVD Silicon Oxide	1.48	4.0	0.3	6.3	103.6	1.7
PECVD Silicon Oxynitride	1.80	4.0	2.0	3.2	102.6	2.2

Note: Surface Carbon contamination range is 6-8 atomic percent

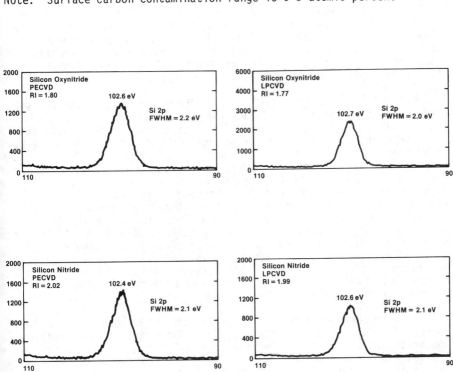

Fig. 7--ESCA Spectra of LPCVD and PECVD Silicon Nitride, Oxide and Oxynitride Films

oxygen, either present in the bulk of the film and/or as a surface oxide contaminant. Unfortunately, ESCA data cannot discriminate between a nitride film which has a thin oxide overlayer from a film which incorporates oxygen into its intrinsic bulk structure, i.e., oxynitride. Hence from this perspective, both nitride and oxynitride films look similar if not identical. It may be noted, however, that silicon nitride should have a Si 2p binding energy about 0.5-1.0 eV lower than that found on these films: values of 101.9 eV [10] and 100.5 to 102.0 eV [11] have been quoted. The discrepancy noted here is probably due to the effect already mentioned: the presence of a surface silicon oxide layer on nitride will not only broaden the peak, but also shifts it to higher binding energy.

As for atom compositions, some clear trends are evident. For the PECVD oxynitride, increasing refractive index correlates with increasing nitrogen and decreasing oxygen in the near surface region (Table 5). A similar trend is seen in the compositions for the LPCVD nitride, oxynitride and oxide (Table 5, also) which may be compared with the bulk Auger compositional data (Table 4). The near surface (ESCA) O/S; ratios range from 0.52 for the nitride to 1.6 for the oxide, compared to the corresponding Auger ratios for the bulk of 0.0 and 1.8, respectively. This comparison suggests that the excess near surface oxygen present in the more nitride-like (higher refractive index) films represents the silicon oxide overlayers, even though this cannot be deduced from the silicon spectra because of the limitation mentioned above. In general, correspondence of ESCA results with both FTIR bonding and Auger compositional data is good.

Nuclear Reaction Analysis (NRA) for Hydrogen: The hydrogen depth profiles of thin (50-100 nm) LPCVD and PECVD films are shown in Figs. 8 and 9. The PECVD films, deposited at substantially lower substrate temperatures, always have more hydrogen in the film's bulk (9-22 atomic percent) as compared to those of LPCVD films (2-4 atomic percent). For both processes, silicon nitride films always contain more hydrogen compared to their oxide and oxynitride counterparts. In oxynitride films, the hydrogen concentration decreases as film refractive index decreases. This is due to thermodynamically favorable formation of Si-O bonds and a reduction of passivated Si-H bonds in the film bulk. For PECVD films, this result is consistent with a previous study by optical emission of Si-O where the reduction of Si-H reactive species in silicon oxynitride plasma deposition glow discharge was noted as the oxygen source reactant increased [12]. This in turn will yield deposited films with more oxygen and less hydrogen in their bulk structure.

We also observed a consistent depletion of hydrogen in all films, especially PECVD films, at the nitride (or oxynitride, oxide)-silicon substrate interfaces. This hydrogen depletion is probably due to the initial transient phenomena of plasma and low pressure chemical vapor deposition processes together with the initial changes in substrate's surface state and composition [8,13]. Details of the variation of hydrogen depth profile with initial transient phenomena and the deposition mechanism have been previously published [13]. The substantial difference in hydrogen concentration between PECVD and LPCVD films can be used to explain differences in etch rates, density, degree of randomness in film structure and possibly film stress.

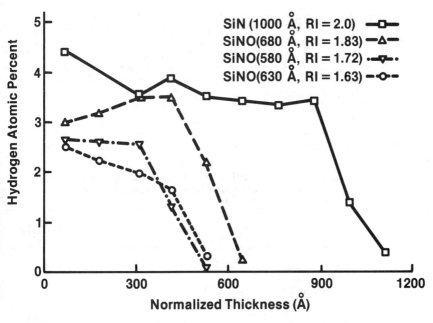

Fig. 8--Hydrogen Depth Profiles of LPCVD SiN & SiNO

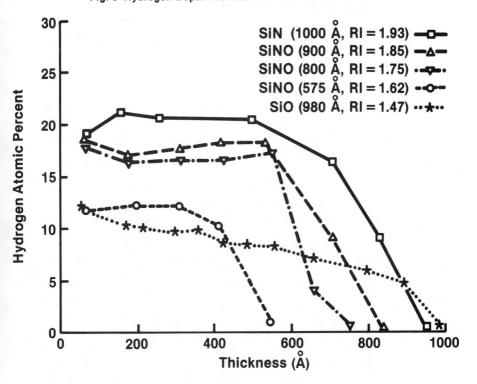

Fig. 9--Hydrogen Depth Profiles of PECVD Silicon Nitride, Oxynitride and Oxide

Electron Spin Resonance (ESR) Analysis: Electron spin density measurement of deposited films shows that both LPCVD and PECVD oxynitride films with a refractive index of 1.75-1.78 have the lowest spin density as compared to other films, Table 6. Figure 10 shows the absorption derivative of typical ESR signals for LPCVD and PECVD silicon nitride and oxynitride films. The LPCVD silicon nitride and oxynitride (RI=1.86) samples show 2 signals each about 2 gauss wide (g=2.0005 and 2.0059) at a magnetic field of 3360 gauss. The PECVD samples also show 2 signals in the same positions except they are broader (2.5-3.0 gauss wide). In general, these ESR signals can be attributed either to silicon or nitrogen dangling bonds. Since the

Table 6--Spin Density Results of Plasma and Low Pressure Chemical Vapor Deposited Silicon Nitride and Oxynitride at 8-9 °K.

Condition	Film Type	Refractive Index	Average Spin Density (Spin/cm^3)
PECVD	Nitride	1.95	1 E16
	Oxynitride	1.85	5 E14
	Oxynitride	1.75	Very Small
	Oxynitride	1.65	6 E16
	Oxide	1.47	Small
LPCVD	Nitride	2.0	3 E16
	Oxynitride	1.86	1 E17
	Oxynitride	1.76	Very Small
	Oxynitride	1.68	Small

Note: Small $\simeq 10^{12} - 10^{13}$ } Can't be determined exactly.
Very Small $< 10^{11}$

nitrogen atoms have nuclear spin of unity, the nitrogen dangling bonds will split the signal into three lines [14], which is not observed in this case. Therefore, the ESR signal must be generated mainly from the silicon. The possible threefold configurations of the silicon atoms have been discussed in previous reports [15,16]. The low spin density value of silicon oxynitride films with RI=1.75-1.78 indicates that the films have less dangling bonds and possibly low-trapped charge thus may result in better electrical properties as compared to other nitride and oxynitride counterpart films. Our previous measurements [3,17,18] indicate that the silicon oxynitride films (RI=1.75-1.78) generally have better electrical properties, lower stress, higher thermal stability and lower pinhole density as compared to other compositions. These previous measurements together with above analysis data suggest the existence of one relatively stable bonding structure of silicon oxynitride films deposited by PECVD and LPCVD processing. It appears that the symmetrical tetrahedral

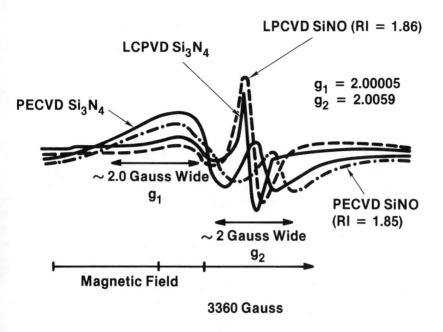

Fig. 10--ESR Spectra of PECVD and LPCVD Si_3N_4 and SiN_xO_y

structure of N_2-Si-O_2 may be the best candidate, in terms of its bonding and structure, as the most stable mixed oxynitride tetrahedron, Table 3. This mixed tetrahedron may also be the most abundant bonding configuration in the PECVD and LPCVD amorphous silicon oxynitride films with RI=1.75-1.78, (Si, N, O and H). This possible stable structure may be formed by a half-hybrid hydrids orbital bonding structure of stable SiO_4 and SiN_4 tetrahedra; i.e., a linear combination of 1/2 SiO_4 + 1/2 SiN_4 bonding orbitals.

CONCLUSION

Silicon nitride, oxide and oxynitride films deposited by PECVD and LPCVD processing were analyzed and compared using FTIR, Auger, XPS, NRA for hydrogen and ESR techniques. The PECVD films exhibit a more random atomic configuration, contain more hydrogen and substantially different atomic compositions and physical properties compared to LPCVD films. However, there are marked similarities in the bonding and its variation with composition in film deposited by both processes. The analysis data and other measurements suggested that these amorphous silicon oxynitride films with RI=1.75, deposited by PECVD and LPCVD processes, may be the most stable oxynitride films where mixed silicon oxynitride tetrahedral N_2-Si-O_2 bonding structures are most abundant.

ACKNOWLEDGMENTS

The authors acknowledge the assistance of Professors W. Lanford of State University of New York at Albany for Nuclear Reaction Analysis of hydrogen, and A. L. Rieger of Brown University for Electron Spin Resonance analysis. Support from D. Dobuzinsky, S. J. Pierce and our management is highly appreciated.

References

[1] M.J. Rand and J.F. Robert, "Silicon Oxynitride Films from NO-NH_3-SiH_4 Reactions", J. Electrochem. Soc., Vol. 120, p. 446, (1973).
[2] S.V. Nguyen, S. Burton and P. Pan, "The Variations of Physical Properties of Plasma Depositied Nitride and Oxynitride with Their Composition", J. Electrochem. Soc., Vol. 131, No. 10, pp. 2348-2353 (1984).
[3] J. Underhill, S.V. Nguyen, M. Kerbaugh, D. Sundling, "Silicon Oxynitride Films as Intermediate Layer for Tri-Layer Resist System", Proceedings of SPIE Meeting, Vol. 539. Advance in Resist Technology and Processing II, pp. 83-89 (1985), Santa Clara, California.
[4] W.A. Lanford and M.J. Rand, "The Hydrogen Content of Plasma Deposited Silicon Nitride", J. Appl. Phys., Vol. 49, p. 2473 (1978).
[5] W.R. Knolle and H.R. Maxwell, Jr., "A Model of SIPOS Deposition Based on Infrared Analysis", J. Electrochem. Soc., Vol. 127, No. 10, pp. 2254-2258 (1984).
[6] W.R. Knolle and J.W. Osenbach, "Structure of Plasma Deposited Silicon Nitride Determined by Infrared Spectroscopy", J. Appl. Phys., Vol. 58, No. 3, pp. 1248-1254 (1985).
[7] G. Lucovsky, "Relation of Si-H Vibrational Frequency to Surface Bonding Geometry", J. Vac. Sci. Tech., Vol. 16, pp. 1225-1228, Oct. (1979).
[8] S.V. Nguyen and P. Pan, "Initial Transient Phenomena in Plasma Enhanced Chemical Vapor Deposition Process", Appl. Phys. Letter, Vol. 45(2), pp. 134-136 (1984).
[9] R. Hezel and W. Streb, "Characterization of Silicon Oxynitride Films Prepared by Simultaneous Implantation of Oxygen and Nitrogen Ions in Silicon", Thin Solid Films, Vol. 124, pp. 35-41 (1985).
[10] J.A. Taylor, "Further Examination of the Si KLL Auger Line in Silicon Nitride Thin Films", Application of Surface Science, Vol. 7, pp. 168-184 (1981).
[11] R.K. Lowry and A.W. Hogrefe, "Application of Auger and Photoelectron Spectroscopy in Characterizing IC Materials", Solid State Technology, pp. 71-75, January 1980.
[12] S.V. Nguyen, "Optical Emission Spectroscopic Study of Silicon Nitride and Oxynitride Deposition Glow Discharge", Proceedings of the 9th International Conference on Chemical Vapor Deposition, pp. 213-232. Electrochemical Society Spring 1984 Meeting, Cincinnati, Ohio, U.S.A.
[13] S.V. Nguyen, W. Lanford and P. Rieger, "Variation of Hydrogen Bonding, Spin Density and Hydrogen Depth Profiles of Plasma Deposited Silicon Nitride and Oxynitride Films with Deposition Mechanisms", Proceeding of 7th International Symposium in Plasma Chemistry; IUPAC, pp. 56-61, July 1-5, 1985. Eindhoven, The Netherlands. J. Electrochem. Soc., Vol. 133, p. 970 (1986).

[14] C.L. Ultee, "Some Observation on the Electron Paramagnetic Resonance Spectra of Gases Free Radicals", J. Phys. Chem., Vol. 64, 1873 (1960).
[15] M. Kastner, D. Alder and H. Fritzches, "Valence - Alternation Model for Localized Gap-States in Lone-Pair Semiconductors", Phys. Rev. Lett., Vol. 37, 1504 (1976).
[16] D. Alder, "Density of States in the Gap of Tetrahedral Bonded Amorphous Semiconductors", Phys. Rev. Lett., Vol. 41, 1755 (1978).
[17] S.V. Nguyen, "Infrared, Auger and Electrical Characterization of Plasma Deposited Silicon Nitride and Oxynitride Films", Ext. Abst. No. B-2,, pp. 24-25, Electronic Material Conference, Burlington, Vermont, June 1983, U.S.A., and Ext. Abst., pp. 52-54, Electronic Material Conference, Boulder, Colorado, June 1985, U.S.A.
[18] P. Pan, J. Abernathy and C. Schaefer, "Properties of Thin LPCVD Silicon Oxynitride Films", Journal of Electronic Material, Vol. 14 (5), pp. 617-632 (1985).
[19] W.A.P. Claasen, H.A.J.Th.v.d.Pol., A.H. Goemans and A.E.T. Kuiper, "Characterization of Silicon Oxynitride Films Deposited by Plasma Enhanced CVD", J. Electrochem. Soc., Vol. 133, pp. 1458-1463 (1986).

E. J. Bawolek

MONTE CARLO SIMULATION OF PLASMA ETCH EMISSION ENDPOINT

REFERENCE: Bawolek, E. J., "Monte Carlo Simulation of Plasma Etch Endpoint," Emerging Semiconductor Technology, ASTM STP 960, D. C. Gupta and P. H. Langer, Eds., American Society for Testing and Materials, 1986.

ABSTRACT: As dry etch requirements become more severe, the trend is toward greater automation of the etch process. Automated endpoint detection is desireable to prevent excessive substrate damage and to increase etch reproduceability. The challenge in designing an automated endpoint detector is to provide a unit with sufficient intelligence to correctly recognize etch completion despite wafer non-uniformity and wafer to wafer variations. In this paper a simple, easy to implement simulation program based on Monte Carlo techniques is described which can be used to study the effect of wafer thickness and etch rate uniformity on the emission endpoint transition. A version of the program which incorporates loading effects is also demonstrated. The loading effect is shown to increase the abruptness of the endpoint transition. The simulation is compared with actual emission data for nitride and polysilicon etching in a commercial single wafer plasma etcher.

KEYWORDS: Plasma etch, emission, endpoint, Monte Carlo, loading effect, uniformity

Plasma etching is extensively used in state-of-the-art semiconductor manufacturing. In production usage it is often desireable to have an automatic method for terminating the etch process at completion. Toward this end, various endpoint determination techniques have been developed, based on emission spectroscopy, residual gas analysis, laser reflectometry, etc. Of these, emission spectroscopy is perhaps the most widely used method [1]. In this technique, the optical emission of a major product or reactant species is monitored as the etch progresses. The variation of emission intensity with time is correlated to the completion of the etch process. In the case where a product species is monitored, the emission is seen to decrease as the etch nears completion.

Mr. Bawolek is a Senior Member of Technical Staff at GTE Laboratories Microelectronics Technology Center, 2010 W. 14th Street, Tempe, AZ 85281.

In this paper we present the results of a Monte Carlo simulation to predict the endpoint signal associated with an etch process. The case studied is that where emission spectroscopy is used to monitor a product species. The simulation is of interest for several reasons: First, the development of a model for predicting the endpoint signal provides information about the etch process. That is, assumptions about the etch process are verified through successful modeling. Next, by predicting variations of the endpoint signal with time the design of automatic shutoff systems is simplified. Finally, the effects of changes in either film thickness or etch rate uniformity can be predicted, as well as their possible deleterious effect on the endpointing system.

DESCRIPTION OF THE SIMULATION

The Monte Carlo simulation is a computer-based modeling technique which derives its name from the fact that random numbers generated by the computer are employed in the computations [2]. The key advantage to Monto Carlo techniques is their ease of implementation. By using random numbers to simulate the statistical behavior of physical systems, the use of complicated analytic expressions is avoided. This simplicity has its cost, however; the accuracy of the simulation depends on the quantity of random numbers employed. Thus, programming simplicity is obtained at the expense of computational volume. Other approaches to the problem are possible. For example, direct integration of two probability distributions is an apparoach which could coverge faster than the Monte Carlo. The issue here is that coding would be more complex, especially in the case of a loading effect. Here one of the functions is time dependent. The Monte Carlo method is able to incorporate this variation with minimal changes in the coding.

Several assumptions were made in the development of the model:

1. The etch process is characterized by an average etch rate and associated standard deviation (uniformity).
2. The film to be etched is characterized by an average thickness and associated standard deviation (uniformity).
3. The product species emission signal is linearly proportional to the product species concentration.
4. The product species concentration is linearly proportional to exposed film surface.
5. The etch power and pressure are constant.

For the purposes of this paper the terms standard deviation and uniformity will be considered interchangeable. The key assumptions involve the etch rate and film thickness. These quantities are treated as gaussian distributed random numbers. That is, the variation in these quantities from point to point on the wafer surface is completely described by an average and standard deviation. Variations having a geometrical symmetry, for example a radially varying etch rate, are simplified in that the spatial variations are absorbed into the gaussian distribution. A further extention of these assumptions is that the standard deviation remains a constant fraction of the instantaneous etch rate. The justification of this assumption is based upon experimental studies of the etch rate as a function of wafer area.

Table 1 presents an outline of the program. Inputs include the film thickness and etch rate and their associated uniformities. In addition, the background and intitial emission intensites are input. The simulation does not attempt to determine these from first principles, since these quantities are easily determined experimentally. Rather, the program is concerned with the transition which occurs as the etch progresses. The primary output of the program is a prediction of the emission intensity vs. time for the process. In addition, where a loading effect is simulated the instantaneous and average etch rates are output, as well as the exposed film area.

TABLE 1--Program outline

Inputs

Average film thickness
Film standard deviation
Average etch rate
Etch rate standard deviation
Background (no load) emission intensity
Initial (full load) emission intensity

Loading effect (etch rate vs. area)[a]

Initial exposed area (initial etch rate)[a]

Outputs

Emission intensity vs. time

Instantaneous etch rate vs. time[a]

Average etch rate vs. time[a]

Exposed film area vs. time[a]

Program Uses

Predict endpoint curves
Study effects of wafer or etch nonuniformity
Study loading effect[a]
Predict slope of endpoint transition
Predict duration of endpoint transition

[a]Version with loading effect

In the simulation, the wafer is conceptually composed of 1000 discrete points. Each point is associated with a film thickness and etch rate. The values of etch rate and film thickness vary randomly from point to point, but the overall distribution is characterized by the corresponding average and standard deviation. The program operates by randomly generating 1000 etch rate and film thickness pairs while incrementing a timekeeping variable. This quantity was chosen empirically to provide a smooth output curve when plotted. (More points could be generated to give additional accuracy, but this would be of dubious value since experimental emission data is typically gathered on chart recorders and is therefore accurate to three significant figures at most.) For each pair of points the system tests whether the etch rate multiplied by the time is greater than the associated film thickness. If this condition is true, the thin film is completely etched away and that point makes no contribution to emission intensity. If the condition is not true, some film remains and that location continues to contribute to the emission signal. The fraction of locations which contribute to the emission signal is determined by totaling the number of points which fail the test criteria. To restate the above in a mathematical program outline:

(FOR EACH POINT)

$$\text{TEST:} \quad E(n, X_e, \sigma_e) * \text{TIME} \stackrel{?}{\geq} F(n, X_f, \sigma_f) \quad (1)$$
(If true K=K-1)

(NEXT POINT)

where
$E(n, X_e, \sigma_e)$ = gaussian random etch rate
$F(n, X_f, \sigma_f)$ = gaussian random film thickness
TIME = time into the etch process
n = index variable for the 1000 points
X_e = average etch rate
σ_e = etch rate uniformity
X_f = average film thickness
σ_f = film thickness
K = number of points contributing to the emission signal (K=1000 initially)

It is a simple matter from this point to determine a proportional variation in the emission signal. This entire procedure is repeated and the time incremented. In this wasy a tabulation of emission intensity vs. time is produced. The version incorporating the loading effect operaties in a similar manner, but with some modifications: Since the etch rate varies with the film area, the program makes use of an instantaneous etch rate. This etch rate is recomputed with each time increment. The instantaneous etch rate is then integrated to obtain an average etch rate and standard deviation. The random generator uses this value in creating the 1000 point pairs. The remaining computations are completely analogous to the previous case.

The original source code for the simulation was written in BASIC and implemented on a Texas Instruments TI 99/4A computer. The elegance and simplicity of the Monte Carlo technique becomes evident when one considers that the source code required less than 100 lines, including comments. An additional version was written in FORTRAN on a Digital Equipment VAX computer to take advantage of increased computational speed.

SIMULATION RESULTS AND DISCUSSION

As previously stated, the principal output of the program is a tabulation of emission intensity vs. time for a set of specified input conditions. Our first simulation studied the effect of film thickness uniformity on the endpoint transition. Figure 1 shows the results of a series of simulations in which the etch rate was assumed to be completely uniform (zero standard deviation) at 1000 Å/min, with no loading effect. The average film thickness was held at 1000 Å, while the standard deviation was computed for 0, 20, 50, 100 and 200 Å. Several important trends are seen in the figure:

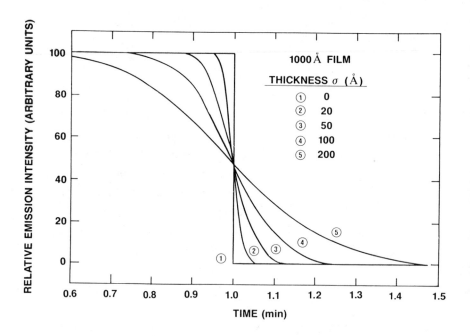

FIG. 1--Endpoint simulations for 1000 Å film with several uniformities. The etch rate was constant at 1000 Å/min. with standard deviation of 0 Å/min.

First, the endpoint transition is instantaneous when the film thickness and etch rate are completely uniform. This is what would be intuitively expected. As the film thickness becomes less uniform, the endpoint transition is spread over an increasing time interval. This is also expected, since a nonuniform wafer will begin to clear earlier in some areas than others. Furthermore, all curves pass through a single point at an emission intensity of 50%. Some thought will show why this is so: the 50% point corresponds to where the average etch rate multiplied by the time equals the average film thickness. This is always true regardless of the associated standard deviations. Finally, all of the curves are symmetrical about the 50% point, with the maximum slope occuring at that point. For this case of no loading effect, the simulation results are completely analogous when the film thickness is held uniform and the etch rate uniformity is varied. Therefore those results will not be repeated here.

The curves presented in Figure 1 should be familiar to those with a background in statistical mathematics: The series of curves is completely analogous to those obtained by the integration of the binomial distribution function. These cumulative distribution curves find use in acceptance sampling and other applications. This similarity follows naturally from the design and assumptions associated with the Monte Carlo simulation. These curves could have been arrived at by strictly analytical expressions; the advantage to the Monte Carlo technique is that the use of higher mathematics is avoided and conceptual simplicity is retained.

The statistical foundation of the simulation and its predictions is further illustrated in Figure 2. Here the interaction between film thickness and etch rate uniformity is illustrated. The results are based on an average film thickness to be etched of 1000 Å, and an average etch rate of 1000 Å/min. Curve 1 shows the simulation results for the case where the film thickness standard deviation is 100 Å, with a completely uniform etch. (Exactly the same results are obtained if the etch rate standard deviation is 100 Å and the film is completely uniform.) Curve 2 shows the results when both etch rate and film thickness standard deviations are 100 Å. Curve 3 shows that the results when the film thickness standard deviation is 200 Å.

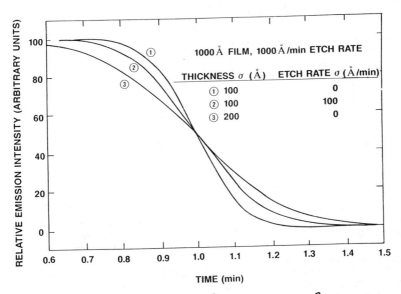

FIG 2--Endpoint simulation for 1000 Å film with 1000 Å/min. etch rate, for three conditions of etch rate and film thickness uniformity. The results show that the effects of nonuniformity are not additive.

Comparison of curves 2 and 3 shows that the effect of standard deviations are not additive. That is, one cannot add the etch rate deviation with the film thickness deviation to obtain an "equivalent" deviation. The reason for this is also statistically based; the "equivalent" standard deviation is actually given by the square root of the sum of the deviations squared. Application of this method to Case 2 would result in an "equivalent" deviation of 141 A. Since curve 2 lies approximately midpoint between the others this can be seen to be true.

The presence of a loading effect alters the predictions. Figure 3 shows the results of a simulation in which a positive loading effect was presumed to behave in accordance with the formula given by Mogab [3]:

$$e = \frac{X}{1 + Y*A} \qquad (2)$$

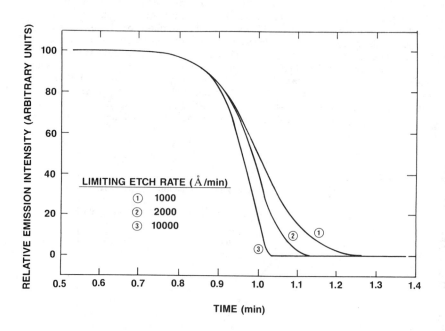

FIG 3--Endpoint simulation for 1000 Å film with 100 Å uniformity in the presence of a loading effect. The initial (blank wafer) etch rate was taken as 1000 Å/min., with zero standard deviation. The etch rate was assumed to increase to the limiting etch rate as the wafer cleared. The loading effect was assumed to behave according to formula (2).

Where e is the etch rate in Å/min, A is the fractional exposed wafer area, and X and Y are constants. This formula defines an etch rate which varies as a function of area. The constant X defines the limiting (no film remaining) etch rate, while Y defines the initial (full wafer) etch rate. For example, take X = 10000 Å/min and Y = 9. Then the etch rate will be 1000 Å/min when A=1 (film completely covers wafer) increasing to 10000 Å/min when A=0 (wafer clear). The simulations presented in Figure 3 assume an initial uniform etch rate of 1000 Å/min on a blank wafer having 1000 Å average film thickness with 100 Å standard deviation. As the loading effect increases (limiting etch rate increases) the endpoint transition is seen to become increasingly abrupt. In addition, the endpoint transition loses the symmetry seen in the absence of the loading effect. It is interesting that increasing the limiting etch rate from 1000 to 10000 Å/min produces only a moderate change in the endpoint slope. The reason for this becomes evident in Figure 4.

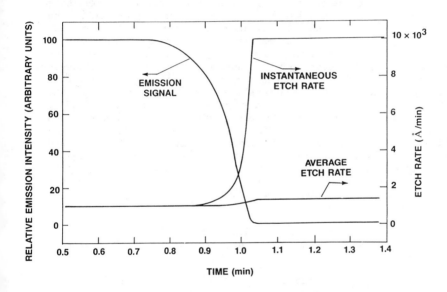

FIG 4--Endpoint simulation for 1000 Å film with 100 Å standard deviation in presence of a loading effect. The initial (blank wafer) etch rate is taken as 1000 Å/min., increasing to 10,000 Å/min. limiting rate (clear wafer). The etch rate standard deviation was taken to be zero. The loading effect was assumed to behave according to formula (2).

Here an endpoint simulation with loading effect is presented with corresponding plots of the instantaneous and average etch rates. Film thickness is 1000 Å, with 100 Å uniformity. The etch rate varies from an initial value of 1000 Å/min to a limiting value of 10000 Å/min. Although the instantaneous etch rate increases rapidly as the wafer clears, the average etch rate increases only slightly. This is because the average is the time integral of the instantaneous etch rate. The instantaneous etch rate rises rapidly, but over such a short time interval that it makes little contribution to the overall average. Thus, even a large loading effect produces only a moderate change in the endpoint transition.

EXPERIMENTAL VERIFICATION

   Experimental verification of the model was obtained by studying the etching of nitride and polysilicon in a commercial 125 mm single wafer plasma etcher using sulfur hexafluoride as the etchant. The first step in this study was to determine the magnitude of the loading effect. This was accomplished by fabricating nitride and polysilicon test wafers having varying degrees of exposed area, the unexposed area being covered with photoresist. The film thickness was measured at 9 locations prior to etching. The wafer was then partially etched and the locations remeasured. The etch rate and etch rate uniformity as a function of area are easily determined by dividing film thickness loss by etch time. The results of this experiment are summarized in Table 2. It can be seen that the nitride etch rate is essentially invariant with respect to exposed film area, while the polysilicon etch rate is a strong function of exposed area. For this reason nitride etching was chosen as a model system for the for the no loading effect case. Polysilicon was chosen as the model system to study endpointing in the presence of a loading effect.

   The data for polysilicon etching was analyzed by nonliner regression to produce an empirical equation for the loading effect. The result was the following equation:

$$e = \frac{547}{1 + 10.66\, A^{1/2}} \qquad (3)$$

Where e is the instantaneous etch rate in Å/s, and A is the fractional exposed area for a 125 mm wafer. This equation was used in the subsequent simulations involving loading effect. It should be reemphasized that this equation represents an empirical "best fit" to the loading effect data and is not based on a physical model. Sample 125 mm wafers of both nitride and polysilicon were measured at 9 locations to determine their average film thickness and uniformity. Theoretical endpoint transitions were computed by Monte Carlo simulation using this data and the etch rate data acquired previously. The wafers were then etched to completion while the emission transition was recorded. The results were compared with those obtained by simulation.

TABLE 2--Etch rates vs. exposed film area[a]

Nitride Etching[b]

Fractional Exposed area	Etch rate, Å/s	Etch rate σ, %
1.00	30.0	6.0
0.50	31.6	4.9
0.08	28.5	2.3

Polysilicon Etching[c]

Fractional Exposed area	Etch rate, Å/s	Etch rate σ %
1.00	47.4	5.4
0.32	75.3	4.3
0.02	228.6	7.2

[a] 125mm dia. wafers (12272 mm² area)
[b] Based on 30s etch at 100 watts 250 mTorr SF_6
[c] Based on 30s etch at 75 mwatts 320 mTorr SF_6

Table 3 compares the results for etching of nitride wafers (no loading effect). Three different comparisons are made: The transition time is the time required for the emission to drop from 90% of its initial value to 10% of its final value. The slope figure specifies the slope of the emission vs. time curve at the midpoint between the initial and final values. The time to endpoint is the time required for the emission value to reach the midpoint between the initial and final values. This latter figure is analogous to the "clear time" which would be computed dividing average film thickness by average etch rate. All of these parameters were obtained by measurements on the experimental endpoint curves and by calculations made using the emission vs. time data from the simulation. Reference to the table shows that the simulation has tended to overestimate the transition time and time to endpoint while underestimating the slope. These systematic errors are possibly due to errors in the estimation of the nitride etch rate and uniformity. This is due in part to the fact that initiation effects were ignored in computations of the etch rate. The initiation time is a period following plasma ignition during which little etching occurs. If greater accuracy were desired in the simulation the initiation period could be incorporated as an interval of zero etch rate. Experimentally, the initiation time can be determined by extrapolating the linear region on a plot of thickness loss vs. etch time to zero thickness. The corresponding time is the initiation time. The overall shape of the predicted emission curve is in good agreement with experiment, as shown in Figure 5, for nitride sample 9A. The shape of the experimental endpoint curve is seen to be symmetric, as predicted by the model for the case of no loading effect. The error between prediction and experiment appears primarily as an offset between the two curves. For all cases shown the disagreement between experiment and simulation is emphasized by the time scale used to record the endpointing event. An expanded time scale was employed to provide the most rigorous test of the program. If a compressed scale typical of production use had been employed it would be difficult to distinguish between experiment and simulation. Consider for example the slope parameter for sample U6: Expressed as an angle the experimental slope of -7.46 is $-82.4°$. The simulated slope of -3.45 corresponds to an angle of $-73.8°$. Now, if the time scale had been compressed by a factor of 60 to express the time in minutes instead of seconds these angles would become $-89.9°$ and $-89.7°$ respectively. The apparent agreement is improved.

TABLE 3--Experimental vs. simulation results for nitride wafers

Sample ID	Transition Time[a], s		Slope[b], %/s		Time to endpoint[c], s	
	Exp.	Sim.	Exp.	Sim.	Exp.	Sim.
U6	4.4	10.0	-7.46	-3.45	39.75	47.0
U7	5.0	8.0	-7.35	-4.65	45.50	51.0
U8	4.2	8.0	-7.69	-4.73	49.50	55.5
9A	5.3	6.5	-6.94	-5.51	42.50	45.5

[a] Time required for emission to decay from 90% to 10% of span between initial and background intensities

[b] Slope when emission is at 50% of span between initial and background intensities

[c] Time to point where emission is at 50% (midway) between initial and background intensities

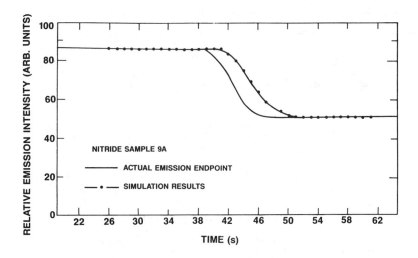

FIG 5--Comparison between experimental and simulation results for nitride etching. The experimental curve shows the symmetry expected for the case where no loading effect exists.

Table 4 compares the simulation and experimental results for polysilicon etching. As previously noted, this process exhibits a pronounced loading effect. The simulation results presented in Table 4 were developed with the loading effect correction applied in the model. Reference to the table shows good agreement for the transition time and time to endpoint. There is a tendency for the simulation to underestimate the slope. Figure 6 presents a comparison between simulation and experimental results. Good agreement is seen in the curve shapes. The offset between them is again attributable to initiation effects. The experimental curve shows a slightly more gradual tail. This may represent a contribution by undercutting to the emission signal. Undercutting is typically present in the etching of polysilicon with sulfur hexafluoride [4]. Also shown are the results obtained when the loading effect is ignored. Clearly, incorporation of the loading effect makes a significant improvement in the model accuracy.

TABLE 4--Experimental vs. simulation results for polysilicon wafers[a]

Sample ID	Transition Time[b], s		Slope[c], %/s		Time to endpoint[d], s	
	Exp.	Sim.	Exp.	Sim.	Exp.	Sim.
3	11.6	9.0	-10.00	-7.84	93.00	97.50
4	8.0	9.8	-8.70	-7.13	94.00	97.75
U5	12.0	11.6	-8.00	-5.89	118.00	120.10
U6	11.5	11.5	-8.89	-7.42	116.75	121.10
U7	11.5	13.5	-8.89	-5.29	119.00	123.00
U8	9.5	11.8	-10.00	-7.81	118.75	122.25

[a] Simulations include corrections for loading effect.

[b] Time required for emission to decay from 90% to 10% of span between initial and background intensities.

[c] Slope when emission is at 50% of span between initial and background intensities.

[d] Time to point where emission is at 50% (midway) between initial and background intensities.

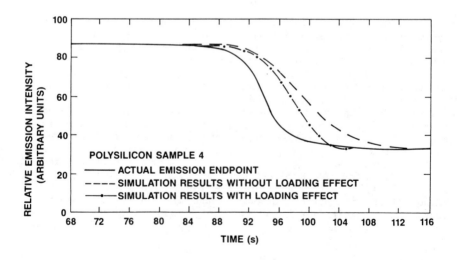

FIG 6--Comparison between experimental and simulation results for polysilicon etching. The simulation results incorporating the loading effect are seen to give a better agreement with experiment.

SUMMARY AND CONCLUSIONS

The effects of wafer and etch rate uniformity on emission endpoint transition have been modeled using Monte Carlo techniques. A simple model based on gaussian descriptions of etch rate and film thickness was proposed. This model predicted spreading of the endpoint transition as the film thickness and etch rate became less uniform. In the absence of loading effects the endpoint curve was seen to posses a symmetry about the transition midpoint. The incorporation of loading effects into the model showed an increase in the abruptness of the transition, and upset the symmetry. The simulation tended to underestimate the transition slope for nitride etching. This was attributed to errors in the estimation of the nitride etch rate, and its uniformity. Good agreement was obtained with experimental results for polysilicon etching.

The reader is cautioned that the simplicity of the model presented here may not extend to all etch processes without modification. Rather, for the cases of nitride and polysilicon etching, the model's assumptions appear to be sufficient to obtain good results. The success of the Monte Carlo technique should encourage its use to model other processes. The model should be easy to modify to include other phenomena, such as variations in batch size, power, and pressure. Thus, additional sophistication can be built into the model where it becomes necessary.

ACKNOWLEDGEMENTS

The author would like to thank Rocky Sanders for writing a Fortran version of the simulation, and Chris Teutsch for performing benchmark tests of program execution times. Thanks also to Tom Fitzgerald and Dr. Steven Mak for reviewing this manuscript, and to Joann Rothbard for the actual typing.

REFERENCES

[1] Weiss, A.D., "Endpoint Monitors," Semiconductor International, Vol. 6, No.9, Sept. 1983, pp. 98-99.

[2] Milikan, R. C., "The Magic of the Monte Carlo Method," BYTE, Vol. 8, No. 2, Feb. 1983, pp. 371-373.

[3] Mogab, C. J., "The Loading Effect in Plasma Etching," Electrochemical Society Journal, Vol. 124, No.8, Aug. 1977, pp. 1262-1268.

[4] Mucha, J. A., "The Gasses of Plasma Etching: Silicon-Based Technology," Solid State Technology, Vol. 28, No. 3, March 1985, pp. 123-127.

Tom Abraham and Robert Theriault

PROFILE CONTROL OF PLASMA ETCHED POLYSILICON USING
IMPLANT DOPING

REFERENCE: Abraham, T., and Theriault R., "Profile Control of Plasma Etched Polysilicon Using Implant Doping," Emerging Semiconductor Technology, ASTM STP 960, D. C. Gupta and P. H. Langer, Eds., American Society for Testing and Materials, 1986.

ABSTRACT: The effect of implant doping using phosphorus on the edge profiles of plasma-etched polysilicon was investigated. Unique rounded profiles were obtained with as-implanted samples. The profile curvature showed a strong dependence on the fluorine to chlorine ratio in the etch process. There was good correlation between the lateral etch rate variation across the polysilicon thickness and the doping profile. Electrical line width measurements and SEM micrographs indicated line width control of ±0.3 μm. The step coverage of as-deposited PSG was studied and enhanced coverage was observed as the edge profile curvature was increased.

KEY WORDS: edge profile, implant doping, doping profile, undercut, Silicon, Boron, Phosphorus, SF_6, C_2ClF_5, polysilicon, PSG.

Plasma etching has become the accepted technique for the patterning of polysilicon gate material in MOS devices. The impetus for this has been the enhanced dimensional control using dry etching. In general, fluorinated and/or chlorinated gases are used [1] with the former resulting in isotropic etching and the latter in anisotropic etching with vertical slopes. Several processes use a combination of the two for higher etch rates and improved oxide selectivities.

Nearly all references to etching with chlorinated gases [2,3] indicate that the anisotropic edge slopes are obtained with little or no undercut. Various models have been postulated to explain this behavior. The surface damage induced anisotropic model explains it on

---

T. Abraham is Senior Development Engineer and R. Theriault is the Manager, IC Development at Northern Telecom Ltd., P.O. Box 3511, Station C, Ottawa, Ontario, Canada. K1Y 4H7

the basis of impinging ions producing lattice damage creating active sites for the formation of the volatile etch product. There is another model which postulates that lateral etching is inhibited by the deposition of a polymer film on the sidewall [4]. Polymer buildup and subsequent inhibition of etching of the exposed surface is prevented by ion bombardment. Chlorinated gas systems tend to show a strong etch rate dependence on n-type doping. In some early work by Mogab et al [1], it was shown that when an ion implanted but non-activated polysilicon (Poly) layer was etched using Cl_2, no etch rate differential was observed. After the sample was annealed however, there was a large difference in etch rate between doped and undoped areas. They postulated a theory based on the assumption that the chemisorption of chlorine atoms was determined by the rate of electron transfer. Hence the high concentration of free electrons in n-type poly resulted in significantly higher etch rates. Boron doped poly showed an etch rate dependence similar to that of undoped poly, lending further credence to this theory.

In this paper, results of doping on the lateral etch rate are presented showing that isotropic-type etching behavior is possible in chlorinated systems. However since this is doping dependent, control of the doping profile using ion implants can result in well defined tapered etching of the polysilicon.

While anisotropic etching gives good dimensional control, the steep 90° slopes or in some cases re-entrant profiles can cause problems with step coverage. This is especially true in double-metal structures where metal lines in close proximity to sharp poly lines can result in deep grooves being formed during planarization of the inter-metal dielectric.

EXPERIMENTAL PROCEDURE

Plasma etching of the samples was carried out in a Drytek Drie 100 parallel plate plasma mode system. The gas chemistry used was C_2ClF_5 and SF_6. This system has a vertical electrode stack capable of etching six wafers at a time using a 13.56 MHz RF generator. Endpoint detection is by means of laser interferometry. Some of the etching was done in a Lam 480 Autoetch system using CCl_4 to investigate the effect of the sidewall film on lateral undercut. All ion implants were carried out in an Eaton-Nova NV-160 high-current implanter.

Scanning electron micrographs were used to analyze the edge profiles and undercut. The edge profiles were correlated to the doping profiles generated using 'SUPREM' modelling. Linewidth control was investigated with electrical probing of specially designed structures and step coverage results were obtained by depositing 7% PSG on top of etched poly features and looking at cross-sectional scanning electron micrographs.

Results and Discussion

The Freon 115 (C_2ClF_5) gas dissociates to form free chlorine and unsaturated fluorocarbon radicals with little or no free fluorine. Thus C_2ClF_5 in effect acts as a chlorinated gas when etching polysilicon. SF_6 on the other hand dissociates to produce high concentrations of atomic fluorine [5]. The chlorine to flourine ratio can therefore be varied by changing the C_2ClF_5 to SF_6 flow rate ratios. The effect of changing this ratio on etch profiles of $POCl_3$ doped polysilicon is shown in Figure 1. The results are typical with increasing undercut as the SF_6 flow is increased. With 100% C_2ClF_5, anisotropic etching with minimal undercut is obtained. The profile is however slightly re-entrant or negative. This has been observed in several instances when etching highly doped polysilicon in chlorinated plasmas and has been explained on the basis of surface migration of the chlorine etch species [6].

When a sample of polysilicon, implanted with phosphorus using a dose of 2×10^{16} at 60 keV, was etched with a 1:1 SF_6 to C_2ClF_5 ratio, the resulting edge profile (Figure 2) showed a completely different characteristic compared to $POCl_3$ doped poly. While the profile is straight at the base, it rounds off towards the top with a significant degree of undercut. At the top of the profile there is a notched step with a slightly re-entrant slope. Several experiments were carried out to investigate whether the undercut was due to amorphization of the top layer during the implant process or whether the dopant type had any effect. Samples implanted with silicon and boron were etched using the same conditions and in neither case was this rounded undercut profile observed (Figure 3).

Koike et al [7] have reported a strong dependence of etch rate on n-type doping when etching in chlorinated gases. The lateral undercut in fact increased when the implant dose was increased, indicating that this feature was caused by dopant dependent lateral etching of the top layer of polysilicon (Figure 4). Some dopant activation seems to occur during the implant doping process. The vertical etch rate also showed a strong dependence on poly depth, decreasing in value as the film was etched. This is shown in the etch rate trace shown in Figure 5.

Simulation of doping profiles in implant doped polysilicon was carried out using 'SUPREM' modelling (Figure 6). The etched edge profile results showed good correlation to the simulated as-implanted doping profile. The re-entrant step at the top of the profile can be related to the increasing doping concentration in the top 1000 Å of the film. The bottom of this step appears to correspond to the peak of the doping profile.

The simulation results showed that annealing the sample at 600°C would have the effect of activating the dopant without driving it in, while a 900°C anneal would redistribute the dopant throughout the film and activate it.

An experimental matrix was carried out by varying the C_2ClF_5 to SF_6 ratio for different annealing conditions. The etch conditions, annealing temperatures and the corresponding resist undercut are listed in Table 1. The profile results are shown in Figure 7. For as-implanted samples, the lateral undercut at the top of the film increases as the chlorine content in the gas mixture is increased. In all three cases, the notch at the top shows a slight negative slope. After a 600°C anneal, this undercut increases with respect to the as-implanted case. Again, the undercut increases as the C_2ClF_5 to SF_6 ratio is increased. The notch at the top of the profile however changes from being re-entrant to a straight edge. This is clearly seen for high C_2ClF_5 flows. The sensitivity of the edge profile to the doping concentration is highlighted at the high C_2ClF_5 to SF_6 flow rate ratio with good correlation to the simulated doping profiles. After a 900°C anneal, the etch characteristics revert back to standard chlorine/fluorine behavior showing increasing undercut as the fluorine content is increased. The dependence of lateral undercut on doping for C_2ClF_5 is clearly illustrated in these experiments.

TABLE 1 -- Resist Undercut for Varying Etch Conditions

Annealing Conditions	$SF_6:C_2ClF_5$ (Flow Rate Ratio)		
	1:1	1:3	1:9
As implanted	0.44 μm	0.67 μm	0.72 μm
600°C anneal	0.62 μm	0.72 μm	1.30 μm
900° anneal	0.39 μm	0.33 μm	0.28 μm

Step coverage and linewidth control were investigated using as-implanted and annealed samples which were subjected to a poly oxidation and an additional anneal prior to PSG deposition. With as-deposited LPCVD 7% PSG, the rounded profile enhances step coverage resulting in smooth curvature of the glass (Figure 8). The top re-entrant notch appears to have little or no adverse effect. On the other hand, straight or slightly sloped edges result in re-entrant PSG profiles with the possibility of void formation for close proximity features.

Electrical linewidth measurements indicate a three-sigma value of ±0.2 μm for as-implanted poly etched in a 3:1 chlorine to fluorine ratio. For $POCl_3$ doped samples, the corresponding three-sigma value was ±0.3 μm. Thus, profile shaping can be carried out without adversely impacting linewidth control. Since the bottom half of the poly is virtually undoped, linewidth control will be similar to anisotropic etching of poly.

Etching as-implanted samples in a CCl_4 plasma showed completely different results (Figure 9). There was little or no undercut with the slopes similar to those of standard POCl$_3$ doped polysilicon. The gas CCl$_4$ is known to be susceptible to polymer formation and Bernacki et al [8] have reported evidence of a sidewall deposit. This film inhibits any lateral etching and may account for the fact that profiles similar to the C$_2$ClF$_5$ case were not seen with CCl$_4$. The former does not generally form polymer films.

Considering the two models for anisotropic etching discussed earlier, the work reported here supports the validity of the sidewall passivant model. The surface damage induced model does not adequately explain the undercut seen here and would have to be modified to account for the n-type doping sensitivity of the lateral etch rate in chlorinated plasmas to be applicable to this type of etching.

Different edge profiles could be obtained by modifying the implant process. Either changing the dopant species to arsenic or implanting through a thin oxide layer and then removing this prior to polysilicon etching could result in rounded profiles without the notch at the top.

Summary

Tapered etching of polysilicon in chlorinated plasmas has been shown to be possible. The strong sensitivity of certain chlorinated gases such as C$_2$ClF$_5$ to n-type doping and the good correlation of edge profiles with doping profiles indicates the possibility of producing tailored edge profiles. Enhanced step coverage can be obtained with rounded profiles. This non-anisotropic etching did not have any adverse effect on linewidth control. Similar slopes were not obtained in CCl$_4$ plasmas indicating that sidewall passivation effectively inhibits any lateral etching.

ACKNOWLEDGEMENTS

The authors would like to thank Pak Leung and Jaan Kolk for many useful discussions. We would also like to thank Brian Tait and Judy Saeki for the SEMs and Jim Fraser for the implant doping of the polysilicon.

REFERENCES

[1] C.J. Mogab et al, J. Vac. Sci. Technology, Vol. 17(3), p721, 1980.
[2] J.W. Coburn et al, J. Vac. Sci. Technology, Vol. 16(2), p391, 1979.
[3] A.C. Adams et al, J. Electrochem. Soc., Vol. 138,2, p366, 1981.
[4] D.L. Flamm et al, J. Vac. Sci. Technology B, Vol. 1(1), p23, 1984.
[5] R.W. Light et al, J. Electrochem. Soc., Vol. 130, p1567, 1983.
[6] T.M. Mayer et al, IEDM Meeting, 25th Tech. Dig., p44, 1981.
[7] A. Koike et al, Abstracts Electrochem. Soc., Montreal, p343, 1982.
[8] S.E. Bernacki et al, J. Electrochem. Soc., Vol. 131, 8, p1927, 1984.

(a)
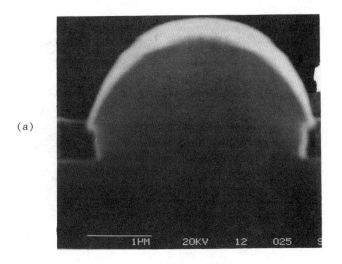

(b)
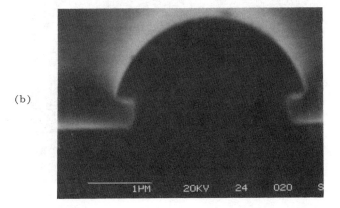

FIG. 1--Edge profile and undercut of $POCl_3$ doped polysilicon for varying $SF_6 C_2ClF_5$ flow rate ratios, (a) 100 sccm C_2ClF_5, (b) 25 sccm SF_6 : 75 sccm C_2ClF_5.

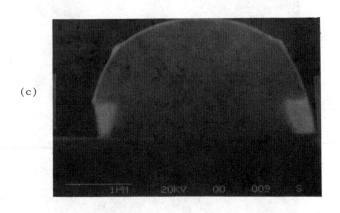

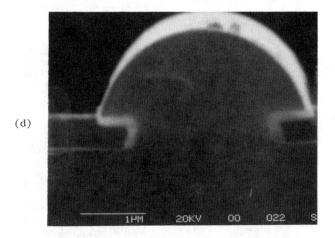

FIG. 1--Edge profile and undercut of $POCl_3$ doped polysilicon for varying $SF_6:C_2ClF_5$ flow rate ratios, (c) 50 sccm SF_6 : 50 sccm C_2ClF_5, (d) 75 sccm SF_6 : 25 sccm C_2ClF_5.

FIG. 2--Edge profile and undercut of as-implanted sample (2×10^{16} #/cm^2 P, 60 keV) after etching with a $SF_6:C_2ClF_5$ flow rate ratio of 1:1 (50 sccm/50 sccm).

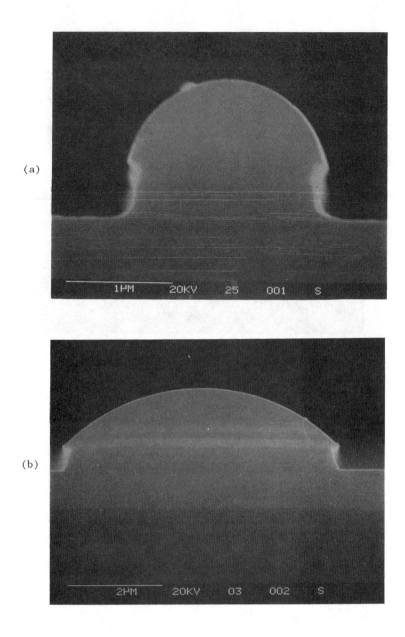

FIG. 3--Edge profiles (a) Profile of silicon implanted sample, (b) Profile of boron implanted sample.

ABRAHAM AND THERIAULT ON PLASMA ETCHED SILICON 213

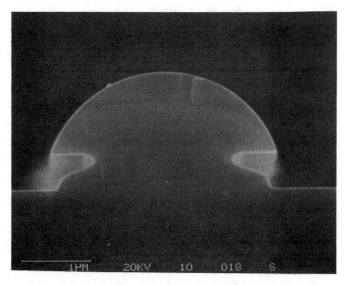

FIG. 4--Edge profile and undercut of as-implanted sample
(3 x 10^{16} #/cm^2 P, 60 keV) after etching with a $SF_6:C_2ClF_5$
flow rate ratio of 1:1 (50 sccm/50 sccm).

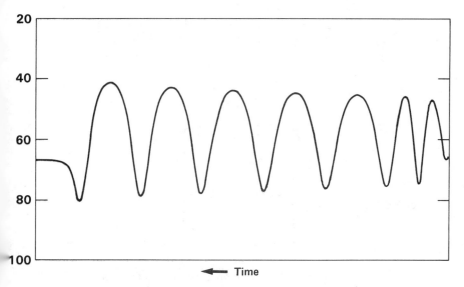

FIG. 5--Laser interferometry etch rate trace of as-implanted sample.

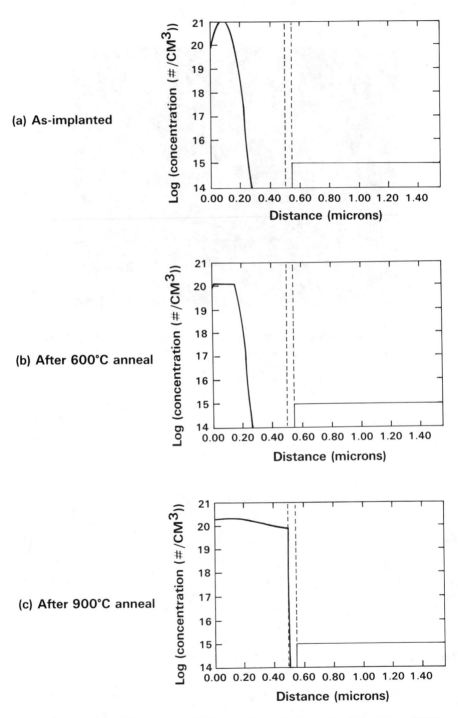

FIG. 6--Simulated doping profiles (a) As-implanted, (b) After 600°C anneal, (c) After 900°C anneal.

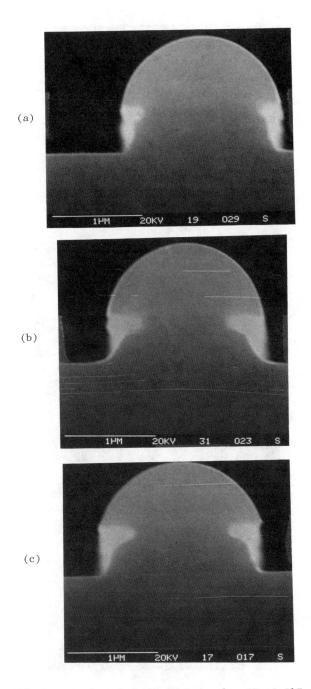

FIG. 7--SEM micrographs of as-implanted poly, $SF_6:C_2ClF_5$ = (a) 1:1 (b) 1:3 (c) 1:9

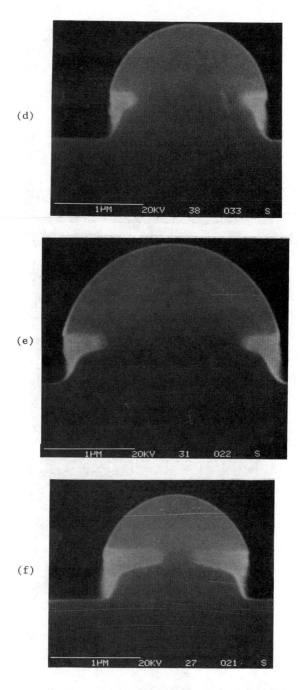

FIG. 7--SEM micrographs of implant doped poly with 600°C anneal, $SF_6:C_2ClF_5$ = (d) 1:1 (e) 1:3 (f) 1:9.

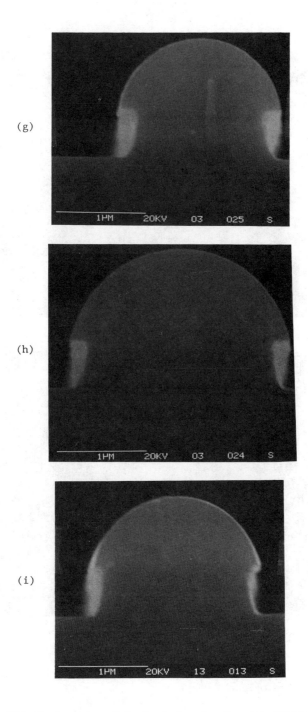

FIG. 7--SEM micrographs of implant doped poly with 900°C anneal, $SF_6:C_2ClF_5$ = (g) 1:1 (h) 1:3 (i) 1:9

(a)

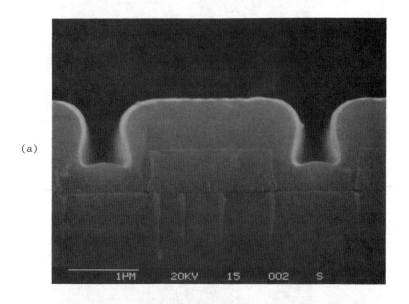

(b)

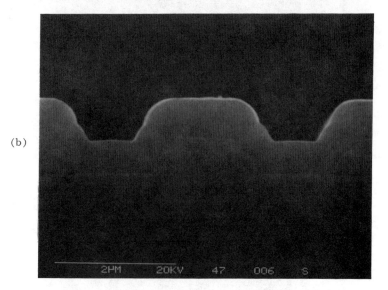

FIG. 8--Step coverage of as-deposited 7% PSG (a) POCl$_3$ doped sample (b) As-implanted sample.

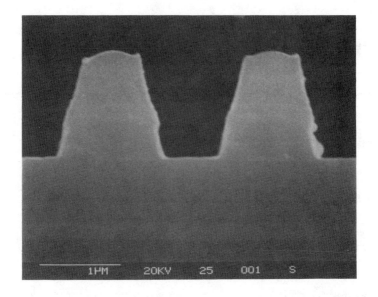

FIG. 9--Edge profile of as-implanted sample etched in CCl_4 plasma.

Keith C. Vanner, John R. Cockrill, James A. Turner

THE EFFECTS OF PLASMA PROCESSING OF DIELECTRIC LAYERS ON GALLIUM ARSENIDE INTEGRATED CIRCUITS.

REFERENCE: Vanner, K.C., Cockrill, J.R., Turner, J.A., "The Effects of Plasma Processing of Dielectric Layers on Gallium Arsenide Integrated Circuits," Emerging Semiconductor Technology, ASTM STP 960, D.C. Gupta and P.H. Langer, Eds., American Society for Testing and Materials, 1986.

ABSTRACT: In the commercial production of GaAs I.C.s it is essential to minimise processing induced damage to the shallow semiconducting layers of the MESFET. Currently used fabrication sequences do not involve the use of highly energetic plasma processes such as RIE under conditions which may cause surface damage as a result of ion bombardment. In the development of new fabrication technologies it is essential to ensure that device performance is not degraded by the use of such plasma processes. The use of such techniques as plasma enhanced CVD to deposit passivation or

---

The authors are research scientists at Plessey Research (Caswell) Ltd., Allen Clark Research Centre, Caswell, Towcester, Northamptonshire, England.

interlayer dielectrics, and subsequent patterning by plasma or reactive ion etching, may adversely affect the active layer as a result of ion or radiation bombardment, or by chemical contamination of the surface.

This paper will present results from a study of the effects of plasma processing of GaAs MESFETs in MMICs. Measurements of device performance show that assessment of FET characteristics can provide a valuable complement to diode measurements and surface analysis.

KEYWORDS: gallium arsenide, plasma damage, plasma etching, RIE.

The development of high yield processes for the commercial manufacture of GaAs I.C.s involves the fabrication of MESFET structures on the surface of a thin semiconductor layer. The quality of this metal-semiconductor interface has an important bearing on device performance, and it is very susceptible to processing induced damage which can cause low yields or unacceptable variations in circuit characteristics.

Subsequent processing includes the fabrication of passive components which require multi-level structures with interlayer dielectric insulation. Since gallium arsenide, unlike silicon, does not form a stable native oxide suitable for use as this insulation layer, thin film deposition techniques are required.

Suitable films such as silicon nitride or silicon dioxide, may be deposited by sputtering or plasma enhanced CVD (PECVD). High temperature dielectric deposition processes such as conventional CVD are not compatible with GaAs which will undergo structural changes at temperatures below those required for satisfactory film deposition.

Although most GaAs MESFET manufacturing processes use vacuum evaporation and "lift off" photoresist techniques to obtain the desired metal patterns, a recent trend has been towards self-aligned structures which require the use of reactive ion etching (RIE) or sputter etching to pattern metal or dielectric films in contact with the wafer surface.

Such processing inevitably causes damage or removes material from the wafer surface and this will adversely affect the characteristics of the FET. The active layer which is usually produced by ion implantation of a semi-insulating GaAs wafer, may extend to a depth of about 500nm in a depletion mode device, or in the case of enhancement mode devices the implant depth may only be about 150nm.

Both enhancement and depletion mode devices are susceptible to surface damage since ohmic and Schottky contacts are made to the upper surface of this layer.

Table 1 shows the plasma processes frequently used to deposit and pattern thin films on the wafer surface.

TABLE 1--Plasma processes applied at the wafer surface

Process	Material	Application
PECVD	Silicon nitride Silicon nitride	Annealing cap Interlayer dielectric
Sputter Deposition	Silicon nitride Metals Metal compounds	Annealing cap Schottky contacts
RIE	Dielectrics Metals GaAs	Via holes for interconnects Self aligned processes Contact definition Localised thinning of active layer
Plasma etching	Dielectrics	Via holes for interconnects Layer removal Photoresist removal
Sputter etching	GaAs	Surface clean prior to film deposition

In most cases the GaAs wafer surface is exposed during an over-etch stage which is necessary to ensure complete removal of films having non-uniform thickness, or in the case of film deposition the wafer is briefly exposed while a continuous film accumulates over the entire surface. A sputter etch is frequently given to clean a surface prior to deposition of a thin film by sputtering.

Gallium arsenide etching by RIE is sometimes deliberately carried out to reduce the thickness of the active layer of the device, in this case material is eroded by the combination of chemical and physical effects of reactive ion bombardment. This process obviously creates some damage to the remaining current carrying layer, but because GaAs etching is in this case the object of the process, this aspect of RIE has not been considered in this investigation.

All the processes considered have in common the creation of ions in an R.F. glow discharge that is maintained between two parallel plate electrodes, one acting as a carrier for the wafers. The essential feature of plasma and reactive ion etching is the generation of reactive species in the plasma which lead to etching of the film, in the case of sputtering the plasma is usually of argon, and bombarding ions are non-reactive. In all cases further exposure of the wafers to electron and radiation bombardment occurs, the extent of this additional flux is often not clearly known.

The work reported here is mainly confined to a study of the effects of plasma processing of silicon nitride dielectric layers. The effects of the deposition of these films has been studied by examination of diode and MESFET structures on GaAs, supplemented by surface analysis and ellipsometric evaluation of the etched surfaces.

In order to provide a contrast between the reactive processes of plasma etching and deposition, some studies were carried out on GaAs surfaces that had been sputter etched in pure argon. This experiment is particularly relevant to GaAs processing in view of the considerable current interest in the deposition of refractory metals and their compounds by sputtering, for use in self aligned processes.

EXPERIMENTAL PROCEDURES

The experiments undertaken were aimed at isolating the degrees of active layer damage in the processes of PECVD, plasma etching and RIE, of silicon nitride thin films on the wafer surface.

The initial experiment was carried out to assess the effect of silicon nitride deposition and plasma etching on a GaAs surface prior to deposition of the Schottky contacts. The assessment was of vertical diode structures. A second experiment utilised planar structures with thin n-doped semiconducting layers in order to accentuate the effect of surface damage, and also to allow measurements to be made on FETs in addition to diodes. In this experiment etching of the dielectric by RIE was also investigated. A further study of the effect of silicon nitride RIE was carried out in a third experiment. In this case the normal device fabrication sequence was modified in order to avoid annealing the samples after the RIE processing. Finally, for comparison, diodes were made on sputter etched surfaces where argon ion bombardment of the GaAs surface had taken place.

Certain aspects of the processes were common to all the experiments, and for brevity are summarised in Table 2. Experimental variations are detailed where relevant.

The initial experiment was intended to look at the effect of nitride deposition and plasma etching by the assessment of diode structures fabricated on the processed surface.

TABLE 2--Process conditions

Silicon nitride film deposition parameters	
System configuration	: Parallel plate electrodes (aluminium)
Chamber pressure	: 133 Pa
Electrode temperature	: 300°C
Gas mixture	: $SiH_4/NH_3/N_2$
R.F. Power density	: 200 W.m^{-2}
R.F. Frequency	: 13.56MHz
Film deposition rate	: 10nm.min^{-1}
Refractive index	: 2.00-2.02
Plasma etching parameters.	
System configuration	: Parallel plate electrodes (aluminium)
Chamber pressure	: 26.6 Pa
Etch gas	: Carbon tetrafluoride (CF_4)
R.F. Power density	: 5×10^3 W.m^2
R.F. Frequency	: 13.56MHz
Etch rate	: 40nm.min^{-1}
RIE parameters	
System configuration	: Parallel plate electrodes (aluminium)
Chamber pressure	: *
Etch gas	: CF_4
R.F. Power density	: *
R.F. Bias voltage	: *
R.F. Frequency	: 13.56MHz
Sputter etch parameters	
System configuration	: R.F.Diode (stainless steel)
R.F. Frequency	: 13.56MHz
Gas	: Argon
Pressure	: 1.33 Pa
R.F. Power density	: 750 W.m^{-2}
R.F. Bias voltage	: 200V

* denotes experimental variable

The wafer was prepared, doped throughout its bulk, with ohmic contacts alloyed onto the rear face. After cleaving into four quadrants one piece was used as a control and was not plasma processed, the other quadrants were treated as follows:-

A: PECVD silicon nitride...HF acid etch...effect of deposition.
B: PECVD silicon nitride...plasma etch...effect of plasma etch.
C: PECVD silicon nitride...plasma etch+over etch...effect of plasma etch.

All three samples were coated with 120nm of silicon nitride in the same deposition cycle. Sample A was then dipped in concentrated (48%) hydrofluoric acid for 5 minutes and then rinsed in ultra-pure water. Sample B was etched in the plasma etching system for 3¼ minutes, which was just sufficient to remove the nitride film. Sample C was then plasma etched under the same conditions but for 7½ minutes, in order to subject the wafer surface to a prolonged exposure to the etching plasma.

After completion of the silicon nitride processing Schottky contacts were prepared on all four samples by evaporation of titanium and use of the "lift off" technique. The Schottky barrier contacts were rectangular, having an area of $48250 \mu m^2$.

Diodes were measured on each quadrant and the data is given in Table 3.

TABLE 3-- Diode parameters

Sample	Ideality factor	Barrier height
Control	1.063 ±0.004	0.86 eV
HF etch	1.063 ±0.004	0.84 eV
Plasma etch	1.050 ±0.009	0.87 eV
Plasma etch (over-etch)	1.066 ±0.008	0.86 eV

Auger and SEM analysis of the surfaces of the etched quadrants showed that no residues or contaminants were present on the surface apart from the normal thin oxide film. Ellipsometric measurements of the wafer surface are reported later in this paper.

It was felt that diode measurements could be usefully augmented by assessment of changes in FET characteristics due to material damage. Bulk doped wafers are not suitable for FET fabrication and subsequent experiments were carried out on shallow ion implanted layers on semi-insulating GaAs wafers. The maximum thickness of these active layers was 150nm, and device characteristics are known to be greatly influenced by any damage in this shallow surface layer.

The next experiment was undertaken with the object of including RIE in the range of plasma processing conditions. Two RIE conditions were used in order to achieve different R.F. bias voltages in successive etching runs.

The nitride processing details were as follows:-

Entire wafer: 135nm. PECVD silicon nitride deposited.
Quadrant A: Silicon nitride etched in conc. HF, rinsed in ultra-pure water.
Quadrant B: Plasma etched under standard conditions, minimal over-etch.

Quadrant C: RIE in CF_4, 13.3 Pa, 1.6×10^3 W.m^{-2}, 160V bias, duration 1.5min.

Quadrant D: RIE in CF_4, 4 Pa, 2.5×10^3 W.m^{-2}, 250V bias, duration 2min.

After the silicon nitride had been etched, FETs were fabricated on the four quadrants. This required the samples to be subjected to an anneal of 430°C for 5 minutes to alloy the ohmic contacts to the semiconductor. Schottky contacts (of titanium) were then prepared without further etching of the surface. It was considered likely that annealing would remove some of the plasma induced damage (1,2), but that this process sequence would be most representative of a typical I.C. fabrication process.

Diode results are shown in Table 4.

TABLE 4--Diode parameters

Sample	Ideality factor	Barrier height
HF etch	1.25 ±0.03	0.86 ±0.02eV
Plasma etch	1.34 ±0.02	0.85 ±0.04
RIE, 160V	1.24 ±0.02	0.85 ±0.01
RIE, 250V	1.42 ±0.02	0.82 ±0.03

Ideality factors are worse than in the previous experiment because measurements were made on FET structures not optimised for diode measurement.

FET measurements were made on 1μm gate length devices and are normalised to the values obtained on the HF etched sample. These results are shown in Table 5. The terms I_{dss}, V_p and g_m normally used to characterise FETs are defined in (3).

TABLE 5--FET characteristics

Sample	I_{dss}	V_p	Transconductance (g_m)
HF etch	1.00	1.00	1.00
Plasma etch	1.09	1.03	1.10
RIE, 160V	0.95	0.87	1.10
RIE, 250V	0.74	0.77	0.94

Auger and SIMS analysis of the samples showed that fluorine and aluminium were present on the RIE quadrants. The aluminium was largely present at the surface and is assumed to have been back sputtered from the chamber and electrodes. The fluorine had penetrated the wafer surface to a depth of approximately 100 nm. and its concentration was highest on the high bias RIE sample.

A further investigation was made with the object of studying in more detail the effect of RIE, and also the effect of annealing the plasma damaged surface after device fabrication. In this study silicon nitride was deposited on three quadrants of a GaAs wafer, with the ohmic contacts already alloyed to the surface. The nitride was removed by RIE using three different bias voltages. A control quadrant was not subjected to any plasma processing. The details of the RIE conditions are as follows:-

Nitride deposition: 135nm deposited using standard conditions.

Sample A: 100V bias voltage, 400 $W.m^{-2}$

Sample B: 225V bias voltage, 1.3 x 10^3 $W.m^{-2}$

Sample C: 350V bias voltage, 2.5 x 10^3 $W.m^{-2}$

All samples: Etched in CF_4 at 2.66 Pa pressure, for a duration estimated to give 20% over etch.

After RIE, Schottky contacts of titanium were fabricated in the usual way.

Measurements were made as described above on diodes and 1 μm gate length FETs. The four quadrants were then annealed at 430°C for 10 minutes. This annealing condition was chosen to simulate the ohmic contact alloying anneal given in the previous experiment before device measurement. No adverse metallurgical effects were expected for this combination of temperature and time on the completed devices. The diode data is given in Table 6.

TABLE 6--Effect of annealing on diode parameters

	Before anneal		After anneal	
Sample	Ideality Factor	Barrier Height	Ideality Factor	Barrier Height
Control	1.45	0.71eV	1.37	0.67eV
RIE, 100V	1.73	0.63eV	1.34	0.65eV
RIE, 225V	1.47	0.68eV	1.26	0.68eV
RIE, 350V	1.46	0.69eV	1.31	0.66eV

The FET data, normalised as before, is given below, in Table 7.

TABLE 7--Effect on annealing on FET characteristics

	Before anneal			After anneal		
Sample	I_{dss}	V_p	g_m	I_{dss}	V_p	g_m
Control	1.00	1.00	1.00	0.98	0.99	1.02
RIE, 100V	0.74	0.82	0.77	0.90	0.89	0.96
RIE, 225V	*	*	*	0.82	0.79	0.93
RIE, 350V	*	*	*	0.84	0.76	0.99

*No working devices could be found on these samples before annealing, after annealing comparable device yields were obtained on each quadrant.

The investigation of surface damage induced by sputter etching was intended not only to assess the effects of ion bombardment by non-reactive ions (in contrast to the reactive etching conditions so far considered), but also because sputter cleaning prior to metal deposition is an important aspect of many self aligned processes now being developed.

For the sputter etch experiment a deeper implanted layer was used and so FET measurements are not directly comparable between experiments. The ohmic contacts were deposited and sintered before sputter etching thus no anneal took place after etching. The results are given in Table 8.

TABLE 8--Device characteristics: sputter etched surfaces

Sample	Ideality factor	Barrier height	I_{dss}	g_m
Control	1.21	0.65 eV	1.00	1.00
200V/30sec.	1.22	0.67 eV	0.78	1.00
200V/1min.	1.48	0.54 eV	0.72	1.10
200V/1.5min.	1.41	0.56 eV	0.72	1.00

ELLIPSOMETRIC MEASUREMENT OF DAMAGED SURFACES

It has been reported that measurement of the complex refractive index yields data that can be correlated with damage created near the surface (4). Such a technique would provide a useful means of determining the degree of surface damage after plasma processing, and measurements have been made on GaAs wafers after various plasma treatments of the type already described. The equipment used was a Gaertner ellipsometer using monochromatic (laser) illumination of wavelength 632.8 nm.

It was found that measurements on wafers that had been exposed to the atmosphere gave very variable results and it was usually necessary to remove the natural surface oxide by a short aqueous ammonia rinse immediately before measurement. This process is known to remove the native oxide layer of about 4-5nm thickness without removing GaAs.

The depth to which the ellipsometer characterises the surface is not clearly established, especially where damage is present. The measured refractive index is a mean value for this surface layer and it is possible that a very thin, highly damaged layer would give a similar ellipsometric measurement to a thicker, less damaged layer. The degree of surface roughness that exists may also influence the values obtained.

If the complex index of refraction is expressed in the form:

$$n = n_s - ik_s.$$

it is found that the extinction coefficient, k_s, changes systematically with plasma processes, and this coefficient has been used to compare the effects of various plasma processes. Details of these measurements are shown in Table 9.

TABLE 9--Ellipsometric measurements on GaAs.

Proces	n_s	k_s	Comment
Clean wafer	3.84	0.28	
PECVD + HF etch	3.84	0.29	
" + plasma etch	3.84	0.29	Typical processing conditions with up to 20% over-etch of nitride in CF_4
" + RIE, 50V	3.84	0.29	
" + RIE, 100V	3.83	0.31	
" + RIE, 225V	3.83	0.31	
" + RIE, 350V	3.84	0.31	
RIE, 225V	3.83	0.32	Wafer exposed to plasma for 5 mins
RIE, 350V	3.83	0.35	Wafer exposed to plasma for 30 mins
RIE, 60V	3.40	1.08	Wafer etched in CCl_2F_2 for 2 mins
RIE, 240V	2.94	1.42	Wafer etched in CCl_2F_2 for 2 mins

It can be seen that the RIE processes induce small change in k_s, when typical process conditions give brief periods of over-etch of the silicon nitride layer. Longer periods of etching in CF_4 at high bias voltages do not cause large increases in k_s. If the GaAs surface is deliberately etched by introducing a chlorine component into the discharge, large changes in both n_s and k_s rapidly occur.

DISCUSSION OF RESULTS

The first experiment using bulk doped wafers and fabricating only diode structures revealed no significant differences between the control sample and the samples that had undergone PECVD and plasma etching. In all cases diode idealities were close to 1.00, and barrier heights exceeding 0.8 eV showed that good Schottky barriers had been achieved. Major changes in the surface characteristics as a result of processing would have degraded these parameters. This indicated that the plasma processing had not caused severe damage to the semiconductor surface on which the Schottky contact was deposited.

The use of thin implanted layers allowing the fabrication of FET devices showed that surface changes as a result of processing could impair the device performance without severe degradation of the diode properties. The data shown in Tables 4 and 5 show that the diode properties of the sample etched by RIE at 160V bias are as good as the control sample, whereas the plasma etched and 250V samples have worse idealities. The 250V bias RIE sample shows reduced FET characteristics consistent with thinning of the active layer, and it is likely that this sample (which was etched for longer than the 160V RIE sample), has been over-etched. The plasma etched sample shows improved FET characteristics, but a poor diode ideality. The ellipsometric measurements of this and similarly etched surfaces indicate degradation of the surface, and this effect may be responsible for the impaired ideality factor. As all these samples were annealed at 430°C after plasma processing it is possible that some damage effects have been removed.

The final experiment to assess PECVD and RIE of silicon nitride avoided the annealing stage after plasma processing, and also attempted to control the degree of over etch during RIE. After processing it is clear that diode measurements do not correlate well with device characteristics. The diode performance of the 225V and 350V RIE samples is close to that of the control sample, whereas the lowest bias RIE sample is degraded in both ideality and barrier height. The FET measurements show that of the RIE samples only in the low bias case can FET action be detected. A possible explanation for this change could be that much of the active layer (less than 150nm. thick) has been sputtered away by the RIE process, but since the over-etch time was carefully restricted to 20% for all the RIE samples this would imply a faster etch rate for GaAs than for silicon nitride in the CF_4 plasma. GaAs is known not to etch chemically in a fluorinated plasma, in the absence of chlorine (5,6) and its etch rate in CF_4 is considerably less than that of silicon nitride. Measurements on these and other GaAs samples lead to an estimate of 10nm. for the maximum likely etch depth under such conditions of RIE. Ellipsometric data does not show any differences between the RIE samples, although it indicates that they have been roughened compared to the control. Surface analysis by SIMS and Auger have also failed to show differences between the samples and no residues or penetration of the GaAs surface could be detected.

After annealing a considerable improvement was observed in the electrical characteristics. Diode idealities improved and good performance was obtained from all the FETs. It was also noticed that the RIE samples became less light sensitive. Reductions in I_{dss} and V_p imply that some reduction in active layer thickness or conductivity has occurred, but this may be due to crystal damage which would be removed by further annealing. Other investigations have also shown that ion bombardment induced damage may be removed by annealing at temperatures between 300°C and 400°C.(1,2).

Since no evidence has been obtained for the introduction of interstitial contaminants by the RIE process, it is proposed that crystal damage has occured under conditions of ion bombardment, this is analogous to the damage created by the much higher energies used in ion implantation. Such damage, caused by recoil effects, may disrupt the GaAs lattice to a sufficient depth to isolate the bulk of the surface layer from the surface contacts.

It has been reported (2) that argon ion bombardment during sputter etching or ion milling can cause changes in surface layers by depleting the lattice of As leading to donor like traps and a reduction in barrier height. In this investigation, no depletion of As could be detected in any of the samples, and the reduction in I_{dss} of the FETs appears to rule out the formation of such an additional n type layer.

The brief investigation of sputter etching damage has also shown the value of combining device measurements with diode evaluation. At bias voltages comparable to those used in RIE, diode properties are degraded when etching times exceed 30 seconds. Even after 30 seconds of sputter etching FET characteristics are significantly degraded.

CONCLUSIONS

Preliminary results of an investigation into plasma process induced device damage have shown the value of combining FET measurements with those on Schottky diodes. The use of surface analytical techniques such as Auger and SIMS, combined with the use of ellipsometry to study surface degradation can provide valuable assessments of the state of etched surfaces but do not give enough information to explain changes in device characteristics. A similar conclusion has been reported by Pang et al (7) from a study of the effects of ion beam etching.

Changes in the implanted layer as a result of momentum transfer from impinging ions may be studied by techniques such as RBS or C/V profiling (8). It is hoped to continue this investigation along such lines in due course.

For processes such as PECVD and plasma etching where the wafers are placed on a grounded electrode, it is concluded that minimal damage occurs under the conditions reported here. Takahashi et al have also shown that prolonged exposure of n type GaAs to a CF_4 plasma in the etch mode does not impair diode performance (9). It is known that the use of lower R.F. frequencies can drastically affect the plasma conditions, and in particular frequencies below about 1 MHz may allow ions to cross the cathode dark space between plasma and grounded electrode within one half cycle (10). Under such conditions ion bombardment of wafers may occur and give rise to additional damage effects. Damage to implanted surfaces by low frequency PECVD is known to occur (11).

The use of ellipsometry to determine the extinction coefficient, k_s, of plasma processed films provides a rapid means of assessing changes after processing. Care must be taken to avoid spurious changes due to the presence of oxide layers on the wafer surface.

Measurements of k_s would be more valuable if the thickness of the layer measured by ellipsometry was known. It might be possible to monitor damage profiles by sequential GaAs etching and ellipsometry as has been reported for ion implant profiles by Kim and Park.

ACKNOWLEDGMENTS.

The authors would like to thank their colleagues at several research establishments within the United Kingdom, as well as those at Caswell, for valuable discussions of the effects of plasma damage on GaAs devices. Thanks are also due to Dr. J. Mun of STL who provided some of the wafers used in this investigation.

REFERENCES

[1] Wu, C.S., Scott, D.M., Wei-Xi Chen, Lau, S.S., Journal of the Electrochemical Society, Vol. 132, No. 4, April 1985, pp.918-922.

[2] Yamasaki, K., Asai, K., Shimada, K., and Makimura, T., Journal of the Electrochemical Society, Vol. 129, No. 12, December 1982, pp.2760-2764.

[3] Williams, Ralph, Gallium Arsenide Processing Techniques, Artech House, 1984.

[4] Kim, Q. and Park, Y.S., Journal of Applied Physics, Vol. 51(4), April 1980, pp.2024-2029.

[5] Smolkinsky, G., Chang, R.P., Mayer, T.M., J. Vac. Sci. Technol., 18(1), Jan/Feb. 1981, pp.12-16.

[6] Donnelly, V.M., Flamm, D.L., Ibbotson, D.E., J. Vac. Sci. Technol. A1(2), Apr.-June 1983, pp.626-628.

[7] Pang, S.W., Geis, M.W., Efremow, N.N., and Lincoln, G.A., Journal of Vacuum Science and Technology, Vol. B3(1), January/February 1985, pp.398-401.

[8] Pang, S.W., Lincoln, G.A., McClelland, R.W., DeGraff, P.D., Geis, M.W., and Piacentini, W.J., Journal of Vacuum Science and Technology, Vol. B1(4), October/December 1983, pp.1334-1337.

[9] Takahashi, S., Murai, F., and Kodera, H., IEEE Transactions on Electron Devices, Vol. ED-25, No. 10, October 1978, pp.1213-1218.

[10] Chapman, Brian, Glow Discharge Processes, John Wiley and Sons, 1980.

[11] J. Mun, private communication.

Marinus Kunst, Andre Werner, Gerhard Beck, Udo Kuppers, and Helmut Tributsch

QUALITY CONTROL AND OPTIMIZATION DURING PLASMA DEPOSITION OF a-Si:H

REFERENCE: Kunst, M., Werner, A., Beck, G., Kuppers, U., Lilie, J., and Tributsch, H. "Quality Control and Optimization during Plasma Deposition of a-Si:H" Emerging Semiconductor Technology, ASTM STP 960, D.C.Gupta and P.H.Langer, Eds., American Society for Testing and Materials, 1986.

ABSTRACT: It is shown that contactless time-resolved microwave conductivity (TRMC) measurements permit an in-process quality control during the plasma deposition of amorphous hydrogenated silicon (a-Si:H) films. The relation between the transient photoconductivity of the amorphous semiconductor and its quality is discussed. The influence of various parameters, such as the substrate temperature and chemical composition on the shape of the signals is explained in terms of charge carrier kinetics. An automated production process to yield quality films is presented to demonstrate the use of these measurements.

KEYWORDS: amorphous silicon, transient photoconductivity, quality control, automated production process, plasma deposition

INTRODUCTION

The properties of hydrogenated amorphous silicon (a-Si:H) depend critically on the production conditions. For applications of this material in photosensors, field effect transistors and solar cells [1] it is important to provide a low defect density, i.e., a low density of state in the mobility gap. Normally the quality of the

Drs. Kunst, Werner, Beck, Kuppers, and Lilie are research scientists, Professor Tributsch is member of the chairmen board of the Bereich Strahlenchemie, Hahn-Meitner-Institut fur Kernforschung Berlin, 1000 Berlin 39, Federal Republic of Germany.

material has been determined by infrared absorption [2], spin density [3] and photoconductivity [4] measurements after the production process.

For an efficient quality control and optimization, however, it is necessary to monitor in-situ, i.e., during the production process, a physical property which reflects the density of states in the mobility gap of the film. For practical reasons in-situ measurements should preferentially be contactless and optical reflectivity [5], photoinduced absorption [6] and time-resolved microwave conductivity [7] (TRMC) measurements can be applied. If the relation between the in-situ signal obtained and the film quality is known, the signal can be used for a computer-aided process optimization.

In this work the TRMC technique is applied to detect the in-situ photoconductivity of intrinsic a-Si:H. This method is based on the proportionality between the excess conductivity and the relative change of reflected microwave power on pulsed illumination. This method allows contactless photoconductivity measurements.

It has been preferred to perform transient measurements as they contain more information than steady state ones. The transient photoconductivity is not simply related to the density of states in the mobility gap and it does certainly not reflect the complete density of states. Various electronic parameters, however, which are essential for the performance of an electronic device have a clear influence on the transient photoconductivity.

In this paper, after a description of the equipment, the relation between the transient photoconductivity and the film quality is discussed. Then a preliminary experiment is presented where during the a-Si:H production one preparation parameter, i.e., the substrate temperature, is varied via a microcomputer in such a direction that after some cycles a film with a relatively optimal photoconductivity will be produced. Finally the possibilities and limitations of an automated production process controlled by optimization of the transient photoconductivity are discussed.

EXPERIMENTAL

The a-Si:H films are produced by the glow-discharge in an equipment as described elsewhere [8]. TRMC measurements are performed in Ka band set-up

(26.5-40 GHz) connected to the glow-discharge system. Excess charge carriers were induced by a 15 ns (FWHM) pulse of a Nd:YAG laser at 532 nm. To avoid light interference effects the glass substrate was roughened.

Fig. 1 shows the plasma deposition chamber, where the microwave set-up is connected via a window to the backside of the substrate. A circulator separates the incident microwave from the reflected ones that are detected with a point contact didode. The signals were transferred to a digitizer and a microcomputer for storage and handling.

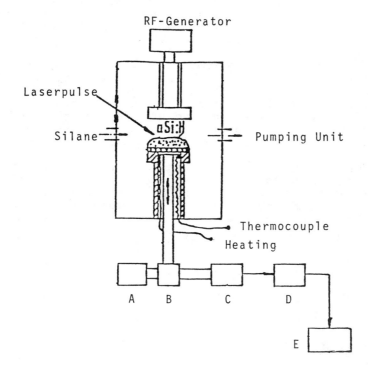

FIG. 1 Plasma deposition chamber allowing in-situ measurements of the transient photoconductivity. A: Gunn diode, B: Circulator, C: Pointcontact detector, D: Digitizer, E: Computer.

TRANSIENT PHOTOCONDUCTIVITY MEASUREMENTS

For small purturbations the relative change of the relected microwave power ($\Delta P(t)/P$) is proportional to the induced conductivity $\Delta\sigma(t)$ on illumination:

$$\frac{\Delta P(t)}{P} = A\Delta\sigma(t) \qquad (1)$$

where A is the sensitivity constant. In intrinsic a-Si:H it has been shown that $\Delta\sigma(t)$ reflects the generation and decay of excess electrons:

$$\Delta\sigma(t) = \Delta n(t)\mu e \qquad (2)$$

where n(t) is the density and μ the mobility of the excess electrons. Recombination and trapping determine the decay. If some of these processes are faster than the width of the exciting laser width then the maximum signal height also gets reduced. The charge carrier transport in a-Si:H devices is determined by these processes which forms the basis for relationship between transient photoconductivity measurements of a-Si:H and the performance of this material in devices. Most charge carrier decay processes involve lattice defects which provide a correlation between the density of states in the mobility gap and the transient photoconductivity.

To derive the relation between the quality and the transient photoconductivity the TRMC signals of films with known different qualities have been investigated. One film is produced at optimal substrate temperature (about 250 C) and another at 100 C. Fig. 2 compares the

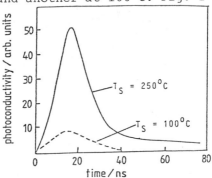

Fig. 2 Comparison of transients in films prepared at different substrate tempertaures Ts. The signals were induced by a 532 nm laser pulse of 1 mJ/cm^2

transients in these films. The film produced at 100 C has very bad properties due to a large number of states in the gap. To avoid temperature corrections both films are measured at room tempertaure (Fig. 2). The TRMC signal induced by 532 nm light in the low temperature film is small and has very high decay rate. The signal in the high temperature film is much larger and shows a slowly decaying tail. The number of defect states in the 100 C film is obviously so high that the decay during the duration of the excitation is already very strong and leads to a reduced maximum signal height. This example shows the relation between the transient photoconductivity and the quality of the material.

The addition of impurities also changes the quality of the produced material. For example a mixture of CH_4 and SiH_4 results in material with a larger band gap than pure a-Si:H. The addition of a small amount of H_2S to the SiH_4 gas during the glow-discharge process leads to a large increase of the decay rate that reduces the amplitude of the signal (Fig. 3). It must be concluded that the a-Si:H:S compound has a larger number of defects which can be related to a higher structural disorder in this material.

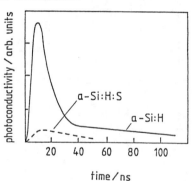

Fig. 3   Influence of a chemical modification (due to H_2S addition during the glow-discharge of SiH_4) on the in-situ transient photoconductivity at 250 C. The signals were induced by a 532 nm laser pulse of 0.3 mJ/cm^2 .

These experiments show epirically the connection between the transient photoconductivity and the quality of a-Si:H films. It can be concluded that for a rough determination of the quality of the material the signal amplitude is a good standard. For more refined quality control and optimization procedures the decay time of the transient photoconductivity must be taken into account but for a preliminary experiment as given here it is sufficient to work with large quality differences giving rise to changes in the signal height.

The intrinsic dependence of the transient photoconductivity mainly due to the activation of the mobility can be accounted for by calibration.

COMPUTER-AIDED PROCESS OPTIMIZATION OF a-Si:H

In-situ measurements of the transient photoconductivity can be used to optimize the process for the production of a-Si:H films during the glow-discharge process. Fig. 4 shows the schematic set-up which is composed in three parts: the process chamber, a

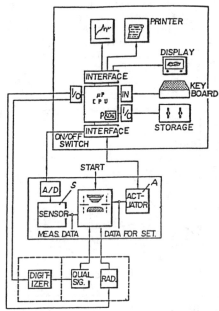

Fig. 4   Diagram of a computer-aided process optimization

quality control unit and a microcomputer. The maximum signal height of the transient photoconductivity signal can be used as sensor (S) of the quality and transferred to a microcomputer which varies the actors (A) with an adequate optimization program seeking after a maximum sensor signal. In principle actors can be all production parameters and a more dimensional optimization must be performed. In the experiment presented only the substrate temperature is changed and all other parameters are fixed. this has the advantage that the result of the optimization can be compared with the optimum substrate temperature between 250 C and 300 C obtained from "trial and error experiments. A correction function adjusting all sensor signals to the same temperature can be used in a more refined experiment. This has been neglected, which is justified in a first approximation.

The used program [9] is so constructed that within a given temperature range, here from 150 C - 300 C, at four different temperatures the transient photoconductivity is determined at the end of a 15 min. glow-discharge. The 15 min duration of the glow-discharge is chosen because it limits the absorption of the exciting 532 nm light pulse mainly to the newly deposited a-Si:H layer (about 0.25 um in 15 min). Selecting the highest sensor signals the program chooses a new smaller temperature range. This range consists of the temperatures at which the highest sensor signals have been measured. The sensor signals obtained at four different temperatures in the new tempertaure range are used in the same way to confine the optimum temperature range. This procedure is stopped if the difference between the four values of the sensor signal within a certain temperature range is smaller than a chosen fixed value. Fig. 5 shows this procedure schematically for our experiment.

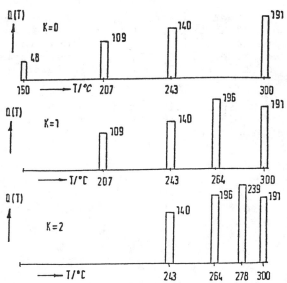

Fig.5 Schematic representation of the sensor signal Q(T) at different temperatures in 3 optimization cycles

In the first interval (150 C - 300 C) the highest values of the sensor signal are situated between 243 C and 300 C. The second interval from 207 C - 300 C confines the optimum temperature around 269 C. The third temperature range yields an optimum temperature interval around 278 C. Without paying too much attention to the exact numerical result of this experiment because of its preliminary nature, it is obvious that the optimization procedure presented here leads to an optimum deposition temperature range that agrees with the range obtained from "trial and error" experiments.

DISCUSSION

The experiments presented and further results to be published [10] show that in principle a fully computerized production of optimal a-Si:H films is possible. With a more refined optimization program and computer control of all actors the quality can be optimized with respect to temperature, deposition rate, gas pressure etc. The experiment described refers to intrinsic a-Si:H but it can be shown [10] that a direct correlation between transient photoconductivity and quality also exists in extrinsic a-Si:H. Therefore, an optimized production can be effectuated in case of extrinsic a-Si:H as well.

The relation between transient photoconductivity and quality is general and not limited to a-Si:H. It can also be applied to the production of other photosensitive materials [11]. Another important application of this procedure is the optimization and quality control of devices, i.e., pn junctions [11], if the dark conductivity of the device is not too large. It is possible to obtain photoinduced microwave absorption characteristics of devices [10], reflecting their optoelectronic properties.

ACKNOWLEDGEMENT

Financial support of the Stiftung Volkswagenwerk is gratefully acknowledged.

REFERENCES

[1] Knights, S.C., Topics in Applied Physics, Vol. 55, Springer Verlag Berlin, 1984.

[2] Amer, N.M., Skumanich, A., *Physica*, Vol. 117/118, 1983, p. 897.

[3] Street, R.A., Biegelson, D.K., Weisfield, R.L., *Phys. Rev. B*, Vol. 30, 1984, p. 5861.

[4] Staebler, D.L., Wronski, C.R., *Appl. Phys. Lett.*, Vol. 31, 1977, p.292.

[5] Pries, W., McLeod, R.D., Card, H.C., Kao, K.C., *Appl. Phys. Lett.*, Vol. 45, 1984, p.734.

[6] Tauc, J, *Semiconductors and Semimetals*, Vol. 21, Academic Press, London, 1984, p.299.

[7] Werner, A., Kunst, M., Beck, G., Lilie, J., Tributsch, H., *Solid State Commun.*, Vol. 56, 1985, p. 127.

[8] Werner, A., Kunst, M., *Appl.Phys. Lett.*, Vol. 46, 1985, p.69.

[9] Kuppers, U., accepted for publication in *IEEE Trans. on Industrial Electronics*.

[10] Publication in preparation.

[11] Tributsch, H., Beck, G., Kunst, M., Kuppers, U., Lewerenz, H.J., Lilie, J., Werner, A., *U.S. Patent Application*, Ser. no. 694932, 1985.

Stephen C. Lee, and Barry L. Chin

EFFECTS OF DEEP UV RADIATION ON PHOTORESIST IN AL ETCH

REFERENCE: Lee, S.C., and Chin, B.L., "Effects of Deep UV Radiation on Photoresist in Al Etch," Emerging Semiconductor Technology, ASTM STP 960, D.C. Gupta and P.H. Langer, Eds., American Society for Testing and Materials, 1986.

ABSTRACT: The photoresist integrity and edge profile can be improved in Al RIE etch by exposing the patterned Al coated wafers to deep UV radiation (λ =220nm). The transmitted DUV intensity on a photoresist coated clear sapphire wafer was measured as a function of the accumulated exposure time. The transmitted DUV intensity reaches a maximum after a period which is proportional to the photoresist thickness. The surface of the DUV exposed photoresist forms a highly polymerized membrane during hard bake and is stressed by the vapor pressure from inside the photoresist. The property of this membrane was studied and reasons for the improved etch integrity are proposed.

KEYWORDS: reactive ion etch, deep UV radiation, positive photoresist, aluminum etch, metal dry etch

This work was performed at Hughes Aircraft Co., Carlsbad, CA 92008 where Drs. Lee and Chin were Members of the Technical Staff. They are presently Staff Engineers at Siliconix Inc., 2201 Laurelwood Road, Santa Clara, CA 95054.

Metal interconnections for VLSI integrated circuits are often delineated by reactive ion etching (RIE) because of its anisotropic property which allows the minimum spacing between metal lines to be reduced. Photoresist erosion is a major concern during RIE of Al in which chlorinated gases are used in the etch process. This is especially true since metal etch is the last critical step in fabricating IC's wherein the surface topology is nonplanar and the photoresist coverage is thinner on metal lines over steps. The process of forming a highly polymerized membrane on top of the photoresist during deep UV (DUV) radiation was studied by transmission DUV measurements. The strength of this membrane was examined at different hard bake temperatures. Reasons for the improvement in photoresist integrity with DUV hardening in Al RIE etch are given.

EXPERIMENTAL METHOD AND RESULTS

The DUV system used in this study was manufactured by Hybrid Technology Group, Inc. and is equipped with a 1000 Watts Mercury-Xenon lamp. The peak DUV spectrum exposed at the wafer surface is in the 200-250 nm range. The metal RIE etch was performed in an Applied Materials AME 8130 system which operates at 13.56 MHz. The rf power is coupled to a floating hexode which holds 24-3" wafers. The etcher was operated with a constant DC self bias of -250 V at the floating hexode resulting in an input power level of 600 watts. The gas composition, flow rate and pressure used in this study were BCl_3 + 10% Cl_2, 80 sccm and 20 mTorr, respectively. The hexode was kept warm (about $40^\circ C$) during etching. The Al-1% Si etch rate was approximately 0.1 um/min.

After exposure to DUV radiation, a thin highly polymerized membrane will form on the top surface of the photoresist. Some of the chemical properties associated with this thin layer were examined and reported in Refs. [1] and [2]. Experiments were performed to examine the process in which this highly polymerized membrane forms. Different thicknesses of Shipley 1400 photoresist were coated onto double side polished transparent sapphire wafers. A photo-detector with the peak response at 220 nm wavelength was placed beneath the photoresist coated sapphire wafer to measure the transmitted DUV intensity. Fig. 1 shows the transmitted DUV intensity for three different photoresist thicknesses of 1.2, 1.5, and 1.8 um as a function of the exposure time. The DUV transmission through a sapphire wafer is about 84% whereas at least three orders of magnitude decrease in transmission was observed for the photoresist coated sapphire wafers. This suggests a strong DUV absorption by the photoresist. The photoresist initially transmits increasing DUV radiation and becomes absorbing to DUV after reaching a maximum level of transmission. The increasing DUV transmission at the begining of exposure may be attributed to a surface-related reaction and the maximum transmission may correspond to the beginning of the membrane formation. The exposure time at which the maximum transmission occurs increases as the photoresist thickness increases. The transmitted intensity is attenuated as the photoresist thickness increases as shown in Fig. 1.

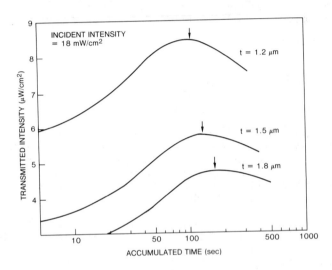

FIG. 1--Transmitted DUV intensity versus accumulated exposure time for photoresist of 1.2, 1.5, and 1.8 um.

The surface morphology of the DUV hardened photoresist was examined after hard bakes for 30 min. at different temperatures between 125 and 250 °C. If the membrane is strong or thick enough it can sustain the vapor pressure build-up inside the photoresist during this temperature bake. This also prevents the photoresist from flowing at this bake temperature. It was observed that the edge profile of the DUV hardened photoresist can be retained even after a 250 °C hard bake temperature whereas the non-DUV hardened photoresist flowed and produced a rounded edge profile. For a membrane of sufficient strength to sustain the differential pressure, the vapor pressure inside the photoresist stresses the membrane and causes it to expand. The membrane is inflated when the wafers are removed from the hard bake oven. A wrinkled surface results for large areas (any linear dimension > 5 um) when the stressed membrane collapses. The surface tension from the sidewalls of the photoresist will inhibit wrinkle formation for fine lines of 5 um or less. No wrinkling was observed when DUV was incident from the sapphire side regardless of the dose level. This would indicate that a free surface is required for membrane formation.

The strength of the membrane can be determined from the minimum DUV dose required to initiate the wrinkled surface at a given hard bake temperature. The vapor pressure inside the photoresist can be expected to rise as the hard bake temperature rises. Fig. 2 shows the minimum DUV dose for the wrinkle to form vs. the hard bake temperature for a 1.5 um thick photoresist. If the vapor pressure within the photoresist is assumed to be a linear function of the hard bake temperature then it may be concluded from Fig. 2 that the strength of the membrane to resist the built-up vapor pressure is a linear function of DUV dose. It was found that the membrane reduced photoresist thickness loss during hard bake. A 0.04 um difference in photoresist thickness after a 30 min bake at $125^{\circ}C$ was measured by surface profilometry between DUV (no wrinkles were formed at this bake temperature) and non-DUV treated photoresist with initial thickness of 1.5 um (see Table 1).

To study the DUV effect on the photoresist erosion rate during Al RIE, both He-Ne laser ($\lambda = 632.8$ nm) interferometry and the reflectance of the He-Ne beam as monitored by a photodetector and power meter were used. Interestingly, the DUV photoresist did not show a measurable reduction in erosion rate over the non-DUV exposed samples for the same hard bake temperature. This was determined by the interference signal I which is modulated by the remaining photoresist thickness over the substrate during the etch. This may be modeled by the following equation:

$$I = A \cos(2\pi X / P)$$

where the period of modulation P is

$$P = \lambda / 2n$$

and A = Maximum amplitude
X = Remaining photoresist thickness
λ = Wavelength of the He-Ne Laser (632.8 nm)
n = Refractive index of photoresist (1.6)

The instantaneous etch rate E is given by

$$E = \frac{dx}{dt} = \frac{dI}{dt} \cdot \frac{P}{2(A^2 - I^2)^{\frac{1}{2}}}$$

It was found that the photoresist erosion rate during Al RIE was related to the hard bake temperature only. Comparison of photoresist erosion rate between a 200 °C hard bake and a 125°C hard bake showed about a 50% slower surface etch rate and 10% slower bulk etch rate (see Table 1).

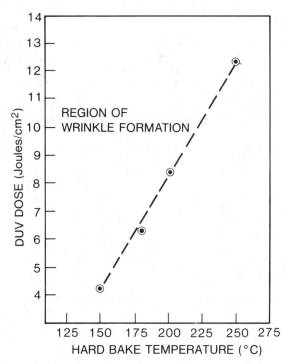

FIG. 2--Minimum DUV dose to cause wrinkle formation versus hard bake temperature.

TABLE 1--Results of experiments at 8.8 Joules/cm²

Hard Bake Temperature (°C)	DUV	Photoresist thickness change after hard bake (Å)	Etch Rate Ratio* (Bulk)	Etch Rate Ratio* (Surface)
125	No	0	1	0.8
125	Yes	+400	1	0.8
200	No	-2000	0.9	0.65
200	Yes	-1500	0.9	0.45

*Etch rate ratio is referenced to the bulk etch rate of photoresist hardbaked at 125°C.

CONCLUSIONS

Improving the etch selectivity of Al to photoresist in Al RIE can be achieved by hardening the photoresist with DUV radiation and subsequently hard baking at a higher temperature. The quantitative reasons for improvement are:

1. The highly polymerized membrane formed on the photoresist surface after DUV radiation minimizes the photoresist thickness reduction in the hard bake cycle.

2. The higher hard bake temperature results in a slower surface and bulk erosion rate. In addition, the sharp edge profile of DUV hardened photoresist patterns is preserved through the Al RIE etch cycle. This is shown in Figs. 3 and 4 which are SEM micrographs of the wafer topology before and after Al RIE etch. The sputtered Al-1% Si was 0.75 um thick with initial photoresist thickness of 1.58 um. The wafer was DUV hardened at a 8.8 Joules/cm^2 dose and hard baked at 200 $^\circ$C for 30 minutes. For a given hard bake temperature, a dose can be determined from Fig. 2 which results in wrinkle formation. The photoresist remaining after RIE was removed using an oxygen plasma strip. The edge profile was transferred to the underlying Al film and resulted in straight wall Al lines which is a requirement for VLSI technology.

REFERENCES

[1] Allen, R., Foster, M. and Yen. Y., "Deep U.V. Hardening of Positive Photoresist Patterns," Journal of the Electrochemical Society, Vol. 129, June 1982, pp. 1379-1381.

[2] Yen, Y. and Foster, M., "Deep UV and Plasma Hardening of Positive Photoresist Patterns," Kodak Microelectronics Seminar, Oct 21-22, 1982, San Diego, CA, pp. 125-130.

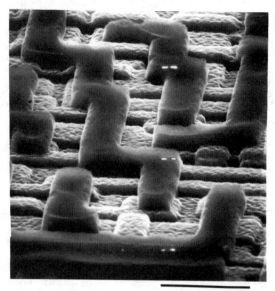

7.5 um

FIG. 3--SEM micrograph of wafer surface before Al etch.

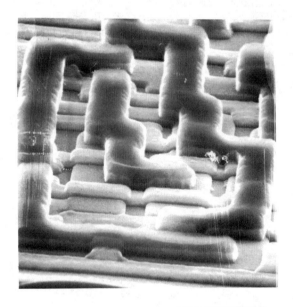

7.5 um

FIG. 4--SEM micrograph of wafer surface after Al RIE with photoresist intact.

Vladimir Starov

INFLUENCE OF X-RAY EXPOSURE CONDITIONS ON PATTERN QUALITY

REFERENCE: Starov, V, "Influence of X-Ray Exposure Conditions on Pattern Quality", Emerging Semiconductor Technology, ASTM STP 960, D.C. Gupta and P.H. Langer, Eds., American Society of Testing and Materials, 1986.

ABSTRACT: We have examined the relationship between x-ray resist sensitivity, spectral characteristics of the x-ray sources, and resist absorption coefficient with respect to our ability to control feature size in x-ray lithography. We have derived an expression relating these properties to the edge-definition quality by estimating the mean distance between the adjacent photon absorption events. It is shown that to ensure feature-size control to 0.05µm necessary to achieve 0.5µm minimum feature size, x-ray resist sensitivity should be no greater than 8 mJ/cm^2 if conventional x-ray sources are used for exposure. The effect of a controlled exposure ambient [1] on edge definition is also demonstrated and discussed.

KEYWORDS: x-ray lithography, shot noise, x-ray resist, exposure ambient, x-ray source for lithography.

INTRODUCTION

X-ray lithography is an emerging manufacturing technique for making integrated circuits with critical dimensions (CD) ≤0.5µm [2]. At this level of miniaturization optical lithographies suffer from diffraction-limited resolution, shallow depth of focus, and poor overlay. X-ray lithography circumvents the diffraction problem by employing very short wavelenghts, λ= 0.1-5nm, and avoiding the use of optical elements such as lenses or mirrors eliminates the difficulties associated with small depth of field and through-the-lens alignment. This inherent simplicity of x-ray lithography makes it particularly attractive for

---

Dr. Starov is manager, Resist and Application, of the X-ray Lithography Program at Varian Associates, Inc., 611 Hansen Way, Mail Stop S-225, Palo Alto, CA 94303.

large-scale commercial applications although some issues need to be addressed before it becomes the preferred manufacturing technique.

High throughput requirement allows only 1-3 min for the exposure of an entire wafer. This creates a need for either brighter x-ray sources or very sensitive, highly-absorbing resists. Accordingly, new, brighter x-ray sources are being developed worldwide [2-5], often primarily for lithographic applications. Significant progress has also been achieved in development of highly-sensitive x-ray resists, notably by researchers at AT&T Bell Labs [6].

Below we will consider the limitations imposed on edge definition by the exposure with short-wavelength x-ray sources. In particular, we will show that the use of highly sensitive resists may lead to an excessive edge roughness. Sensitivity control via exposure in a reactive gaseous ambient suggested by Moran and Taylor [1] will also be discussed.

X-RAY SOURCES AVAILABLE FOR LITHOGRAPHY

There are three classes of sources used in x-ray lithography: (1) electron bombardment of metal targets, both stationary and rotating, (2) hot-plasma x-ray emitters, and (3) synchrotron radiation emitted by charged particles in a storage ring. Advantages and limitations of these x-ray sources have been reviewed previously [2,7]. Some development work has been reported on transition-radiation x-ray sources for lithography but no applications have been demonstrated as yet [8]. The major characteristics of the three types of x-ray sources are listed below in Table 1.

TABLE 1--Characteristics of x-ray sources

Source type	[a]Flux,mW/cm^2	Spot size,cm	[a]Wavelength,nm
Electron bombardment	0.03-1.5	0.15-0.3	0.1-1
Hot plasma	1-10	0.002-0.1	0.6-1
Storage ring	100-1000	(b)	0.3-1

[a] The x-ray flux and wavelength range are given at the resist surface assuming attenuation by 5μm thick BN membrane most commonly used as the mask substrate.

[b] Synchrotron radiation is used as a nearly collimated beam and not as a point source.

In the case of a quasi-point source, the flux at the resist surface is determined (for a given total power emitted) by the distance between the source and the resist and the absorption characteristics of the media (vacuum, Be window, He column, mask substrate). It is therefore advantageous to keep this distance as short as possible. On the other hand, penumbral blur, p, increases as the distance to the source decreases:

$$p = \frac{sg}{d} \; (\mu m) \tag{1}$$

where
- s = the source size in cm,
- g = the mask-to-resist gap in μm, and
- d = the source-to-mask distance in cm.

Thus, for a typical gap in the range $g=20-40\mu m$, and a typical spot size 0.2-0.3cm, d has to be 20-40cm to ensure $p=0.15-0.3\mu m$ for a conventional, electron-bombardment source. This limits the available flux at the resist to the values given in the table above.

Spectral distribution of the power reaching the resist surface is also very important. Although for conventional sources, only characteristic emission wavelength is most often given (e.g. Al_K, Pd_L, W_M), one must remember that up to 60% of the emitted power may be contained in the continuum portion of the spectrum depending primarily on the target material and on the kinetic energy of the bombarding electrons. In addition, bremsstrahlung-rich spectrum reduces the effective contrast of the mask since any absorber material is, in general, more transparent to the higher energy photons. Finally, differences in the incident spectra affect the effective absorption coefficient (R in Eq. 13 divided by the resist thickness x) of the resist and the actual number of x-ray photons absorbed as shown in the next section.

CALCULATION OF PHOTON ABSORPTION EVENTS IN THE RESIST

Let us first consider the case of monochromatic x-ray radiation characterized by a flux at the resist surface I_o (W/cm^2). The flux absorbed in the resist I_{abs} is

$$I_{abs} = I_o - I_t = I_o(1-e^{-ax}) \tag{2}$$

where
- I_t = flux transmitted through the resist,
- a = absorption coefficient of the resist in cm^{-1} at a wavelength λ, and
- x = resist thickness in cm.

The average absorbed power density in W/cm^3, $\bar{I}_{abs}$, is simply

$$\bar{I}_{abs} = I_{abs}/x . \tag{3}$$

Since for polymeric resists $a \sim 1000$ cm^{-1} and $x \sim 10^{-4}$ cm, we have $a \cdot x \sim 0.1$ and $\exp(-a \cdot x) \simeq 1 - a \cdot x$. Hence, Eq. 2 can be written

$$I_{abs} \simeq I_o a x \tag{4}$$

and Eq. 3 becomes

$$\bar{I}_{abs} = I_o a \quad . \tag{5}$$

The corresponding dose absorbed in the resist, $\bar{D}_{abs}$, in J/cm^3 is given by

$$\bar{D}_{abs} = I_o \cdot a \cdot t \tag{6}$$

where

t = exposure time in sec.

For radiation of wavelength λ(nm) there are P_J photons per joule, with P_J given by

$$P_J = \frac{\lambda}{hc} = 5.03 \cdot 10^{15} \cdot \lambda . \tag{7}$$

Thus for the number of photons absorbed per cm^3 of the resist, N_{abs}, we have

$$N_{abs}(\lambda) = \bar{D}_{abs} P_J = 5.03 \cdot 10^{15} \cdot I_o \cdot \lambda \cdot a \cdot t \tag{8}$$

or

$$N_{abs}(\lambda) = 5.03 \cdot 10^{15} \cdot \lambda \cdot a \cdot S = 5.03 \cdot 10^{15} \cdot \lambda \cdot \bar{D}_{abs} \tag{9}$$

where

$$S = I_o t \tag{10}$$

is the resist sensitivity in J/cm^2 and $\bar{D}_{abs}$ is the sensitivity in J/cm^3. Note that $\bar{D}_{abs}$ is a fundamental definition of sensitivity while S is a more accepted quantity usually quoted as resist sensitivity because only incident dose is known for most systems.

As an example, let us consider a well characterized case of DCOPA resist exposed to Pd_L x-rays [1]. In this case, considering only characteristic line emission, λ = 0.437nm and at this wavelength a = 1203cm^{-1} [9]. The reported sensitivity for this resist [1] ranges between 10 mJ/cm^2 for pure N_2 exposure ambient to 22 mJ/cm^2 for the mixture of N_2 with 0.3% oxygen. Taking S=10 mJ/cm^2 and using Eq. 9 we obtain N_{abs} (0.437nm)=2.65·10^{16} photons/cm^3. From Eq. 6 we calculate $\bar{D}_{abs}$ = 12J/cm^3 in this case.

We now compare the result for monochromatic radiation to the case of the full x-ray spectrum. Several such spectra, including that of Pd, have been analyzed by M. Sogard at Varian [10]. He has also developed a computer model which keeps track of the number of photons of different wavelengths transmitted through the various media on the way to the resist and those absorbed in the resist. In this case, a numerical summation for each wavelength present in the spectrum is employed. It is similar to Eq. 8 with I_0 replaced by the incident flux in a small portion of the spectrum, I_0 $(\lambda, \lambda + \Delta \lambda)$. Then the total number of the photons absorbed, N_{abs}, is

$$N_{abs} = \sum_\lambda N_{abs}(\lambda) \qquad (11)$$

and it can be calculated for any resist sensitivity S given the following condition

$$\overline{D}_{abs} = \sum_\lambda N_{abs}(\lambda) \frac{hc}{\lambda} \qquad (12)$$

where $\overline{D}_{abs}$ (J/cm^3) is proportional to the incident dose S (J/cm^2). Since the model also calculates the ratio

$$R = \frac{D_{abs} (J/cm^3)}{S (J/cm^2)} \qquad (13)$$

we can calculate $\overline{D}_{abs}$ for any given S:

$$\overline{D}_{abs} = \frac{SR}{x}. \qquad (14)$$

Then we can scale the coefficients $N_{abs}(\lambda)$ by performing the summation in Eq. 12; using Eq. 11 we compute the desired N_{abs}.

We have estimated N_{abs} for the full spectrum of a Pd target bombarded with the 20 keV electrons using the computer model of Sogard as outlined above. The result, $N_{abs} = 2.42 \cdot 10^{16}$ photons/cm^3, is in good agreement with the corresponding number for the monochromatic case, N_{abs} (0.437nm) = 2.65×10^{16}, calculated using Eq. 9 for the characteristic Pd$_L$ radiation only. The difference between the two cases is in the right direction: for the same energy density absorbed $\overline{D}_{abs}$ = 12 J/cm^3, fewer photons are needed to deliver the dose when higher energy bremsstrahlung photons are accounted for. The small difference between the two numbers, ca.10%, indicates, however, that only a few of the continuum photons contribute to the exposure due to the high transparency of the resist at the shorter wavelengths.

Now we can estimate the mean distance between the adjacent photon absorption events, ℓ, given by

$$\ell = N^{-1/3} \quad (cm) \qquad (15)$$

or, using Eq. 9,

$$\ell = \frac{5.84 \times 10^{-2}}{(\lambda \cdot a \cdot s)^{1/3}} \quad (\mu m) \qquad (16)$$

where λ is in nm, a in cm^{-1}, and S in J/cm^2. For the case of DCOPA exposed to Pd radiation considered above, Eq. 15 gives ℓ = 0.034 µm for S = 10mJ/cm^2. Considering that each photon absorbed in the resist delivers up to 2.8 keV of energy, many chemical events can be initiated in the close proximity to the absorbing atom. The reactive range will be determined by the larger of (1) the ejected electron energy-loss distance or (2) the chemical reaction propagation distance. A typical photoelectron scattering range is 0.1μm for the x-ray energies and materials in question and the reaction can extend over a typical polymer molecule length (e.g., catalytic unzipping of a polymer) or even larger distances for the polymerization or grafting-type negative resists. It is difficult to give a universal estimate of these distances since many factors contribute to each individual case. However, even for very small photoelectron scattering range and reaction propagation distance, the exposed feature edge definition becomes poor when the mean distance between photons absorbed increases to some critical value ℓ_{crit} determined by the resolution desired. For example, if the resolution element is 0.5μm, we require that ℓ_{crit} be $\leq$0.05μm, i.e., below one tenth of the critical dimensions. In this case a 0.5 x 0.5μm^2 dot in 0.5μm - thick resist will be exposed by at least 1000 photons. Since in this case the feature edge is exposed by 100 photons the probability of finding a significant number of unexposed one-photon pixels is small [7].

Using Eq. 16 we have calculated the characteristic values of ℓ for the wavelengths of x-ray sources λ = 0.4, 0.8, and 1.2nm, the resist absorption coefficients a = 500, 1000, and 1500cm^{-1}, and sensitivities S = 1, 8, 125mJ/cm^2. Results are summarized in Table 2.

TABLE 2--Average distance between photons absorbed ℓ (μm)

a, cm^{-1}	S, mJ/cm^2	$\bar{D}_{abs}$, J/cm^3	λ=0.4nm	λ=0.8nm	λ=1.2nm
500	1	0.5	ℓ = 0.10	0.08	0.07
	8	4	0.05	0.04	0.03
	125	62.5	0.02	0.02	0.01
1000	1	1	ℓ = 0.08	0.06	0.06
	8	8	0.04	0.03	0.03
	125	125	0.02	0.01	0.01
1500	1	1	ℓ = 0.07	0.06	0.05
	8	12	0.03	0.03	0.02
	125	187.5	0.01	0.01	0.01

From the table we see that 1 mJ/cm^2 resist approaches the desired values of ℓ only for $\lambda \geq$ 1.2nm and a $\geq$ 1500 cm^{-1}. This means that a resist with sensitivity of a few mJ/cm^2 is likely to have poor edge definition when irradiated by sources in the 0.4-1nm wavelength range most commonly used for lithography. On the other hand, the 8mJ/cm^2 resist satisfies the criterion $\ell \leq$ 0.05μm for most conditions, being marginal only for $\lambda = 0.4$nm in the case of weekly absorbing material, $a \leq 500$ cm^{-1}. Any other combinations of the source wavelength, λ, resist sensitivity, S, and absorption coefficient, a, can be evaluated using Eq. 16. Our conclusion is that 0.5µm resolution and 0.5µm minimum feature size requirement implies that resist sensitivity should be no greater than 8mJ/cm^2, i.e. $S \geq 8$mJ/cm^2, if conventional sources are to be used. For many resists, it translates in to the requirement $\overline{D}_{abs} \geq 8$J/cm^3 for the more fundamental volume sensitivity. X-ray sources with a larger portion of their spectrum at longer wavelengths, such as hot-plasma or synchrotron radiation, may be used with more sensitive resists.

ROLE OF EXPOSURE AMBIENT

Exposure ambient may have a very pronounced effect on x-ray resist sensitivity. In particular, negative-working cross-linking resists are affected most as reported by Moran and Taylor [1]. They showed that by adding oxygen to the N_2 exposure ambient, resolution of DCOPA resist used in their work improved substantially. At the same time, the sensitivity decreased from 10mJ/cm^2 to 17-22 mJ/cm^2.

In working with very sensitive proprietary resists, $S=1-3$mJ/cm^2, we have observed excessive edge roughness as shown in Figure 1a. From the microphotograph

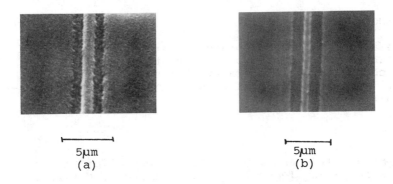

5µm
(a)

5µm
(b)

FIG.1--Proprietary negative x-ray resist exposed to 1.5mJ/cm^2, part (a), and 12mJ/cm^2, part (b).

we estimate the mean roughness of 0.05-0.1µm. As we have shown above, the observed poor edge definition may stem from the large distance between the photons absorbed close to the feature edge. Taking $S = 1.5$ mJ/cm^2, $a = 1000$ cm^{-1}, and $\lambda = 0.437$nm (Pd source was used) we estimated $\ell = 0.07$ µm in close agreement with observed structure at the line edge. In figure 1b we show the same resist exposed in the presence of a small amount of O_2 so that the chemical reactions caused by x-rays are scavenged to some extent. In this case the resist sensitivity dropped to 12mJ/cm^2 and substantial improvement in edge definition is achieved. For this value of S we calculate ℓ = 0.03μm while the observed improvement is even greater. This is because oxygen intercepts the propagating chemical chains resulting in tiny filaments spreading into the unexposed region, Figure 1a. Thus, in addition to the larger number of photons exposing the resist when it is used in the lower sensitivity mode, we have the additional effect of O_2 eliminating the chemical amplification effect responsible for the excessive edge roughness (and the high sensitivity) of the resist, compare Figures 1a and 1b.

SUMMARY AND CONCLUSIONS

We have examined the role played by the x-ray source wavelength, resist sensitivity, and absorption coefficient in obtaining high-resolution, high-quality patterns. We have derived a useful criterion for estimating a potential for good edge definition of an x-ray resist. It is shown that for typical exposure conditions with conventional x-ray sources (energies ~2keV), the resist sensitivity should not exceed 8mJ/cm^2 for good definition of 0.5μm features. Using the controlled ambient technique suggested by Moran and Taylor [1], we have also demonstrated a substantial improvement in edge definition and resolution of a highly sensitive x-ray resist.

ACKNOWLEDGEMENT

The author thanks M. Sogard of Varian Associates for his collaboration and numerous stimulating discussions on the subject of x-ray lithography. I also thank J.M. Moran and G.N. Taylor of AT&T Bell Labs for many helpful discussions on x-ray resists.

REFERENCES

[1] Moran, J.M. and Taylor, G.N., Journal of Vacuum Science and Technology, Vol. 16, No.6, Nov./Dec. 1979, pp. 2020-2024.
[2] Garrettson, G., Karnezos, M., Sogard, M., and Starov, V., "X-ray Lithography: A Primer and Status Report"; technology review for Ruddell and Associates, Inc., Semiconductor and Photo process Engineering, Trinity Center, CA 1985.
[3] Lammers, D., Electronic Engineering Times, August 1985.
[4] Yaakobi, B., Kim, H., Sources, J.M., Deckman, H.W., and Dunsmuir, J., Applied Physics Letters, Vol. 43, No. 7, 1983, pp. 686-688.

[5] Gohil, P., Kapoor, H., Ma, D., Pekerar, M.C., McIlrath, T.J., and Ginter, M.L., *Applied Optics*, Vol. 24, No. 13, 1 July 1985, pp. 2024-2027.
[6] Taylor, G.N., *Solid State Technology*, June 1984, pp. 124-131.
[7] Spiller, E. and Feder, R., "X-Ray Lithography", in *X-Ray Optics* H-J Queisser, Ed., Springer-Verlag, New York, 1977; also see Ch. 13 of *Handbook on Synchrotron Radiation*, Vol. 1, ed. by E.E. Koch, North Holland Publishing Co. 1983.
[8] Piestrup, M., Adelphi Technology, unpublished
[9] Taylor, G.N., Coquin, G.A., and Somekh, S., *Polymer Engineering and Science*, Vol. 17, No. 6, June 1977, pp. 420-429.
[10] Sogard, M., Varian Associates, Inc., unpublished; also see Ref. 2, pp. 80-81 and Fig. 18a,b.
[11] Moran, J.M., and Taylor, G.N., *Journal of Vacuum Science and Technology*, Vol. 16, No. 6, Nov./Dec. 1979, pp. 2014-2019.

R.N. Singh

Palladium Silicide Contact Process Development for VLSI

REFERENCE: Singh, R.N., "Palladium Silicide Contact Process Development for VSLI", *Emerging Semiconductor Technology, ASTM STP 960*, D.C. Gupta and P.H. Langer, Eds., American Society for Testing and Materials, 1986.

ABSTRACT: A process for forming self-aligned small-area palladium silicide ohmic contacts to shallow junction silicon is described. A number of processing steps such as thin-film deposition, thermal annealing for silicide formation, and etching of unreacted palladium are developed in conjunction with modern materials characterization techniques such as SEM, SIMS and RBS. The results indicate that palladium films of ≤500Å thickness have good adherence to silicon and silicon dioxide. The metal-rich, dense, and continuous Pd_2Si readily forms on reaction of silicon with palladium upon thermal annealing. The siliciding reaction disrupts the native oxide and forms an intimate silicide-silicon interface. Unreacted palladium is etched in an aqueous etching solution, but extreme care is necessary to avoid palladium oxide formation during annealing, which leads to etching difficulties. Based on these results an optimum palladium silicide contact process for VLSI is proposed. Excellent contact resistance values between 11 and 25Ω are obtained for palladium silicide contacts of 1.25μm diameter using the optimum process.

KEYWORDS: palladium silicide-silicon contacts, VSLI contacts, VSLI processing

Dr. Singh is a staff research scientist at General Electric Company, Corporate Research and Development, PO Box 8, Schenectady, New York 12301.

INTRODUCTION

An ohmic contact with low resistance and shallow junction characteristics is required for very large scale integrated (VLSI) circuits to achieve device miniaturization and performance goals [1]. Metal silicides have attracted considerable attention for low-resistivity gates, interconnections and ohmic contacts because of metal-like resistivities, stability against processing chemicals, and high temperature stability [2]. Silicides of platinum, palladium, molybdenum, titanium, tantalum and tungsten are being considered for gate metallization and ohmic contacts to active regions of VLSI circuits. Palladium silicide is potentially attractive from the viewpoint of shallow junction formation because palladium forms a metal-rich silicide, Pd_2Si, after reaction with the single-crystal silicon upto 700°C. Other metal silicides such as Mo, Ti, Pt, Ta, and W form silicon-rich silicides thereby consuming significantly more silicon substrate and producing relatively deeper metal-semiconductor junctions. In contrast to platinum silicide contacts, the palladium silicide forms at a low temperature of ~270°C, can be selectively etched at the room temperature, and is compatible with some of the processing chemicals such as HF and HCl acids. Therefore, palladium silicide contacts are attractive for VLSI applications.

Considerable amount of work has already been published on the palladium silicide system of interest to microelectronics industry [3-10]. Kircher [3] studied the metallurgical properties and electrical characteristics of palladium silicide-silicon contacts and showed that palladium silicide readily forms with Si at 200°C. The electrical resistivity of the silicide was reported as $40 \mu\Omega$-cm and the barrier height of the $Pd_2Si/Si(n)$ contact was 0.745 V. Contact resistance to a relatively large size opening ($25 \mu m$ dia.) was 4.5Ω. Palladium was vacuum-evaporated onto a silicon substrate heated to 200°C. The unreacted palladium was etched away with an aqueous solution of $KI + I_2$, an etchant which does not attack Pd_2Si. Kircher [4] also studied the contact metallurgy of shallow junction Si devices and obtained low leakage currents of $~10^{-12}$ amp for $115 \mu m$ diameter contacts on n^+-Si. Buckely and Moss [5] also measured contact resistance using $20 \mu m$ diameter contact holes on p(111)-silicon, but they formed silicide via thermal annealing the palladium film at 260°C in an inert gas. They also suggested that thermal annealing may be conveniently performed at 260°C for 10 min. in air in a laboratory oven without detrimental effect on the contact characteristics. In contrast, this author observed significant sensitivity of annealing environment on the etching and electrical characteristics of palladium silicide contacts.

Previous investigations utilized large area contacts, which are relatively easy to form and may not be sensitive to processing environment. In comparison, small-area contacts (≤ 1-$2 \mu m$ dia.) are required for emerging VLSI technology which demands a clear understanding of the influence of processing environments on electrical characteristics of devices so that an optimum process can be designed for best performance. Therefore, the object of this investigation was to develop a palladium silicide contact process for small area (~$1 \mu m$ dia.) VLSI circuits. Influence of processing conditions on palladium deposition, annealing for silicide formation, and etching of unreacted palladium were studied. An optimum palladium silicide contact process for VLSI is proposed.

EXPERIMENTAL

The palladium silicide contact process was developed on 7.6cm diameter silicon wafers of (100) orientation. Most of the work was performed on plain silicon wafers, but a few wafers with half-oxide and half-bare silicon and patterned wafers with different size contact holes through the oxide were also utilized. All the wafers were cleaned in sequence with Karos (50:50::H_2SO_4:H_2O_2) for 5 minutes, hot concentrated HCl for 5 minutes at 60°C, and deionized (D.I.) water rinsed for 5 minutes. Unless specified otherwise most wafers were also etched for 60s in 1% HF in water to remove the native oxide and then spun dried.

Immediately after cleaning, the samples were loaded in a Perkin-Elmer model 2400 diode sputtering machine equipped with RF power and cryopumping system for thin-film deposition. The sputtering chamber was evacuated to a pressure of 2.7×10^{-4} Pa prior to deposition. Palladium was sputter deposited from a high-purity target at a rate of 11Å/s using high-purity argon gas at a pressure of 0.53 Pa. A wafer to target distance of 0.5 cm was utilized. These sputtering conditions produced palladium film with excellent adherance to silicon and silicon oxide, and low sheet resistance of 3.5Ω/□ for a 500Å-thick film. This corresponds to a film resistivity of 17$\mu\Omega$.cm. The film thickness was measured by a Sloan DEKTEK-II.

Palladium films were heated to higher temperatures for silicide formation. Two types of annealing methods were evaluated. In one case, wafers were annealed in a tube furnace in a gas atmosphere of either argon, nitrogen, argon-hydrogen, or hydrogen gas. High-purity gases were used in all experiments. In addition, these gases were further purified to remove trace amounts of oxygen. In the second case, wafers were annealed in the high vacuum of a sputtering chamber using quartz heating lamps. Influence of annealing method and annealing environment on the silicide formation are presented in the next section.

Unreacted palladium was selectively etched at room temperature using an aqueous solution of KI + I_2. Initial experiments utilized an etching solution with 4g KI, 1g I_2, and 40 ml H_2O [11]. Subsequently, an etching solution with 10g KI, 1g I_2, and 40ml H_2O showed superior etching characteristics and, therefore, was used for most of the process development. Etching was done in an ultrasonic cleaner for periods ranging from 10 to 20 minutes.

Influence of deposition, annealing, and etching conditions on thin-film properties and silicide formation were studied. Each process step was characterized via sheet resistance measurements, SEM, RBS, SIMS, and x-ray diffraction techniques. X-ray diffraction techniques utilized the Cu-K_α radiation and the RBS technique used 2.0 MeV-$^4He^+$ ions at normal incidence, Ω = 141 msr, ϵ = 2KeV, and λ = 2μC. The SIMS elemental profiles are normalized with respect to the silicon background in the substrate and oxygen concentration in the background is also set equal to the silicon intensity for self consistency.

RESULTS AND DISCUSSION

Adherence of Palladium Film

Good adherence of palladium film to silicon and silicon oxide is necessary to achieve pattern definition via photolithographic techniques. Adequate adherence of palladium to silicon and silicon oxide is also required in the silicide contact process so that an intimate palladium-silicon interface is formed for siliciding via interdiffusion. It is well known that palladium has low adherence to silicon oxide [12,13]. Therefore, adherence is determined as a function of film thickness on silicon and silicon oxide.

Palladium films ranging in thickness from 500 to 2500Å are deposited on silicon wafers with half-oxide and half-silicon surface. Excellent adherence, as determined by the tape test, is obtained for all the samples on bare silicon and silicon oxide. Adequate adherence is also observed for samples annealed in argon atmosphere, but lowered adherence is seen for hydrogen-annealed palladium film on oxide. Generally, thicker films ($\geq$1000Å-thick) showed poor adherence to oxide while thinner films ($\leq$500Å-thick) showed good adherence to oxide even after annealing in the hydrogen gas. Thinner palladium films consume less silicon on silicide formation and form shallow-junctions. Therefore, palladium films of $\leq$500Å thickness are chosen for contact process development.

Palladium Silicide Formation

The influence of deposition and annealing conditions on palladium silicide formation are studied. In addition, the effects of native oxide and a chemical oxide on silicon formation are investigated. The results are given below.

Palladium readily reacts with silicon and forms palladium silicide at temperatures as low as 200 to 250°C [3,4]. Even during sputter deposition, the substrate temperature can rise to a sufficiently high level for palladium silicide formation. To study this in situ siliciding behavior, 2000Å of palladium was sputter deposited on a silicon wafer and the thin film was characterized via SEM, RBS, SIMS, and x-ray diffraction analysis. The x-ray diffraction results, shown in Fig. 1 indicate that palladium silicide of Pd_2Si stoichiometry forms during deposition of the film. But, the silicide is not well-crystallized as evidenced by some of the broad x-ray diffraction peaks. RBS analysis of this sample, as given in Fig. 2a, shows plateau in the Si and palladium edges, indicating formation of the palladium silicide. Simulated RBS spectrum [14] is also shown in Fig. 2a and indicates that ~750Å of Pd_2Si forms during deposition of 2000Å of palladium on silicon. SIMS profile (Fig. 2b) also confirms this behavior and shows a palladium silicide thickness of ~500 to 600Å. In addition, SIMS data indicate lower oxygen in the vicinity of silicon-silicide interface and somewhat higher oxygen in the silicide. The presence of palladium silicide in 1.25μm diameter contact hole, during sputter deposition of 2000Å of palladium, is clearly shown in Fig. 3. The SEM micrographs shown in Fig. 3 are taken after etching away the unreacted palladium with the KI + I_2 etchant. These micrographs show that a continuous layer of palladium silicide does not form during the sputtering process. This behavior is inferred from dark holes in Fig. 3, which are the locations where palladium

270   EMERGING SEMICONDUCTOR TECHNOLOGY

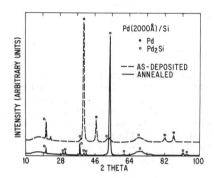

FIG. 1   X-ray diffraction spectra of an as-deposited and annealed sample of Pd on silicon. Palladium silicide is evident in the as-deposited sample.

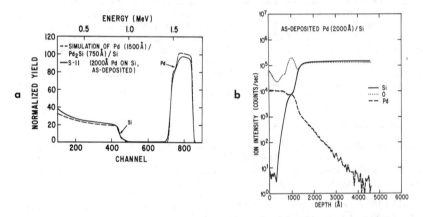

FIG. 2   (a) 2.0 MeV-^{4}He$^+$ backscattering spectra and (b) SIMS profiles from a sample of Pd(2000Å)/Si showing the formation of palladium silicide during sputter deposition.

silicide has not formed. This situation is not desirable for making good contacts to silicon junctions because it will result in unreliable and high-contact resistance value.

In comparison to the above behavior, thermally annealed samples form a dense and continuous layer of palladium silicide as shown in Fig. 4. This sample was furnace-annealed in a gettered argon atmosphere at 430°C for 0.5h and etched with the KI + I$_2$ etch to remove unreacted palladium. A summary of results for samples annealed in vacuum, gettered argon, and argon-10% hydrogen mixture is given in Table 1. The sheet resistance of vacuum annealed samples increased from 4.3 to 7.3Ω/□ and those for furnace annealed samples increased to 5.9Ω/□ after siliciding. Increased sheet resistance for annealed samples is indicative of silicide formation. Gettered argon and argon-hydrogen annealed samples produced similar sheet resistance values; therefore, either can be used as an effective annealing atmosphere.

A thin layer of silicon oxide (native oxide) forms on silicon wafers exposed to the ambient conditions. This is an insulating layer that provides a barrier to good metal-semiconductor contacts. Invariably, this native oxide is etched away in a dilute HF solution prior to the metal deposition. Influence of native oxide and a chemical oxide interposed between silicon wafer and palladium film on silicide formation is studied. The native oxide is grown in air for 48h on clean silicon wafers, whereas a chemical oxide (~50Å) is

FIG. 3   Scanning electron micrographs showing the formation of palladium silicide during sputtering of 2000Å of palladium in 1.25μm contact window.

FIG. 4   Scanning electron micrographs showing the formation of dense palladium silicide after annealing 500Å palladium on silicon at 430°C for 0.5h, (a) top view and (b) side view.

formed by boiling the silicon wafer in concentrated nitric acid at 50°C for 20 minutes. Palladium film (~1000Å-thick) is then sputtered on these wafers and siliciding done for 0.5h at 400°C in an annealing furnace containing Ar-10% H_2 atmosphere. A summary of sheet resistance values for samples with native oxide and chemical oxide is given in Table 2. Samples with native oxide show silicide formation as evident from the characteristic black color of palladium silicide. Sheet resistance values are also typical for Pd and Pd_2Si films of 1000Å and 1700Å thickness, respectively. In contrast, samples with chemical oxide do not show the black color characteristic of the palladium silicide. Furthermore, significant increase in the sheet resistance after annealing is shown by chemical oxide. These samples are characterized via RBS and SIMS techniques to determine the silicide formation and chemistry near the palladium-silicon interface.

TABLE 1 -- Influence of Annealing Environment on Silicide Formation

Annealing Atmosphere	Annealing Conditions		Sheet Resistance ($\Omega/\square$)	
	Temp (°C)	Time (h)	Before	After
Vacuum (in situ)	270	0.25	4.3	7.3
Argon (gettered)	430	0.25	4.09	5.9
Argon-Hydrogen	430	0.25	4.30	5.9

TABLE 2 -- Influence of Thin Oxides on Silicide Formation

Type of Oxide	Annealing Conditions			Sheet Resistance ($\Omega/\square$)	
	Temp (°C)	Time (h)	Gas	Before	After
Native	400	0.5	Ar-H_2	1.6	2.27
Chemical	400	0.5	Ar-H_2	1.8	7.9

Backscattering results from samples with palladium on the native oxide in the as-deposited and annealed states are shown in Fig. 5. A small amount of silicide is evident in the as-deposited spectrum of palladium in Fig. 5a. The annealed spectrum, as shown in Fig. 5b, shows complete conversion of palladium to palladium silicide. Good agreement is observed between experimental RBS data and the simulated RBS spectrum for 1800A Pd_2Si. This suggests that 1A of palladium forms about 1.8A of palladium silicide. It also indicates that the native oxide is disrupted during the silicide formation. SIMS profiles from as-deposited and annealed samples with native oxide are shown in Fig. 5c,d. These show an oxygen peak due to the native oxide towards silicon side of the palladium-silicon interface (Fig. 5c). This peak has moved towards the palladium side (Fig. 5d) after annealing at 400°C for 0.5h. This is also indicative of disruption of native oxide during the silicide formation. Plateau in the palladium and silicon spectra (Fig. 5d) for annealed samples also indicates that a silicide of a uniform stoichiometry forms throughout the original palladium film. A lowered oxygen level at the palladium silicide-silicon interface, as shown by the larger dip in oxygen profile in Fig. 5d, is an indication that extremely clean silicide-silicon interface is formed on reaction of silicon with the palladium. In contrast, samples with the chemical oxide show no evidence of silicide formation in the as-deposited and annealed states as shown by the RBS spectra in Fig. 6a and 6b, respectively. SIMS profiles from samples with the chemical oxide are shown in Fig. 6c,d which indicates that oxygen along with silicon are diffused into the palladium film on annealing at 400°C for 0.5h. The presence of large amounts of oxygen inhibits the palladium-silicide formation and is responsible for the high sheet resistance values (Table 2). The palladium-silicon interface is saturated with oxygen and will lead to significant increase in contact resistance. These results indicate that native oxide can be disrupted during palladium silicide formation whereas a chemical oxide is stable against such disruption.

Etching Behavior of Palladium Film

The palladium silicide contact process requires deposition of palladium film on patterned wafers, siliciding in contact windows via thermal annealing, and selective removal of unreacted palladium via an etching process. A chemical etching process was investigated for selective removal of palladium. Initial experiments utilized an etching solution with 4g KI, 1g I_2, and 40 ml H_2O [11], however, a majority of the work in this investigation was done with an etching solution with 10 g KI, 1g I_2, and 40 ml H_2O. The etching was performed at room temperature. Both of these etchants worked well on as-deposited palladium films, but considerable difficulty was encountered when etching was attempted on thermally annealed samples of palladium. Therefore, the influence of annealing conditions on the etching behavior of palladium was investigated.

Etching behavior is studied on samples with 1000Å of palladium on thermally grown oxide annealed at 400°C for 0.5h. Four types of high-purity annealing gases (argon, nitrogen, hydrogen and argon-10% hydrogen) are used. Initial experiments were done in high-purity nitrogen because Buckley and Moss [5] had suggested that thermal annealing can be performed in air. However, annealing of palladium films in nitrogen at 400°C led to great difficulty in etching. The palladium film discolored after annealing and could not be etched with the KI + I_2 etching solution. Similarly, palladium films annealed in high-purity argon gas showed purple discoloration and could not be etched by the etching solution. In contrast, samples annealed in pure hydrogen and argon-10% hydrogen gas mixture showed no discoloration of palladium film and are completely etched by the KI + I_2 solution. These behaviors suggest that palladium film may be reacting with impurities such as oxygen in the argon and nitrogen gases. Nitrogen gas is therefore purified via passing through a liquid nitrogen trap, but a similar etching difficulty is observed. In contrast, samples annealed in argon gas purified by passing it over a hot titanium-sponge showed excellent etchability of palladium by the KI + I_2 solution. This suggests that oxygen impurity in nitrogen and argon gases is responsible for poor etching behavior of annealed palladium films. The source of etching problems is traced to palladium oxide (PdO) formation which leads to purple discoloration of palladium films. An x-ray diffraction spectrum from discolored palladium film is shown in Fig. 7. It clearly shows a PdO phase along with unreacted Pd and palladium silicide. The presence of PdO film on palladium metal is expected to provide a barrier to

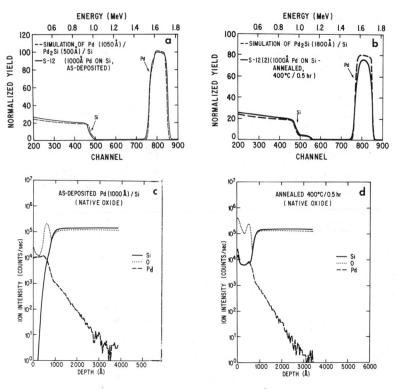

FIG. 5   Backscattering spectra for 2.0 MeV-^{4}He$^+$ ions incident on (a) as-deposited, and (b) annealed samples of Pd(1000Å)/Si. SIMS profiles from (c) as-deposited and (d) annealed samples of Pd(1000Å)/Native Oxide/Silicon.

etching. The palladium oxide film does not form on annealing the samples in either gettered argon, hydrogen, or argon-hydrogen gases; therefore, wafers annealed in these gases can be completely etched via KI + I$_2$ etchant. No residue of palladium is left on oxide after this selective etching process as determined by an SEM examination. Furthermore, the KI + I$_2$ etching solution does not attack palladium silicide. Excellent selective etching characteristics of the KI + I$_2$ etchant for palladium is clearly evident in Fig. 4, which shows palladium silicide formation in the 1.25μm diameter contact window.

These results have established that the thermal annealing environment is extremely important for making clean palladium silicide in contact windows and for good etching characteristics of the palladium film. Thermal annealing should not be performed in air as suggested by Buckley and Moss [5].

Thermal Stability of Palladium Silicide

Thermal stability of palladium silicide (Pd$_2$Si) is investigated via annealing 2000Å thick palladium film on a clean silicon wafer as a function of temperature between 400 and 700°C. The palladium sili-

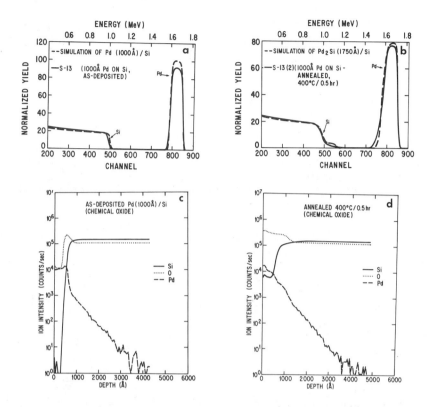

FIG. 6 Backscattering spectra for 2.0 MeV-^{4}He^{+} ions incident on (a) as-deposited, and (b) annealed samples of Pd(1000Å)/chemical oxide/Si. SIMS profiles from (c) as-deposited and (d) annealed samples of Pd(1000Å)/Chemical Oxide/Si.

cide is formed during thermal annealing at 400, 500, 600, and 700°C for 0.5h in an atmosphere containing argon-10% hydrogen.

The sheet resistance of palladium silicide depends on the siliciding temperature as shown in Fig. 8. A sheet resistance value of 1.75Ω/□ is obtained for samples annealed at 400°C, whereas a value of 1.24Ω/□ is observed for those annealed at the 700°C. These samples were annealed for 0.5h at each temperature. An additional annealing period of 5 hours at each of the 500, 600, and 700°C temperatures leads to insignificant changes in sheet resistivity values which indicates good thermal stability of the palladium silicide films. The x-ray diffraction and RBS analyses of samples annealed at the high temperature of 700°C for 5 hrs show Pd$_2$Si as the only silicide phase. This behavior is in agreement with an earlier investigation of Hutchins and Shepela [6], who showed that silicon-rich PdSi phase did not nucleate in Pd$_2$Si films below a temperature of 735°C. Therefore, palladium silicide contacts are expected to be fairly stable up to ~600°C in the absence of interactions with other thin films.

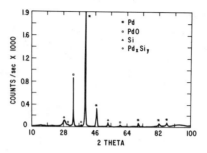

FIG. 7   X-ray diffraction spectrum of a sample annealed in an ungettered argon gas at 400°C for 0.5h. Formation of palladium oxide is evident.

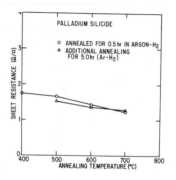

FIG. 8   Sheet resistance of Pd(2000Å)/Si after annealing between 400 and 700°C for 0.5h. Additional annealing for 5h did not significantly influence the sheet resistance.

Preferred Palladium Silicide Contact Process

The palladium silicide contact process, as described in previous sections, is developed with the help of modern materials characterization techniques such as x-ray diffraction, SIMS, RBS, and SEM. This is a potentially useful methodology for understanding the influence of processing conditions on thin film properties, silicon-metal interface chemistry, and for rapid development of small area contact metallization process for VLSI circuits of the future. An optimum palladium silicide contact process, as described below, is developed using this methodology.

Palladium films of 400Å thickness is deposited on patterned wafers for making shallow junction contacts. These films are then silicided for 15 minutes in either a gettered argon gas or in the high-vacuum at ~300°C. Siliciding in the high-vacuum is preferred because it is potentially simpler and can be readily performed in the sputtering chamber subsequent to sputter deposition process using quartz heating lamps. The unreacted palladium is selectively etched using the $KI+I_2$ etchant. Contact resistance values between 11.6 and 14Ω for n^+-silicon, and between 21 and 22.5Ω for p^+-silicon are obtained for palladium silicide contact of 1.25μm diameter using the optimum process. These low contact resistance values are comparable to CVD-tungsten [15] and indicative of excellent contact forming ability of palladium silicide contact process as described in this paper. More detailed results on contact resistance of palladium silicide to shallow n^+- and p^+-silicon junctions is described in another publication [16].

SUMMARY

A process for forming small area ($\leq 1.25 \mu m$ diameter) palladium silicide ohmic contacts to silicon is described. A number of processing steps such as thin film deposition, thermal annealing for silicide formation, and etching of unreacted palladium are developed. These processing steps are characterized via SEM, SIMS, RBS, and x-ray diffraction techniques. A summary of results is given below.

Palladium films of $\leq 500 A$ are found to have excellent adherence to silicon and silicon dioxide. Metal-rich palladium silicide, Pd_2Si, forms upon the reaction of palladium with silicon. This phase is discontinuous when it forms during sputter-deposition. In contrast, dense palladium silicide forms upon subsequent thermal annealing at temperatures $\geq 300°C$ in a gas atmosphere or in a high vacuum. Native oxide on silicon gets disrupted on palladium silicide formation and leads to intimate silicide-silicon interface. A chemical oxide, on the other hand, leads to a poor silicide/silicon interface, and inhibits the formation of stoichiometric Pd_2Si. The metal rich silicide phase, Pd_2Si, is stable up to $700°C$ for annealing periods of 5 hrs.

As-deposited palladium films are readily etched in a solution with 10 g KI, 1g I_2, and 40 ml H_2O. Similarly, vacuum-annealed films are also easily etched with the above etchant. In contrast, considerable etching problems are encountered for gas-annealed samples. Annealing in an ungettered gas leads to palladium oxide formation which cannot be etched with KI + I_2 solution in water. Therefore, the annealing environment should be extremely low in oxygen content to avoid etching problems. For these reasons, annealing in high vacuum is a preferred procedure.

Excellent contact resistance values of 11.6-14Ω for n^+-Si and 22-25Ω for p^+-Si is obtained for palladium silicide contact of $1.25 \mu m$ diameter. These results are obtained with a preferred palladium silicide contact process as described in this paper.

ACKNOWLEDGEMENTS

The author wishes to thank PA Piacente and RH Wilson, for supplying clean wafers; CD Robertson, CI Hejna and GA Smith of Materials Characterization Operation for SEM, x-ray diffraction, SIMS and RBS analyses; K. Borst and RE Ostranger for sputtering; and CA Markowski for preparation of this paper. Encouragement and interest of DM Brown, KW Browall and FN Mazandarany in this investigation are appreciated.

REFERENCES

[1] Vossen, J. L., "VLSI Metallizations: Some problems and trends", *Journal of Vacuum Science and Technology*, Vol. 19, No. 3, 1981, pp. 761-765.
[2] Murarka, S. P., "*Silicides for VLSI Applications*", Academic Press, Inc., 1983.
[3] Kircher, C. J., "Metallurgical Properties and Electrical Characteristics of Palladium Silicide-Silicon Contacts", *Solid-State Electronics*, Vol. 14, 1971, pp. 507-513.
[4] Kircher, C. J., "Contact Metallurgy for Shallow Junction Si Devices", *Journal of Applied Physics*, Vol. 47, No. 12, 1976, pp. 5394-5399.
[5] Buckley, W. D. and Moss, S. C., "Structure and Electrical Characteristics of Epitaxial Palladium Silicide Contacts on Single Crystal Silicon and Diffused P-N Diodes", *Solid-State Electronics*, Vol. 15, 1972, pp. 1331-1337.
[6] Hutchins, G. A. and Shepela, A., "The Growth and Transformation of Pd_2Si on (111), (110) and (100) Si", *Thin Solid Films*, Vol. 18, 1973, pp. 333-363.
[7] Shepela, A., "The Specific Contact Resistance of Pd_2Si Contacts on n- and p-Si", *Solid-State Electronics*, Vol. 16, 1973, pp. 477-481.
[8] Grinolds, H. and Robinson, G. Y., "Study of Al/Pd_2Si Contacts on Si", *Journal of Vacuum Science and Technology*, Vol. 14, 1977, pp. 75-78.
[9] Fertig, D. J. and Robinson, G. Y., "A Study of Pd_2Si Films on Silicon Using Auger Electron Spectroscopy", *Solid-State Electronics*, Vol. 19, 1976, pp. 407-413.
[10] Chu, W. K., Lau, S. S., Mayer, J. W., Muller, H. and Tu, K. N., "Implanted Noble Gas Atoms as Diffusion Markers in Silicide Formation", *Thin Solid Films*, Vol. 25, 1975, pp. 393-402.
[11] Ghandhi, S. K., "*VLSI Fabrication Principles*", John Wiley and Sons, 1983.
[12] Cunningham, J. A., "Expanded Contacts and Connections to Monolithic Silicon Integrated Circuits", *Solid-State Electronics*, Vol. 8, 1965, pp. 735-745.
[13] Srinivasan, M. S. and Svensson, C. M., "Control of Palladium Adherence to Silicon Dioxide for Photolithographic Etching", *Journal of the Electrochemical Society*, Vol. 123, No. 9, 1976, p. 1258.
[14] RBS Simulation Program RUMP, Cornell University, Ithaca, New York.
[15] Wilson, R. H., Private communication, General Electric Research and Development Center, Schenectady, New York 12301.
[16] Singh, R. N., Skelly, D. W. and Brown, D. M., "Palladium Silicide Ohmic Contacts to Shallow Junctions in Silicon," General Electric Report No. 85CRD132, 1985.

Material Defects, Oxygen and Carbon in Silicon

H. Ming Liaw, John W. Rose, and Ha T. Nguyen

CHARACTERIZATION OF SILICON SURFACE DEFECTS BY THE LASER SCANNING TECHNIQUE

REFERENCE: Liaw, H. M., Rose, J. W., and Nguyen, H. T., "Characterization of Silicon Surface Defects by the Laser Scanning Technique," Emerging Semiconductor Technology, ASTM STP 960, D. C. Gupta and P. H. Langer, Eds., American Society for Testing and Materials, 1986.

ABSTRACT: The increase of circuit complexity in an IC chip has greatly increased the cost of processed wafers. The cost can be reduced if the defective wafers which cause the poor process yields can be screened prior to the wafer processing. Recent advanced in laser optics and data processing have made automatic laser scanners available. The laser scanners can be potentially used for 100% of the screening of incoming wafers and as a process control tool for certain steps of wafer processing. This paper reports (1) the correlation of defects detected by the laser scanner with the physical nature of defects, (2) utilization of the laser scanner for the characterization of process-induced defects, and (3) utilization of the laser scanner for monitoring substrate cleaning and epitaxial growth processes.

KEYWORDS: wafer inspection, laser scanner, surface defects, epitaxial silicon.

INTRODUCTION

Evaluation of surface defects in silicon wafers is a critical step in silicon wafer processing. Surface defects include contaminants as well as surface damage. The former could arise from particulate contamination and/or residue of liquids used for wafer cleaning. The latter could be caused by incomplete polishing, improper handling of wafers, or the growth-induced cyrstallographic

Dr. Liaw is a senior member of the technical staff and section manager at the Semiconductor Research and Development Labs (SRDL). Mr. Rose is principal staff engineer and a manager for the epitaxial development program of the Electronic Materials Operation. Ms. Nguyen is a process engineer at SRDL, Motorola, Inc. Semiconductor Products Sector, 5005 East McDowell Rd., Phoenix, Arizona 85008.

defects such as epitaxial stacking faults. Qualitative and quantitative analyses of the surface defects are difficult because of the small defect size and low defect density in current silicon wafers. Traditionally, the qualitative surface defect evaluation is made by visual inspection using an intense light. The amount of randomly scattered reflected light is proportional to the density of surface defects. Quantitative evaluation is made with the aid of an optical microscope by counting the number of each defect type under a given field of view. However, reproducibility of both evaluation methods are poor.

It has been well established that the presence of surface defects in silicon wafers can reduce device manufacturing yields [1-4]. Surface defects due to particulate contamination can lead to the formation of photolithographic defects. The defects due to surface damages or crystallographic imperfections can degrade device performances which in turn reduce the device manufacturing yields. As device dimensions continue to shrink and in the meantime the number of components in a single silicon chip is steadily increased, the device yields are becoming more sensitive to the presence of surface defects in silicon wafers.

Increase in circuit complexity in an IC chip has greatly increased the cost of wafer processing. Device manufacturing costs can be reduced if one does not need to process the wafers which contain high percentages of rejected devices caused by the defects in silicon wafers. Therefore, there is a great incentive to screen the poor wafers before wafer processing. Recent advances in laser optics and data processing technologies have made laser wafer scanners available. The industry has begun to use this technique for incoming wafer screening and possibly for the process monitoring [5-8]. Few reports have been made regarding (i) correlation between the detected defects and their actual physical nature, and (ii) use of the laser scanning technique as a process control tool. The purpose of this paper is to report our study in these two areas.

PRINCIPLE USED IN WAFER INSPECTION

Manual inspection of silicon wafers by light back-scattering has been well established and used by the industry, particularly by silicon wafer manufacturers. The light source used for back-scattering is generally a high intensity tungsten lamp. Procedures for the visual inspection of wafers have been given in ASTM standard-F523. In essence the wafer is illuminated by an intense light beam with an incident angle approaching 90 degrees as shown in Fig. 1(a). When the wafer surface is flat and free of defects, no light scattering will be observed in the reflected beam (whic is called the specular beam). The entire surface shows complete darkness when one views from an angle other than that of the specular beam. The presence of defects on the surface causes a fraction of light to scatter in random directions. Observation of the scattered beam provides information on the defect location and size. These defects can be recorded by using a photographic or TV camera.

Laser wafer scanners use the same principle as the visual inspection with the exception of replacing the human eyes (or photographic camera) by light detectors. Some commercially available scanners use only one detector per scanner to collect the scattered beam [6]. In this case, the light collector preferably encompasses a large fraction of the reflected solid angle, since detecting sensitivity increases proportionally with the solid angle of light collection. Fig. 1(b) shows that the incident laser beam is approximately normal to the wafer main surface and the scattered beam is collected from a wide angle of the solid cone which is also normal to the main surface of the wafer. Other scanners use two detectors, one for the scattered beam and the other for the specular beam. However, neither collectors encompass a solid angle. Fig. 1(c) shows that the specular beam collector is located in the path of the reflected beam, which is 15 degrees from the surface normal, while the scattered beam collector is placed at the perpendicular axis to the main surface of the wafer. In this case, the scattered beam collector will detect defects which reflect light isotropically such as particulates and the specular beam collector will detect defects which reflect light anisotropically such as crystallographic defects (e.g. stacking faults).

The laser scanner shown in Fig. 1 (b) can differentiate defects into haze, point, line and area defects. The haziness of the surface leads to the gradual signal change in the output of the photomultiplier which receives the signals from the light collector. The point defects scatter the laser beam to a greater extent than haze. Therefore, the output of the photomultiplier produces a pulse signal whenever the laser beam encounters a point defect. The amplitude of

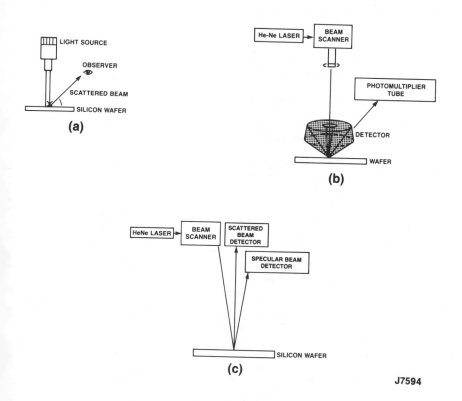

Fig. 1-- (a) Manual surface inspection using high intensity light. (b) Laser scanner with single detector. (c) Laser scanner with two detectors.

the pulse is proportional to the size of the point defect. A strip of adjacent point defects is assigned as a line defect. A cluster of adjacent point defects is identified as an area defect. Some laser scanners can also provide information on the extent of line and area defects in terms of the number of unit area. Some scanners can classify the point defects into different categories according to their sizes. Table 1 shows an example of a scanner which groups the particle sizes into four categories by channel numbers.

TABLE 1-- Particle sizes detected by the laser scanner at threshold setting No. 2

Channel No.	Particle Size (in micrometer)
1	0.26 - 0.42
2	0.43 - 1.60
3	1.60 - 2.60
4	> 2.6

TABLE 2-- Suggested flaw types detected by a scanner with a 2-beam collector.

Flaw #	Flaw type suggested by the manufacturer	Collector
1	Haze	s *beam
2	Small pits and particulates (0.2 - 2 micron).	s beam
3	Large pits and particulates (2 - 20 micron).	s beam
4	Totalization of particulates.	s & sp* beams
5	Scratch (line defects), a minimum of four adjacent pits or particulates.	s beam
6	Area defects.	s beam
7	Light Orange Peel.	sp beam
8	Distortion with abrasion such as an unpolished out saw mark.	s & sp beams
9	Distortion defects of higher frequency such as small dimples, grooves, and mounds.	sp beam
A	Large particulates.	s & sp beams
B	Heavy orange peel and related defects.	sp beam
C	Severe dirt, abrasions, or films.	s & sp beams

*Legend: s stands for scattered, sp stands for specular.

The laser scanner shown in Fig. 1(c) can differentiate the defects into 12 categories. The scattered beam or specular beam collector can detect several types of defects (or flaw types) according to the amplitude of pulses recorded by the photomultiplier. Other categories of the defect types are formulated by combining the signals from both collectors. Table 2 lists the suggested flaw types detected by a laser scanner with two beam collectors.

EXPERIMENTAL PROCEDURES

In this study, a laser scanner with two beam collectors was used as a primary surface defect inspection tool. Scanners with single beam collectors were also used only for the correlation study on different types of scanners. Experiments were made with the following purposes: (i) to find a correlation between the detected defects by a scanner with a single detector and the same defects as detected by a scanner with two detectors, (ii) to examine the physical nature of defects for each flaw type detected by a laser scanner, (iii) to monitor the surface defects introduced by various wafer processing steps, and (iv) to demonstrate the relationship between the defects detected by a laser scanner and the electrical defects calculated from the yield curves of MOS capacitors.

Wafers used for this study were either 3 inch or 100 mm diameter, heavily-doped with Sb or B, and were (100) orientation. The experiments were focused on the study of the introduction of surface defects by the epitaxial growth process. Other processes studied include the substrate cleaning methods and thermal oxidation.

The density of electrical defects was evaluated by using Murphy's Law [9] assuming that the defect distribution follows a delta function. Under this assumption, the relation between the device yield (Y) and defect density (D) is $D=(-\ln Y)/A$, where A is the area of the devices. The devices which we fabricated for this work are MOS capacitors. The capacitors were fabricated with four different sizes (25, 625, 5,000 and 20,000 mil^2) in a given wafer. The yield is defined as the percentage of capacitors which have oxide breakdown fields of greater than 2 MV/cm. The D is evaluated by plotting $\ln(Y)$ vs. A.

EXPERIMENTAL RESULTS

Correlation of Defects Detected by Different Laser Scanners

Comparison of the surface defects detected by a scanner of a single beam detector as shown in Fig. 1(b) was made with those detected by a scanner of two beam detectors as shown in Fig. 1(c). Fig. 2(a) shows the plot of the number of type 1 defects (i.e. haze) detected by the two beam collectors (scanner A) versus the number of defects with 0.26 to 0.42 micrometer sizes detected by the single beam collector (scanner B). Fig. 2(b) shows the plot of defect count of type 4 defects (.2 to 20 micrometer) detected by the scanner A versus the defect count of > 2.6 micrometer detected by the scanner B. Fig. 2(c) shows the plot of the total number of defects detected by the scattered beam collector of Scanner A versus the total number of defects detected by Scanner B. These plots have shown that a weak correlation exists for surface defects detected by these two laser scanners. The correlation curves are somewhat skewed. This is due to the fact that the range of the two laser scanners when viewing the size of defects are only approximately equal. Nevertheless, these plots show that the laser scanners of these two different types are generally agreeable to each other qualitatively. Quantitatively, they are also agreeable within the same order of magnitude.

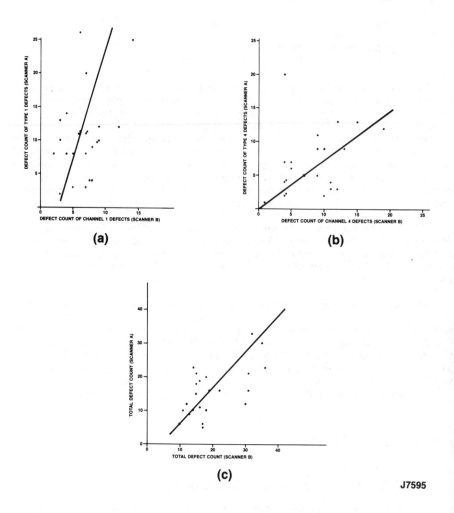

Fig. 2-- Defect counts of Scanner A vs. Scanner B. (a) Comparison of small defects and haze. (b) Comparison of larger defects. (c) Comparison of total defects detected.

Correlation Between the Photo-Backscattering vs. Laser Scanning Techniques

Our preliminary study has shown that both crystallographic and non-crystallographic defects can be detected by an automatic laser scanner. Figs. 3(a) and 3(b) compare the laser surface defect map and the light back scattered photograph of an 80 micrometer thick epitaxial wafer. A good correlation can be observed along the periphery at the right hand side of both figures. The defects existing in the other portions of the wafer are revealed by the laser scanner but not by the photographic camera. It is partly due to the fact that the photographic camera was placed at a fixed angle with respect to the reflected light and was not able to collect the entire reflected beam scattered by the surface defects. The correlation is even harder to make for thin epitaxial wafers in which the defect size is smaller and defect density is lower. However, this study has clearly shown that the laser scanners are certainly advantageous over the photographic or TV camera in terms of better resolution for detecting small defects.

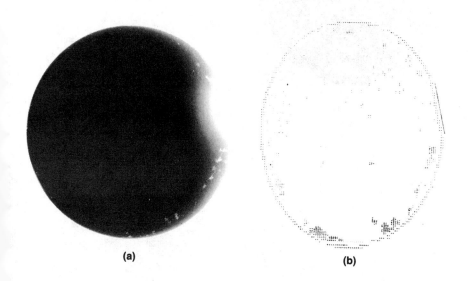

Fig. 3-- Comparison of defects detected by photo-backscattering and laser scanning. (a) Photograph. (b) Laser defect map.

A fairly good agreement has been found between these two techniques for the detection of defects in oxidized wafers. Fig. 4(a) shows the photograph while Figs. 4(b) through (d) are the laser scanning maps of the same wafer, but were taken at diffeent threshold values (i.e. sensitivity) for the collector. The laser scanning maps show that a high sensitivity setting is needed to give a good correlation with the light back-scattered photograph for revealing swirls.

Each dot shown in the laser scanning maps is an alpha or numerical symbol representing the flaw type shown in Table 2. We have correlated the flaw type with the actual defects by visual inspection under an optical microscope. Fig. 5 shows the photographs of the surface defects that were detected by the scattered beam collector. The flaw types 2 or 3 in this case are found to be particulates (5a) or light scratch marks (5b). Flaw type 5 is found to result from tweezer marks (5c) or heavily scratched lines (5d). Fig. 6 shows the photographs of surface defects that were detected by the combination of both scattered and specular beam collectors. Flaw type A includes HCl etch pits (6a and b), large particulates (6c) and epitaxial spikes (6d). Flaw types 8 and 9 include epitaxial stacking faults (e) or etched mesa (6f).

288   EMERGING SEMICONDUCTOR TECHNOLOGY

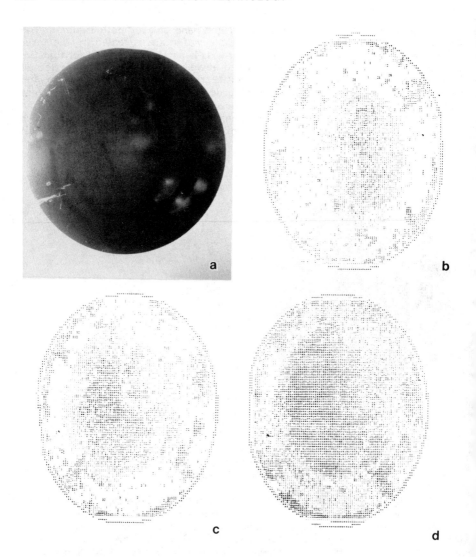

Fig. 4-- Comparison of defects detected by (a) photo-backscattering and (b-d) laser scannings with increasing sensitivity, (b) low sensitivity, (c) medium sensitivity, and (d) high sensitivity.

Laser Scanner Used as a Process Control Tool

Evaluation of pre-epitaxial substrate cleaning methods: The effectiveness of various substrate cleaning methods has been evaluated. Types of cleaning methods include (i) hot nitric cleaning, (ii) DI wafer scrubbing, (iii) hot nitric cleaning followed by DI water scrubbing, (iv) hot nitric cleaning under ultrasonic field followed by DI water scrubbing and (v) jet cleaning which forms a piranha solution by mixing acids from jet streams on wafer surfaces. The results show

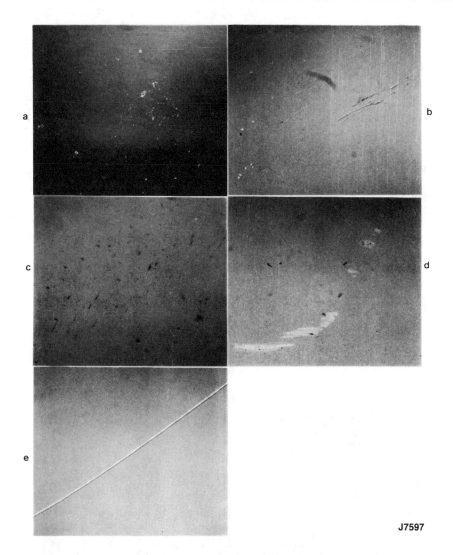

Fig. 5-- Correlation of flow types detected by the scanner to visual inspection. Flaw types 2 and 3 are represented as particulates (a,b), or light scratch marks (c). Flaw type 5 is seen as tweezer scratchs (d), or a scratch (e) under visual inspection.

that the as-received wafers taken from a vacuum sealed package are the cleanest. Hot nitric cleaning alone is not only ineffective but also adds particulates on the wafer surfaces. However, hot nitric cleaning followed by the scrubbing improves the surface cleanliness. Scrubbing alone is also effective.

The new cleaning technique (FSI) using the combination of several jet streams of different chemicals also seems reasonably good. Table 3 lists the average number of surface defects, counted by the laser scanner, of 25 wafers which have been subjected to various wafer cleaning steps.

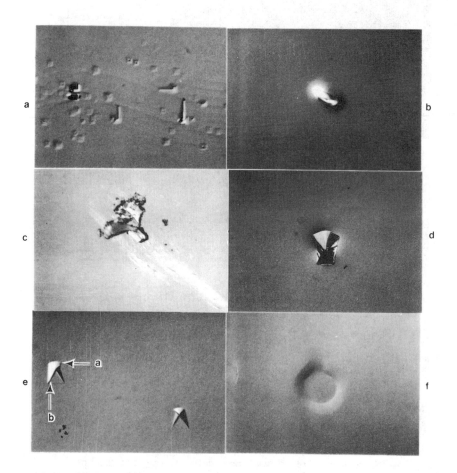

Fig. 6-- Correlation of flaw types detected by the scanner to visual inspection. Flaw type A is seen as HCl etch pits (a,b) large particulates (c), and epitaxial spikes (d). Flaw types 8 and 9 were seen as epitaxial stacking faults (e), and etched mesa (f).

Monitoring epitaxial growth: The laser scanner was used to monitor the defects introduced by the epitaxial growth process. The epitaxial silicon layers were grown using a AMC 7800 reactor at 1150°C at atmospheric pressure. The prebake cycle was carried out at 1180°C with the presence of HCl to remove approximately 0.5 micron from the substrate surfaces. The growth rate was 1 micrometer/min and epitaxial layer thickness was 1.6 micrometer. Fig. 7(a) shows the plot of the defect count of each flaw type before and after epitaxial

TABLE 3-- Average counts of surface defects on wafers and their standard deviation (in parentheses) at each step of substrate cleaning.

Defects	Type 1	Type 2	Type 3	Type 4	Type 5
1. As-polished wafers	15.5 (8.1)	5.76 (6.1)	1.92 (4.9)	8.36 (12.9)	1.60 (6.2)
2. Hot HNO_3 clean	2997 (375)	1313 (38)	37.8 (15)	1351 (390)	0 -
3. Scrub Clean	172 (375)	31 (8)	2 (2)	34 (10)	28 (17)
4. Hot HNO_3 clean and scrub	136 (32)	25 (6)	1.7 (1.5)	27 (9)	17 (19)
5. Hot HNO_3 with ultrasonic followed by scrub	447 (135)	148 (145)	23 (30)	173 (174)	61 (58)
6. FSl clean	248 (126)	37 (27.9)	4.9 (7.2)	24.3 (32.3)	47.0 (31.8)
7. Scrub	120 (27.1)	19.6 (9.9)	3.4 (4.3)	23.6 (14.2)	7.2 (12.6)

growth using B-doped substrates. Fig. 7(b) shows the same kind of plot for Sb-doped substrates. From the results of the plot shown in these two figures we can suggest the following: (1) The defects present in the substrates seem to propagate into the epitaxial layers, (2) The predominate defects on the substrates and epitaxial layers are flaw types 2, 3, and 4, which correspond to particulates, light scratch marks and surface damage caused by tweezers, and (3) The surface defect density is increased by the epitaxial growth process.

The increases in the surface defects by the epitaxial growth process as shown in Fig. 7(a) and (b) are fairly typical for most of the epitaxial films deposited on clean substrates even under different growth parameters. The typical clean substrates are those where the total surface defect counts are generally less than 100 for 3" diameter wafers. However, when the surface defect counts are greater than one thousand with the majority of them being types 1 and 2 flaws, the epitaxial growth can reduce the defect counts. These defects are most likely removed during the prebake cycle in which pure H_2 or a few percent HCl in H_2 is used. The reduction of types 1 and 2 flaws causes the appearance of additional type 9 flaws (Fig. 6(f)); however the amount of type 9 flaws remain much less than the reduced number of types 1 and 2 flaws. Flaw type 9 is an etched mesa resulting from the uneven etching of HCl in the prebake cycle.

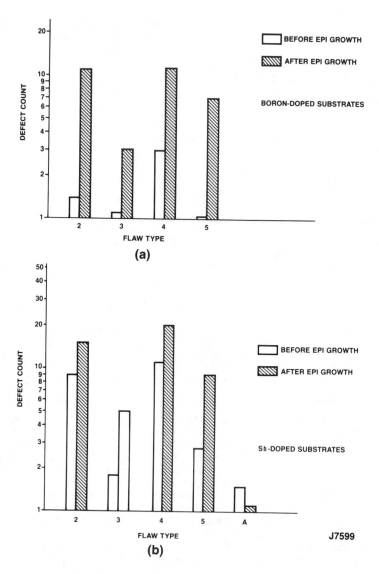

Fig. 7— Plot of defects before and after epitaxial growth. (a) Boron-doped substrate. (b) Sb-doped substrate.

Detection of oxidation-induced defects: The characterization of incoming substrates or epitaxial silicon wafers includes the evaluation of oxidation-induced defects. These defects are revealed by the growth of a few thousand Å thick thermal oxide film followed with a preferential etch. The defects generally consist of shallow etch pits and oxidation-induced stacking faults. A visual evaluation of the oxidation-induced defects is easier than an evaluation of raw wafers because of the existence of a higher defect density (typically from 10^2 to $10^5 cm^{-2}$ as compared to $< 10^2 cm^{-2}$). The laser scanners were also used for the evaluation of these defects. In this case the etching time for revealing the defects was reduced to less than 30 seconds. Table 4 lists the number of flaw types on the epitaxial films deposited on the Sb and B-doped substrates. It is interesting to point out that the epitaxial defect density is higher with the Sb-doped substrate than with the B-doped substrate. This is consistent with the finding of Secco d'Aragonna et. al. [10]. Furthermore, Table 4 also shows that the epitaxial defects in the Sb-doped substrate are predominantly flaw type 2, while those in the B-doped substrate are predominantly flaw type 1.

TABLE 4-- Flaw count of the oxidation-induced epitaxial defects on the Sb-doped and B-doped substrate.

	Flaw Type											
	1	2	3	4	5	6	7	8	9	A	B	C
L	1937	4079	127	4264	0	0	6	2	7	47	0	1
G	147	5599	222	6094	0	0	67	3	89	163	10	0

Legend: L=B-doped substrate; G=Sb-doped substrate.

Correlation of scanner detected defects with electrical defects: Figs. 8(a) and (b) compare the wafer map of the gate breakdown voltage of MOS capacitors and surface defects detected by a laser scanner. The minus sign indicated the capacitors which experienced breakdown at < 2MV/cm. The dot indicates breakdowns occurring between 2 and 6 MV/cm, while the plus sign indicates breakdowns at greater than 6MV/cm. Good correlation between the breakdown field and the defect map (as shown in Fig. 8) is observed only for large capacitors (20,000 mil^2). The small capacitors occupy only a very small fraction of the area in the process control chip while the laser scanner records the whole area of the chip. This is possibly the main reason why a good correlation is hard to observe.

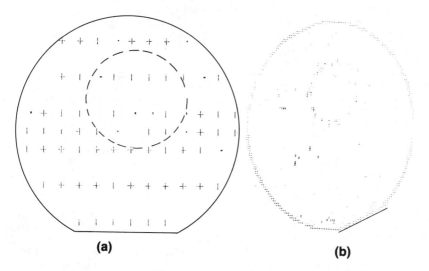

Fig. 8-- Comparison of (a) oxide breakdown voltage map to (b) defect map.

The defect densities calculated from the capacitor yield curves were obtained and compared with the defect counts obtained from the laser scanner. Table 5 lists these results. This table shows that the best correlation was observed between the type 4 flaw and the defect density claculated from electrical measurements.

TABLE 5-- Defect densities calculated from MOS capacitor yield curves and defect counts obtained from a laser scanner.

WAFER NUMBER	MOS MEASUREMENT DEFECT DENSITY	LASER SCANNED DEFECT COUNT TYPE					
		1	2	3	4	5	6
3G11-02	2.36/sq cm	175	92	20	115	45	25
3G11-03	2.52/sq cm	196	95	17	112	46	17
4G11-03	6.00/sq cm	190	95	24	122	64	25
5G11-02	0.80/sq cm	205	73	12	88	46	29
5G11-03	6.83/sq cm	355	157	23	182	0	0
5G11-04	2.00/sq cm	175	93	13	110	53	29

DISCUSSION AND SUMMARY

The current laser scanner used for automatic wafer inspection can be broadly classified into two categories: (i) those which contain a single detector (or collector) with a wide detection solid angle, and (ii) others which contain double detectors with a narrow detection solid angle. Both categories of the scanner are fairly agreeable in terms of differentiating poor wafers from good ones. However, the quantitative correlation is somewhat weak at the present time. This is partly due to the differences in the detector design of the scanners, and partly due to the differences in the computer software which converts the electrical signals into defect sizes. Nevertheless, the defect counts from various scanners are agreeable within the same order of magnitude.

We have found that the surface defects detected by the laser scanners are in fairly good correlation to visual inspection under an intense light or microscope. Using a scanner with two collectors we have found that the nature of surface defects can be classified into those that scatter light isotropically, and those that scatter light anisotropically. The first type of defects includes small particulates, saucer pits, light scratched marks, and oxidation-induced stacking faults. The latter includes large etch pits, pyramids, and epitaxial stacking faults. The defects which scatter light isotropically are most sensitive to the scattered beam collector, while those that scatter anisotropically are most sensitive to the specular beam collector. It has to be pointed out that the orientation of crystallographic defects are directional.

The laser scanner can count the total number of apparent defects from which the defect density can be calculated. In the inspection of raw wafers which generally contain less than 100 defects, the defect count provided by the laser scanners are fairly representative of the true value. However, the defect density in the oxidized wafers is generally much higher, and the defect count obtained from a laser scanner is always lower than the true value. This is not due to lack of detection resolution in the laser scanners but due to the computer software of the scanner. Most scanners may count only one defect within a predefined surface area even though they might detect a multiple number of defects.

We have shown that laser surface scanning can be used for the process control of substrate cleaning and epitaxial growth. This may lead to significant improvement in the device manufacturing yields since certain types of surface defects are correlated to the MOS capacitor yields as demonstrated in this work.

ACKNOWLEDGEMENT

The authors would like to acknowledge Dr. Kent W. Hansen for his encouragement of this work. His critical reading of the manuscript is also greatly appreciated.

REFERENCES

(1) Fejes, P.L., Liaw, H.M., and Secco d'Aragona, F., "Structural Characterization of Processed Silicon Wafers", IEEE Transaction on Componenet, Hybrids, and Manufacturing Technology, Vol. CHMT-6, No. 3, 1983, pp. 314-322.

(2) Muraka, S.P., Seidel, T.E., Dalton, J.V., Dishman, J.M., and Reed, M.H., "A study of Stacking Faults During CMOS Processing: Origin, Elimination and Contribution to Leakage", Journal of Electrochemical Society, Vol. 127, No. 3, 1980, pp. 716-724.

(3) Parrillo, L.C., Payne, R.S., Seidel, T.E., Robinson, M., Reutlinger, G.W., Post, D.E. and Field, Jr. R.L., "The Reduction of Emitter-Collector Shorts in a High-Speed All-Implanted Bipolar Technology", IEEE Transaction of Electron Devices, Vol. ED-28, No. 12, 1981, pp. 1508-1514.

(4) Laneuville, J., Marcoux, J., Orchard-Webb, J., "Defect Characterization of a Silicon Gate CMOS Process", Semiconductor International, May 1985, pp. 250-254.

(5) Gara, A.D., "Automatic Microcircuit and Wafer Inspection", Electronic Test, May, 1981, pp. 60-70.

(6) Galbraith, L.K., "Automated Detection of Wafer Surface Defects by Laser Scanning", Silicon Processing, ASTM STP 804, O.C. Gupta, ed., 1983, pp. 492-500.

(7) Logan, C., "Analyzing Semiconductor Wafer Contamination", Microelectronic Manufacturing and Testing, March, 1985, pp. 9-10.

(8) Burggraaf, P., "Wafer Inspection for Defects", Semiconductor International, July 1985, pp. 56-65.

(9) Murphy, B.T., "Cost Size Optima of Monolithic Integrated Circuits", Proc. IEEE, Vol. 52, Dec., 1964, pp. 1537-1545.

(10) Secco d'Aragona, F., Rose, J., and Fejes, P.L., "Outdiffusion, Defects and Gettering Behavior of Epitaxial N/N+ and P/P+ Wafers Used for CMOS Technology", Proceeding of 3rd International Symposium on VLSI and Technology, W.M. Bullis, S. Broydo, eds., The Electrochemical Society, 1985, pp. 106-117.

Lawrence D. Dyer

DAMAGE ASPECTS OF INGOT-TO-WAFER PROCESSING

REFERENCE: Dyer, L. D., "Damage Aspects of Ingot-to-Wafer Processing ," Emerging Semiconductor Technology, ASTM STP 960, D. C. Gupta and P. H. Langer, Eds., American Society for Testing and Materials, 1986.

ABSTRACT: Intensive efforts have been put into the growth of silicon crystals to suit today's solar cell and integrated circuit requirements. Each step of processing the crystal must also receive concentrated attention to preserve the grown-in perfection and to provide a suitable device-ready wafer at reasonable cost.

A comparison is made between solar cell and I.C. requirements on the mechanical processing of silicon from ingot to wafer. Specific defects are described that can ruin wafers or can possibly lead to device degradation. These include grinding cracks, saw exit chips, crow's-foot fractures, edge cracks, and handling scratches.

KEYWORDS: silicon damage, ingot fracture, wafer cracks, mechanical processing

INTRODUCTION

There is abundant literature on the growth and annealing of silicon crystal ingots to provide the proper resistivity, oxygen concentration, and oxygen clustering for electronic devices. In contrast, there is less published on the mechanical processing of the ingot into slices, although this processing affects the cost and survival potential of the wafers and may affect the device quality. The purpose of the present report is to explain how certain fracture defects arise in ingot-to-wafer processing, how their recognition can lead to preventive measures, and how trends in wafer requirements for the I.C. and solar cell industries suggest the adoption of some new equipment options and concentration of future development work in selected areas of mechanical processing.

Dr. Dyer is a Senior Member of Technical Staff in the Silicon Slice Products Department at Texas Instruments Inc., Sherman, Texas, 75090.

DISCUSSION

First, some remarks will be made on the nature of mechanical processing damage and the effect of chemical etching on the resistance of the slice to breakage. Second, solar and I.C. processes will be compared in their exposure and sensitivity to damage. Third, specific defects that occur step-by-step through ingot-to-wafer processing will be described and discussed. Fourth, what is needed for gains in the mechanical processing of both solar cell and I.C. wafers will be assessed.

Nature of Damage and Effect of Chemical Etching

Mechanical damage to crystalline silicon from abrasive processes consists essentially [1,2] of two superimposed types of defects: 1) a thin layer containing dislocations and debris, and 2) microcracks extending 10 to 20 times deeper, which may have dislocation cracks extending somewhat more deeply [3,4].

Resistance of silicon to chipping or breakage during mechanical processing theoretically depends on the size and sharpness of the microcracks that have been introduced. According to the Griffith formula [5,6], fracture stress depends on the inverse square root of the crack size. If σ_f is the tensile stress for crack propagation, c the crack length, E Young's modulus, and γ the surface energy,

$$\sigma_f \cong \sqrt{\frac{E\gamma}{c}} . \qquad (1)$$

Other formula show (Cottrell [7]) that, for an excavated notch whose root has the radius ρ, the fracture stress depends directly on the square root of the ratio between this radius and the atomic radius a:

$$\sigma_f \cong \sqrt{\frac{E\gamma}{c} \cdot \frac{\rho}{a}} . \qquad (2)$$

This means that if an etchant were to widen cracks laterally at the same rate that the surface is attacked, and if the crack tip were made perfectly semicylindrical in so doing, the fracture strength of an abraded material would be increased greatly.

A common misconception in the semiconductor industry regarding the removal of mechanical damage is that the layer containing the damage must be removed completely to make the material strong enough for further processing. For example, if the maximum depth of cracks after grinding and flatting a silicon crystal amounts to 3.5 mils, it is thought that the etching step has to remove at least this much from the crystal radius to "get rid of the damaged layer". After a slight safety margin is added, the etching might call for 10 mils to be removed from the diameter. On the other hand, according to the understanding that was expressed above, a very brief etching that widened the crack to 2 microns and rounded the crack tip would improve the bending fracture strength by $(10000/1)^{0.5}$ = 100. At present the effect of various etchants in this regard is not known. One may not even need in all cases to remove silicon until an optimum strength is obtained (0.7 mil per side for a sawed slice according to Guidici [8]).

Comparison of Solar and I.C. Processes

Table I shows the main steps in I.C. and solar cell mechanical processing, as well as some possible crack defects that may occur in those steps. Causes of some of these defects have been identified by fracture tracing techniques [9,10]. Solar cell processing has fewer steps, hence less exposure to damage. In addition, solar cell requirements are less severe from a cosmetic standpoint; defects such as the exit chip in sawing are not necessarily a cause for slice rejection in this case. On the other hand, since solar cell wafers are sliced thinner than I.C. wafers, edge and crow's-foot fractures are more likely.

TABLE 1. STEPS IN INGOT-TO-WAFER PROCESSING vs. POSSIBLE FRACTURE DEFECTS

FOR SOLAR CELLS	FOR INTEGRATED CIRCUITS	FRACTURE DEFECT
CROP	CROP	LARGE CHIPS
	SAMPLE	SAME AS WAFERING DEFECTS
GRIND OR SLAB	GRIND	EXCESS CRACK DEPTH
	ORIENT	
flat (opt.)	FLAT	EXCESS CRACK DEPTH
	ETCH	
INGOT HANDLING	INGOT HANDLING	EDGE DAMAGE SUCH AS INGOT FLAKE CHIPS
WAFER (18 mils)	WAFER (24 mils)	EXIT CHIPS, EDGE FRACTURES DEEP MICROCRACKS, EDGE CHIPS
	LASER MARK	
edge gr. (opt.)	EDGE GRIND	CROW'S-FOOT FRACTURES
	HEAT TREAT	EDGE CHIPPING AND CRACKS
	LAP	CROW'S-FOOT FRACTURES & SCRATCHES
ETCH	ETCH	
	BACKSIDE DAMAGE	
	POLISH	EDGE CRACKS
FURNACE PROCESSING	FURNACE PROCESSING	WAFER FRACTURES, EDGE CHIPS, CRACKS, PARTICLES

Specific Defects

Cropping chips.-Various types of fracture losses can occur in cropping the top and tang ends of a crystal or in cutting it to standard lengths. Figure 1 shows a fracture created from using too much blade force in a cropping saw. The problem is similar to exit chip formation in slicing, described below; and therefore is found mainly on <100> ingots. A slab of silicon as thick as the chip is lost immediately at this work station. Other fractures may not show up until the slicing stage as was the case in Figure 2. These problems can obviously be solved by limiting the force at the cropping saw and the severity of handling the crystal at each step.

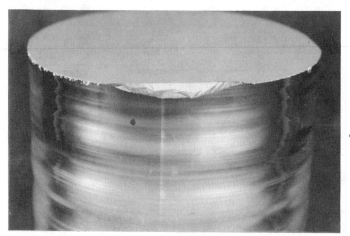

Figure 1. Cropping Chip on 125mm <100> Silicon Ingot

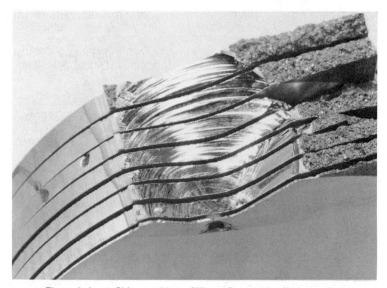

Figure 2. Ingot Chip on 100mm Silicon, Detected at Wafering Step

<u>Grinding cracks</u>.--Figure 3 shows grinding cracks made by a centerless-type crystal grinder some years ago. The defects were found at wafer epitaxy, where they had caused slip. The crack pattern is that of the Hertzian crack [11,12], modified by the cylindrical geometry. Figure 4 shows a schematic diagram of slip generation from excessively deep cracks formed at the crystal grinding step. Figure 5 shows grinding cracks made by a center-type crystal grinder. Clearly there is a need to limit the severity of the grinding process to mimimize such cracks; one such means is provided on the Siltec crystal grinder [13], i.e., by measuring the current drawn by the drive motor [14].

Figure 3. Cracks Produced by Centerless Crystal Grinder and Subsequent Slip in Epitaxial Processing

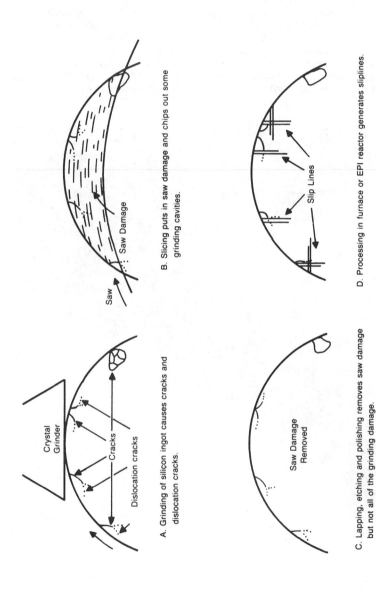

Figure 4. Schematic of Slip Generation from Excessively Deep Crystal Grind Cracks A. Grinding of Silicon Ingot Causes Cracks and Dislocation Cracks. (B) Slicing Puts in Saw Damage and Chips out Some Grinding Cavities. (C) Lapping, Etching, and Polishing Removes Saw Damage But Not all of the Grinding Damage. (D) Processing in Furnace or EPI Reactor Generates Sliplines.

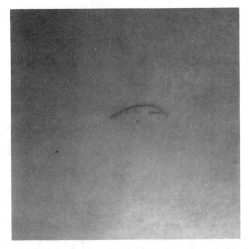

Figure 5. Cracks Produced by Center-type Crystal Grinder

Flake Chips on Ground Ingot Surfaces.--Figure 6 shows a type of breakage that creates very heavy losses relative to the volume disturbed. It consists of a shallow chip almost parallel to the ingot axis, and is caused by end contact with other heavy objects after grinding. One way of preventing these losses is to bevel or round the ends of the ingot.

Figure 6. Flake Chip on Ground Ingot

Exit Chips/Saw Fractures.--Figure 7 shows a saw exit chip that was made with the I.D. saw. This defect is a cosmetic one for the semiconductor industry, but a slice cannot be sold with an exit chip, partly because of the potential of generating increased particle counts in the wafer fabrication facility [15]. In the solar wafer the exit chip can, if severe, constitute a "saw fracture" [17]. In the light of studies by Chen and Leipold, [16], and Dyer [17], the exit chip/saw fracture is probably the major limitation on lower slice thickness for solar-cut (100) wafers [17]. The causes of exit chips have been outlined previously [17]; they fall basically into two categories--bending forces [16] and wedge forces [17]. Since there are numerous ways to cause the exit chip, it remains a perennial problem.

Deep Saw Damage ("Sparkle").--Figure 8 shows an area on a (100) sawed slice that reflects light preferentially at the same angle as the exit chip. This appearance is sometimes called "sparkle" and often accompanies the exit chip. The appearance is due to deeper-than-usual microcracks [17] that are partially revealed by an intersecting crack and show up under the microscope as small, scattered facets. This collection of defects probably has one of the same causes as exit chips, but occurs earlier in the cut where the crystal still has enough breadth to resist complete crack propagation.

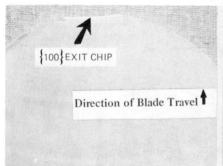

Figure 7. Saw Exit Chip

Figure 8. Deep Saw Damage ("Sparkle") on (100) Silicon

Saw Edge Fractures.--Figures 9 and 10 show a type of defect related to exit chips and which is cause for immediate slice rejection. It is a fracture that initiates at the edge of the wafer and is associated with a depression in the slice caused by saw blade removal. It occurs more frequently in thin slices. Deep saw damage is often evident elsewhere on the slice. It occurs when the I.D. saw blade has lost some of its tension. On (100) slices, the cracks are parallel to saw marks; on (111) slices, they are usually located within 20 degrees of the mounting strip. The mechanism postulated for this defect is that the blade becomes too flexible, all or in part, and is deviant before entering the cut. As it enters the cut, it applies a bending stress to the wafer at its edge. In the (111) case, the edge is kept from fracturing as far as the mounting strip extends. In the (100) case this protection is not enough because of the ease of making the exit chip-like fracture.

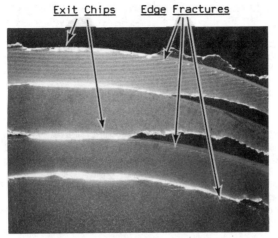

Figure 9. Saw Edge Fractures on (100) Silicon

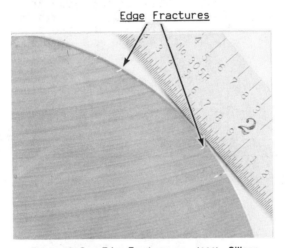

Figure 10. Saw Edge Fractures on (111) Silicon

Figure 11. Crow's-foot Fracture from Burr on Vacuum Chuck

Crow's-Foot Fractures: Figure 11 shows a typical crow's-foot fracture that occurred in the edge-grind process because of a 1.3-mil burr on a vacuum chuck. Figure 12 shows the mechanism by which such fractures occur; an analysis was given previously [18]. Thin slices are more susceptible to this type of fracture. Another operation in which crow's-foot fractures originate is lapping. Note that the mechanism is different from that of Fig. 12; the edge-grind fracture occurs because of bending stresses, the lap fracture is thought to arise from wedging by abrasive particles [19].

Edge Cracks from Heat Treating. Sometimes an annealing operation is carried out to stabilize resistivity. Figure 13 shows a crack formed at a slice heat-treat operation. These cracks can form either from the impact of transfer into the quartz boats or from boat slots that pinch the slice edges. In the instance of Figure 13, the temperature was high enough for dislocation generation. Such defects can also occur during furnace processes that are used in the fabrication of wafers into devices.

Edge Cracks at Polish: Figure 14 shows edge cracks that developed during the polish operation. These are characterized by nearly radial development and by polish cloth beveling. They are envisioned to form in response to the tangential tensile stresses that occur when the slice friction is too high and the slice edge is relatively cool.

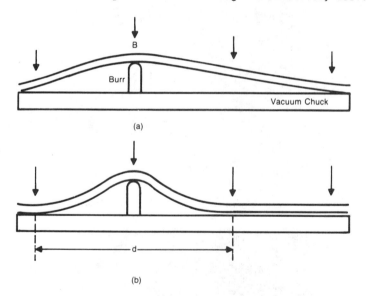

(a) Slice is bent over burr: max. tensile stress at B. Weak vacuum.

(b) Strong vacuum flattens outer regions against plate and increases curvature and stress at "B".

(c) (Not shown). Crack starts near "B" but stops as reverse curvature area is reached. (Within "d").

Figure 12. Mechanism of Burr-induced Fracture on Vacuum Chuck

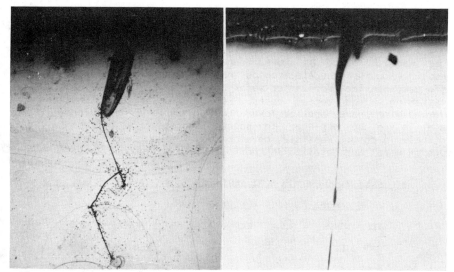

Figure 13. Edge Crack from Heat-treating Silicon Slice in Quartz Boat

Figure 14. Edge Crack at Polish

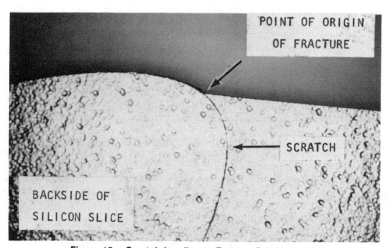

Figure 15. Scratch Leading to Furnace Breakage (30X)

Scratches: Figure 15 shows an example of a backside scratch that caused wafer breakages in furnace operations. The fracture origin [10] in each case coincided with the point of tangency of the scratch. This breakage is postulated to occur as follows: Furnace operations cause wafers to undergo transient stresses that bow the wafer one way or the other. If a scratch is located on the convex side of the wafer, it is subject to tensile stress that can exceed the crack propagation stress for the scratch. Such scratches must be avoided in wafer processing not only for the survival of the individual slice, but for avoiding loss of other devices from particles generated in the fracture process.

Improvements for Solar Cell and I.C. Mechanical Processing

Better mechanical processing methods and equipment are desired by both I.C. and solar cell industries, but the needs are divergent in some ways. Table 2 shows trends in wafer requirements from both industries. For the electronics industry the trends dictate a broad effort to provide for larger wafer sizes, slice traceability, better flattening, and better edges. For the solar cell industry the overwhelming needs are for lower cost and greater efficiency; if any gain is to be achieved in solar ingot-to-wafer processing, it is entirely in cost. Possible developments for achieving gains for each type of wafer are detailed in the following sections.

TABLE 2.--TRENDS IN SLICE REQUIREMENTS, I.C. AND SOLAR CELL

REQUIREMENT	SOLAR	I.C.
SIZE, INCHES	HOLDING AT 4-5	INCREASING TO 8
TRACEABILITY	--	DESIRED
LOWER COST	GREAT PRESSURE	SLIGHT PRESSURE
FLATNESS	O.K.	FLATTER
BOW	O.K.	LESS BOW
HIGHER QUALITY		
o EDGE	O.K.	REDUCED FRACTURES & PARTICLE GENERATION
o SURFACE	O.K.	GREATER SMOOTHNESS, NO BLEMISHES, FRONT & BACK SIDES
o INTERNAL	IMPURITY CONTROL	IMPURITY CONTROL

Needed Developments in Mech. Processing for Solar Cells: The lower cost requirement on solar cell ingot-to-wafer processing puts emphasis on less kerf loss, thinner slices, faster processing, etc., always with the trade-off of more chips and breakage. Increasing wafer size appears to have no benefits for solar cells because of higher I^2R losses, and because the attendant increased metalization to avoid such losses causes sufficient shadowing to negate the benefit.

Considering Tables 1 and 2 together, the hopes for improvements in solar cell ingot-to-wafer mechanical processing lie mainly in two areas: <u>crystal preparation</u> and <u>sawing</u>. In crystal preparation it is clear that limits on sawing force and grinding drag will reduce the defects to a low level; some benefit may be obtained from bevelling the ingot and from some amount of crystal etch after grinding. A brief summary of possible advances in sawing for solar cells follows.

Recent advances in substrate sawing for the electronics industry have been reviewed in 1984, namely large capacity saws, slice retrieval, fault systems and productivity increasers [13]. Among new features called for by users were in-situ blade tension and drag measurement. For the solar cell industry, all but the first of these advances will be useful. Unfortunately, most present equipment does not include the advances; even the blade tracking systems are mostly retrofits. Any of the new large saws may be used, but the larger blade size usually means that a thicker blade core is necessary to obtain the same stiffness, and the kerf loss is increased. The new saws may, of course, be ordered with the smaller head sizes if desired. It is the author's opinion based on the foregoing discussion that some development of new saws especially for solar cell slicing should be done, aiming more at thinner slices and thinner cuts than larger size. A suggested list of studies or developments follows.

o Improvements in blade core strength so that thinner cores may be used.

o In-situ blade tension measurement.

o A method to maintain fairly constant blade tension without sacrificing blade concentricity.

o A method of assessing the degree of damage to the slice while the slice is freshly off the saw. This would allow the operator to take corrective action before many other slices are ruined.

o A study of why rotary crystal slicing doesn't work and the design of a saw specifically for rotary crystal. This type of cutting would permit smaller blades, thinner cores, and less kerf.

o Autobalancing. Silicon Technology Corporation has an autobalancing attachment, presently available for 22- and 34-inch saw heads, that rebalances the saw while running. This feature increases productivity by avoiding shutting down the saw for the imbalance that comes with broken slices, and may be a refinement that permits lower slice thicknesses.

Additional developments needed for I.C.-type wafers with respect to fracture: Considering Tables 1 and 2, improvements in ingot-to-wafer mechanical processing for I.C.s lie not only with crystal preparation and sawing, but with larger equipment, laser marking, and lap development. Possible improvements in these areas to reduce fractures and solve other problems are given as follows:

o For larger ingot diameters, larger equipment is required from crystal grind through polish. Since new equipment is required, it may be possible to accomplish more than upsizing. For example, chipping losses of various kinds can be reduced by combining some of the operations to minimize moving the heavy ingots from one station to another. A crystal grinder can be equipped to machine the crystal diameter, determine the position to be flatted, and grind the flat.

o Laser marking: This new processing step allows identity of the slice to be established and provides traceability back to crystal and polysilicon source. It apparently has no lasting deleterious effect so long as the damage is relieved by a chemical etch [20,21].

o Wafer flatness: Basic flatness of the wafers is generated at the lapping operation. Several recent developments that have promise are:) pressure control through load cell sensing; this feature reduces the chances for breakage and scratches due to over-pressuring [22],

2) thickness control by a frequency gauge that senses the thickness of sacrificial quartz chips, and 3) automatic plate reversal and tachometer speed readout. These allow the lapping plates to be kept flatter in the production environment.

o **Edge Quality:** For I.C. use, the quality of wafer edge is of considerable importance in avoiding chips and particles in wafer fabrication and edge build-up in epitaxial deposition. Factors such as the edge profile, edge finish, and size and shape of flat or notch are all involved in edge quality. For the larger wafer sizes, SEMI has noted increasing interest in replacing the flat with a notch [23]. What is lacking for a rational decision on these factors are data on the strength and particle-generation tendencies for various conditions of edge. An early attempt at edge strength measurement was done by Guidici [8], with a ball-on-ramp impact test. While results in this apparatus are only comparable to each other in short time-frames, they are still valid in a relative sense. For example, Guidici found in this type of edge test that (111) slices were 10 to 23 % stronger than (100) slices, and that CZ wafers were 25 to 53 % stronger than FZ. A more complete study of edge quality factors is clearly needed.

SUMMARY

The nature of mechanical damage and chemical etching to relieve such damage are discussed with regard to ingot-to-wafer mechanical processing of I.C. and solar cell silicon material. Various examples of fracture defects are described and explained. I.C. and solar mechanical processes are compared as well as trends in requirements in slice parameters. These considerations combine to suggest that the main area for innovative improvement in the mechanical processing of solar cell slices lies in aiming at lower cost in the wafering step. For integrated circuits a broad approach is indicated to provide for larger wafer sizes, traceability, and improved flatness and edge quality. Various studies or desired features that would benefit the integrated circuit and solar cell efforts are suggested.

ACKNOWLEDGMENTS

The author would like to express appreciation to Sam Rea and Jim Rinehart for helpful discussions.

REFERENCES

[1] Stickler, R. and Booker, G. R., "Surface Damage on Abraded Silicon Specimens", Philosophical Magazine, Ser. 8, Vol. 8, No. 5, 1963, pp. 859-876.

[2] Kuan, T. S., Shih, K. K., Van Vechten, J. A. and Westdorp, W.A., "Effect of Lubricant Environments on Saw Damage in Si Wafers," Journal of the Electrochemical Society, Vol. 127, No. 6, 1980, pp. 1387-1394.

[3] Allen, J. W., "On the Mechanical Properties of Indium Antimonide," Philosophical Magazine, Vol. 2, No. 24, 1957, pp. 1475-1481.

[4] Allen, J. W., "On a New Mode of Deformation in Indium Antimonide," Philosophical Magazine, Vol. 4, No. 45, 1959, pp. 1046-1054.

[5] Griffith, A. A., "The Phenomena of Rupture and Flow in Solids," Philosophical Transactions of the Royal Society of London, 221, 1921, pp. 163-198.

[6] Griffith, A. A., "Theory of Rupture," in Proceedings of the First International Congress for Applied Mechanics (Delft), 1924, p. 55.

[7] Cottrell, A. H., "Theoretical Aspects of Fracture," in Fracture, ed. by B.L. Auerbach, D.K. Felbeck, G.T. Hahn, and D.A. Thomas, The Technology Press of Massachusetts Institute of Technology and John Wiley & Sons, New York, 1959, pp. 20-53.

[8] Guidici, D. C., "Using Wafer Fracture Strength to Evaluate Orientation and Method of Crystal Growth," Insulation/Circuits --June 1978, p. 27.

[9] Dyer, L.D., "Fracture Tracing in Silicon Wafers," The Electrochemical Society Extended Abstracts, Vol. 83-2, 1983, pp. 553-554.

[10] Dyer, L. D., "Fracture Tracing in Semiconductors Wafers," Semiconductor Processing, ASTM STP 850, Dinesh Gupta, Ed., American Society for Testing and Materials, 1984, pp. 297-308.

[11] Frank, F. C. and Lawn, B. R., "On the theory of Hertzian fracture," Proceedings of the Royal Society of London, Vol. 299A, No. 1458,1967, pp. 291-306.

[12] Lawn, B. R., "Hertzian Fracture in Single Crystals with the Diamond Structure," Journal of Applied Physics, Vol. 39, No. 10, 1968, pp. 4828-4836.

[13]   Siltec Corporation, Product Bulletin--Model 540/590 Grinders.

[14]   Dyer, L. D., "Recent Advances in Substrate Sawing Technology," Industrial Diamond Review, Vol. 44, No. 501, 1984, pp. 74-76.

[15]   Dyer, L. D., "Exit Chipping in I.D. Sawing of Silicon Crystals," The Electrochemical Society Extended Abstracts, Vol. 81-1, 1981, pp. 785-786.

[16]   Chen, C.P. and M.H. Leipold, "Analytical Calculation of Thickness Versus Diameter Requirements of Silicon Solar Cells," The Electrochemical Society Extended Abstracts, Vol. 80-2, 1980, pp. 712-713, and Chen, C. P., "Minimum Silicon Wafer Thickness for I.D. Wafering," Journal of the Electrochemical Society, Vol. 129, No. 12, 1982, pp. 2835-2837.

[17]   Dyer, L. D., "Exit Chipping in I.D. Sawing of Silicon Crystals," Proceedings of the Low-Cost Solar Array Wafering Workshop, 8- 10 June 1981, Phoenix, Ariz. Jet Propulsion Laboratory Publication No. 5101-187, (DOE/JPL-1012-66), 1982, pp. 269-277.

[18]   Dyer, L. D. and J.B. Medders, "Defects Caused by Vacuum Chuck Burrs in Silicon Wafer Processing," in VLSI Science and Technology/1984, ed. by K.E. Bean and G. Rozgonyi, The Electrochemical Society, Pennington, New Jersey, 1984, pp.48-58.

[19]   Maruyama, S. and O. Okada, "Crow Track Formed by Mechanical Force on Silicon Crystal Wafer," Japanese Journal of Applied Physics, Vol. 3, No. 5, 1964, pp. 300-301.

[20]   Scaroni, J. H., "Using Lasers to Mark Identification Data on Silicon Wafers," Microelectronic Manufacturing and Testing, April 1982, p. 16.

[21]   Singer, P., "Wafer Marking and Reading", Semiconductor International, Dec. 1982, p. 35.

[22]   Sherer, J. B., cited by W. R. Runyan, Semiconductor Technology, McGraw-Hill Company, New York, 1965, p. 232.

[23]   "SEMI Standard for 200 mm Polished Monocrystalline Silicon Slices (Proposed)", M1 STD.9-84, Semiconductor Equipment and Materials Institute, Inc., 625 Ellis St., Suite 212, Mountain View, California, 94043-2295.

Toshio Shiraiwa and Shoji Inenaga

HYDROGEN IN SILICON AND GENERATION OF HAZE ON SILICON SURFACE IN AGING

REFERENCE: Shiraiwa, T and Inenaga, S.,"Hydrogen in Silicon and Generation of Haze on Silicon Surface in Aging", Emerging Semiconductor Technology, ASTM STP 960, D.C. Gupta and P.H. Langer, Eds., American Society for Testing and Materials, 1986

ABSTRACT: Hydrogen in silicon wafers was analysed by the gas analysis method of metals. Silicon wafers made by routine process contain a few ppm hydrogen. Diffusive hydrogen was observed in p-type silicon. It was proved that these hydrogen is absorbed into the wafer from water containing acid. The present results agree with current reports that hydrogen diffuses into p-type silicon and acts as a compensator of an acceptor.
Haze which is observed on mirror polished surface of p-type silicon was investigated by SEM and ESCA, and they revealed that haze is due to scattering of light by nodules of silicon oxide which grow on the wafer sarface in density of $10^4/cm^2$. The present anthors suggest that oxidation is due to water formed by hydrogen which diffused out to the surface of wafer as well as moisture in the atmosphere.

KEYWORDS: Silicon Wafer, hydrogen analysis, haze, acceptor compensator, ESCA

It has been reported in the recent studies that hydrogen in p-type silicon acts as a compensator of an

Dr. Shiraiwa is an executive technical counselor of Osaka Titanium Co., Amagasaki, Japan: S. Inenaga is a research scientist of Kyushu Electric Metals Co., Kohoku, Saga, Japan.

acceptor forming a bond between a boron and a hydrogen ion or a hydrogen complex.(1-4) In these studies hydrogen was introduced into the silicon wafer from plasma (1-3) and from chemical reagent (4). The behavior of hydrogen in silicon has been investigated by electrical resistance and it has been suggested that the hydrogen related donor is H(1,3,4) or OH(2).

In the steel industries, hydrogen in steel is one of the most important subject of study because it causes brittle fracture. The major origine of hydrogen in steel is solute hydrogen in molten steel or absorbed hydrogen from water containing acid in the pickling process. The analytical method of hydrogen in steel has been progressed for total hydrogen and diffusive one.

The present authors have tried to apply the analytical method used for steel to silicon and ascertained existence of hydrogen in silicon.

On the other hand, it is well known that the surface of a mirror polished silicon wafer is very active and haze appears on the surface when it is stored for a long time even in the desirable environment. In the present study, microscopic investigation has revealed the origin of the haze, and the relation between the haze occurence and hydrogen in the wafer has been investigated.

ANALYSIS OF HYDROGEN

Method of Analysis

<u>Fusion method</u>: This method is widely used in hydrogen analysis of steel. Fig.1 shows the fusion method. The silicon sample is put in a carbon crucible together with iron and tin powder and it is heated up to 1900 c by an impulse electrical current in argon gas flow. The dissociated hydrogen from the sample is detected by a thermal conductivity detector(TCD).

<u>Mass spectrum method</u>: Discharged hydrogen from the sample heated up to 650°C from room temperature in vacuum is analysed by a mass spectrometer. Fig.2 shows the analyser.

Result

<u>Treatment of Surface</u>: The total hydrogen of silicon which was cleaned by RCA reagent (NH_4OH and H_2O_2) and dried by a centrifugal dryer was a few tens ppm and it decreased to a few ppm when the water on the cleaned surface was replaced by acetone and dried. This difference may be due to the adsorbed water on the hydrophylic surface.

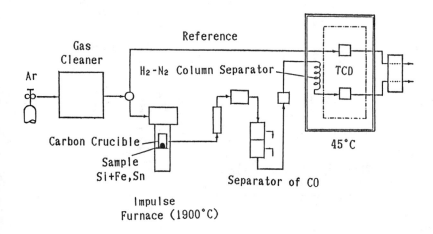

Fig. 1--Block diagram of total hydrogen analyzer.

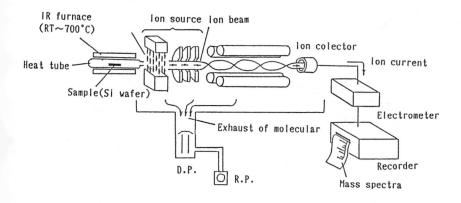

Fig. 2--Block diagram of discharged hydrogen analyzer.

316   EMERGING SEMICONDUCTOR TECHNOLOGY

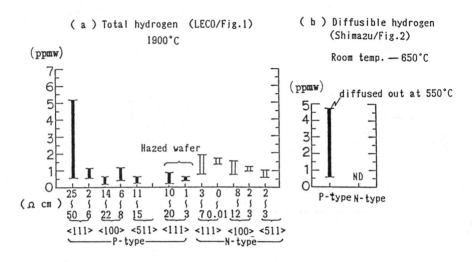

Sample wafer : Polishing --> Cleaning --> Spin dry --> Anhydride by acetone

Fig. 3--Hydrogen in wafers.

Although there is a possibility that remaining acetone on the surface affects the hydrogen analysis, replacing the remaining water by acetone has been adopted for treatment of the surface of samples in order to reduce the adsorbed water.

Total hydrogen and discharged hydrogen: Fig.3 shows the results of hydrogen analysis. The length of the orthogonal bars show the scattering of measured results. The samples are finished mirror polished wafers from routine process. Total hydrogen is observed in both p-type and n-type silicon, but discharged hydrogen up to 650 °C is observed only in p-type silicon. The peak of discharging is at 550 °c at the heating rate of 25 °c/min. The present results show hydrogen is absorbed into silicon and hydrogen in p-type silicon is diffusive,but it does not diffuse in n-type silicon.

Origin of hydrogen: Total hydrogen of p-type silicon after various process of wafer making is shown in Fig.4. It clearly shows that the hydrogen is absorbed from water containing chemical reagent as reported (4). In the present experiment the silicon wafer after each process

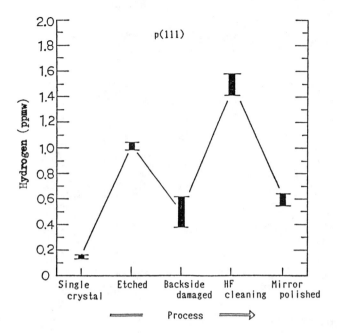

Fig. 4--Hydrogen after various process.

was cooled by liquid nitrogen in order to suppress discharge of diffusive hydrogen before analysis.

Electrical resistance: Fig.5 shows change of electrical resistance of p-type silicon in aging, 80°C×2hrs. It is higher than the estimated value from dopant concentration immediately after the last cleaning process of wafer and it decreases after aging. The time needed for aging depends on temperature.

Conclusion of Hydrogen Analysis

The experimental results agree with the current reports that hydrogen (or OH) in p-type silicon composes bond with boron and electrical resistivity increases and the hydrogen can also diffuse in the silicon. In these reports the existence of hydrogen was suggested from preparation process of specimen, such as exposing the wafer to hydrogen plasma or acid etchant. The present analysis has directly ascertained the existence of hydrogen. The hydrogen exists in wafers made by routine process and the major origin is suggested to be in

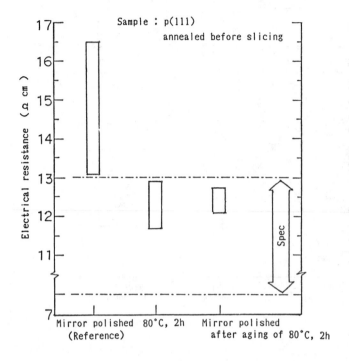

Fig. 5--Aging effect for Electrical resistance of p-type silicon.

etching process and hydrogen is absorbed from wafer containing acid. The hydrogen in p-type silicon is diffusive and it affects electrical property. In the present experiment, hydrogen behavior has been examined mainly for p-type silicon. Study of hydrogen in n-type silicon is remained in future.

HAZE ON WAFER SURFACE AFTER AGING

Haze or fogging on a wafer surface sometimes occurs when the wafer has been stored for a long time. The occurence is accelerated by coexistence of a small quantity of water in the packing case, and this fact suggests that haze occurence is water related phenomena. However it happens in dryed ambient atmosphere and the reason has been investigated related to hydrogen in the wafer, where hydrogen diffuses out to the surface and generate water combining with oxygen.

(a)

(b)

Fig. 6--SEM images of haze.

Microscopic Observation

SEM images: Fig.6 shows SEM images of hazed surface of silicon wafer. Fig.6 (a) is a image of low magnitication and many nodules (small bright spots) are observed in the density of $10^4/cm^2$. Fig.6 (b) shows an enlarged image of a nodule. Besides a nodule, extended structure of lower height is observed on the surface.

X-ray and ESCA analysis: In the x-ray spectrum analysis of the nodule, a solid state detector detected Sik x-rays only, which can detect elements heavier than sodium.

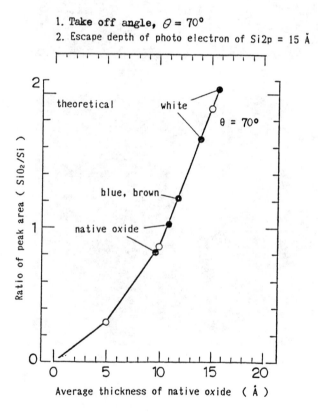

Fig. 7--ESCA analysis of hazed surface.

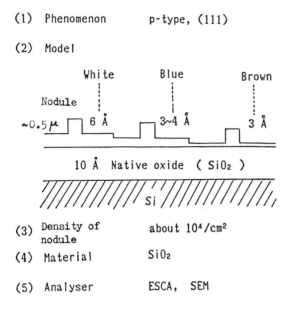

Fig. 8--Schematic illustration of haze

ESCA analysis shows silicon and oxygen on the wafer surface and oxygen intensity increases on hazed surface as shown in Fig.7. In Fig.7 the theoretical intensity ratio of the Si2p photo electron from SiO_2 and Si for the thickness of SiO_2 is shown, where the take off angles is 30 . The calculation is carried on the assumption that the oxide layer is uniform and escape depth of Si2p photo electron is 15A. The black circles are the experimental results of the hazed wafers. The analysed area is 2mmϕ. The color, white and blue, is the color of haze observed. The nodules grow up to nearly 0.5 μ as shown in Fig.6 (b), but contribution of nodule to intensity of photo electron in ESCA may be very small because the escape depth is only 15Å and the surface ratio of nodules to observed surface is very small. Then the estimated thickness of oxide layer in Fig.7 means the thickness of uniform oxide layer grown on the surface.

The haze appearance is due to the light scattering by nodules. The change of the color may be due to the density and dimension of the nodule.

Origin of haze: These experimental results suggest that the haze appearance is due to the oxidation of the surface, where oxide nodules grow on the surface and scattering of light by them gives the appearance of haze. Fig.8 shows a schematic illustration of haze.

Haze and Water

As mentioned already, haze is induced by water. For example if 20mg/l of water is put in the packing case of wafers, haze occurs on the mirror polished surface of all kinds of silicon wafers after one week aging. When the wafer is stored in dry ambient atmosphere, haze occurence is scarce and it is limitted to p-type silicon.
Haze induced by water is thought to be due to oxidation by existence of water, and present authors sugest that the same phenomen has occured on the p-type silicon on which haze is observed after aging. That is, hydrogen is absorbed into wafers in making process and if it remains after finishing, there is a possibility that it diffuses out in p-type silicon and makes water reacting with OH in natural oxide of hydrophilic surface or oxygen in atmosphere. If it is assumed that the hydrogen concentration is 1ppmw in the wafer of 0.5mm thickness and this hydrogen diffuses out to the surface and makes water, the surface of the wafer is covered by the water film of 50Å thickness.
Diffusive hydrogen is observed only in p-type silicon and this is the explanation why haze is observed only in p-type silicon in a dry atmosphere.

CONCLUSION

The gas analysis method of metals has been applied to silicon wafers and the existence of hydrogen in the wafers has been ascertained. The hydrogen is absorbed from water containing acid. The hydrogen in p-type silicon is diffusive, and it may induce the haze by making water at the surface in aging.
SEM and ESCA investigations revealed that this haze appearance is due to growth of SiO_2 nodules in density of $10^4/cm^2$.

ACKNOWLEDGMENT

The authors would like to express their sincer thanks to Dr. N. Fujino of Central Research Laboratories and Mr. Y. Yoshihara of Kokura Steel Works of Sumitomo Metal Ind.

tor their assistance to our works by analysis of hydrogen.

REFERENCE

(1) Pankove, J.I. and Berkeyheisei, J.E. "Neutralization of Acceptors in Silicon by Atomic Hydrogen", Applied physics Letters, Vol.45, No.10, 15 Norvember 1984, pp.1100-1102.
(2) Hansen, W.L., Pearton, S.J. and Haller, E.E., "Bulk Aceptor Compensation produced in p-type silicon at near-ambient Temperatures by H_2O Plasma", Applied Physics Letters, Vol.44, No.6, 15 March 1984, pp.606-608.
(3) Tavendale, A.J., Alexiev, D. and Williams, A.A., "Field Drift of the Hydrogen-related, Acceptor-neutralizing Defect in Diodes from Hydrogenated Silicon", Applied Physics Letters, Vol.47, No.3, 1 August 1985, pp.316-318.
(4) Vieweg-Gutberlet, F.G. and Siegesleitner, P.F., "A Model to Explain the Electrical Behavior of p-type silicon Surface after a Chemical Treatment". Journal of Electrochmical Society, Vol.126, No.10, October 1979, pp.1792-1794.

Margareth C. Arst

IDENTIFYING GETTERED IMPURITIES IN SILICON BY LIMA ANALYSIS

REFERENCE: Arst, M. C., "Identifying Gettered Impurities in Silicon by LIMA Analysis," Semiconductor Processing, ASTM STP 960, D. C. Gupta and P. H. Langer, Eds., American Society for Testing and Materials, 1986.

ABSTRACT: LIMA - Laser Induced Mass Analysis - was used to identify impurities captured in defect structures formed by three different backsurface damage systems. Analysis data indicate that a poly layer and wet sandblast damage provide greater getter efficiency than abrasive SVG backsurface damage.

In this study oxidized CZ and FZ silicon samples were evaluated. Impurities such as Na, K, Al, Cr, Cu as well as H, C, O and Cl could be identified in the backsurface structures and the oxide layer, but not in the silicon bulk beyond the getter related defects. The so identified gettered impurities were directly correlated to the defect density observed on the front surface, while also considering the impact of the interstitial oxygen concentration in the CZ wafer samples.

The LIMA technique uses a Nd:Yag laser to vaporize and ionize portions of the sample. The generated ions are then injected into a "time of flight" mass spectrometer, which displays a complete mass spectrum within microseconds. This capability to vaporize and characterize microvolumes of sample materials in relatively short times make this an attractive technique for bulk analysis.

KEYWORDS: gettering, laser ionization mass analysis, backsurface damage, evaporate, ionize, contaminants

Dr. Margareth C. Arst is Engineering Manager for Fab Process Materials Engineering, CQ and R at Signetics, 811 E. Arques Avenue, Sunnyvale, CA 94086.

INTRODUCTION

Extrinsic and intrinsic gettering is generally applied in IC processing to assure both low metallic contamination and reduced structure imperfections in and near the device active areas.

Extrinsic gettering shows getter activity from the beginning of the first processing step onward, whereas intrinsic - oxygen related - gettering requires an incubation period until it becomes functionally effective [1]. During these first processing steps, the getter capability depends nearly exclusively on the extrinsic backsurface damage related structures/getter sinks to capture metallics and other trace impurities. These to be gettered impurities, may be contained in the incoming wafer material, result from surface contamination or be added during processing in the fabrication line [2].

It was the intent of this study to find and apply a technique which can qualitatively and possibly quantitatively identify gettered impurities in backsurface damage structures of processed wafers.

The more established material analysis techniques like secondary ion mass spectrometry (SIMS) and AUGER Electron Spectrometry (AES) per se, are not well suited to analyze bulk or layer structures to a depth of 5 um and larger. Therefore, we investigated the application of a recently introduced technique, the Laser Ionization Mass Analysis (LIMA) technique.

The principle of the LIMA technique is the mass and intensity analysis of the ionic components formed by high power laser irradiation of a material surface. The analysis area in the LIMA technique ranges from 1 to 5 um in diameter and the analytical depth can range from several monolayers to approximately 1 um per laser shot. The analysis depth depends primarily on the laser power density absorbed by the sample. Samples can be loaded into the analyzer chamber within minutes and a complete mass spectrum, identifying both elemental and molecular species, can then be produced in less than a second. Due to this rapid mass analysis process of the ion plume created by a single laser pulse, repeated pulses of the laser energy onto the same area of the sample accomplish rapid depth profiling. This feature of the laser microprobe analysis technique provides a distinct advantage compared to other established analysis methods when investigating residual impurities in successive layer structures which are present on or in wafers during the various stages of IC processing.

INSTRUMENTATION - TECHNIQUE

LIMA analysis of the samples evaluated in this study was performed by Charles Evans Associates*, on a LIMA 2A reflection mode laser Microprobe [3]. The schematic diagram of this instrument is shown in Figure 1. The operational details of this technique can be best described by the flow of events during sample analysis:

- Insert sample - without specific preparation - into the main analysis chamber. The precise analysis point on the sample is indicated by a He : Ne spotting laser and displayed on a TV camera or by a high power microscope.

- A short pulse of the Q-switched Nd : YAG laser, focused on the selected sample spot evaporates and ionizes a microvolume of the sample.

- The ionization process produces both positive and negative ions which can be accelerated and focused into the time-of-fight (TOF) mass analyzer. Only one ion polarity can, however, be analyzed for each laser shot.

- The ions are separated according to their velocity (mass) as they travel along the drift tube. The ion reflectron located at the end of the flight tube provides second order energy focusing which improves the mass resolving power of the TOF spectrometer.

- The mass separated ions are sequentially detected by the electron multiplier and the electron multiplier output signal is amplified, digitized and stored on a transient recorder for further data manipulation by an on-line computer.

- The total acquisition time for one spectrum is typically less than 1/10 of a second.

Another unique feature of the LIMA technique is its ability to analyze for both the molecular and elemental composition of the same sample material. The microanalysis for molecular constituents is achieved by operating the laser at relatively low power densities (less than 10^8 W/cm^2) which minimizes the heating of the vaporized sample and therefore significantly minimizes decomposition of certain classes of molecular compounds. The higher laser power density of $10^8 - 10^{12}$ W/cm^2 is more sensitive for elemental analysis. Independent of the laser power density, the ionization process can produce either positive or negative ion mass spectra. The choice of the detected polarity primarily depends upon the efficiency with which the elements (or molecules) of

---

*Charles Evans Associates, 1670 South Amphlett Blvd., San Mateo, CA 94402.

interest form positive or negative ions [3, 4]. For example, in elemental analysis the alkalis and alkaline earths form positive ions more efficiently, while hydrogen, the halogens, oxygen, sulfur, etc., more readily form negative ions.

The elemental detection sensitivities are in the ppm range. The simplicity and speed of this technique make it time effective and economical to measure multiple spots on a surface, improving the accuracy of the analysis.

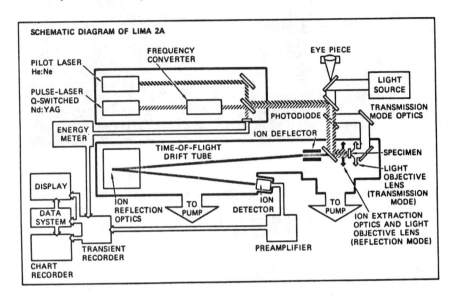

Figure 1 Schematic diagram of LIMA 2A reflection mode laser microprobe

Experimental Aspects

a) Wafer samples with three different types of back- surface damage were used in this study (Figure 2). Type one had a 1.3 um poly layer on the wafer backsurface; type two had an abrasive damage application - SVG damage - and type three had "wet sandblast" damage. During the first thermal cycle damage types two and three develop defect structures due to the mechanical stress induced during the damage applications. Figures 2 A, B and C identify these structures through TEM cross sections.

b) Czochralski (CZ) and Floatzone (FZ) grown silicon wafers, each with one of the above described types of backsurface damage, were processed through a high (1100°C) or high-low (1100/800°C) or low (800°C) temperature oxidation cycle.

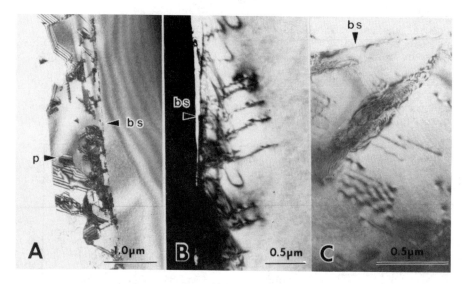

Figure 2 TEM Cross Sections

    A) 1.3 um poly layer annealed at 1150°C/100 min.
    B) SVG abrasive BSD annealed at 1150°C/30 min.
    C) Wet sand blast BSD annealed at 1000°C/150 min.
    p = poly, bs = backsurface

c) Sample testing by LIMA analysis was designed to first analyze the oxide layer, then the poly layer or the backsurface damage defect structures and finally the monocrystalline wafer bulk (Figure 3). Multiple areas (1-3 um diameter) were analyzed on each wafer sample to identify the local distribution of gettered impurities in the layers of each specimen.

d) Complementing the LIMA data, SIMS analysis using an oxygen primary beam was then applied to display the lateral distribution of selected impurities. These secondary ion images obtained at different depth levels in the backsurface getter structures showed the distribution - and concentration - variations of these impurities with increasing distance from the backsurface.

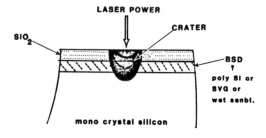

Figure 3 Schematic of sample testing by LIMA

RESULTS AND DISCUSSIONS

Twenty samples were evaluated by the laser microprobe technique and the results were correlated to the defect/S-pit density on the front surface.

Sample characteristics and LIMA data are summarized in Table 1. LIMA spectra typical for specific backsurface damage systems and wafer types are displayed in Figures 4 and 5.

IMPURITIES DETECTED BY LIMA ANALYSIS
(WAFER BACKSURFACE)

TABLE 1

SAMPLE TYPE BSD	IMPURITIES DETECTED IN THE			INTERSTITIAL OXYGEN CONCENTRATION RANGE	HAZE ON FRONT SURFACE
	SiO_2	POLY LAYER	BULK		
CZ POLY	Na, K, Al, Cr, C, N, O	Al, K, Ca, Cr, Cu, C, H, O, F, Cl	-	MED	ZERO
CZ POLY	Al, Na, K, Ca, Mn, Cu	Na, K, Ca, Mn	-	LOW	5%
CZ SVG	Al, Na, K	-	Al, K, C, O	MED	ZERO
CZ SVG	Na, K, Mn	-	-	HIGH	ZERO
CZ WET SAND BLAST	Al, C, N, O	-	Al, K, Ca, Cr, Cu, C, O, H, F, Cl	MED	ZERO
FZ POLY	Na, K, C, Cl	-	-	B. D. L.	15-60%

ALL SAMPLES WERE PROCESSED THROUGH A HIGH THERMAL CYCLE.
B. D. L. - Below Detection Limit

Detailed information of the getter behavior of the various wafer types, as they can be correlated with the laser microanalysis data, can be listed as follows:

o   The elemental detection sensitivity of the laser microprobe mass analysis technique showed the capability to identify contaminants in intentionally created extrinsic getter sites (backsurface damage) in thermally processed CZ wafers (Figures 4 and 5). Trace elements Al, K, Ca, Cr, Cu, C, H, O, F and Cl could be detected in the poly layer and the defect structures related to the sand blast backsurface damage (BSD). Wafers with abrasive damage - SVG - showed only Al, K, C and O in the bulk next to SiO_2/Si interface (Table 1).

o   However, no contaminants could be identified in the backsurface damage structures of FZ wafers.

o   Many elements detected in the backsurface getter centers could also be found in the SiO_2 layer. Additional mass peaks, observed in the positive and negative ion spectra, were Na, K, C and Cl. These impurities could indicate surface and/or furnace contamination.

o   Microanalysis of the wafer bulk, beyond the getter structures, did not reveal trace elements as found in the layers above.

o   Only wafers processed through a high temperature step showed gettered impurities in the SiO_2 and backsurface damage layer mass spectra.

o   Impurities detected by LIMA were verified by SIMS analysis and their lateral distribution in randomly selected areas in the backsurface getter structures were displayed by secondary ion images.

At the present time, laser microprobe mass analysis technique is primarily a qualitative technique producing the mass and intensity of the various components in a material. Quantitative analysis with this technique will require the development of suitable microanalytical standards from which the ion intensity can be correlated with elemental (or molecular) concentration. However, approximate (semi-quantitative) analysis can be performed using internal standard reference specimens which ideally have chemical, optical and thermal properties similar to those of the sample and are analyzed at the same laser power density [3]. Evaluating the positive ions spectrum of a "wet sandblast" backsurface structure (Figure 5) the Cu peak intensity represents approximately 7 ppm.

In general the main success in applying the LIMA analysis technique in this study was the qualitative identification of gettered impurities. Increased activities in silicon and IC product applications should force the need for creating more standards and establishing standard/sample test conditions which should add the capability to better quantify the identified ionic components.

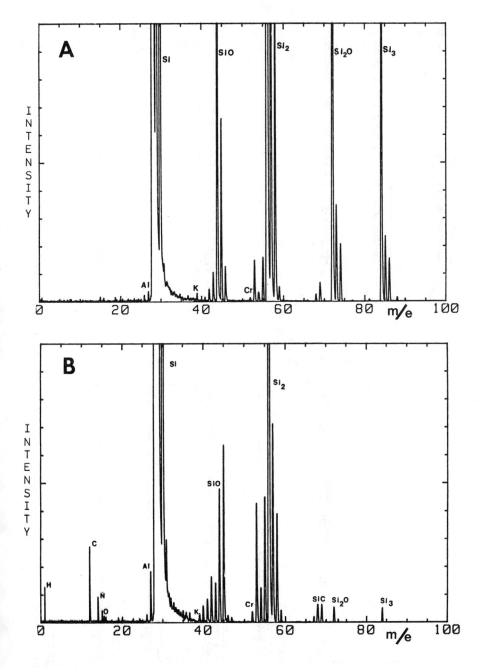

Figure 4 CZ wafer with poly BSD

    A) Positive ion spectrum of SiO_2 layer
    B) Positive ion spectrum of poly layer

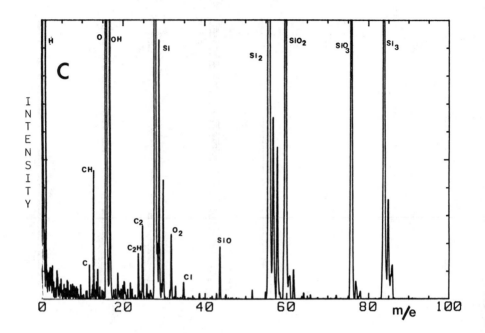

Figure 4 continued
C) Negative ion spectrum of SiO₂ layer

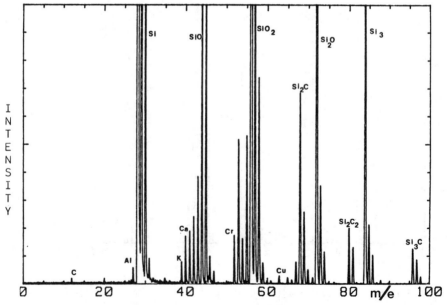

Figure 5 CZ wafer with wet sand blast BSD, positive
ion spectrum of defect structure
(See Figure 2C)

Table 1 lists the contaminants gettered in the backsurface damage regions of CZ wafer samples processed through thermal oxidation at 1100°C/4h. During oxidation, silicon is consumed (approximately 42% of the SiO_2 thickness) and the gettered impurities seem to move with the Si atoms into the SiO_2 layer. Etch removing this contaminated SiO_2 layer during processing reduces the gettered impurities within the extrinsic getter centers. This reduces the quantity of contaminants possibly available for redistribution when the extrinsic getter defect structure out anneals during high temperature processing steps. LIMA data of the SVG samples seemed to indicate getter defect out annealing as the spectra revealed less trace elements than observed for the other two getter systems. It is well established that the thermal stability of these defects, a function of the intensity of the abrasive damage, is much lower than that of the wet sand blast or poly layer related defect structures and grain boundaries.

The data in Table 1 show that LIMA analysis could not detect contaminants in backsurface getter structures in FZ material. Due to the growth method, FZ crystals, incorporate a minimum of residual contaminants. It seems that the impurity levels accumulated from the wafer bulk and from thermal processing do not reach a level that could be detected by this analysis method. The same negative result was obtained when incoming wafers with backsurface damage were analyzed to see if the damage application caused wafer degradation.

In general, LIMA analysis identified impurities in 40-60% of the test areas only and showed a considerable variation of impurity types in the various test spot related mass spectra. This inconsistency of detecting previously identified impurities can be explained by the non uniform distribution of the getter defect structures and the small test spot size, 1 to 3 um in diameter. SIMS analysis helped to verify this assertion through secondary ion image displays of selected elements as shown in Figure 6, A through D. Images A and B show the distribution and relative concentration of Na at two levels of the backsurface defect structure. While images C and D identify the distribution of Cu and Ca at two specific distances from the backsurface.

Wafer cleanliness and getter effectiveness can be evaluated by observing the S-pit/haze density (associated with metallic impurities) on the front surface of oxidized samples [5]. We applied this method to the various wafer types used in this study and the results - by percent of wafer area covered with haze - are listed in Table 1. Comparing the wafer characteristics like backsurface damage and interstitial oxygen concentration to the area covered by haze it is apparent that a medium or medium-high oxygen concentration is needed to obtain a haze free wafer (Table 1). This substantiates the benefit of the complementary action of extrinsic and intrinsic (oxygen related) gettering in high temperature processing.

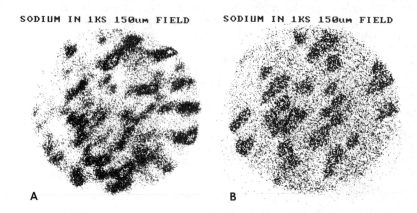

A) Approximately 500 Å into the backsurface structure.

B) Approximately 100 Å below area displayed in A.

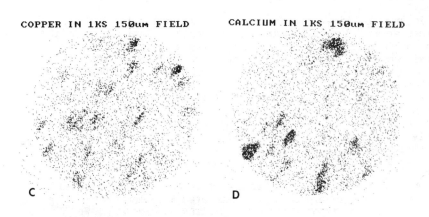

C) Aproximately 500 Å.

D) Approximately 600 Å into the backsurface structure.

Figure 6 Secondary ion images displaying the lateral distribution of impurities in the backsurface getter structures. (Black/grey shades indicate signal intensity and can be related to impurity concentration.)

CONCLUSION

Laser Ionization Mass Spectrometry has been shown to be an effective technique to identify elemental and molecular components (in the ppm range) in silicon backsurface structures and layers. Gettered contaminants, like Al, Cr, Cu, etc., could be detected in the backsurface damage and the SiO_2 layer. The SiO_2 showed in addition contaminants Na, K, N and Cl, most likely introduced during thermal oxidation.

Overall, LIMA analysis data indicates that a poly layer or wet sandblast backsurface damage is a more effective extrinsic getter system than the abrasive SVG system.

The special advantages of the LIMA technique are the high sensitivity and the easy sample analysis as well as the rapid depth profiling. However, more development work is needed to make this analysis method an effective qualitative and quantitative microanalytical tool.

ACKNOWLEDGEMENTS

The author would like to thank C. Hitzman and R. Odom for performing the LIMA and SIMS analysis, C. Vorst for wafer processing, W. Stacy and K. Ritz for the TEM sections, R. Gong and B. Hamre for their encouragement and support, and C. Dominguez for typing the manuscript.

REFERENCES

[1] Huff, H. R. and Shimura, F., Solid State Technology March 1985, pp. 103

[2] M. J. J. Theunissen et al, in house Philips Nat. Lab Report #5829

[3] Hitzman, C. J. and Odom, R. W., Internal Technical Report, Charles Evans & Associates, 1670 S. Amphlett Blvd., Suite 120, San Mateo, CA, 94402

[4] T. Dingle, B. W. Griffiths, Microbeam Analysis 1985, edited by J. T. Armstrong, pp 315.

[5] Stacy, W. T., Allison, D. F. and Wu, T. C., Semiconductor Silicon/1981 edited by H. R. Huff, R. J. Kriegler, Y. Takeishi, pp. 406

Hisaaki Suga and Koji Murai

Effect of Bulk Defects in Silicon on SiO_2 Film Breakdown

REFERENCE: Suga, H., and Murai, K., "Effect of Bulk Defects in Silicon on SiO_2 Film Breakdown," Emerging Semiconductor Technology, ASTM STP 960, D. C. Gupta and P. H. Langer, Eds., American Society for Testing and Materials, 1986.

ABSTRACT: The origins of degrading the dielectric breakdown strength of silicon dioxide on silicon were analysed by controlling surface and bulk lattice defects and back side damage. As a result, two different mode of breakdown were observed. One is caused by lattice defects emerging on the silicon/silicon dioxide interface, but the degree of degradation in breakdown strength was only as much as 1 MV/cm. The other is due to low density of point like contamination. This kind of contamination was controlled with the gettering by lattice defects nearby the surface region.

KEYWORDS: defect in silicon, silicon oxide, dielectric breakdown, gettering, contamination

INTRODUCTION

The dielectric breakdown for silicon dioxide film in metal-oxide-semiconductor (MOS) device is thought to depend on the quality of the thin film [1], defects in silicon substrate [2], contamination in oxidation process [3-6] and so on. It is expected that some of origins of the degrading of breakdown strength are excluded by defect control through the intrinsic gettering process [7-9]. Surface denudation through out-diffusion of oxygen and impurity trapping by getters in the bulk [9] would be effective in

Dr. Suga is manager at semiconductor division, Central Research Institute, Mitsubishi Metal Corporation, 1-294, Kitabukuro, Omiya, Japan 330; Murai is associate manager at technical division, Japan Silicon Corp., 314, Kanauchi, Nishisangao, Noda, Japan 278.

keeping the soundnesses of silicon dioxide after various thermal processes.

Intrinsic gettering (IG) is one of the most reliable technique to utilize the defects associated with the precipitation of supersaturated oxygen in Czochralski grown silicon for the sink of impurity. While, extrinsic gettering (EG) is the technique to utilize the high density of dislocations which are mostly given on the back side of wafer to capture the unwanted impurity. Further, combining these techniques ensures the reliability of the device performance against the degradation by contamination.

The purpose of our research is to analyse the origin degrading the breakdown field strength by bulk defect control and extrinsic gettering via backside damage.

EXPERIMENTAL PROCEDURE

Polished Czochralski grown wafers of 125 mm diameter were prepared. Their initial oxygen content was ranged between 0.72 to 0.91 × 10^{18} atoms/cm^3. The resistivity was in the range of 9 to 10 ohm·cm.

Polished wafers were annealed for introducing desired lattice defects after RCA cleaning[10]. The thermal processes employed are in the following; 1st step: annealing at 1100°C for surface denudation, 2nd step: the one of the process for defect formation was ramp annealing from 550°C to 950°C, and the other was composed of two consecutive annealing after ramp annealing mentioned above as shown in Fig. 1. We define these two IG processes as weak IG and strong IG, respectively. Then a backside damage by fine silica blasting was given on the some of wafers in order to provide the external or extrinsic gettering. Before oxidation of the wafer, the gront surface of 5 micron in thickness was polished off for excluding the effect of surface damage in the blasting process.

Thin film of silicon dioxide was formed to get thickness of 590 A by annealing at 900°C for 16 min under dry oxygen ambient and for 27 min under wet oxygen ambient, and then the sample was annealed under dry nitrogen ambient for 30 min.

The type, density and distribution of defects were examined by analytical electron microscopy (Hitachi H800LB) and x-ray topography (Rigaku RU 200). Impurity analysis was done by the SIMS (Cameca ims-3f) and content of oxygen dissolved in silicon was examined by FTIR (JEOL JIR-40X).

Breakdown, it was determined, would be a substantial leakage current of greater than 5 µA through 5 mm diameter

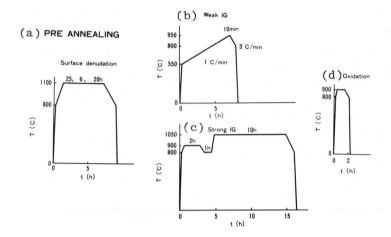

Fig. 1 Thermal treatments for IG process and oxidation process. (a) Preannealing for surface denudation (annealing time: 2.5h, 6h or 20h) (b) Weak IG process (c) Strong IG process (includes weak IG process) (d) Oxidation process.

Al gate. A constant ramp rate of 0.05 Volt/sec was used in all cases. IV characteristics of 30 gates for each wafer was measured.

RESULTS AND DISCUSSION

Defect analysis

We applied two processes of weak IG and strong IG to produce the different defect condition. Fig. 2 shows that in case of weak IG process there was little interstitial oxygen change after the annealing treatment, but in case of strong IG process significant reduction of interstitial oxygen content which depended on the preannealing time, was observed.

According to TEM observation, in weak IG processed wafers, a lot of microprecipitates of 200 to 300 Å in size were confirmed and in the strong IG processed wafers, punched dislocations and stacking faults in addition to grown precipitates of a few hundred nanometer in size were observed as shown in Fig. 3. The density of these defects depended on the preannealing time, as shown in Table 1. The results seen here for shorter, low-temperature ("weak IG") vs. longer, high-temperature ("strong IG") anneals

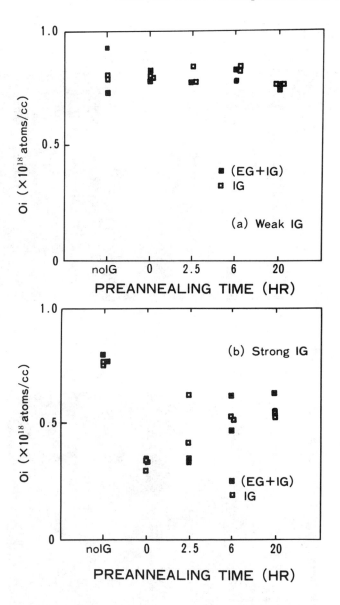

Fig. 2 Disolved oxygen content in silicon after IG thermal process.

WEAK IG

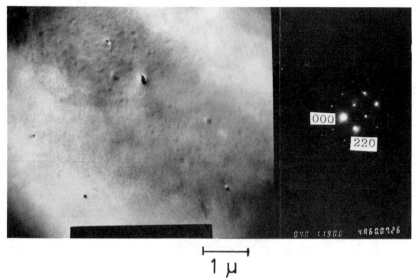

STRONG IG

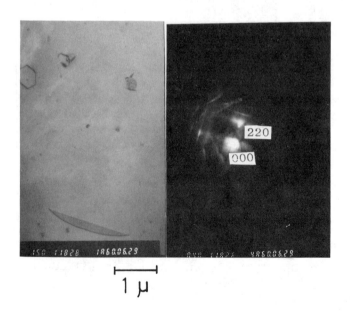

Fig. 3 TEM photographs showing the defects in silicon introduced by IG process.

Table 1 Microdefect density in IG processed wafers.

(a) WEAK IG

	PREANNEALING TIME			
	0 HR	2.5HR	6 HR	20HR
DISLOCATION, 10^5 lines/cm^2	0	0	0	1.1
STACKING FAULT, 10^4/cm^2	0	0	0	0
OXIDE, 10^{10}/cm^3	10.7	∼0	< 0.01	< 0.01

(b) STRONG IG

	PREANNEALING TIME			
	0 HR	2.5HR	6 HR	20HR
DISLOCATION, 10^5 lines/cm^2	6.8	10.0	7.6	7.6
STACKING FAULT, 10^4/cm^2	27.0	5.3	3.5	1.8
OXIDE, 10^{10}/cm^3	19.6	0.70	0.17	0.45

would be expected from earlier works [11][12][13]. It seems that the preannealing retarded the oxygen precipitation, as far as this preannealing condition was concerned.

In weak IG processed wafers, only a defect observed was oxygen precipitates except in the wafers annealed for 20 hr, where in the preannealing process a limited number of precipitates would grow up to evolve the punched dislocations.

The preannealing significantly affected not only the defect condition but also the defect distribution, which is shown in the optical microphotographs for the sectional view of the wafer (Fig. 4). Longer preannealing makes thicker surface defect-free (denuded) zone [14]. Thickness of denuded zone are 20, 40 and 60 μm for preannealing time of 2.5, 6 and 20 hr, respectively. But for the IG processed wafer without high temperature preannealing the high density of lattice defects was emerged on the wafer surface and this had been confirmed by thin foil observation by transmission electron microscopy. Long preannealing often generate the low density of microdefects in the denuded zone, which are caused by heterogeneous nucleation.

342 EMERGING SEMICONDUCTOR TECHNOLOGY

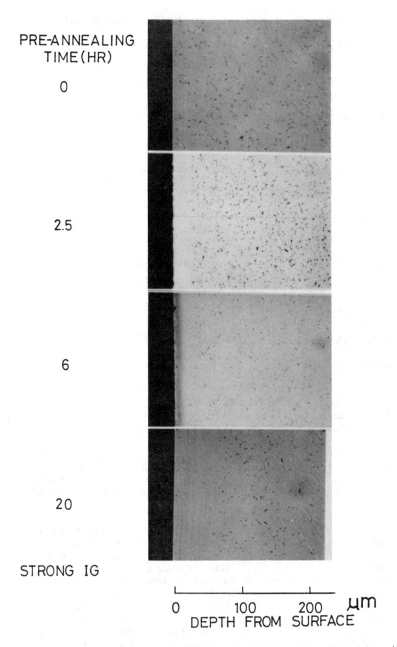

Fig. 4 Optical microphotographs showing the sectional view of various IG processed wafers.

Breakdown of thin film

Fig. 5 shows the relationship between average of breakdown field and preannealing time for various IG processed wafers. Their breakdown field changes depending on the preannealing time. Introduction of more defects in the wafers by prolonged annealing may degrade more prominently the breakdown strength of the thin films. The IG processed wafers without preannealing have the lowest field and no IG processed wafers have the highest average breakdown field. This phenomenon can be explained by the effect of the degradation by the introduction of thermal defect on the surface layer.

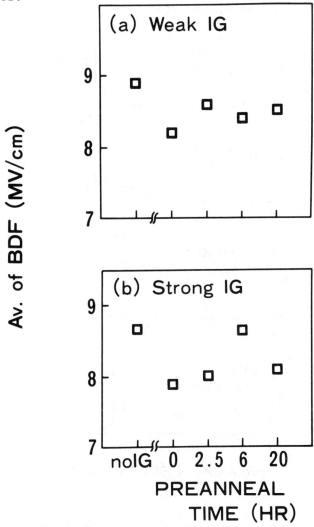

Fig. 5 Average of breakdown field (BDF) strength for IG processed wafers.

The histograms of the breakdown field give us the other information in the breakdown behaviors. Fig. 6(a) and (b) show the histogram of the no IG processed wafers. Two distinct peaks of distribution are observed in Fig. 6(b). Let us define the higher peak and lower peak as I mode and P mode, respectively. I mode and P mode correspond to the intrinsic breakdown and the weak spot like breakdown [15]. I mode and I + P mode were frequently observed in no IG processed wafers.

In the wafers where such defects as oxygen precipitates, dislocations and stacking faults in the surface region of the wafers were introduced on the interface between the film and the substrate, the specific histogram of breakdown

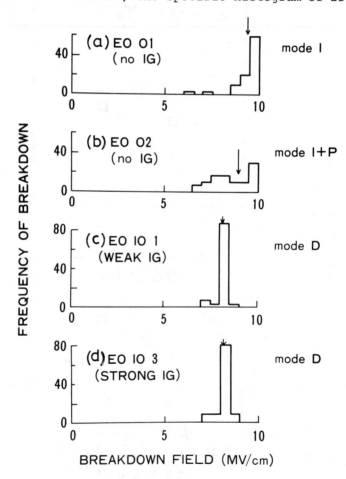

Fig. 6 Histograms of breakdown field for wafers with and without IG process.

field was appeared as shown in Fig. 6(c) and (d). The introduction of lattice defects without preannealing shows the these sharp peaks ranged between 8 to 9 MV/cm, and this breakdown was only 1 MV/cm in degradation compared with I mode breakdown. Let us define this histogram of breakdown as D mode.

Backside damage

For the enhancement of the gettering efficiency, backside damage was given on the IG processed wafers. Fine silica blasting on the backside of the wafers introduced the damage which produced a long range elastic strain field through the wafer thickness as shown in the x-ray section topograph (Fig. 7(a)). Black and white contrast imply the strain distribution through thickness [16]. TEM photographs (Fig. 7(b) and (c)) clearly shows the change of blasted damage into the densed dislocation colony by annealing at 900°C. This colony would play a role of effective getter.

Effect of backside damage on breakdown

In no IG processed wafers with backside damage, both I mode and I + P mode were observed, which is similar to the case without the damage. In the wafers where lattice defects were introduced near surface region, D mode of breakdown was mostly observed. Overall view of the histogram of thin film breakdown field in the damaged wafers did not change with the one in the wafers without damage. Average of breakdown field of the damaged wafers seems to be higher than that of the wafers without backside damage, shown in Fig. 8 and thus this extrinsic gettering by the backside damage is considered to be effective to assist the intrinsic gettering action. But a few cases which indicated an abrupt lowering of breakdown field were observed. Note the cases of 20 hr preannealing in weak IG processed wafer and of 6 hr preannealing in the strong IG processed wafer. This lowering would be caused by contamination introduced by the blasting process. This conclusions are not very convincing from the SIMS impurity analysis. But from the following classification, the appearance of P mode could imply the particulate contamination. There was no correlation of precipitation behaviors with the lowering.

Classification of histogram of breakdown

Fig. 9 shows the three characteristic modes of breakdown; I mode, D mode and P mode. Their ranges are 9 to 10 MV/cm, 8 to 9 MV/cm and 5 to 8 MV/cm, respectively. On the basis of this definition, the breakdown mode of heat treated wafer is classified as Table 2. In no IG processed wafers, I mode and I + P mode are dominant, and in IG processed wafers without preannealing D mode of breakdown is dominant. While in the weak IG processed wafer with pre-annealing I + P mode of breakdown occurs frequently, in the strong IG processed wafers P + D mode of breakdown is dominant.

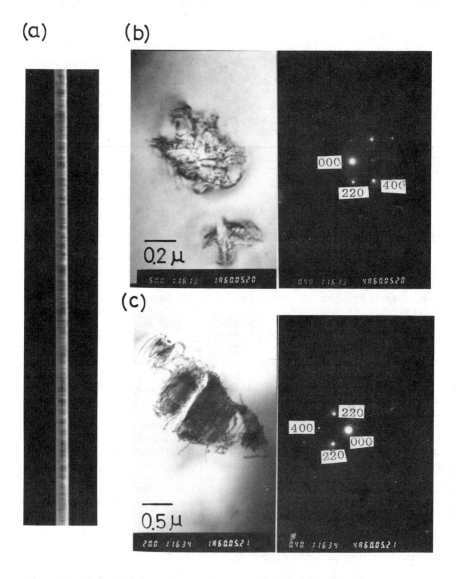

Fig. 7 Defect structure due to backside damage.
(a) X-ray section topograph showing a long range strain field emanating from the backside of wafer. (b) TEM photograph of the defect before annealing. (c) TEM photograph of defect after annealing.

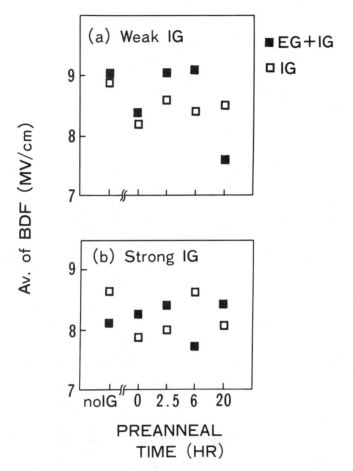

Fig. 8 Effect of backside damage on the average of breakdown field (BDF).

In the weak IG processed wafers with preannealing, localized contamination on the wafers occurs and the getter introduced in the wafer can not maintain the soundness of the wafer surface because of the relatively low gettering capacity of the getter in the weak IG processed wafers. Width of the denuded zone should be so narrow that getters can act as a effective site for collecting unwanted impurities near the surface.

In the strong IG processed wafers, the degradation of breakdown is significant. This might come from the defect formation inside the denuded zone judged from the TEM observation, considering that D mode is frequently observed and impurity decorated defects inside the denuded zone deteriorate the breakdown strength. But as far as we used analytical electron microscopy with energy dispersive x-ray analyser to examine microsegreagation, the decoration of impurity on the lattice defects could not be confirmed, unlike the other literatures [17-19]. This might come from the difference of contamination level.

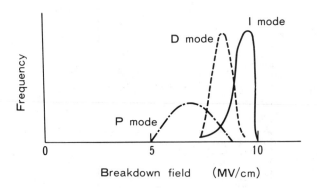

I mode : Intrinsic breakdown (9-10MV/cm)
D mode : Defect induced breakdown (8-9MV/cm)
P mode : Contamination induced breakdown (5-8MV/cm)

Fig. 9 Three characteristic breakdown modes.

Table 2 Classification of histograms of breakdown in IG and/or EG processed wafers. Oxidation was done separately for weak IG processed wafers and strong IG processed wafers.

WEAK IG

	NO IG	PREANNEALING TIME (HR)			
		0	2.5	6	20
NO EG	I / I+P	D / D	I / I+P	I+P / I+P	I+P / I
EG	I / I	D / D	I+D / I+P	I+P / I	D / P+D

STRONG IG

	NO IG	PREANNEALING TIME (HR)			
		0	2.5	6	20
NO EG	I / I+P	D / P+D	D / P+D	I+P / D	P+D / P+D
EG	I+P / I+P	D / D	P+D / P+D	P+D / P+D	P+D / P+D

Origin of P mode breakdown

While D mode of breakdown can be considered to be the result of the lattice defects on the substrate surface, the origin of P mode breakdown is somewhat difficult to identify. Because the distribution of breakdown strength was broad, we could not distinguish the statistic variation from experimental error. But some key to solve the problem is the fact that the existence of surface defects changed P mode breakdown into D mode breakdown. In the other words, surface defects eliminated the low field breakdown below 8 MV/cm. One of possible explanation for this is the improvement of the interface by the gettering of microcontamination by the defects on the thermal process. In this case the degree of degradation of breakdown is only as much as 1 MV/cm compared with I mode of breakdown which corresponds the intrinsic breakdown of the film. So the aggregation of contamination which would cause a local field enhancement was spread over the dislocation line or stacking fault plane and the dilution of contaminant thus results at these lattice defects. This kind of aggregate must be small in volume because the decoration on defects did not produce the substantial degradation of breakdown although the impurity aggregates would be easily captured by the defects.

The distribution of the aggregates on the wafer would be dispersive. Fig. 10(a) shows the I + P mode of histogram of breakdown field, which is composed of clearly split peaks; one at 9 MV/cm and one below 8 MV/cm. The gate positions of P mode breakdown on that wafer are not preferentially grouped, but are randomly distributed on the wafer as shown in Fig. 10(b). This mode of breakdown can be caused by point-like imperfection for each gate, not by film

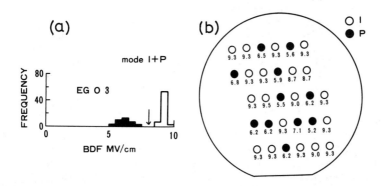

Fig. 10 Histogram and distribution on the wafer showing P mode breakdown. This wafer was processed by no IG and EG.

or layer contaminant. The density of the imperfection would be about 200 points for the wafer, if statistically uniform distribution of point-like imperfection is supposed. This kind of imperfection may be some micro-particulate contamination which happens while the wafer cleaning process goes on. But as far as SIMS analysis concerns, we could not detect any specific contaminant on that degraded wafer. The size of these particulate contaminants could be very small compared with oxide thickness, because serious deterioration of breakdown below 5 MV/cm were not observed.

CONCLUSION

We tried to maintain the high breakdown strength of silicon dioxide film on silicon wafer by controlling the bulk lattice defects and by introducing dislocation colonies at the backside of wafers. As a result, we obtained the possible cause of degradation for breakdown strength and also a few important knowledges about the improvement of the breakdown strength as follows.

(1) Three distinct breakdown modes of thin film were observed; I mode ranged between 9 to 10 MV/cm due to intrinsic breakdown, D mode ranged between 8 to 9 MV/cm due to surface lattice defects and P mode ranged between 5 to 8 MV/cm due to point-like contamination. Surface defects can maintain the high breakdown field even after the gettering of unwanted impurities because of extremely low contamination on wafers.
(2) Oxygen microprecipitates which were not reveled by chemical etching and of which volume could not almost be detected by an infrared spectroscopy had the effective capacity to maintain the breakdown strength.
(3) Narrow width of denuded zone took advantage in high efficiency to maintain the high breakdown strength. But such a long thermal process to introduce the punched dislocations in the wafer degraded the breakdown strength.
(4) Addition of EG to IG processed wafer improved the breakdown characteristics, but a degradation due to the contamination in EG process might happen in some cases.

REFERENCES

[1] Bhattachryya, A., Vorst, C., and Carim, A., "A Two-Step Oxidation Process to Improve the Electrical Breakdown Properties of Thin Oxides," Journal of the Electrochemical Society, Vol. 132, No. 8, August 1985, pp. 1900-1903.
[2] Yamabe, K., Taniguchi, K., and Matsushita, Y., "Thickness dependence of dielectric breakdown failure of thermal SiO_2 films," The Electrochemical Society

Extended Abstracts, San Francisco, May 1983, Abstract No. 309, pp. 482-483.
[3] Itsumi, M., and Kiyosumi, F., "Origin and Elimination of Defects in SiO_2 thermally Grown Czochralski Silicon Substrate," Applied Physics Letters, Vol. 40, No. 6, March 1982, pp. 496-498.
[4] Duffalo, J. M., and Monkowski, J. R., "Particulate Contamination and Device Performance," Solid State Technology, Vol. 27, No. 3, March 1984, pp. 109-114.
[5] Monkowski, J. R., "Chemistry, Physics, and Defect Source in Semiconductor Gas Processes," Microcontamination, Vol. 2, No. 1, February/March 1984, pp. 37-43.
[6] Honda, K., Ohsawa, A., and Toyokura, N., "Breakdown in Silicon Oxides-Correlation with Cu precipitates," Applied Physics Letters, Vol. 45, No. 3, August 1984, pp. 270-271.
[7] Ikuta, K., Nakajima, S., and Inoue, N., "Intrinsic Gettering Effects on Reactive Ion Etching Damage and Thin Oxide Film Breakdown," Extended abstract of the 16th Conference on Solid State Device and Materials, Kobe, 1984, pp. 483-486.
[8] Tan, T. Y., Gardner, E. E., and Tice, W. K., "Intrinsic Gettering by Oxide Precipitate induced Dislocations in Czochralski Si," Applied Physics Letters, Vol. 30, No. 4, February 1977, pp. 175-176.
[9] Tice, W. K., and Tan, T. K., "Nucleation of CuSi Precipitate Colonies in Oxygen-rich Silicon," Applied Physics Letters, Vol. 28, No. 9, May 1976, pp. 564-565.
[10] Kern, W., and Puotinen, D. A., "Cleaning Solution Based on Hydrogen Peroxide for use in Silicon Semiconductor Technology," RCA Review, Vol. 31, June 1970, pp. 187-264.
[11] Suga, H., Shimanuki, Y., Murai, K., and Endo, K., "Imhomogeneous generation of OSF in silicon single crystal with 4 inches diameter (III)," Preprint of Spring meeting of Japanese Society of Applied Physics, April 1983, p. 661.
[12] Suga, H., Shimanuki, Y., and Kainuma, M., "Correlation of secondary defects with oxygen precipitation in Cz-silicon," Preprint of Fall meeting of Japanese Society of Applied Physics, October 1984, p. 606.
[13] Shimura, F., and Tsuya, H., "Multistep Repeated Annealing for Cz-Silicon Wafers: Oxygen and Induced Defect Behavior," Journal of Electrochemical Society, Vol. 129, No. 9, September 1982, pp. 2089-2095.
[14] Huber, D., and Reffle, J., "Precipitation Process Design for Denuded Zone Formation in Cz-Silicon Wafers," Solid State Technology, Vol. 26, No. 8, August 1983, pp. 137-143.
[15] Chou, N. J., and Eldridge, J. M., 'Effects of Material and Processing Parameters on the Dielectric Strength of Thermally Grown SiO_2 Film," Journal of Electrochemical Society, Vol. 117, No. 10, October 1970, pp. 1287-1293.

[16] Weissmann, S., and Saka, T., "Characterization of Strain Distribution and Annealing Response in Deformed Silicon Crystals," *Advances in X-ray Analysis*, Ed. Murdil, H. F., Barrett, C. S., Newkirk, J. B., and Rudd, C. O., Plenum Publishing Company, Vol. 20, 1977, pp. 237-244.

[17] Suga, H., Shimanuki, Y., Murai, K., and Endo, K., "Precipitation Behavior of Deposited Metals in Cz-Silicon," *Semiconductor Processing, ASTM STP 850*, Gupta, D. C., Ed., American Society for Testing and Materials, 1984, pp. 241-256.

[18] Lin, P.S.D., Marcus, R. B., and Sheng, T. T., "Leakage and Breakdown in Thin Oxide Capacitors-Correlation with Decorated Stacking Faults," *Journal of the Electrochemical Society*, Vol. 130, No. 9, September 1983, pp. 1878-1883.

[19] Stacy, W. T., Allison, D. F., and Wu, T. C., "Metal Decorated Defects in Heat-Treated Silicon Wafers," *Journal of the Electrochemical Society*, Vol. 129, No. 5, May 1982, pp. 1128-1132.

Warren K. Gladden and Aslan Baghdadi

FREE CARRIER ABSORPTION AND INTERSTITIAL OXYGEN MEASUREMENTS

REFERENCE: Gladden, W. K., and Baghdadi, A., "Free Carrier Absorption and Interstitial Oxygen Measurements," *Emerging Semiconductor Technology*, ASTM STP 960, D. C. Gupta and P. H. Langer, Eds., American Society for Testing and Materials, 1986.

ABSTRACT: The infrared (IR) absorption of n- and p-type silicon samples was measured over the concentration range $\sim 10^{15}$ to 6×10^{17} atoms/cm^{-3}. The free carrier absorption exhibited a power-law dependence on wavenumber. The data were fit to a logarithmic function with this dependence, and these results were applied to the determination of the baseline from which to compute the corrected net IR absorption at 1107 cm^{-1} due to interstitial oxygen. Application of this correction for the free carrier absorption results in an improvement of 3% to 30% in the accuracy of the oxygen content determination at dopant concentrations above 10^{16} atoms/cm^{3}.

KEYWORDS: free carrier absorption, interstitial oxygen, infrared absorption, FT-IR, fourier transform infrared spectroscopy, intrinsic gettering, semiconductors, silicon, n-type silicon, p-type silicon

INTRODUCTION

It is well known that the presence of interstitial oxygen affects both the mechanical stability of processed silicon wafers and, via the intrinsic gettering process, the minority carrier lifetime of the wafer. In both bipolar and MOS processes, the oxygen content needs to be known to better than about 2 ppma in order to control intrinsic gettering [1]. One of the problems in interstitial oxygen measurements has been the choice of a baseline from which to determine the net absorption of the oxygen local vibrational mode at 1107 cm^{-1}. For high-resistivity material, the

---

Mr. Warren K. Gladden is an electronics engineer and Dr. Aslan Baghdadi is a physicist in the Semiconductor Electronics Division at the National Bureau of Standards, Gaithersburg, Maryland 20899. Contribution of the National Bureau of Standards. Not subject to copyright.

spectrum of a float-zone wafer can be subtracted from that of the sample, and a relatively flat baseline is obtained. Although some absorption due to dopant atoms is present even in high-resistivity materials, its effect is overshadowed by lattice absorption and reflectance losses at the surfaces of the wafer. In the case of low-resistivity material, free carrier absorption becomes a more significant contribution to the baseline and is highly frequency-dependent.

Fourier Transform Infrared Spectroscopy (FT-IR) is a quick, nondestructive technique for the evaluation of oxygen in silicon. The limits of usefulness for FT-IR measurements are determined by the background absorption characteristics of the material being analyzed. With epitaxial layers grown on heavily doped substrates becoming one widely used solution for latch-up problems in CMOS devices [2-4], extending these limits becomes increasingly important.

In this paper, the effect of free carrier absorption on the determination of interstitial oxygen in silicon wafers is shown. The classical theory is outlined, and its limitations discussed. Relevant research on the free carrier absorption in both n- and p-type silicon is reviewed. The measurement technique is described, followed by the results of absorption measurements on silicon wafers with carrier concentrations ranging up to 6×10^{17} atoms/cm^3. These are discussed with particular emphasis on the implications to oxygen measurements.

THEORY

From the classical Drude model [5], the free carrier absorption cross section, or the absorption coefficient α per impurity N, is given by

$$\frac{\alpha}{N} = \frac{4\pi}{nc} \cdot \frac{e^2 \tau}{m^*(1+\omega^2\tau^2)}, \quad (1)$$

where n is the refractive index of silicon, c is the speed of light, e is the electronic charge, m^* is the carrier effective mass, ω is the frequency of radiation, and τ is the electron relaxation time which accounts for the scattering mechanisms necessary for intraband absorption. Over the carrier concentration and frequency range considered in this paper, the refractive index is assumed to have a constant value of 3.42 [6]. From knowledge of the carrier concentration and electrical conductivity σ, the relaxation time can be determined from the relation

$$\tau = \frac{m^*\sigma}{Ne^2}. \quad (2)$$

The use of a constant relaxation time for all the carriers in the system is a rough approximation at best. Dumke [7] has shown that for the case of a classical distribution of carriers, the classical theory is only applicable if $\hbar\omega \ll kT$. This requirement limits the validity of the Drude model to wavenumbers much less than 200 cm^{-1} for room temperature measurements. Since the interstitial oxygen peak is observed at $\sim$1000 cm^{-1}, agreement between the classical theory and measured results, for either dopant type, is not expected. A more rigorous treatment of the scattering mechanisms involved in the absorption process is required to fully describe the phenomenon.

In the case of n-type silicon, free carrier absorption is proportional to the carrier concentration, and the dominant momentum conserving mechanisms involved are acoustic-mode lattice scattering and ionized-impurity scattering [8]. The absorption cross section varies as $\omega^{-1.5}$ when lattice scattering is the dominant mechanism, and as ω^{-3} for impurity scattering. The functional forms of the absorption cross sections are well known in both cases [9-11].

Explanation of the absorption phenomenon in p-type silicon has proven to be a more difficult problem. In measurements dating back to 1950, the free carrier absorption in p-type silicon was found to be proportional to the concentration of holes and varying smoothly as ω^{-2}. Researchers also found discrepancies between the magnitude of the experimental absorption data and that predicted from the classical Drude theory as shown in Equations (1) and (2). Since 1950, the only comprehensive experimental analysis of the free carrier absorption in p-type silicon was presented by Leung [12] in 1971. The measured absorption qualitatively followed that predicted by the classical theory, but quantitatively disagreed with theoretical results available at that time. Using the results of Kahn [13] for free carrier absorption due to direct interband transitions, Leung found them not to be prominent contributors to the observed absorption.

In summary, absorption due to free carriers consists of contributions from an intraband process involving some scattering mechanism so as to conserve momentum, and an interband process. This is true regardless of the type of dopant. Thus, despite the incompleteness of the theory for p-type silicon, the free carrier absorption in silicon can generally be described as a weighted sum of three processes:

(a) scattering by acoustic phonons; α proportional to $\omega^{-1.5}$

(b) scattering by ionized impurities; α proportional to ω^{-3}

(c) direct interband transitions; the carrier concentration and frequency dependence of this term are as yet undetermined for p-type silicon.

This can be written as

$$\alpha_{\text{free carrier}} = A\omega^{-1.5} + B\omega^{-3} + C_{\text{interband}}. \tag{3}$$

EXPERIMENTAL RESULTS

Thirteen samples (8 n-type and 5 p-type) were used in this study. Tables I and II list the sample no., thickness, resistivity, carrier concentration, and dopant type for the n- and p-type samples, respectively. All p-type samples were doped with boron. Both sides of the samples were polished to avoid complications due to back-surface scattering. The polish used was a commercially available colloidal silica. Sample thickness was measured by a micrometer. The resistivity was determined from spreading resistance and four-point probe measurements [14,15].

In past work, the carrier concentration has been determined from the resistivity using Irvin's curves [16]. Thurber et al. [17,18] have shown significant differences between their measurements and Irvin's data. As a result, they have given new empirical relationships for carrier concentration and mobility as a function of resistivity for boron- and phosphorus-doped silicon. Similar relationships are available for arsenic-doped silicon [19]. These relationships were used to determine the carrier concentrations from the measured resistivities in Tables I and II.

Table I Electrical data for n-type samples

sample no.	thickness (mm)	resistivity (Ω-cm)	dopant density (atoms/cm^3)	dopant type
N-1	0.53	0.7	7.38×10^{15}	phosphorus
N-2	0.584	0.65	8.0×10^{15}	phosphorus
N-3	0.519	0.5	1.07×10^{16}	phosphorus
N-4	0.507	0.41	1.34×10^{16}	phosphorus
N-5	0.560	0.3	1.94×10^{16}	arsenic
N-6	0.451	0.19	3.40×10^{16}	arsenic
N-7	0.538	0.18	3.61×10^{16}	phosphorus
N-8	0.573	0.029	6.03×10^{17}	arsenic

Table II Electrical data for p-type samples

sample no.	thickness (mm)	resistivity (Ω-cm)	dopant density (atoms/cm^3)
P-1	0.581	10.8	1.28×10^{15}
P-2	0.518	3.8	3.7×10^{15}
P-3	0.542	1.0	1.51×10^{16}
P-4	0.509	0.6	2.68×10^{16}
P-5	0.385	0.33	5.5×10^{16}

A NICOLET* 8000 Fourier transform spectrometer was used to measure the infrared transmittance of the samples. A Globar* was used as the infrared source, a KBr/Ge beamsplitter, and a DTGS (deuterated triglycine sulfate) thermal detector. All measurements were taken with the spectrometer chamber evacuated so as to eliminate significant atmospheric absorption. The transmittance through a double-side polished wafer is given by

$$T = \frac{(1-R)^2 e^{-\alpha d}}{1 - R^2 e^{-2\alpha d}}, \qquad (4)$$

where d is the sample thickness, R is the surface reflectivity of silicon, and α is the absorption coefficient. Since T is measured, Equation (4) can be solved for α

---

* Certain commercial equipment, instruments, or materials are identified in this paper in order to adequately specify the experimental procedure. Such identification does not imply recommendation or endorsement by the National Bureau of Standards, nor does it imply that the materials or equipment identified are necessarily the best available for the purpose.

yielding

$$\alpha = -\frac{1}{d}\ln\left\{\frac{-0.49 + \sqrt{0.2401 + 0.36T^2}}{0.18T}\right\}, \qquad (5)$$

where the silicon reflectivity, R, has been set equal to 0.3, its approximate value over the carrier concentrations and frequencies considered here [20]. Consider that the sample transmittance is ratioed to that of a reference. Since the transmittance T is related to the absorption coefficient through a logarithm, the ratio of transmittances implies taking the difference of the absorption coefficients. In general, the absorption coefficient can be expressed as the sum of several contributions,

$$\alpha = \alpha_{\text{oxygen}} + \alpha_{\text{lattice}} + \alpha_{\text{free carrier}}. \qquad (6)$$

If oxygen-free high resistivity float-zone silicon is used as the reference material, then $\alpha_{\text{free carrier}}$ will be negligible because of the high resistivity and α_{oxygen} will be zero. Thus

$$\alpha_R \simeq \alpha_{\text{lattice}}. \qquad (7)$$

So

$$\alpha_S - \alpha_R \simeq \alpha_{\text{oxygen}} + \alpha_{\text{free carrier}}, \qquad (8)$$

where the subscripts S and R refer to the sample and reference, respectively, and the absorption coefficients on the right-hand side of the above equation only refer to the sample. Figure 1 shows a typical transmittance spectrum. It is often difficult to determine the correct baseline absorption due to free carriers because of the presence of lattice and interstitial oxygen absorption. However, by ratioing the transmittance of the sample to that of a float-zone reference (i.e., applying Equations 7 and 8), the lattice absorption can be removed thus allowing the free-carrier baseline to be more readily determined. The result of applying this procedure to the transmittance spectrum of Figure 1 is shown in Figure 2.

Following the analysis of Pankove [21], the absorption cross section is assumed to be proportional to $\tilde{\nu}^m$ where $\tilde{\nu}$ is the wavenumber of the IR radiation, and the value of m indicates the dominant scattering mechanism for the observed absorption. The measured absorption cross section of the n-type samples was then fit to the equation

$$\ln\frac{\alpha}{N} = m\ln\tilde{\nu} + b, \qquad (9)$$

where b is the intercept. All data were fit over the wavenumber range 1000 cm^{-1} to 1300 cm^{-1}, omitting data between 1060 cm^{-1} and 1200 cm^{-1} so as to avoid the oxygen peak. This truncated portion of the absorption spectrum was used because only the interstitial oxygen peak at 1107 cm^{-1} was of interest, and this region is also used in the proposed ASTM test method for the automatic determination of oxygen in silicon [22]. Table III shows the values of the fitted parameters for each of the n-type samples.

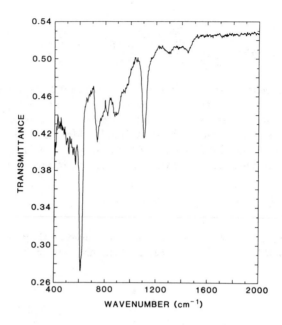

Figure 1 Measured transmittance of the arsenic-doped sample N-5, 1.94×10^{16} atoms/cm^3.

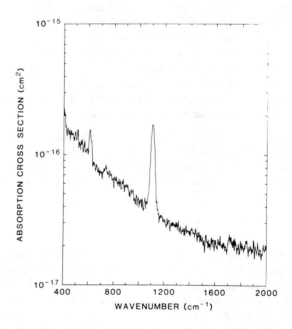

Figure 2 Measured absorption cross section of n-type sample N-5 after application of Equations (5-8).

Table III Fit coefficients for n-type samples

sample no.	dopant density (atoms/cm^3)	slope	intercept
N-1	7.38×10^{15}	-1.96 ± 0.11	-23.6
N-2	8.0×10^{15}	-1.90 ± 0.12	-24.3
N-3	1.07×10^{16}	-2.01 ± 0.09	-23.5
N-4	1.34×10^{16}	-2.18 ± 0.08	-22.5
N-5	1.94×10^{16}	-1.81 ± 0.07	-25.2
N-6	3.4×10^{16}	-1.67 ± 0.04	-26.0
N-7	3.61×10^{16}	-1.78 ± 0.04	-25.5
N-8	6.03×10^{17}	-1.97 ± 0.004	-23.4

The slopes of the fits were found to be approximately two for the majority of the samples measured. This agrees with the dependence published by Weeks [23]. There is a quantitative difference between his calculations and the data published here, roughly a factor of 2 to 3, owing to the difference in the choice of τ. Weeks uses the τ as calculated from Equation (2). For the truncated spectral region used in this analysis, the n-type measurements presented here agreed with the theoretical calculations for acoustic-mode lattice scattering where τ was calculated from the acoustic-mode scattering mobility [10]. Due to the incompleteness of the theory for p-type silicon, Equation (9) was used to fit the p-type data as well. These results are shown in Table IV.

Table IV Fit coefficients for p-type samples

sample no.	dopant density (atoms/cm^3)	slope	intercept
P-1	1.28×10^{15}	-2.11 ± 0.09	-21.1
P-2	3.7×10^{15}	-2.41 ± 0.09	-19.5
P-3	1.51×10^{16}	-1.96 ± 0.03	-22.9
P-4	2.68×10^{16}	-1.86 ± 0.01	-23.5
P-5	5.5×10^{16}	-1.84 ± 0.01	-23.6

A similar wavenumber dependence of roughly slope two was found for the p-type data as well. Again, this is consistent with the findings of Leung and others as was previously stated. The measured cross sections differed from the Drude results, Equation (1), by an order of magnitude and the theoretical calculations from lattice scattering differed by roughly half that amount. The explanation for these differences is as yet unknown.

DISCUSSION

To determine the effect of free carriers on the accuracy of the oxygen content determination, the IR transmittance data was reduced using two methods:

(1) The measured absorption coefficient spectrum, using Equations (5-8), was determined from 1000 to 1300 wavenumbers. The baseline contribution was determined by a linear fit to the data, omitting data between 1060 cm^{-1} and 1200 cm^{-1} to avoid the oxygen peak. This baseline was subtracted from the measured absorption coefficient spectrum yielding the corrected spectrum. The corrected absorption data about the oxygen peak, from 1090 to 1123 wavenumbers, was fit to a fourth-order polynomial. A fourth-order polynomial rather than a lorentzian or gaussian function was used because it provided a best fit to the data. From the first derivative of the fit, the wavenumber at which the peak occurred, $\tilde{\nu}_{peak}$, was determined. The 4th-order polynomial was evaluated at $\tilde{\nu}_{peak}$ yielding α_{peak}. The oxygen content $[O]_i$ is related to α_{peak} by the multiplicative factor 9.63 cm·ppma [24].

(2) As outlined in method (1), the measured absorption coefficient spectrum is obtained. The baseline contribution is determined from Equation (9) using the appropriate fit coefficients listed in Tables III and IV. As before, the baseline is subtracted from the measured spectrum yielding the corrected spectrum. α_{peak} and $\tilde{\nu}_{peak}$ were determined as described in method (1). There are small differences between the fit for the free carrier absorption and the measured absorption coefficient. Therefore, a linear fit is done through the baseline data. The evaluation of the linear fit at $\tilde{\nu}_{peak}$ yields α_{base}. In this case, $[O]_i$ is related to the net absorption peak height (i.e. $\alpha_{peak} - \alpha_{base}$) by the multiplicative factor 9.63 cm·ppma.

Both methods attempt to correct for the effect of free carrier absorption. Method (1) assumes a linear baseline and Method (2) uses the curved baselines as determined from Equation (9). The differences between these two methods are shown in Tables V and VI where the superscripts (1) and (2) refer to oxygen content determined by methods (1) and (2), respectively.

Table V Interstitial oxygen content data for n-type samples

sample no.	dopant density (atoms/cm^3)	$[O]_i^{(1)}$ (ppma)	$[O]_i^{(2)}$ (ppma)	Δ (ppma)
N-1	7.38 × 10^{15}	32.9	33.1	+ 0.2
N-2	8.0 × 10^{15}	24.0	24.1	+ 0.1
N-3	1.07 × 10^{16}	24.6	24.8	+ 0.2
N-4	1.34 × 10^{16}	21.1	21.2	+ 0.1
N-5	1.94 × 10^{16}	25.0	25.2	+ 0.2
N-6	3.4 × 10^{16}	20.7	20.9	+ 0.2
N-7	3.61 × 10^{16}	28.0	28.3	+ 0.3
N-8	6.03 × 10^{17}	19.1	29.3	+10.2

Table VI Interstitial oxygen content data for p-type samples

sample no.	dopant density (atoms/cm^3)	$[O]_i^{(1)}$ (ppma)	$[O]_i^{(2)}$ (ppma)	Δ (ppma)
P-1	1.28×10^{15}	31.1	31.2	+0.1
P-2	3.7×10^{15}	31.1	31.3	+0.2
P-3	1.51×10^{16}	38.4	38.9	+0.5
P-4	2.68×10^{16}	29.1	29.9	+0.8
P-5	5.5×10^{16}	29.3	31.0	+1.7

The observed differences can be explained with the help of Figure 3. The IR absorption after correction for silicon lattice absorption is shown in the solid curves, and after correction for both lattice absorption and free carrier absorption in the dashed curves. Figure 3a shows the computed absorption coefficient for the phosphorus-doped sample N-1. For both curves, the baseline can be determined from a straight-line fit. Figure 3b shows the curves for the arsenic-doped sample N-8. Because of the high carrier concentration (6×10^{17} atoms/cm^3), there is definite curvature in the baseline of the solid curve because of free carrier absorption. Once this contribution is subtracted from the solid curve (i.e., the dashed curve), a straight line can be fit to the baseline. This baseline curvature accounts for the observed differences in Tables V and VI.

CONCLUSIONS

It has been shown that the absorption by free carriers has a significant influence on the interstitial oxygen determination. Uncertainties of 3 to 5% were observed in the p-type data and as high as 30% in the n-type data. Significant differences in oxygen content because of background absorption were seen at concentrations above 10^{16} atoms/cm^3. Over the carrier concentration range studied and the limited wavenumber range, 1000 cm^{-1} to 1300 cm^{-1}, the free carrier absorption varied smoothly with wavenumber roughly as $\tilde{\nu}^{-2}$. Although the sample set is small, there is no reason to believe that more low-resistivity samples would show any frequency-dependent behavior which is different than that already observed in this data set. The variation of free carrier absorption with frequency is a smooth function and therefore should exhibit no sharp spectral features at lower resistivities.

From this analysis, two recommendations can be made for the measurement of the oxygen content of silicon wafers with resistivities down to approximately 0.03Ω-cm for n-type silicon and down to approximately 0.3 Ω-cm for p-type silicon:

(1) All IR transmittance measurements should be referenced to that of a high-purity float-zone sample.

(2) The free carrier absorption of the sample should be determined assuming a power dependence on wavenumber, and the IR absorption spectrum corrected for this contribution, using a fitting procedure as outlined here.

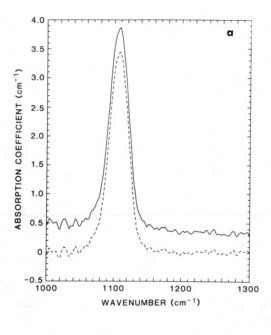

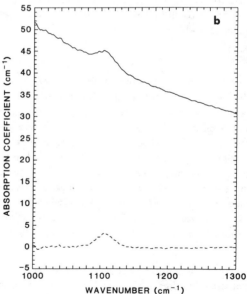

Figure 3 Absorption spectra of n-type samples N-1 (a) and N-8 (b) where the solid curves represent the absorption spectrum after correction for silicon lattice absorption, and the dashed curves represent the absorption spectrum after correction for lattice absorption and absorption due to free carriers.

The application of these steps will significantly improve the accuracy of oxygen content determinations for resistivities below 0.1 Ω-cm for n-type and 0.5 Ω-cm for p-type silicon.

ACKNOWLEDGMENTS

The authors would like to express their appreciation to J. Ehrstein for providing the samples used in this study, to D. Ricks for the resistivity and thickness measurements, to M. Thomas for the sample preparation, and to M. I. Bell for helpful discussions of the theory discussed here.

REFERENCES

[1] Jastrzebski, L., "Origin and Control of Material Defects in Silicon VLSI Technologies: An Overview," *IEEE Transactions on Electron Devices* Vol. ED-29, 1982, p.475-487.

[2] Payne, R. S., Grant, W. N., and Bertram, W. J., "Elimination of Latch Up In Bulk CMOS," IEDM-80, section 10.2, 1980, pp. 248-251.

[3] Parillo, L. C., Payne, R. S., Davis, R. E., Reutlinger, G. W., and Field, R. L., "Twin-Tub CMOS – A Technology for VLSI Circuits," IEDM-80, section 29.1, 1980, pp. 752-755.

[4] Borland, J. O., and Deacon, T., "Advanced CMOS Epitaxial Processing for Latch-Up Hardening and Improved Epilayer Quality," *Solid State Technology* Vol. **27**, No. 8, 1984, pp. 123-131.

[5] Seitz, F., *Modern Theory of Solids* (McGraw-Hill Book Company, Inc., New York, 1940), pp. 638-640.

[6] Schumann, P. A., Keenan, W. A., Fong, A. H., Gegenwarth, H. H., and Schneider, C. P., "Silicon Optical Constants in the Infrared," *Journal of the Electrochemical Society* Vol. **118**, 1971, pp. 145-148.

[7] Dumke, W. P., "Quantum Theory of Free Carrier Absorption," *Physical Review* Vol. **124**, 1961, pp. 1813-1817.

[8] Spitzer, W., and Fan, H. Y., "Infrared Absorption in n-Type Silicon," *Physical Review* Vol. **108**, 1957, pp. 268-271.

[9] Fan, H. Y., "Effects of Free Carriers on the Optical Properties," *Semiconductors and Semimetals*, Willardson, R. K., and Beer, A. C., Eds., Academic Press, New York, 1967), Vol. 3, pp. 406-408.

[10] Fan, H. Y., Spitzer, W., and Collins, R. J., "Infrared Absorption in n-type Germanium," *Physical Review* Vol. **101**, 1956, pp. 566-572.

[11] Debye, P. P., and Conwell, E. M., "Electrical Properties of N-Type Germanium," *Physical Review* Vol. **93**, 1954, pp. 693-706.

[12] Leung, P. C., "Infrared Absorption by Free Carriers in p-type Silicon," Ph.D. thesis, University of Southern California (1971).

[13] Kahn, A. H., "Theory of the Infrared Absorption of Carriers in Germanium and Silicon," *Physical Review* Vol. **97**, 1955, pp. 1647-1652.

[14] F84-84a, "Standard Method for Measuring Resistivity of Silicon Slices with a Collinear Four-Probe Array," *Annual Book of ASTM Standards*, Volume

10.05, American Society for Testing and Materials, Philadelphia, Pennsylvania, 1984, pp. 191-207.

[15] F525-84, "Standard Method for Measuring Resistivity of Silicon Wafers Using a Spreading Resistance Probe," *Annual Book of ASTM Standards*, Volume 10.05, American Society for Testing and Materials, Philadelphia, Pennsylvania, 1984, pp. 455-471.

[16] Irvin, J. C., "Resistivity of Bulk Silicon and of Diffused Layers in Silicon," *Bell System Technical Journal* Vol. **41**, 1962, pp. 387-410.

[17] Thurber, W. R., Mattis, R. L., and Liu, Y. M., "Resistivity-Dopant Density Relationship for Boron-Doped Silicon," *Journal of the Electrochemical Society* Vol. **127**, 1980, pp. 2291-2294.

[18] Thurber, W. R., Mattis, R. L., Liu, Y. M., and Filliben, J. J., "Resistivity-Dopant Density Relationship for Phosphorus-Doped Silicon," *Journal of the Electrochemical Society* Vol. **127**, 1980, pp. 1807-1812.

[19] Masetti, G., Severi, M., and Solmi, S., "Modeling of Carrier Mobility Against Carrier Concentration in Arsenic-, Phosphorus-, and Boron-Doped Silicon," *IEEE Transactions on Electron Devices* Vol. **ED-30**, 1983, pp. 764-769.

[20] Graupner, R. K., "Analysis of Infrared Spectra for Oxygen Measurements in Silicon," *Silicon Processing, ASTM STP 804*, Gupta, D. C., Ed., American Society for Testing and Materials, 1983, pp. 459-468.

[21] Pankove, J. I., *Optical Processes in Semiconductors* (Prentice-Hall, Inc., Englewood Cliffs, N. J., 1971), pp. 74-76.

[22] Proposal P-112, "Proposed Test Method for Interstitial Oxygen Content of Silicon Slices by Computer-Assisted Infrared Spectrophotometry," *Annual Book of ASTM Standards*, Volume 10.05, American Society for Testing and Materials, Philadelphia, Pennsylvania, 1984, pp. 691-698.

[23] Weeks, S. P., "Influence of Electrically Active Impurities on IR Measurements of Interstitial Oxygen in Silicon," *Semiconductor Processing, ASTM STP 850*, Gupta, D. C., Ed., American Society for Testing and Materials, 1984, pp. 335-342.

[24] F121-79, "Standard Test Method for Interstitial Atomic Oxygen Content of Silicon By Infrared Absorption," *Annual Book of ASTM Standards*, Part 43, American Society for Testing and Materials, Philadelphia, Pennsylvania, 1979, pp. 519-521.

Naohisa Inoue, Toshihiro Arai, Tadashi Nozaki, Kazuyoshi Endo and Kiichiro Mizuma

HIGH RELIABILITY INFRARED MEASUREMENTS OF
OXYGEN AND CARBON IN SILICON

REFERENCE: Inoue, N., Arai, T., Nozaki, T., Endo, K., Mizuma, K.,"High Reliability Infrared Measurements of Oxygen and Carbon in Silicon," Emerging Semiconductor Technology, ASTM STP 960, D. C. Gupta and P. H. Langer, Eds., American Society for Testing and Materials, 1986.

ABSTRACT: High reliability infrared measurements of oxygen and carbon in silicon single crystals are presented. One is the preparation and distribution of standard sample sets with known oxygen contents. Oxygen content range is 5 to 11×10^{17} atoms/cm^3 and the accuracy is estimated to be within 4×10^{16} atoms/cm^3. The other is the establishment of a standard infrared measurement procedure for carbon performed by round robin infrared measurement and charged particle activation analysis. From the procedure the conversion coefficient is determined to be $(8.5 \pm 0.9) \times 10^{16}$ atoms·cm^{-3}/cm^{-1}.

KEYWORDS: silicon, oxygen, carbon, infrared absorption, conversion coefficient, charged particle activation analysis

INTRODUCTION

Oxygen and carbon are two main impurities found in Czochralski silicon. Oxygen in silicon has both detrimental and beneficial effects on device characteristics. In particular, oxygen precipitation is successfully utilized for intrinsic gettering, which requires precise oxygen content control. This means that accurate measurement of the oxygen content must be carried out. Carbon in silicon plays some role in oxygen precipitation. As a result, accurate measurement of the carbon content is also necessary. Infrared absorption measurement is commonly used to determine the oxygen and carbon contents in silicon.

This work was performed by the Japan Electronic Industry Development Association (JEIDA) committee. Dr. Inoue is a senior research engineer at the NTT Electrical Communications Laboratories, Atsugi, Kanagawa, 243-01; Professor Arai is from the Institute of Applied Physics, Tsukuba University, Niihari, Ibaragi, 305; Dr. T. Nozaki is a principal scientist at the Institute of Physical and Chemical Research, Wako, Saitama, 351; and Mr. K. Endo and Mr. K. Mizuma are with Japan Silicon Co. Ltd., Noda, Chiba, 278.

However, the current measurement procedures are insufficient for the necessary accuracy: The infrared measurement procedure is not determined in detail. Moreover different spectrometers give different absorption coefficients due to instrument characteristics even if the same measurement procedure is employed. In addition, non-absolute infrared measurement requires calibration to achieve oxygen content. However, the calibration constant obtained by various absolute measurement method includes large uncertainties and is not yet established.

In order to establish an accurate measurement method of oxygen, round robin infrared measurement as well as charged particle activation analysis have previously been conducted in the Japan Electronic Industry Development Association (JEIDA) committee (1). A standard infrared measurement procedure has been proposed and a conversion coefficient of 3.03×10^{17} atoms·cm^{-3}/cm^{-1} has been obtained. However there is still a problem that the difference between absorption coefficeints obtained by different laboratories easily exceed 10% even when the same procedure is employed. Another project has therefore been conducted; the preparation and distribution of standard samples with known oxygen content used for calibration. 50 sets of standard samples having 4 samples with oxygen content near 6, 8, and 10×10^{17} atoms/cm^3, and a reference (oxygen-free), were prepared. The accuracy is estimated to be within 4×10^{16} atoms/cm^3.

A round robin infrared measurement of carbon in silicon has been conducted, also combined with a charged particle activation analysis. A standard infrared measurement procedure is then established. And a new conversion coefficient is determined to be $(8.5 \pm 0.9) \times 10^{16}$ atoms·cm^{-3}/cm^{-1}. These results should prove helpful for the accurate determination of oxygen and carbon contents in silicon. In this paper, the standard infrared measurement procedures, results, and accuracy evaluation are given.

OXYGEN STANDARD SAMPLES

Procedure

Procedure adopted for standard sample preparation is shown in Fig.1. "Primary standards" with known oxygen content were used for calibration. They are those used in the previous round robin measurement, and their oxygen content was previously determined by the charged particle activation analysis on samples cut adjacent to them. Linearity between absorption coefficient and oxygen content has already been confirmed. "Standard samples" whose oxygen content to be

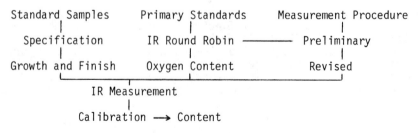

FIG. 1--Oxygen standard sample preparation procedure.

determined and reference wafers without detectable oxygen were prepared. Infrared absorption coefficients of both the primary standards and the standard samples were determined using the same spectrometer and method (standard measurement procedure). Standard sample oxygen content was determined using the calibration line obtained from the primary standards.

Standard Sample Preparation

Standard sample oxygen content was determined to cover the range 5 to 11×10^{17} atoms/cm^3 (using 3.0×10^{17} atoms·cm^{-3}/cm^{-1} conversion coefficient). This commonly used range consists of 3 oxygen content levels: high($\sim 10 \times 10^{17}$ atoms/cm^3), medium(~ 8), and low (~ 6), in addition to oxygen-free. 50 sets of oxygen-containing samples were prepared by the Czochralski method (2), as shown by the sample specifications listed in Table 1. Sample resistivity was selected to be above 10Ω cm so that free carrier absorption does not affect the accuracy. Reference wafers were cut from a crystal grown by float zone method. The samples were finished into 2 mm thick, 50 mm diameter (40 mm for reference), double side mirror polished wafers. These specifications were found in the previous round robin measurement to give most accurate results.

TABLE 1--Standard sample specification.

Thickness	2 mm ± 10 µm
Surface	Both Sides Mirror Polished
Resistivity	$> 10 \Omega$ cm
Carbon Content	$< 2 \times 10^{16}$ atoms/cm^3
Diameter	50 mm (ref. 40 mm)

Primary standards with known oxygen content were the ones used in the previous round robin infrared measurement. Their oxygen content was determined by charged particle activation analysis using the reaction:

$$^{16}O\ (^3He,\ p)\ ^{18}F\ (^{18}F:\ 110\ min.,\ \beta^+)$$

on 1 mm thick samples which were cut adjacent to them. Good linearity was obtained between the absorption coefficient and activation data (1). 15 samples of 22 were selected and used for the primary standards.

Standard Infrared Measurement Procedure

Standard measurement procedures, ASTM F121 (3) and DIN 50438/1 (4), are currently in use. However, several important factors are not given uniquely, such as baseline procedure, resolution, and reflection correction. A different procedure, although within the tolerance in the standards, gives a different result. An example of infrared transmission spectrum is shown in Fig. 2. And measurement procedures adopted here are summarized in Table 2 along with the ASTM. In the procedure adopted here, the difference method with an oxygen-free reference sample was employed for accurate baseline determination. Figure 2 shows that there are two subsidary absorption peaks on both sides of the main peak. The baseline with start/end points inside or outside the subsidary peaks gives different absorption coefficient. It has been confirmed by the previous round robin measurement that baseline outside subsidary peaks (between 1300 and 900 cm^{-1}) yields a good results (1). Therefore the same method is adopted here. Next, correction for multiple reflection is performed, which gives a few % more accurate result than the correction for single reflection. The parameters used for calculation are also the same as before (lattice vibration absorption coefficient is 0.85 cm^{-1} and reflectivity is 0.3).

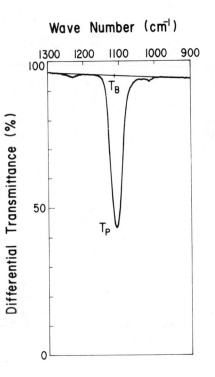

FIG.2-- Differential transmission spectrum of oxygen. Resolution = 4 cm^{-1}. Baseline is drawn between 1300 and 900 cm^{-1}. Oxygen content = 8×10^{17} atoms/cm^3. T_P is the transmittance at the absorption peak and T_B is that on the baseline.

TABLE 2--Standard infrared measurement procedure for oxygen.

Item	JEIDA	ASTM
Measurement Method	Difference Method	Dif. or Air Refer.
Resolution	4 cm^{-1}	< 5 cm^{-1}
Baseline	Between 1300 and 900 cm^{-1}	not given
Reflection Correction	Multiple Reflection	Single Refl.
Instrument Check	100% and 0% Accuracy	ibid.
Measured Value	Peak Absorption Coefficient	ibid.
Lattice Absorption	0.85 cm^{-1}	0.4 cm^{-1}
Reflectance	0.30	0.30
Conversion Coefficient	3.03×10^{17}	2.45×10^{17}

Results and Discussion

Infrared Measurement Accuracy: Infrared measurement was performed by 6 organizations who commercially supply infrared spectrometers (5). Three groups consisting of two laboratories each measured 50 standard samples and 5 primary standards. They are denoted by A, B and C groups in the following. One laboratory in each group used a dispersive type and the other a Fourier Transform type spectrometer. The accuracy was first checked by measuring the primary standards. The results are shown in Fig. 3. The absorption coefficient obtained by 6 laboratories agree well, within 10%. And linearity between the absorption coefficient and the activation data is good. Conversion coefficients obtained ranged between 3.03 and 3.29, which is close to the 3.03 value obtained in the previous work. Furthermore, no systematic difference was observed between the two types of spectrometers.

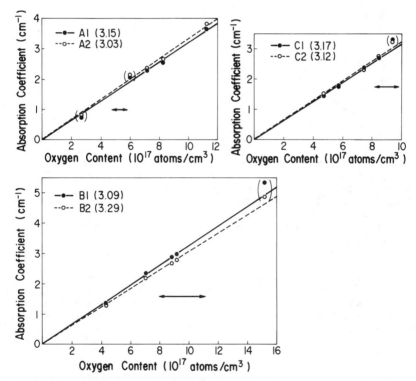

Fig. 3--Measurement accuracy estimated with primary standards. Number 1 laboratory used the dispersive type and the second used the Fourier Transform type spectrometer. Figures in parentheses are conversion coefficients. Arrows indicate the oxygen content ranges of 50 standard samples measured by the three groups.

Two absorption coefficient data sets of standard samples in each group are compared in Fig.4. Straight line is the correlation for the primary standards. Scatter of the standard sample data around the correlation line is very small (within 2%).

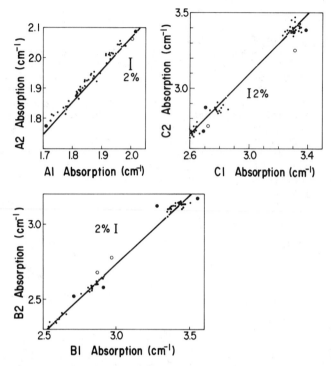

Fig. 4--Correlation of the two infrared absorption data sets for the primary standards (open circles, those outside arrows in Fig. 3 are not shown here) and the standard samples (dots and filled circles). Straight line is the correlation for the primary standards except those shown in the parentheses in Fig. 3. Filled circles are standard samples whose oxygen content was re-determined for accuracy evaluation (see text and Tab. 3).

Oxygen Content Determination: The oxygen content of the standard samples was determined as follows. The oxygen content for each laboratory was calculated using each calibration line shown in Fig.3. Then their averaged value on the two data was assigned as the sample oxygen content. The high content sample range is 10.2 to 10.7×10^{17} atoms/cm^3, the medium content range is 7.8 to 9.1, and the low content range is 5.4 to 6.3.

Oxygen Content Accuracy: Figure 5 shows the comparison of the two data sets for oxygen content in each group. This figure shows that the oxygen content data sets of 144 samples agreed to within 0.4×10^{17} atoms/cm^3, and that the maximum difference was only 0.6×10^{17} atoms/cm^3. Assuming that the two data sets do not include errors on the same side, the error due to infrared measurement is estimated at below 0.2×10^{17} atoms/cm^3.

Errors including those due to the primary standard oxygen content were estimated by measuring all primary standards and selected standard samples (shown by filled circles in Fig.4, those seem to include large discrepancy) at NTT. Figure 6 and Tab. 3 show the results. Linearity

Fig. 5--Comparison of the two sets of oxygen content data obtained by the two laboratories in each group using dispersive type and Fourier Transform type spectrometers.

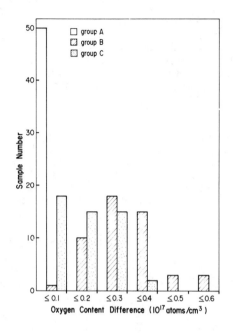

between the oxygen content and the absorption coefficient is again very good for the primary standards, which suggests that the oxygen content accuracy is good. It should be noted, however, that the primary standard sets for the 3 groups give slightly (2%) different calibration

Fig. 6--Evaluation of the primary standard oxygen content accuracy. Four calibration lines are shown, with all primary standards except those shown by "•" and with primary standard sets for A, B, and C groups (the upper three lines are displaced upward by 1 cm^{-1} each). The attached figures are corresponding conversion coefficients.

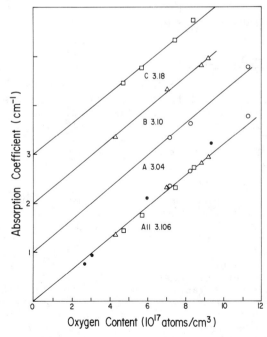

lines, which would result in some systematic errors. The mean values for two laboratories in each group agree with the contents re-determined by using the calibration line shown by "All" in Fig. 6 to within 0.2×10^{17} atoms/cm^3 (Tab. 3). The results show that the systematic error due to the primary standard sets is very small, and that the total error is also small.

The standard sample oxygen content error is the sum of the errors due to infrared measurement (2%) and those due to primary standard oxygen content (2%), i.e., 4% or 0.4×10^{17} atoms/cm^3. 50 sets of standard samples were prepared containing three levels of oxygen content, selected randomly from those measured by the three groups. Measurement accuracy in case of using the standard sample set may be better than 4% because 3 standard samples are used for calibration. Thus accurate measurement free from inter-laboratory discrepancy is possible, compared to the measurement without standard samples.

TABLE 3 - Standard sample oxygen content accuracy evaluation (10^{17} atoms/cm^3).

Group	Sample	Lab.1	Lab.2	Average	Re-determined
A	N34	5.4	5.4	5.4	5.4
	N25	6.4	6.3	6.3	6.4
B	SB7	9.0	8.5	8.7	8.7
	SA24	11.0	10.4	10.7	10.7
C	OB20	9.0	8.6	8.8	8.6
	OA18	10.5	10.4	10.5	10.4

CARBON MEASUREMENT

Sample Preparation

Carbon containing crystals were grown by the Czochralski or float zone method (effect of oxygen coexistence was examined). Two 2 mm thick double side mirror polished samples were prepared from the same part of the crystals. Other sample specifications are nearly identical to those in Table 1. The carbon content of 36 pairs of samples ranges from below the detection limit (10^{16} atoms/cm^3) to about 3×10^{17} atoms/cm^3.

Measurements

<u>Standard Infrared Measurement Procedure</u>: An example of the differential transmission spectrum of carbon in silicon is shown in Fig. 7. The carbon absorption peak is located at 16.5 μm, which is the shoulder of the strong lattice vibration peak. Full width at half maximum is about 3 cm^{-1}, which is very narrow. These features represent a serious problem in achieving accurate measurement. So, the standard measurement procedure was determined first. The procedure is listed in Table 4 along with the ASTM F123-83 standard procedure (6) for comparison. The difference method was adopted to compensate lattice vibration peak. The resolution was selected as 2 cm^{-1}, since 1 cm^{-1} yields poor reproducibility and 4 cm^{-1} yields a reduced value of absorption coefficient. Full width at half maximum reported by

all laboratories did not exceed 6 cm^{-1}, satisfying the ASTM standard requirement. Baseline is drawn outside the subsidary absorptions, i.e. between 640 and 560 cm^{-1}. Correction for multiple reflection is not necessary because of the large absorption.

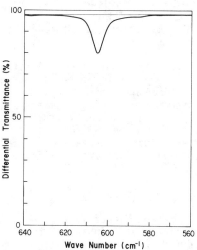

FIG.7-- Differential transmission spectrum of carbon in silicon. Resolution = 2 cm^{-1}. Full width at half maximum is 5.5 cm^{-1}. Baseline is drawn between 640 and 560 cm^{-1}. Carbon content = 5×10^{16} atoms/cm^3.

TABLE 4--Standard infrared measurement procedure of carbon in silicon.

ITEM	JEIDA	ASTM
Method	Difference Method with Reference	ibid.
Resolution	2 cm^{-1}	< 2 cm^{-1}
Baseline	Between 640 and 560 cm^{-1}	not given
Reflection Correction	Neglected	ibid.
Instrument Check	100% and 0% Accuracy	ibid.
Measured Value	Peak Absorption Coefficient	ibid.
Conversion Coefficient	8.5×10^{16}	1.0×10^{17}

Round Robin Infrared Measurement: Infrared measurement was performed by 18 laboratories. Half of these used dispersive type and the other half used Fourier Transform type spectrometers. Each laboratory measured two sets of samples which were cut adjacent to each other with an interval of several months to improve the experimental accuracy.

Charged Particle Activation Analysis(CPAA): After round robin infrared measurement, 22 samples were analyzed by the reaction

$$^{12}C\ (^3He, \alpha)\ ^{11}C\ (^{11}C: 20.4\ min.).$$

A reference wafer was also analyzed. The procedure is as follows. Each sample was bombarded with ^{3}He particles for 20 min. through an aluminum foil (80 um thick) in the Institute of Physical and Chemical Research cyclotron. ^{3}He energy at the sample surface was set from 11.5 to 12 MeV to eliminate interference by the $^{16}O(^3He, 2\alpha)^{11}C$ reaction. The bombarded sample was then dissolved in a NaOH solution containing the Na_2CO_3 carrier. After the addition of $KMnO_4$, the solution was heated in a microwave oven to first vaporize the water and to then give dull red melt(700-800°C). Finally, the ^{11}C was converted into CO_2 gas and fixed as Li_2CO_3. Positron annihilation radiation from the Li_2CO_3 was measured by either a well-type NaI detector or a pair of BGO scintillators operating in coincidence, and the decay was followed to ascertain that the radiochemical purity was that of ^{11}C. The carrier recovery for correction of the counting result was determined by weighing and was found to be 60-80%. A graphite plate was used as the activation standard.

Secondary Ion Mass Spectroscopy(SIMS): Secondary ion mass spectroscopy was performed to estimate the accuracy and sensitivity of the method and to confirm the accuracy of CPAA. Between infrared measurement and CPAA, 5 samples were analyzed. The primary ion was Cs^+ (1 μA) which was accelerated at 14.5kV. A carbon ion implanted sample with a peak concentration of 9×10^{16} atoms/cm^3 was used as a standard. Surface contamination was removed by Cs ion bombardment and a 60 μm diameter area was analyzed.

Results and Discussion

Infrared Measurement: Most laboratories obtained nearly equal absorption coefficients for the same sample. In addition, measurements performed with an interval also agreed to within 10% in many laboratories. These results confirm that the standard measurement procedure adopted is quite effective. However a few laboratories gave extremely small absorption coefficients compared to those from majority as shown later. Baseline problem (wavy) arose in the dispersive type spectrometers.

CPAA: Good reproducibility was generally obtained. Carbon content in the reference sample was determined to be 0.04×10^{17} atoms/cm^3.

SIMS: Linearity between CPAA and SIMS results was good for samples above 10^{17} atoms/cm^3, confirming the accuracy of both methods.

Conversion Coefficient: The conversion coefficient without any data exclusion is 9.3×10^{16} atoms·cm^{-3}/cm^{-1}. This was obtained by simple averaging of $k = (c)/\alpha$ for all samples and laboratories, where (c) is the carbon content obtained by CPAA and α is the absorption coefficient. Figure 8 shows the conversion coefficient histogram for the laboratories. The values of this histogram were obtained by averaging k on all samples in a set (thus, one laboratory gave two conversion coefficients). Most of the data lie between 7 and 10×10^{16} atoms·cm^{-3}/cm^{-1}. Three of the laboratories which all used the dispersive type spectrometers gave extremely large coefficient (small absorption). Except for this problem, there is no systematic difference between the data obtained by the two types of spectrometers. Averaging k on laboratories except the 3 laboratories which gave extremely large values yields 8.7×10^{16} atoms·cm^{-3}/cm^{-1}.

Figure 9 shows the conversion coefficients obtained for the samples. These coefficients were obtained by averaging k on all laboratories except three. 21 samples gave coefficient between 7 and 10×10^{16} atoms·cm^{-3}/cm^{-1}, but only one with a very low carbon content gave an extremely large one. Averaging k on samples except one yields 8.9×10^{16} asoms·cm^{-3}/cm^{-1}. It should be noted that no systematic shift is observed toward the low content samples, although

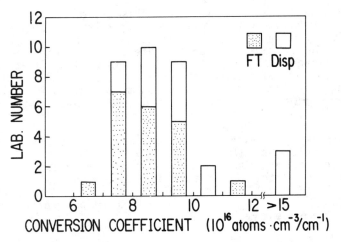

Fig. 8-- Conversion coefficient histogram for laboratories. All laboratories except one gave two conversion coefficients for the two sample sets.

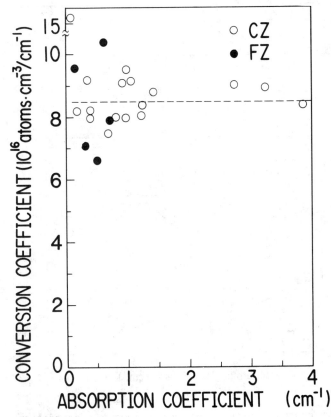

Fig. 9--Conversion coefficients for various carbon content samples.

coefficient scatter tends to be large. This assures that the simple averaging of k on samples is acceptable regardless of sample carbon content. Averaging on all samples except one (and 3 laboratories) yields 8.5×10^{16} atoms·cm^{-3}/cm^{-1}. There was no systematic difference between CZ and FZ samples which shows that oxygen coexistence does not affect the value.

Obtained conversion coefficients are summarized in Table 5 with the data exclusion rule and standard deviation. The relationship between mean absorption coefficient and activation data obtained for sample and laboratory exclusion is shown in Fig. 10. It is observed that good linearity is obtained even in the low carbon content samples. In addition, a calibration line can be drawn so that it crosses the origin. These results suggest that the obtained conversion coefficient is very accurate. Furthermore, the obtained conversion coefficient is a little smaller than the one given in the ASTM, 1.0×10^{17} atoms·cm^{-3}/cm^{-1} (6). That is, a larger absorption coefficient was obtained in the present work.

TABLE 5--Conversion coefficient obtained from various data exclusions.

Data Exclusion	Conversion Coefficient (10^{16} atoms·cm^{-3}/cm^{-1})	Standard Deviation
None	9.3	1.84
Laboratory	8.7	1.00
Sample	8.9	0.90
Lab.& Sample	8.5	0.86

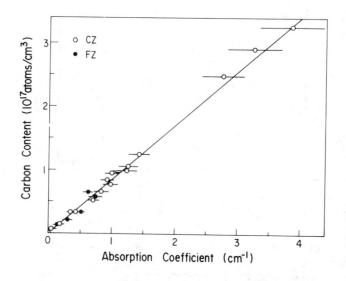

FIG.10--Carbon content versus infrared absorption coefficient after lab.& sample data exclusion. The bar represents standard deviation.

SUMMARY

Oxygen standard samples were prepared and their oxygen contents were accurately determined by infrared measurement using primary standards with known oxygen contents. A sample set consists of 3 oxygen content levels: high ($\sim 10 \times 10^{17}$ atoms/cm^3), medium (~ 8), low (~ 6) and an oxygen-free reference. They cover a content range widely used today. The content accuracy was estimated to be within 0.4×10^{17} atoms/cm^3. 50 sets were prepared and distributed to organizations including ASTM and DIN. It is assured that the so far high level of measurement errors due to instrument or measurement procedure will be much reduced by using these standard samples.

Carbon round robin infrared measurement was performed along with charged particle activation analysis. 36 pairs of samples were measured by 18 laboratories and 22 samples were analyzed. A standard measurement procedure was established and an accurate conversion coefficient of $(8.5 \pm 0.9) \times 10^{16}$ atoms·cm^{-3}/cm^{-1} was obtained. This is 15 % smaller than the ASTM standard. Data correction for the activation analysis, statistical examination and measurement on additional samples are now underway, proving only a slight change must be made. The result will be reported in the near future.

ACKNOWLEDGEMENTS

This work has been greatly assisted by the continuing efforts of K. Higuchi and S. Furukawa of JEIDA. The authors are grateful to S. Nakamura of Matsushita Technoresearch for SIMS measurement. We would also like to acknowledge discussions with T. Iizuka and M. Tajima of the Optoelectronics Joint Research Laboratory and M. Watanabe of the Toshiba Company.

REFERENCES

(1) Iizuka, T., Takasu, S., Tajima, M., Arai, T., Nozaki, T., Inoue, N., and Watanabe, M., J. Electrochem. Soc., 132 (1985) 1707.
(2) Oxygen standard samples, carbon containing samples, and reference oxygen-carbon-free samples were prepared by Japan Silicon, Shin-Etsu Handotai, Osaka Titanium, Komatsu Electronic Metals and Toshiba Ceramics. In addition, a few samples were provided by Wacker Chem. and Monsanto.
(3) ASTM F 121-83, Annual Book of ASTM Standards, Philadelphia, 1984, Vol 10.05, p. 240.
(4) DIN 50438/1, Teil 1, 1978, Beuth Verlarg CmbH, Berlin 30.
(5) Infrared measurement of oxygen standard samples was performed at ASCO(D), JEOL(F), Shimadzu (D), Digilab (F), Hitachi (D), and Nicolet (F). D denotes the dispersive type and F denotes the Fourier Transform type spectrometer.
(6) ASTM F123-83, Annual Book of ASTM Standards, 1984, Vol 10.05 p. 245.

Field Enhancement and Contamination Control Aspects

Samares Kar and Manju Tewari

NATURE OF PROCESS-INDUCED Si-SiO$_2$ DEFECTS AND THEIR INTERACTION WITH ILLUMINATION

REFERENCE: Kar, S., Tewari, M., "Nature of Process-Induced Si-SiO$_2$ Defects and their Interaction with Illumination" Emerging Semiconductor Technology, ASTM STP 960, D.C. Gupta and P.H. Langer, Eds., American Society for Testing and Materials, 1986.

ABSTRACT: Transparent gate structures were fabricated by electron beam evaporation of Sn-doped In$_2$O$_3$ on oxidized p-Si substrates. The samples were oxidized at 1100°C in dry oxygen. No post-oxidation or post-electrode-deposition annealig was carried out. Admittance-voltage-frequency measurements were made under optical illumination. Interface state density distributions and hole and electron capture cross-sections were obtained using the recently developed optical metal-oxide-semiconductor (MOS) admittance technique. The experimental interface state density profile contained two peaked distributions, one near the valence band-edge E_v, and the other near the conduction band-edge E_c, overlying a concave background. The peak near E_c was sharper and the peak density was higher than in the case of the peak near E_v. The capture cross-section vs band-gap energy profile also displayed a peaked distribution for interface states under each of the peaks. With increasing illumination, the state density at the peak increased, the peak energy location moved closer to the respective band-edge, and the capture cross-section decreased. The experimental results show the presence of defects at unpassivated Si-SiO$_2$ interfaces, which exchange electrons/holes with silicon bands under illumination.

KEYWORDS: optically-activated states, radiation damage, interface defects, silicon-oxide interface, MOS admittance

Ms. Tewari is a senior research assistant and Professor Kar is a member of the Electrical Engineering faculty, Indian Institute of Technology, Kanpur-208016, India.

INTRODUCTION

Inspite of a very large number of investigations carried out to study the Si-SiO$_2$ interface states, several aspects of this very important subject are still not well understood. Perhaps the most crucial among these is the origin of the interface states, specifically those that are process-induced. Another aspect, which can have important bearing on the operation of optoelectronic devices, but has received scant attention so far, is the behaviour of the Si-SiO$_2$ interface defects under optical illumination. [An interface defect is a physical or chemical imperfection at the interface, while an interface state is an allowed state in the silicon bandgap at the interface].

As increasing use is being made of processes, such as, electron beam evaporation, sputtering, and plasma deposition /etching, a better understanding becomes necessary of the nature of defects induced by these processes. It is also important to know the characteristics of the interface states generated by the process-induced defects.

There are some reasons to suspect that the interface defects behave differently under light than in the dark, when the usual electrical signals are applied, and to believe that optically-activated interface defects exist. The voltage-axis intercepts of the reciprocal-squared-capacitance vs voltage, $C^{-2}(V)$, characteristics of metal-oxide-semiconductor (MOS) solar cells have been observed to decrease with increasing illumination intensity [1-3]. This observation was made in the case of each of a large number of MOS solar cells examined on p-type and on n-type silicon and with various types of front contact materials such as different metals and In$_2$O$_3$ and SnO$_2$. The significant decrease of the zero-bias silicon band-bending, φ_i^o, with increasing light intensity could not be explained by any fact other than an increase in the interface state charge. More recently, Henderson [4] while examining the electron spin resonance signal from trivalently-bonded-silicon with a dangling bond directed towards the oxide, observed a large enhancement of the signal amplitude under optical illumination. This increase was attributed to an optical modification of the charge state of the dangling bonds. Recently, an optical MOS admittance technique [5-7], has been applied to the investigation of various types of MOS structures, fabricated on both p-type and n-type silicon, with oxide thickness varying between 70 and 800 Å, and with Au, Al, In$_2$O$_3$, or SnO$_2$ front contacts deposited either by filament or electron gun evaporation, or rf sputtering, or spray hydrolysis. In the case of each of the structures investigated, additional interface states were found to respond to the applied ac signal under optical illumination. The increse in the MOS admittance with illumination intensity could only be due to an increase in the interface state density.

In order to investigate further the response of the interface defects to the applied ac signal under optical illumination, p-Si/SiO$_2$/In$_2$O$_3$ transparent gate structures, were fabricated by dry oxidation of silicon wafers at 1100°C, followed by electron beam evaporation of Sn-doped In$_2$O$_3$. These structures are most suitable for the optical MOS admittance technique because of good transparence (higher than 90 percent) of the In$_2$O$_3$ layer, low series resistance (the In$_2$O$_3$ resistivity was ~10^{-4} Ohm. cm), and excellent mechanical and chemical stability. The admittance measurements were made over a wide frequency range in the dark and under various intensities of tungsten lanp illumination. Measurements were carried out on a large number of unpassivated p-Si/SiO$_2$/In$_2$O$_3$ structures to verify reproducibility. Measurements were also repeated on the same structure at intervals of time. No permanent photon damage was detected. The measured data were analyzed to obtain the interface state density distributions and the state cross-section vs band-gap energy profiles for capture of holes as well as of electrons.

EXPERIMENT

Sample Fabrication

Fabrication was carried out in a class 100 laminar flow clean room. The starting materials were single crystal p$^+$p epitaxial silicon wafers of 50-mm dia with (100) surface orientation, 0.015 Ohm.cm substrate resistivity, and 1.0 Ohm.cm layer resistivity. The epitaxial layer thickness was about 8 microns. Epitaxial wafers were chosen to reduce the bulk silicon series resistance. The wafers were degreased in warm trichloroethylene, degreased in warm acetone, cleaned ultrasonically in acetone, and finally degreased in warm methanol. They were subsequently etched in HF, rinsed in 14-16 MOhm.cm deionized water, and dried in dry filtered nitrogen gas. Preoxidation was then carried out in a Thermco resistance-heated furnace in dry oxygen at 1100°C for 30 min. at atmospheric pressure. The wafers were then etched in HF, rinsed in deionized water, ultrasonically cleaned in acetone, etched in HF, rinsed in deionized water, and then dried in dry nitrogen. Final oxidation was carried out in dry oxygen at 1100°C at atmospheric pressure for 30 min. The purpose of preoxidation was to remove the surface layer damaged during mechanical polishing. No post-oxidation annealing was carried out.

Immediately after oxidation, the wafers were introduced into the vacuum chamber of a Varian VT-112B ultrahigh vacuum system fitted with sorption, sublimation, and sputter-ion pumps. After the vacuum chamber was pumped down to 1.0×10^{-7} Torr, the substrate temperature was brought up to 300°C. Filtered dry oxygen was then introduced and the partial pressure of oxygen was adjusted to 1.0×10^{-5} Torr. Layers of tin-doped indium oxide were deposited on the substrates by upward electron-beam evaporation of tin-containing (10.0 at.%) indium oxide tablets, Initially, the substrates were protected from outgassing of the indium oxide tablet by a shutter arrangement. During electron-beam evaporation, the cryopanel was chilled by liquid nitrogen.

Evaporation was carried out using a Varian 2-kW, 3- crucible electron gun with stainless-steel crucibles, cooled by chilled recirculating water. Substrates were located about 17 cm above the tablet, and were supported on molybdenum shadow masks in a substrate holder. Circular dots of 1.0-mm dia were deposited and the indium oxide layer thickness was about 2000 Å.

In the next step, oxide from the back of the wafer was dissolved in HF after the front had been protected by apiezon wax. Subsequently, the Au back contact was formed by filament evaporation in a separate Varian VT-112A system. No post-metallization annealing was carried out.

Measurements

For electrical measurements, the samples were kept in a dry box, well shielded against electrical fields. Contact was made to the transparent gate with a fine tipped (0.2-mm tip diameter) telescopic spring probe. Admittance-voltage measurements were made in the dark as well as under four intensities of illumination, L_1, L_2, L_3, and L_4, over the frequency range 60 Hz - 13 MHz, using a General Radio 1616 precision capacitance system or a Hewlett Packard 4192A LF impedance analyzer, a Keithley 616 digital electrometer, and a finely adjustable dc power supply. The General Radio 1616 precision capacitance system was employed to measure the capacitance-voltage characteristics at 60 and at 120 Hz. The Hewlett Packard 4192A LF impedance analyzer was employed to measure the capacitance-voltage characteristics at 1 and 3 MHz and the conductance-frequency and the capacitance-frequency characteristics at various values of the bias. The frequency range for the admittance-frequency characteristics was 60 Hz - 13 MHz. The illumination intensity L_1 was the lowest, and the illumination intensity L_4 was the highest. Tungsten lamp illumination was used and the light was admitted into the sample box through a tiny slit. The area of the transparent gate was measured under a WILD M8 stereo zoom microscope.

Data analysis

The method of analysis and the reduction of the illuminated MOS capacitance and conductance data, to obtain the interface state density distribution and the hole and electron capture cross-sections, has been described elsewhere [5-7].

RESULTS AND DISCUSSION

A number of p-Si/SiO$_2$/In$_2$O$_3$ structures were examined to ascertain the reproducibility of the electrical characteristics. Further, the electrical measurements were repeated at different times to ascertain the sample's stability. Figure 1 presents the measured capacitance-voltage characteristics at different frequencies and illuminations, such as, 60 Hz, 120 Hz, and 1 MHz and the illumination levels of 0, L_1, L_2, L_3, and L_4. For each illumination condition, the C-V characteristics were recorded at 60 and 120 Hz using the General Radio 1621 precision capacitance system, and at 100 kHz, 1MHz, 3 MHz, and 5 MHz using the Hewlett Packard 4192A LF analyzer. Even under the lowest illumination level L_1, the 60 Hz C-V characteristic qualified as the low frequency plot. However, in the dark, the 60 Hz characteristic exhibited no dispersion from the 120 Hz characteristic from -5.0 V to 0.25 V, above which the former deviated from the latter (cf. Fig. 1). It was not possible to carry out capacitance measurements below 60 Hz, because of noise in the supply line. The voltage range used for the admittance measurements was -5.0 to 2.0 V.

The 1 MHz C-V characteristic was an optimal high frequency plot. At frequencies higher than 1 MHz, the errors introduced by the leads became significant, although, care was taken to neutralize the effects of the induced magnetic field on the impedance of the leads between the HP analyzer terminals and the sample. The lead-induced error was much larger in the case of the MOS capacitance rather than MOS conductance.

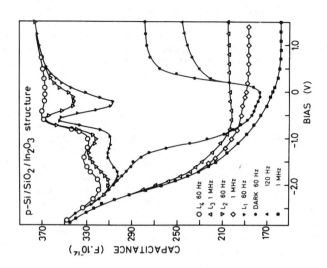

Fig.1: Capacitance-voltage (C-V) characteristics of a typical p-Si/SiO$_2$/In$_2$O$_3$ structure, measured in the dark and under the light intensities L$_1$, L$_2$, L$_3$ and L$_4$, at 60 Hz, 120 Hz and 1 MHz.

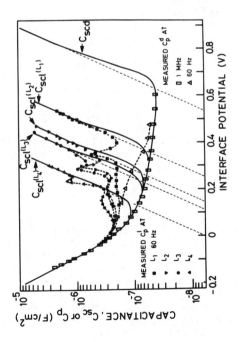

Fig.2: Experimental parallel capacitance, C$_p^i$, vs. interface potential, ψ_i, and calculated space charge capacitance, C$_{sc}$, vs interface potential characteristics, corresponding to the dark condition and to the illumination levels L$_1$, L$_2$, L$_3$ and L$_4$.

Fig. 1 shows that the constant high frequency minimum capacitance in the strong inversion regime increases with increasing light intensity. This is because at a higher illumination level, strong inversion sets in for a smaller silicon band-bending, ψ_i [6].

It is quite apparent from the 60 Hz characteristics that two interface state density peaks are present, one near the valence band-edge, E_v, and the other near the conductance band-edge, E_c. Measured characteristics also show that the densities of both the peaks increase with the illumination intensity.

Figure 2 depicts experimental and calculated characteristics between the interface potential ψ_i and the calculated space charge capacitance C_{sc} or the experimentally-obtained parallel capacitance C_p. C_p is the sum of the interface state capacitance C_{is} and C_{sc}, and is obtained from the measured low frequency MOS capacitance C [5]. The solid lines in Fig. 2 represent the dark space charge capacitance, C_{scl}, under intensities L_1, L_2, L_3 and L_4, calculated as functions of ψ_i, using the experimental values of the doping density and the quasi-Fermi level separation, δE_F [5]. The broken lines in Fig. 2 represent the experimentally obtained parallel capacitance in the dark, C_p^d, and under various illumination intensities, C_p^l, as functions of ψ_i. The experimental ψ_i was obtained from the integration of the measured low frequency C-V characteristic.

Figure 2 shows that the measured C_p^d at 1 MHz is very close to the calculated C_{csd}, indicating that for the given sample, the doping density is fairly uniform and most of the interface states are not able to follow the 1 MHz signal. One can observe from the differences between the C_p^l (ψ_i) and the respective C_{scl} (ψ_i) characteristics that the interface state density peak near E_c is much larger in magnitude and is also closer to the band-edge than the peak near E_v.

Figure 3(a) presents the experimental interface state density as a function of the bandgap energy, obtained by using the optical MOS capacitance technique, while Fig. 3(b) depicts the same obtained by using the optical MOS conductance technique. Figure 3(a) shows that the interface state density peak near E_c is not only higher in magnitude and is closer to the band-edge, but the peaked distribution is also sharper than the peaked profile near E_v. Under the dark condition, the peaked distribution near E_v is very broad. With increasing light intensity, in the case of both the peaked distributions, the peak magnitude initially increases and then tends to saturate, the peak energy location moves closer to the nearest band-edge, and the peaked profile becomes sharper.

Figure 4 represents the MOS conductance/frequency, G/f, vs f characteristics, measured under the illumination intensity L_3, for various values of the applied bias in the accumulation/depletion regime. In most cases, accurate values of the conductance could be obtained from the HP 4192A LF analyzer over the frequency range of 700 Hz to 3 MHz. Below 700 Hz, the HP analyzer was not sensitive enough due to noise, and above 3 MHz, the leads gave rise to errors. Figure 5 contains the parallel conductance/angular frequency, G_p/ω, vs f characteristics, obtained from the measured plots of Fig. 4. One can easily recognize, by noting the change, in the values of the frequencies, f^{max}, for which the maxima of G_p/ω occur, with the applied bias, that the source of the

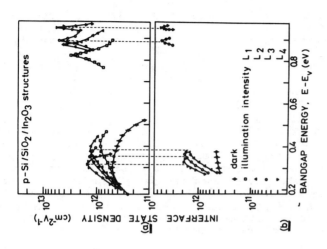

Fig.3: Experimental interface state density, N_{is}, as a function of the band-gap energy, $E-E_v$, obtained from (a) the MOS capacitance data, and (b) the MOS conductance data, corresponding to the dark condition and the illumination levels L_1, L_2, L_3 and L_4.

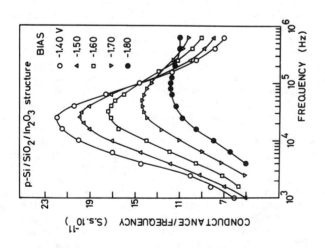

Fig.4: Measured conductance/frequency, G/f as a function of the frequency, f, for different values of the applied device bias, V, in the accumulation/depletion regime, corresponding to illumination intensity L_3.

measured conductance is the loss involved in the exchange of holes with the valence band [7].

Figure 6 presents the G/f vs f characteristics, measured under illumination intensity L_3, for various values of the applied bias in the inversion regime, while Fig. 7 contains the G_p/ω vs f characteristics, obtained from the plots of Fig. 6. Figure 7 shows that the G_p/ω maxima occur within a very narrow range of the frequency f. From the shift in the values of f^{max} with the applied bias, it can be determined that the source of the measured conductance is electron exchange with the conductance band (cf. Section II).

For light intensities below level L_3, G_p/ω vs f characteristics did not exhibit any clean maxima in inversion. This is related to the fact that the rate of supply of the minority carriers to the interface was not high enough to allow the inversion layer to follow the higher frequency signals. In other words, when the light intensity is not high enough, the inversion layer as well as the interface states are cut off from the bulk silicon at higher frequencies. Thus, the minimum light intensity necessary for the optical MOS conductance technique is higher than that for the optical MOS capacitance technique.

The halfwidths of the G_p/ω vs f characteristics were examined to determine which of the three models, namely: the single level state, the state continuum, and the statistical fluctuation models, best represented a certain characteristic. With increasing light intensity, the halfwidth was found to decrease. The G_p/ω vs f characteristics in the inversion regime displayed single level characteristics, while those in the accumulation/depletion regime, initially exhibited statistical form, but with increasing ligh intensity, assumed single level character.

The interface state density distributions obtained from the conductance data have been presented in Fig. 3(b), and the corresponding data on the hole and the electron capture cross-sections have been presented in Fig. 8. The vertical broken lines in Fig. 3. indicate the positions of the state density peaks obtained from the capacitance data, for comparison with the peak energy locations obtained from the conductance data. Figure 3 indicates good agreement between the results obtained from MOS capacitance and those obtained from MOS conductance. The peak energy locations tally very well, except for the peak near E_v under the dark condition. The peak magnitudes also agree in most cases.

Figure 8 shows that the hole capture cross-sections, σ_h, are much higher than the electron capture cross-sections, σ_e. This may partly be related to the fact that the values of σ_e belong to states located very close to E_c, while the values of σ_h belong to states located closer to the midgap, cf. Fig. 8. It is also apparent from Figs. 8 and 3 that the capture cross-section vs bandgap energy profile goes through a peak and that this peak occurs at about the same energy at which the state density also goes through a maximum. In Fig. 8, the arrows indicate the corresponding energy locations of the state density peaks. With increasing light intensity, the electron capture cross-sections can be seen to decrease strongly. Again, this may partly, be a consequence of the energy location moving closer to the band-edge. A strong variation of the capture cross-section with the energy may be behind the observation of a narrower peak from the conductance technique than from

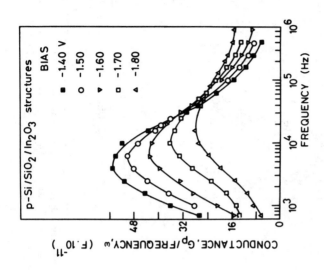

Fig.5: Experimental parallel conductance/angular frequency, G_p/ω, vs frequency characteristics, for different values of the applied bias, V, in the accumulation/depletion regime, corresponding to illumination intensity L_3.

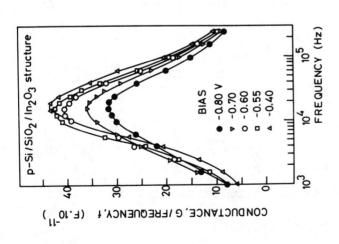

Fig.6: Measured conductance/frequency, G/f as a function of the frequency, f, for different values of the applied device bias, V, in the inversion regime, corresponding to illumination intensity L_3.

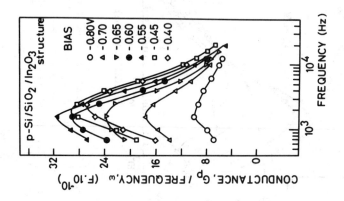

Fig. 7: Experimental parallel conductance/angular frequency, G_p/ω, vs frequency characteristics, for different values of the applied bias, V, in the inversion regime, corresponding to illumination intensity L_3.

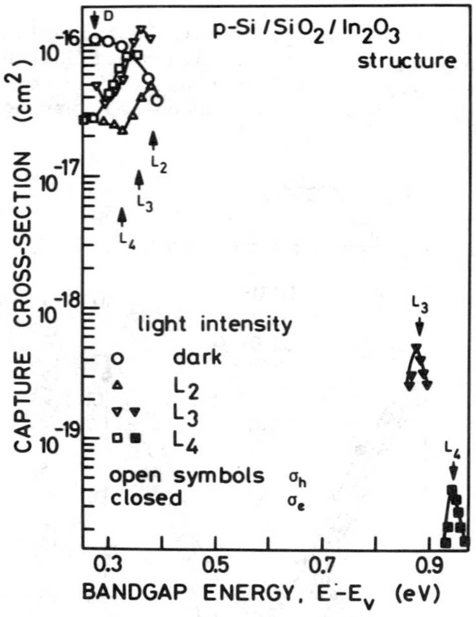

Fig. 8: Experimental hole and electron capture cross-section, σ_h and σ_e, as function of the bandgap energy, $E-E_v$, corresponding to the dark condition and the illumination levels L_2, L_3 and L_4.

the capacitance technique, cf. Fig. 3(b). For, even with a strongly varying capture cross-section, all the states will contribute to the capacitance, if the frequency is low enough. However, in the case of the conductance, only those states will contribute, for which $\omega\tau$ is close to 1.0.

CONCLUSIONS

From the experimental observations, the following conclusions may be made. The nature of the peak near E_c appears to be different from that of the peak near E_v. The former is closer to the band-edge, its peak density is many times larger, its capture cross-section is orders of magnitude smaller, and its profile is narrower than that of the latter. Hence, the peak near E_c might have a different physical origin than that of the peak near E_v. It can be assumed that the lower bandgap peak is generated by bonding defects [8,9]. Poindexter et al [8] had assigned the two peaks observed by them, in unpassivated MOS structures, at $E_v + 0.31$ eV and at $E_v + 0.83$ eV, to trivalently bonded silicon with a dangling bond pointed towards the oxide. For both these peaked profiles, the state densities at the peak were about the same on the (111) surface.

The peak, observed in the upper bandgap, could be due to defects induced by the electron beam processing. Earlier observatins have indicated the interface states, generated by an electron beam to have very small capture cross-sections and to be located very close to the conduction band-edge [10]. The mechanism by which electron beam processing could induce interface states has not been well understood. Two possibilities have been suggested, namely, electron trapping and X-ray damage. X-rays are believed to disrupt the chemical bonds, while the electron traps are believed to be located in SiO_2 in the proximity of the Si-SiO_2 interface.

Furthermore, the large change in the state density at the peak, the energy location of the peak, and the capture cross-section, with the illumination intensity are worth noting. With increasing optical radiation, the peak density increases, the peak energy location moves closer to the band-edge, and the state capture cross-section decreases. In addition, the state density profile becomes narrower, and assumes more of a single level character. The optical effect was more pronounced in the case of the peak near the conduction band edge. The results, obtained in this investigation, clearly show that defects exist at or near the silicon-oxide interface, which do not respond to the applied ac signal in the dark, but are activated in the presence of optical radiation. No indication of permanent photon damage was seen. The process of optical activation of the Latent defects could involve optical modification of the charge state of the dangling bonds [4], or other optically induced changes in the bonding parameters, in the case of the bonding defects. The theoretical calculations of Sakurai and Sugano [9] show that changes in bonding parameters can be expected to bring about variations in the most important parameters, that characterize an interface state. The electron trapping defects are believed to be located in the oxide, but very close to the interface. The optical radiation, perhaps, facilitates transport of the electrons across the oxide barrier.

ACKNOWLEDGEMENT

This work has been carried out under a cooperative research project between Indian Institute of Technology, Kanpur and the Pennsylvania State University, University Park, supported by the National Science Foundation, USA.

REFERENCES

[1] S. Kar, "On the role of interface states in MOS solar cells", J. Appl. Phys. 49, 5278 (1978).
[2] S. Kar, "Effects of interface states, tunneling, and metal in silicon MOS solar cells", Technical Digest of the IEEE International Electron Devices Meeting, Washington, D.C. (1977), p. 56 A.
[3] S. Kar, S. Varma, and S. Bhattacharya, "I-V, I_D-V, and C-V characteristics of MOS and PN junction solar cells under concentration", Conference Record of the IEEE Photovoltaic Specialists Conference, Washington, D.C. (1978), p. 667.
[4] B. Henderson, "Optically induced electron spin resonance and spin-dependent recombination in Si/SiO_2", Appl. Phys. Lett. 44, 228 (1984).
[5] S. Kar, S. Varma, P. Saraswat, and S. Ashok, "Interface investigation using transparent conductor-oxide-silicon structures", J. Appl. Phys. 53, 7939 (1982).
[6] S. Kar and S. Varma, "Determination of semiconductor quasi Fermi level separation under illumination", J. Appl. Phys. 54, 1988 (1983).
[7] S. Kar and S. Varma, "Determination of silicon-silicon dioxide interface state properties from admittance measurements under illumination", J. Appl. Phys. 58, 4256 (1985).
[8] E.H. Poindexter, G.J. Gerardi, M.E. Rueckel, P.J. Caplan, N.M. Johnson, and D.K. Biegelsen, "Electronic traps and P_b centers at the SiO_2 interface: Band-gap energy distribution", J. Appl. Phys. 56, 2844 (1984).
[9] T. Sakurai and T. Sugano, "Theory of continuously distributed trap states at $Si-SiO_2$ interfaces", J. Appl. Phys. 52, 2889 (1981).
[10] E. Rosencher, A. Chantre, and D. Bois, "Electron beam induced defects at $Si-SiO_2$ interface", Proceedings of the International Topical Conference on the Physics of SiO_2 and its Interfaces, (edited by G. Lucovsky, S.T. Pantelides, and F.L. Galeener (Pergamon, New York, 1980), p. 331.

Eric C. Maass

A STRATEGY FOR REDUCING VARIABILITY IN A PRODUCTION SEMICONDUCTOR FABRICATION AREA USING THE GENERATION OF SYSTEM MOMENTS METHOD.

REFERENCE: Maass, E. C., "A Strategy for Reducing Variability in a Production Semiconductor Fabrication Area Using the Generation of System Moments Method," Emerging Semiconductor Technology, ASTM STP 960, D.C. Gupta and P. H. Langer, Eds., American Society for Testing and Materials, 1986.

ABSTRACT: Tightening the distributions of key device parameters is increasingly important in the enhancement and prediction of yields, and in improving the quality of the products of a semiconductor fabrication area. Extensive experimentation is not always feasible in a production area. Statistical tools and knowledge of the effects of processing factors on the device characteristics are used to help identify the most significant causes of variability with the fewest experiments. The Generation of System Moments method, combined with reliable mathematical models or simulation programs which relate device parameters to processing factors partitions the device parameter variance to the causes according to partial derivatives obtained from the model. In identifying the most significant factor, this method is an effective tool in evaluating methods and tradeoffs in efforts to improve process control. The most significant factors involved in threshold and breakdown voltage variability were identified using this approach.

KEYWORDS: Parametric variability, statistical modelling, device sensitivity, generation of system moments, error propagation, Pareto analysis.

Reduction of the variability of key device parameters generally results in improvements in the quality, yield, and the predictability of the yield of the integrated circuits produced. The probability of producing integrated circuits with characteristics which do not conform to specifications is reduced, thereby improving quality (defined as conformance to specifications). Furthermore, in reducing the variability of key device parameters, the integrated circuits produced are more nearly uniform. Hence, the yield is more uniform, resulting in greater

E. C. Maass is device engineering section manager for Motorola, 2200 West Broadway, Mesa, Arizona 85202.

predictability of yield, with attendant benefits to on-time delivery of quality product.

Tools currently available towards this goal of reduced parametric variability include statistical analysis of process and device information, process control techniques such as control charting, and experimentation. The costs and time constraints involved place limitations on the desirability of extensive experimentation in a production semiconductor fabrication area. While factorial experimental designs have proven beneficial, the multiple splitting and re-merging of groups in such a design is a drawback. Ideally, the possible candidates for factors in the factorial design should be screened to achieve the goal of using the fewest and simplest possible experiments which can result in the desired tightening of the key device parameters.

The variability of the device parameters important to the functioning of integrated circuits can be partitioned in two ways, each of which provides insight into approaches to reduce the variance:

> 1) the hierarchical levels of components of the variance of the device parameter: across (within) wafer, wafer-to-wafer, and lot-to-lot.

> 2) the sources of the variance, in terms of the processing factors to which the device parameter is sensitive.

In partitioning the variance, one level generally dominates. This "family of variation" provides a clue as to the cause of the variation; for example, the dominance of wafer-to-wafer variance might indicate that the variation derives from the position of wafers in a furnace, or the sequence of processing through a single-wafer operation. The commonly dominant lot-to-lot variance, on the other hand, may indicate variation involving batch operations.

Similarly, in partitioning the variance according to the process factors causing the variance, one process factor also generally dominates. Better control of the dominant factor will result in less variability of the device parameter affected. Methods for determining this dominant cause include linear regression techniques, which assume a linear model: the parameter is assumed to be a linear, additive function of the independent (process) factors.

In a variety of instances, however, there are equations and simulation programs available which provide more exact models to describe the sensitivity of device parameters (such as threshold voltage, current gain, breakdown voltages, and sheet resistances of layers) to the processing factors. Sensitivity analysis, in conjunction with statistical information from the process area and test devices on the wafers produced, can implicate the dominant processing factor causing the variability of the device parameter. The generation of system moments method, also known as statistical error propagation [1, 2, 3] provides a means to use the wealth of theoretical work and simulation programs available in the semiconductor field to partition variance according to the processing factor causes in a manner analogous to regression techniques.

METHOD

One strategy for reducing variance uses both partitioning of variance according to level of variation, and partitioning of variance according to the process factors to which the device parameter is sensitive. this strategy is illustrated by the flowchart in Figure 1.

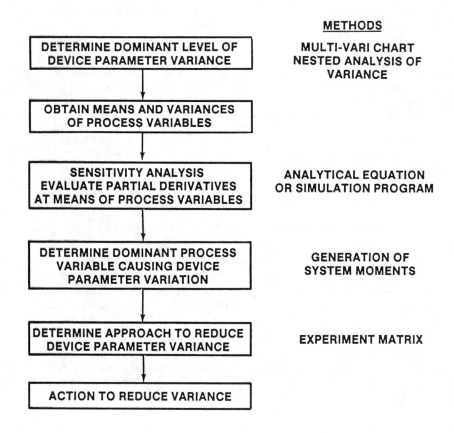

Figure 1. Flowchart describing a strategy to reduce parametric variance.

The first step involves identifying the dominant level of variance, whether across (within) wafer, wafer-to-wafer, or lot-to-lot. Techniques to achieve this include the Multi-vari method [4]; however, a more direct approach is possible if a suitable database of parametric information exists, from which an estimate of the nested variance at each level can be extracted by Variance Components analysis [5].

Partitioning of variance according to the process factors involves obtaining the sensitivity of the device parameter to the process factors, in terms of a set of partial derivatives. The partial derivatives can be obtained from the theoretical equation, or by means of multiple runs of a suitable simulation program. The partial derivative must be evaluated in the vicinity of the means of each processing factor.

In the generation of system moments method, the portion of the device parameter variance which can be attributed to the variance of each processing factor "cause" is estimated by the square of the product of the standard deviation of the processing factor and the partial derivative of the device parameter with respect to the processing factor, evaluated at the mean of the processing factor. This relationship is summarized by equation 1, which is derived by expanding the functional relationship about the means of the process factors by a multivarate Taylor series expansion [1, 2]. Retaining terms up to third order, and assuming that process factors are uncorrelated:

$$[S(P)]^2 = \sum_{i=1}^{n} \left[\frac{\partial P}{\partial X_i} \cdot S(X_i) \right]^2 + \sum_{i=1}^{n} \left(\frac{\partial P}{\partial X_i} \right) \left(\frac{\partial^2 P}{\partial X_i^2} \right) \mu_3(X_i) \quad (1)$$

WHERE: $S(P)$ = STANDARD DEVIATION OF DEVICE PARAMETER P

$S(X_i)$ = STANDARD DEVIATION OF PROCESS FACTOR X_i

$\mu_3(X_i)$ = THIRD CENTRAL MOMENT OF PROCESS FACTOR X_i

The last term can be neglected under certain circumstances (for example, symmetrically distributed process factors). From this simplification, the variance of device parameter P can be partitioned into the variance due to each process factor:

$$[S(P_i)]^2 = \left[\frac{\partial P}{\partial X_i} \cdot S(X_i) \right]^2 \quad (2)$$

Once the variance has been partitioned according to the dominant level of variation and the dominant processing factor causing the variation, alternative strategies for reducing variance can be considered and an effective experiment designed with a better understanding of the problem being addressed.

THRESHOLD VOLTAGE VARIANCE EXAMPLE - A CASE STUDY

In recent years, a few papers have been published which deal with the sensitivity of threshold voltage to processing factors [6, 7]. In this case study, the threshold voltage variance and sensitivity was analyzed as discussed above. Since the production process utilized geometries not considered subject to short channel effects, the sensitivity analysis was concerned with the gate oxide thickness, oxide/interface charge density, and the concentration of the doping in the channel region.

Variance Components analysis of the levels of variance indicated that the dominant level of variation was from lot-to-lot (Figure 2). This has been found to be typical of many device parameters, which are generally dependent on batch operations (such as hot processing). The statistical analysis, using the generation of system moments method, subsequently used lot-to-lot variance estimates.

The sensitivity of threshold voltage to gate oxide thickness, oxide/interface charge density, and concentration of doping was obtained by means of the equation:

$$V_t = \phi_{ms}(N_d) + 2\phi_f(N_d) - \frac{Q_b(N_d)}{C_o(t_{ox})} - \frac{Q_i}{C_o(t_{ox})} \tag{3}$$

WHERE

$\phi_{ms}(N_d)$ = THE METAL-SEMICONDUCTOR WORK FUNCTION DIFFERENCE,

$\phi_f(N_d)$ = THE FERMI POTENTIAL,

$Q_b(N_d)$ = THE CHARGE PER UNIT AREA IN THE SURFACE DEPLETION REGION AT INVERSION,

$C_o(t_{ox})$ = THE GATE OXIDE CAPACITANCE PER UNIT AREA

N_d = THE DOPING CONCENTRATION IN THE CHANNEL,

t_{ox} = THE GATE OXIDE THICKNESS, AND

Q_i = THE OXIDE/INTERFACE CHARGE PER UNIT AREA.

Using the lot-to-lot variance estimates for the device parameter, V_t, and the three processing parameters, gate oxide thickness, oxide/interface charge density, and concentration of doping, the sensitivities to each processing parameters are illustrated by Figure 3. By normalizing each factor in terms of its lot-to-lot standard deviation, the relative importance of each parameter in determining the variability of threshold voltage is depicted graphically.

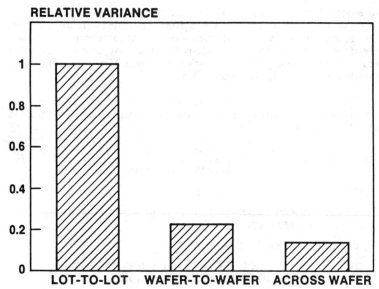

Figure 2. Pareto chart of relative variance of threshold voltage at each nested level of variance.

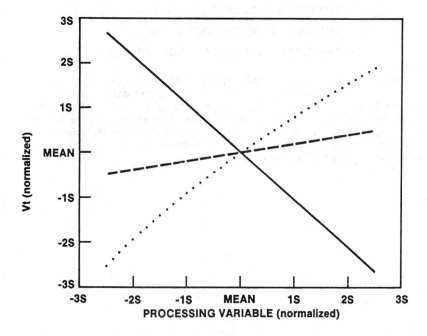

Figure 3. Normalized threshold voltage as a function of the normalized processing factors: Gate oxide thickness ― ― ―
Oxide/interface charge density ⎯⎯⎯
Concentration of doping in the channel region ....

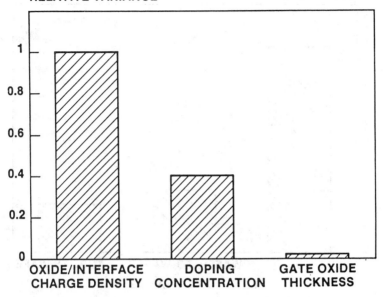

Figure 4. Pareto chart of relative variance of threshold voltage attributed to each processing factor source of variance.

A Pareto chart was constructed using the lot-to-lot variance estimates and the partial derivatives evaluated at the means of each processing factor, as described in equation 1. Figure 4 indicates that oxide/interface charges are the dominant source of threshold voltage variability in this case study.

BVCEO VARIANCE EXAMPLE - SECOND CASE STUDY

As a second case study, the collector-emitter breakdown voltage, BVceo, was analyzed using the same approach. The sensitivity analysis was concerned with epi doping concentration, the effective "intrinsic" epi layer thickness (the distance from the base junction depth to the top of the outdiffused subcollector), and the current gain (Hfe) of the NPN.

Variance Components analysis indicated that the lot-to-lot level of variation was again dominant (Figure 5). In this instance, analysis of a fourth level of variance - from epi run to epi run within a lot - could have provided additional insight. The statistical analysis, using the generation of system moments method, subsequently used lot-to-lot variance estimates.

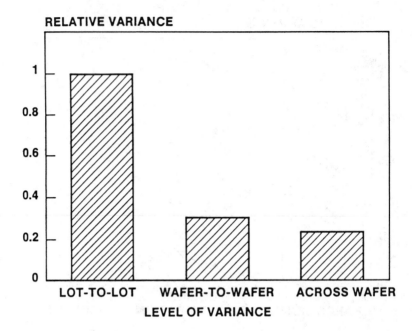

Figure 5. Pareto chart of relative variance of BVceo at each nested level of variance.

The sensitivity of BVceo to the epitaxy doping concentration, the epitaxy layer thickness, and the current gain (Hfe) was obtained by means of an analytical equation [8]. The sensitivities to each processing factor are illustrated by Figure 6.

The Pareto chart constructed using the lot-to-lot variance estimates and the partial derivatives evaluated at the means of each processing factor is shown in Figure 7. The intrinsic epitaxial layer thickness is found to be the dominant source of BVceo variability in this case study.

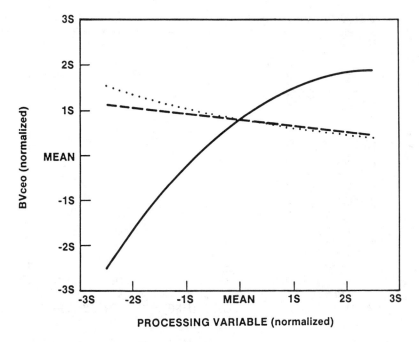

Figure 6. Normalized BVceo as a function of the normalize processing factors:

⋯⋯ NPN current gain (Hfe)
——— Intrinsic epitaxy layer thickness
– – – Concentration of doping in the epitaxial layer.

This example also serves to illustrate a caution in using the generation of system moments method. As shown in Figure 6, the relationship between BVceo and intrinsic epitaxy layer thickness reaches a point at which the slope is zero, corresponding to the transition from thickness to non-thickness limited breakdown for the collector-base junction. Had the mean of the epi layer thickness been just sufficient to reach non-thickness limited breakdown, the generation of system moments method would have suggested no contribution of this process factor to the variability of BVceo, due to a value of zero for the partial derivative evaluated at the mean intrinsic epi layer thickness. However, the intrinsic epi layer thickness variance would in fact contribute to the variability of BVceo by virtue of the standard deviation being such that thickness-limited breakdown would be observed at some frequency.

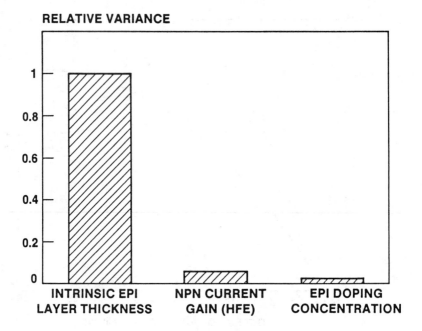

Figure 7. Pareto chart of relative variance of BVceo attributed to each processing factor source of variance.

By plotting the relationship between the processing factors and the device parameter, as in Figure 6, situations like that described above can be anticipated. Comparing the changes in the device parameter at plus or minus two standard deviations from the mean of the process factor, squared, might be a useful alternative for the generation of system moments in such situations.

SUMMARY

The parametric variances of threshold voltage, and of collector - emitter breakdown voltage were each analyzed in terms of the relative contribution attributed to the process factor causes, using the generation of system moments method in conjunction with device models. In each case, the dominant level of variance and the dominant process factor causing the variance was identified, facilitating effective experimentation and corrective action to reduce the variability of the device parameter.

ACKNOWLEDGEMENTS

The author would like to thank A. R. Alvarez and H. Weed for valuable technical discussions, G. Stickney for stimulating interest in the use of process simulation as a process control tool, K. Williams for BVceo calculations, F. Lamy for his encouragement and S. Metzger for help in the preparation of this paper.

REFERENCES

(1) Hahn, G. J. and Shapiro, S.S., "Drawing Conclusions About System Performance from Component Data," in Statistical Models in Engineering, John Wiley and Sons, New York, 1967, pp. 225 - 259.

(2) Tukey, J.W., Technical Reports 10, 11, and 12: (10) "The Propagation of Errors, Fluctuations and Tolerances - Basic Generalized Formulas"; (11) "The Propagation of Errors, Fluctuations and Tolerances - Supplementary Formulas", (12) "The Propagation of Errors, Fluctuations and Tolerances - An Exercise in Partial Differentiation," Statistical Techniques Research Group, Princeton University, Princeton, New Jersey.

(3) Deming, W. E., Some Theory of Sampling, John Wiley and Sons, New York, 1950.

(4) Seder, L.A., "Diagnosis with Diagrams," Industrial Quality Control, Vol. 6, No. 4, January and March, 1950.

(5) Box, G., Hunter, W., and Hunter, J., "Study of Variation," in Statistics for Experimenters, John Wiley and Sons, New York, 1978, pp. 571-583.

(6) Selberherr, S., Schutz, A., and Potzl, H., "Investigation of Parametric Sensitivity of Short Channel MOSFETS," Solid State Electronics, Vol. 25, No. 2, 1982, pp. 85-90.

(7) Yokoyama, K., Yoshii, A., and Horiguchi, S., "Threshold-Sensitivity Minimization of Short-Channel MOSFET's by Computer Simulation," IEEE Transactions on Electron Devices, Vol. ED-27, No. 8, August 1980, pp. 1509-1514.

(8) Roop, R., "Trends in High Voltage Integrated Circuit Technology," Technical Report 43, Motorola Bipolar Technology Center, Mesa, Arizona, 1984.

Clark H. Beck

COMPUTERIZED YIELD MODELING

REFERENCE: Beck, C. H., "Computerized Yield Modeling", *Emerging Semiconductor Technology, ASTM STP 960*, D. C. Gupta and P. H. Langer, Eds., American Society for Testing and Materials, 1986.

ABSTRACT: A computer program has been developed which extends the existing technique for calculating global circuit yield factor of lateral geometric design rules by process simulation. The current paper describes the inclusion of resistor limits as part of the general yield model. The program depends upon a reliable data base including: 1) geometric tolerances, 2) skews or offsets of critical dimensions during processing, 3) tolerable minimum separation of circuit element edges, 4) film tolerances of resistivity and thickness, and 5) resistor design limits.

The program incorporates conventional means of original design rule creation.

KEYWORDS: process simulation, yield analysis, modeling, design rules

Progressively smaller profit margins in the semiconductor manufacturing industry require correspondingly more accurate product yield estimates at wafer sort. Three failure modes generally describe losses: 1) Statistically controlled processes from which the global circuit yield factor, Yproc, is derived, 2) statistically occurring random defects which define the random yield factor, Ydef, and 3) operational or design mis-

Mr. Beck is a process engineer at Sierra Semiconductor, 2075 N. Capitol Ave, San Jose, Ca 95132.

takes which are beyond the scope of this paper.

Overall yield, Y, will be defined as the product of two calculable independent yield factors:

$$Y = Yproc \times Ydef \quad (1)$$

GLOBAL PROCESS CONTROL YIELD FACTOR

<u>Lateral Geometric Design Rule Creation</u>

There are six types of lateral geometric design rules [1] as shown in Fig. 1. We will discuss a generalized two layer type.

A process architect understands that the designed separation of circuit element edges must be greater than the process actually accommodates at the end of processing because of inevitable factory tolerances. The general case of a minimum layout rule, Lr, is therefore the sum of two components:

$$Lr = Mr + Tr \quad (2)$$

where
    Tr = stochastically summed individual processing tolerances of registration and circuit element dimensions, and

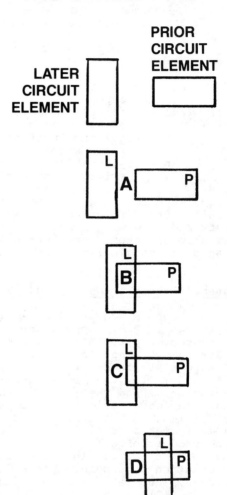

Fig. 1--Design rule types.

M_r = "margin limit" or minimum tolerable physical separation of circuit element edges at the end of processing as defined by experiment or physical simulation.

$$T_r = (T_{cdp}^2 + T_{cdl}^2 + \sum_{n=1}^{N} R_n^2)^{1/2} \qquad (3)$$

where
T_{cdp} = critical dimension tolerance of prior layer edge,
T_{cdl} = critical dimension tolerance of later layer edge,
N = total number of alignments involved in a rule, and
R_n = registration tolerance of nth alignment involved in the rule.

While the margin limit is fixed by the process architecture, tolerances depend upon the factory equipment and the confidence level (or number of standard deviations) chosen to obtain a particular yield with respect to margin limit violation.

Lateral Geometric Design Rule Yield Calculation

A process composed of a single design rule may be expected to yield as a function of margin limit violation according to the confidence level expressed in the tolerances. That is, if the tolerances are composed of a single standard deviation, the yield should be about 65%. Or, if the tolerances are twice the standard deviation, then the yield should be about 94%. If more than one design rule are involved the yield will probably be less. The upper yield limit will be the be the yield factor of a single rule and the lower yield limit, the product of all individual rule yield factors. An analysis of rule interlacing might provide a direct calculation of probable yield for a multiple design rule process, but a simpler approach has been described by Heavlin [2] using simulated processing and will be extended in this paper.

Multiple process simulations may be used to independently discover process yields of any number of rules. We will show here the general case of a single rule yield calculation and describe the extension to many rules.

Process Simulation: The factory processing engineer knows the standard deviations of his critical dimensions and registrations. A table of those tolerances permitted us to calculate original design rules. Process simulation is achieved in two steps:

1. A random normal generator in the program creates an assemblage of numbers whose collective characteristic is a histogram describing the normal distribution of the familiar "bell curve". That is, approximately 65% of the numbers are within the range of + or - unity; approximately 94% of the numbers are within the range of + or - 2; etc.

2. Different random normals are multiplied on a one for one basis by each of the standard deviations of critical dimensions and orthogonal registrations involved in the process. These products are the simulated process excursions which may be used to test whether or not a simulated process will yield.

Yield Test: The algorithm presumes yield failure if the margin limit of a design rule is violated in any radial direction. The simulated process excursions created above apply to individual layers and alignments. In order to test for margin limit violation we must sum the individual excursions involved in a design rule. Since critical dimension excursions are physically limited to radial symmetry, we use the scalar sum of the two critical dimension excursions plus the vector sum of the registration excursions involved. That is, the rule yields if the following inequality holds:

$$(Lr - Mr) > (RN1 \times Tcdp) + (RN2 \times Tcd1) + \overline{(RN3 \times R1) + (RN4 \times R2) + \ldots + (RNn+2 \times Rn)} \qquad (4)$$

where
  RN = random normals taken in sequence, and
  R = registration excursions.

If the design rule margin limit is violated during a simulated process then the process fails to yield. For a multiple rule process the algorithm simply repeats the above test for each of the rules in the process and requires zero failures of any margin limit in order to tally a yielding process simulation.

When all rules of a process have been tested on the first process simulation another set of simulated excursions are created from a fresh set of random normals and all rules again tested. After several hundred such sets of process excursions have been tested the yield factor, Yproc, is found from the ratio of yielding simulations to total simulations.

Heavlin found Yproc for lateral geometric design rules in this fashion. The current paper will now extend the principle to include resistor yields as a function of process excursions and show how other physical or electrical circuit limitations may be similarly included in a single statistically controlled global yield factor.

Resistor Creation

An ideal straight resistor is described by its width, length, thickness and resistivity. It may have maximum and minimum resistance limits imposed by process design. The cross section of an actual resistor may resemble Fig. 2, whose conformation to underlying topology and processing characteristics may require a distortion factor which sums with the width variable before employing the resistance equation:

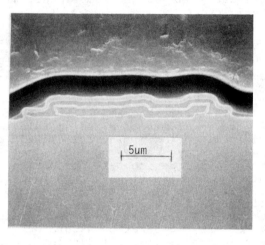

Fig. 2--Two superimposed polysilicon resistors.

$$RescX = Resyf \times LeX/[(WX + DX) \times Tf] \qquad (5)$$

where

    RescX = resistance of resistor X,
    Resyf = resistivity of film f,
      LeX = length of resistor X,
       WX = nominal width of resistor X,
       DX = distortion factor of resistor X, which is determined from processing representative structures, and
       Tf = thickness of film f.

We will calculate only simple straight resistors with uniform resistivity, width and thickness.

Resistivity, thickness and distortion factor are defined by the process. Product design defines resistance limits. Layout limits usually define resistor length. Therefore, the width is typically the adjustable parameter which is calculated to achieve the average between limiting tolerable resistances.

Resistor Yield Calculation:

Process simulation may be extended to include any combination of process variables whose tolerances are defined from random normal distributions. We have already shown how random normals may be multiplied by standard deviations of critical dimensions and registrations to find simulated process excursions for testing lateral geometric design rules.

Resistor Process Simulation: In order to accommodate resistor yields we find similar excursions of resistivity and resistor film thickness. The excursion of resistor width is defined by one of the same six types of lateral geometric design rules defined in Fig. 1. The length of a resistor is similarly governed but for simplicity, we will assume resistor length is much greater than process excursions and therefore may be treated as a constant unless the process is shrunk. The resistor distortion factor is constant.

Yield Test: As in the case of lateral geometric design rules we assume the straight resistor may be aligned in any radial direction and will correspondingly seek a worst case yield condition. First, we recall and rewrite equation (4):

$$M_r < L_r - (C_r + R_r) \qquad (6)$$

where
 C_r = scalar sum of critical dimension excursions, and
 R_r = vector sum of registration excursions.

We observe that to find worst case yield we sought the maximum value of the term $(C_r + R_r)$ which corresponded to the worst case edge excursions. However, a resistor with both maximum and minimum resistance limits has no fixed margin limit and must be evaluated with respect to all combinations of simulated excursions of the relevant parameters. The second step then, is to ask: What factors affecting resistance can change with resistor orientation on the circuit? Inspection of the factors involved in equation (5) shows all are independent of angular orientation except the width which can vary between $W + (C_r + R_r)$ and $W + (C_r - R_r)$. We can write the test inequalities for a single yielding straight resistor:

$$\text{Rescmax} > \frac{[\text{Resyf} + (\text{RN1} \times \text{Tresf})] \times \text{LeX}}{\{[(\text{WX}+\text{Cr}-\text{Rr})+\text{DX}] \times [\text{Tf}+(\text{RN2} \times \text{Tthkf})]\}} \quad (7)$$

where
      Rescmax = maximum tolerable resistance,
        Tresf = resistivy tolerance of film f, and
        Tthkf = thickness tolerance of film f.

and

$$\text{Rescmin} > \frac{[\text{Resyf} + (\text{RN1} \times \text{Tresf})] \times \text{LeX}}{\{[(\text{WX}+\text{Cr}+\text{Rr})+\text{DX}] \times [\text{Tf}+(\text{RN2} \times \text{Tthkf})]\}} \quad (8)$$

where
      Rescmin = minimum tolerable resistance.

If any margin limit or resistor fails, the simulated process fails. The yield factor, Yproc, after several hundred sets of excursions have been tested, is the ratio of yielding simulations to total simulations.

Simulated Process Example:

A set of CMOS design rules was created and their probable yield, Yproc, for margin and resistance violation calculated. The program was then used to calculate a succession of yields based on the same rules and resistors, but with progressively smaller shrinks. Fig. 3 demonstrates the expected rapid decline of yield with shrink factor. Also plotted is the maximum good dice per wafer which is the product of Yproc and the number of candidate dice per wafer and defines the optimum design shrink for the product.

Discussion of Process Simulation:

The design rules most sensitive to shrinkage are those whose margin component is proportionately greatest. For any percentage shrink of a rule, the margin limit remains unviolated only if the "tolerance" component of the rule can absorb the processing excursions imposed on it, since margin limit is fixed by the process. Resistors, on the other hand, will yield independently of shrinks until their reduced widths are dominated by unshrinkable process excursions.

This flexible program accommodates several applications:

1. Some users need to create design rules from known margin limits of existing processes before calculating probable global yields.
2. Others have existing rules to be tested with

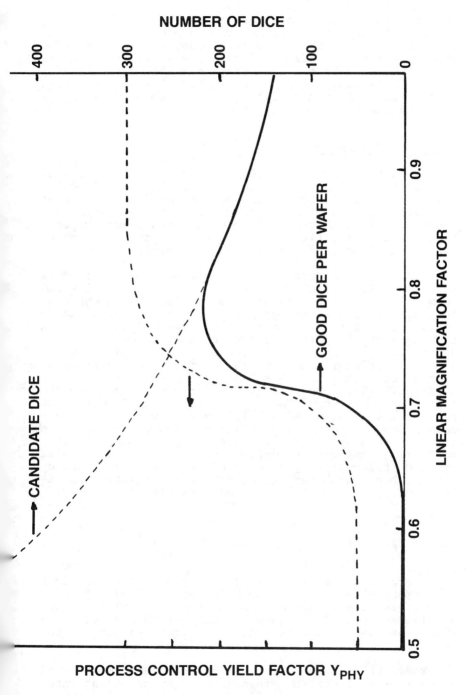

Fig. 3--Global Yield Plot

new processes and factory tolerances.
3. Still others have existing products which require optimal shrinks. In this case, the user may also want to alter the sizing of individual layers in order to better utilize his existing and tested circuit data base. By sizing, we mean the change of line widths and separations on individual layers while maintaining a constant pitch, or magnification.

Future Process Simulation Work:

Simulated process yield models share two common similarities with actual production yields: 1) In general, an entire process may be simulated with random excursions of all process parameters including geometric edges and film thicknesses, resistivities and capacitances, 2) All critical parameters may then be tested for survival against the simulated excursions.

The algorithm developed here provides a usable tool for predicting yields based on resistor and lateral geometric design rule margin violations. Although the lateral geometric design rule concept is fully developed, the current algorithm is limited to straight resistors whose widths are constrained by two layer rule definitions of Fig. 1. Future work will provide analysis of resistors whose widths change with direction, such as serpentines. One layer rule resistors, without a variable width due to registration excursions, are the most common type. They are fully treated with the current algorithm. Also, future work will incorporate capacitors, which include all isolating, nonconducting films, just as resistors include, in principle, all long conducting circuit elements. Later extensions will involve calculating yields based on electrical parameters such as punch through as substrate resistivity affects diffusion lengths and depletion widths.

DISTINCTIONS BETWEEN RANDOM AND GLOBAL YIELDS

Sierra experience suggests that yield distributions of extreme clustering as shown in the yield plot Fig. 4 are indicative of global process control failures such as plasma etching "bullseyes", whereas random distributions as in yield plot Fig. 5 indicate random defect losses.

The single common factor relating random defect and global process control yield losses is the shrink factor. Therefore, the yield factor of each failure mode can be calculated independently so long as they share a common shrink factor.

SUMMARY

We have extended a simulated global process yield model to include resistors. We have also indicated how other electrical and processing parameters may be incorporated in a general global yield factor.

REFERENCES

[1] Beck, C. H., "Design Rule Verification", presented at Kodak Interface, San Diego, Ca, Nov 1983; available from Kodak.
[2] Heavlin, W. D. and Beck, C. H., "On Yield-Optimizing Design Rules", <u>IEEE Circuits and Devices</u> Magazine, Vol. 1, No. 2, March 1985, pp. 7-12.

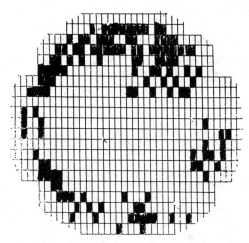

Fig. 4--Clustering Failures

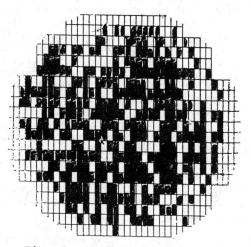

Fig. 5--Random Defects

Michael D. Brain

PARTICLE AND MATERIAL CONTROL AUTOMATION SYSTEM FOR VLSI MANUFACTURING

REFERENCE: Brain, M. D., "Particle and Material Control Automation System for VLSI Manufacturing",*Emerging Semiconductor Technology, ASTM STP 960*, D. C. Gupta and P. H. Langer, Eds., American Society for Testing and Materials, 1986.

ABSTRACT: This paper describes a practical automation system for integrated circuit fabrication combining the SMIF (Standard Mechanical InterFace) concept with a new approach to material control. SMIF implementation is shown to provide a cost effective clean environment and a unique opportunity for information management automation. Each SMIF port acts as a material "gateway", controlling lot processing via data attached directly to the SMIF pod. Misprocessing is eliminated, particulate contamination is reduced, and productivity is improved.

KEYWORDS: Standard Mechanical Interface (SMIF), material control, contamination control, clean room technology, automation

In response to the need for cost effective particle control in integrated circuit manufacturing, the Standard Mechanical InterFace (SMIF) concept was developed. The SMIF concept is based on the realization that a small volume of still, particle free air, with no internal source of particles, is the cleanest possible environment for wafers [1]. The current implementation of the SMIF concept makes use of a small sealed, ultra-

Michael D. Brain is a project manager responsible for material control projects at Asyst Technologies, Inc., 46309 Warm Springs Blvd., Fremont, California 94539.

clean box (or "Pod") for a cassette of wafers (fig. 1).

The SMIF Pod serves as a standardized means of storing cassettes of wafers and transporting the cassettes among processing equipment while isolating the wafers from the ambient environment. This reduces the need for clean space to the processing equipment and the Pods themselves. The Pod is only opened to a clean environment, and then sealed to prevent the leakage of particles and to reduce the possibility of air movement inside the Pod during storage and transportation.

Equipment isolation is a technique used to improve process cleanliness by reducing the possibility of cross-contamination between process steps. SMIF provides a means of transferring material between isolated process environments. In a facility not constructed for equipment isolation, each piece of equipment may be enclosed within a canopy made of plexiglass or other clean material (fig. 2). If the equipment generates particles, small HEPA filtration units are used on the canopies to maintain cleanliness. Each equipment enclosure contains a "SMIF port" which mates to the bottom of a SMIF Pod, enabling a robotic mechanism to transfer a cassette of wafers from a SMIF Pod onto the input elevator of the processing equipment. This is accomplished such that the wafers never come into contact with the ambient environment.

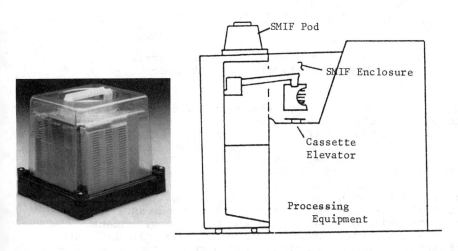

Figure 1: SMIF Pod

Figure 2: Process Equipment with SMIF enclosure and robotic cassette loader

In Figure 3, the level of airborne particulate inside a SMIF equipment enclosure within a class 100,000 room is compared to a peopleless class 10 clean room. The results show that a SMIF enclosure provides significant improvement over a peopleless clean room for protection against large particles (over a micron) and SMIF is shown to be roughly equivalent to a peopleless clean room in protection against sub-micron particles.

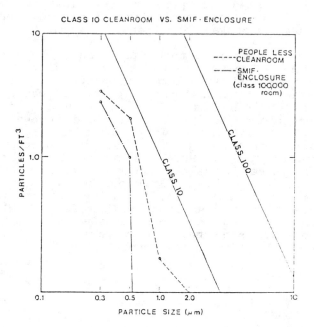

Figure 3: Clean room versus SMIF Enclosure, Airborne particles per cubic foot

In figure 4 we see the results of experiments [2] comparing the cleanliness of wafer transport and handling in five different cases: (1) SMIF system in a class 10 clean room, (2) SMIF system in a class 1000 clean room, (3) SMIF system in a class 20,000 clean room, (4) conventional handling of cassettes in "blue-box" type cassette carriers, and (5) conventional handling of open cassettes. In these experiments, the initial wafer surface contamination was measured with an Hitachi DECO HLD-200C in a class 10 environment. The wafers were then transported 30 meters, loaded into a simulated process station, then transported back to the wafer inspection system and measured again. This was repeated one hundred (100) times for each of the five different experimental conditions. Ten wafers (150 mm) were measured in each cassette. In all tests, the operators were fully gowned and followed careful clean room procedures.

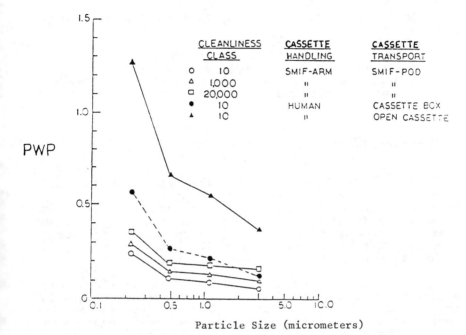

Figure 4: Particles Per Wafer Per Pass (PWP),
SMIF versus conventional wafer handling.

The data for each of the five cases was gathered and plotted. The nominal slope of the data was determined, thus providing a rate of particle gain termed <u>particles per wafer per pass</u> or PWP. The average PWP of ten wafers were plotted against particle size ranges for each of the five test conditions (figure 4). These results show that SMIF is significantly more effective in controlling particulate contamination than conventional cassette handling, particularly with respect to small particles (less than 0.5 microns).

In a SMIF system people never enter the wafer processing environment. This greatly increases the cleanliness of the process and eliminates many variables in clean space design. As a result, the working conditions enjoyed by the operations staff can be less restrictive and hence more productive.

It has been stated [3] that a SMIF implementation makes it possible to maintain a Class 10 or better processing environment; the general environment is much less stringent, perhaps Class 1000. The same author estimated that energy expenses may be reduced by as much as 75% due to the decreased clean air requirements.

418   EMERGING SEMICONDUCTOR TECHNOLOGY

In summary, a SMIF implementation provides major benefits:

- reduction of particulate contamination.
- co-existence with people in the fab.
- simple application to new and existing fabs.
- low cost, low risk.

SMART-SMIF: MATERIAL CONTROL AUTOMATION

In an IC facility using the SMIF concept, all material must pass through a SMIF port to move from a SMIF Pod into the processing equipment. For this reason, the transfer of material from a SMIF pod is an ideal location to perform material control and tracking. Material control may be automated by the same robotic mechanism that performs the material transfer.

For material control, we have added an electronic memory circuit to the SMIF Pod (fig. 5). This memory is programmed with the identity of the material inside the Pod, plus a description of the process as a sequence of steps (or operations) to be performed by specific equipment. At the time of processing, a user places a SMIF Pod onto a SMIF port, at which time the transfer robot reads the tag <u>before</u> opening the SMIF Pod. No danger exists of improper processing because the robot has been previously configured to know what equipment it is attached to. The Pod will simply not be opened if it is placed on the wrong port.

Figure 5: SMIF Pod with Smart-Tag memory.

Between process steps, the Pods are stored in a Smart-Storage area (fig. 6). When a Pod is placed in Smart-Storage, the Pod memory is automatically read. Operators may query the Smart-Manager (an IBM PC compatible) using a light pen menu to determine what material is waiting in storage for processing.

Using the Smart-Manager utilities, all decisions may be made locally, based on information obtained from a distributed database. Complete information is stored exactly where it is needed. Because the lot-related data is stored on the carrier of the material, flow of information is physically tied to the processing of material.

A run card or lot traveller system is currently used in most IC facilities. Under this system, a deck of cards is attached to each material carrier containing the process description. The operators check this card to determine what operation is to be performed next. When an operation has been completed, the operator notates its completion so that the process step will not be inadvertently repeated.

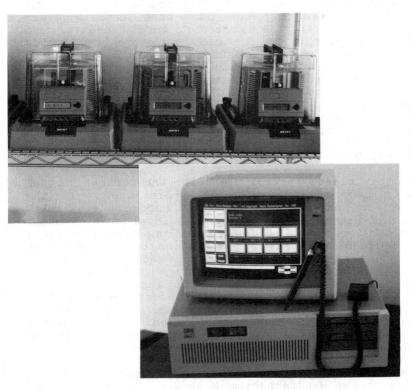

Figure 6: Smart-Storage Area

The run card system is effective for several reasons: (1) the pertinent information is easily and reliably retrieved by the operator in a timely fashion; (2) all the information required to perform the next operation is available locally and is under local control; and (3) progress is recorded without having to use a terminal or keyboard. In short, the run card system is reliable, user friendly, efficient, and extremely inexpensive.

However, using run cards to control process flow has several disadvantages. Occasionally, a run card is misread, or an operator may forget to make a mark on the card after completing a process step. When a batch of wafers is placed into a proces station, the run card is removed from the cassette. Misprocessing is likely if the correct card is not returned to the cassette after each step.

The "smart tag" method of material control is conceptually an automated run card system, eliminating the element of human error.

Other methods of verifying process sequence and automating lot tracking have been considered. Wafers sealed inside a SMIF Pod are not easily examined to determine identity from bar code or optical character recognition (OCR) [4]. A machine readable code could be placed on the exterior of the SMIF Pods, and readers added to the robotic transfer mechanisms at each processing station. The robot would read the code and communicate with the global facility host system to request permission to process the material contained inside that particular SMIF Pod. The host would maintain a table identifying what material is in which pod at any given time.

In this global control scenario, the central control computer becomes vital to fab operations. If the computer or communications systems become unavailable, processing cannot continue. As the level of automation in the fab increases, so does the volume of transactions with the facility host. Each station could require up to fifty transactions to perform each step in the process sequence. Real time transactions would be required to route the material to the next station, identify the material, initiate transfer from the SMIF Pod, download the correct process recipe, record parameters used and measurements taken. In the highly distributed system we propose, communications with the global host would be significantly reduced.

Many IC facilities have already installed central Computer Aided Manufacturing systems. With access to facility wide information, these systems are capable of optimizing throughput while minimizing inventory.

Reporting capabilities provide invaluable information for fab management. Many of these systems also acquire, store, and retrieve engineering data for improved process control. Operators are usually asked to record the information required by these global information systems using a terminal located in the clean room. In some facilities, paper is used to record fab operations and a staff of "keypunch" operators are given a stack of paper every day to enter into the central system.

The intelligent storage area described above automatically has the information required to update the central data base. Information is automatically acquired and sent to the central system, eliminating the burden of manual data entry and the associated potential for errors. This reduces the need for costly and space consuming terminals in the facility.

While most information systems provide useful services to the operation of the facility, they often place a burden on operations staff. The "Smart-SMIF" system relieves this burden and provides a environment conducive to a productive IC fabrication facility.

FUTURE

As processes become more complex and circuit geometries become smaller, the limitations of processes and process equipment become critical. A finer degree of control is required to maintain yields at a profitable level. Feed forward and backward of process parameters will be required at the wafer and die site level. This requires the manipulation and timely delivery of a large volume of data for each cassette of wafers. Multiplied by the total number of cassettes in the facility, this becomes a massive database. It is our opinion that distribution of this database to the lot level (ie. lot tags) is the first practical solution to this problem.

Once communications has been established between the process equipment and the material handling robotics, the smart tag will become available to the equipment to store and retrieve information. This information will automatically be available to each piece of downstream equipment (feed forward). Smart-Storage areas could be programmed to capture and interpret this information and alter the process parameters of later lots to reflect the results of previous lots (feed back).

CONCLUSION

SMIF improves the die yield through contamination control. The material control aspects of Smart-SMIF improves wafer yield. Together, the Smart-SMIF system improves the productivity and profitability of the integrated circuit manufacturing process.

ACKNOWLEDGMENTS

The author would like to acknowledge with appreciation the contributions of Joseph Nardini and Cindy Brain in the completion of this paper.

REFERENCES

[1] M. Parikh, U. Kaempf,"SMIF: A Technology for Wafer Cassette Transfer in VLSI Manufacturing," Solid State Technology, July 1984, pp. 111-115.

[2] H. Harada, Y. Suzuki, "SMIF System Performance at 0.22 um Particle Size," Solid State Technology, December 1986

[3] M. Parikh, A. Bonora, "SMIF Technology Reduces Clean Room Requirements," Semiconductor International, May 1985, pp. 32-36.

[4] S. Gunawardena, U. Kaempf, B. Tullis, J. Vietor, "SMIF and Its Impact On Cleanroom Automation," Microcontamination, September 1985, pp. 54-62,108

Nelda D. Casper and Bradley W. Soren

SEMICONDUCTOR YIELD ENHANCEMENT THROUGH PARTICLE CONTROL

REFERENCE: Casper, N. D. and Soren, B. W., "Semiconductor Yield Enhancement through Particle Control," Emerging Semiconductor Technology, ASTM STP 960, D.C. Gupta and P. H. Langer, Eds., American Society for Testing and Materials, 1986.

ABSTRACT: Contamination control has always been an important part of semiconductor processing. Until recently, it has been fairly easy to control particles at the level necessary to obtain satisfactory LSI device yields, by requiring personnel to wear protective clothing and by filtering the room air. However, device geometries have shrunk to the point where a more sophisticated means of identifying and controlling particles is needed in order to fabricate working VLSI devices. We have developed a method for identifying, measuring, monitoring, and controlling particles in a cleanroom. This method involves the characterization of wafer fabrication defects with respect to process equipment, measurement of particles with an automatic wafer inspection system, defect monitoring over time through control charts, and the establishment of regular cleaning procedures.

KEYWORDS: Contamination control, particles, cleanroom, plasma etcher.

Ms. Casper is Senior Engineer for maskmaking operation. Westinghouse Electric Corporation, Box 1521, MS-3240, Baltimore, Maryland 21203. Mr. Soren is an Engineer in plasma etch process development. Westinghouse Electric Corporation, Box 1521, MS-3974, Baltimore, Maryland 21203.

INTRODUCTION

Particle control must be considered a major factor in maintaining profitable manufacturing yield. Contamination from the lab environment, lab personnel, process equipment, and all materials used in wafer fabrication are major concerns. Typical technology expected in microcircuit manufacturing is shown in Table 1 [1].

TABLE 1--Typical micromanufacturing technology: past, present, and future.

Year	Feature Size(um)	Complexity (critical levels)	Die Size (mils)	Typical Product
1982	3	6	250	64K DRAM (VLSI)
1984	2	7	275	
1986	1.5	8	300	256K DRAM (Early ULSI)
1988	1.25	9	325	
1990	1.0	10	350	1024K DRAM (Complex ULSI)

We at Westinghouse have developed a method of reducing particles in our advanced MOS product.

ANALYTICAL PROCEDURES

An automatic laser scanning particle measuring system was set up to detect particles on bare bulk silicon wafers in the 2 micron to 100 micron range. (All particle counts referenced in this paper are given in counts per 4 inch wafer and are averaged over a sample size of at least three wafers). It divides the particle counts into 13 different defect categories based on size and type of defect (e.g., scratch, pit, bump, film irregularity). The information is available in two reports: a printed list of counts in each category for each wafer (see Fig. 1) and a wafer map that shows flaw position on the wafer (see Fig. 2). Electrical measurements were made using an automatic tester on resolution test patterns having VLSI geometries in an attempt to predict device yields.

Wafer		1	2	3	4	5	6	7	8	9	A	B	C	D
0101	ADJ	0	25	4	35	0	0	19	0	8	6	1	0	0
	INT	19	25	4	0	0	0	19	0	8	6	1	0	0
0102	ADJ	0	20	1	23	0	0	0	0	1	2	0	0	0
	INT	14	20	1	0	0	0	0	0	1	2	0	0	0
0103	ADJ	0	31	1	34	0	0	1	0	0	1	1	1	0
	INT	17	31	1	0	0	0	1	0	0	1	1	1	0
0104	ADJ	0	28	0	30	0	0	0	0	0	2	0	0	0
	INT	19	28	0	0	0	0	0	0	0	2	0	0	0
0105	ADJ	0	28	10	40	9	0	45	7	0	2	0	0	0
	INT	21	31	16	0	0	0	45	7	0	2	0	0	0
0106	ADJ	0	23	1	25	0	0	0	0	0	1	0	0	0
	INT	9	23	1	0	0	0	0	0	0	1	0	0	0
0107	ADJ	0	70	10	82	15	0	1	0	0	2	0	0	0
	INT	20	84	10	0	0	0	1	0	0	2	0	0	0
0108	ADJ	0	46	8	60	0	0	2	0	2	6	0	0	0
	INT	12	46	8	0	0	0	2	0	2	6	0	0	0

FLAWS

FIG. 1--Inspection system printout.

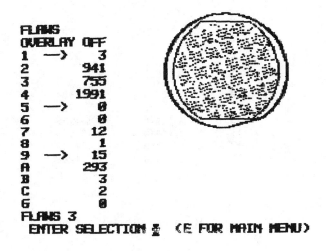

FIG. 2--Wafer map.

The following equations were used to relate particle counts, defect density/unit area, and VLSI yield.

$$Y = e^{-AD} \quad (1)$$

where
- Y = yield
- A = area
- D = defect density per unit area

and

$$D = \frac{\text{Particle count}}{\text{pixels per wafer}} \quad (151 \text{ defects/cm}^2) \quad (2)$$

where

151 defects/cm^2 = maximum number of counts detectable

Figure 3 shows the relationship between particle count, defect density/unit area, and VLSI yield (based on one masking level). These yield calculations give only a rough approximation of actual device yield, but they are helpful in determining maximum allowable particle counts for individual process steps.

YIELD (%)	D/CM²	3" COUNT	4" COUNT
95	0.73	23	45
90	1.51	48	92
85	2.32	74	142
80	3.19	101	195
75	4.11	130	251
70	5.1	162	311
65	6.15	195	375
60	7.3	232	445
55	8.54	271	521
50	9.9	314	604
45	11.8	375	720
40	13.1	416	799
35	15	476	915
30	17.2	546	1049
25	19.8	629	1208

FIG. 3--Yield as a function of defect density and particle count.

PARTICLE SOURCES

The following six potential sources of particulate contamination were identified (see Fig. 4), and each was analyzed to determine its contribution to the particle problem.

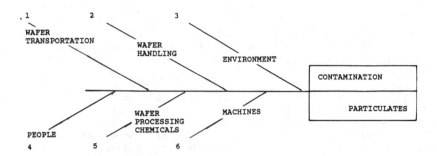

FIG. 4--Potential contamination sources.

1. Wafers get transported in their processing boxes between cleanrooms through "non-clean" hallways. These boxes were checked to see if particles could leak in during this inter-lab movement by storing one box in a class 100 clean room and another box in a "non-clean" area for 19 days. The average particle counts were 3.2 and 6.4, respectively. The delta of +3.2 is an insignificant difference in counts. Therefore, wafer transportation was eliminated as a major yield detractor.

2. Wafer handling that occurred during a typical process sequence was broken down into four categories (see Fig. 5). Best-case and worst-case handling procedures were examined for changes in particle counts. There was no significant increase in counts for either procedure (+0.6 and +1.8, respectively). Since this experiment was conducted, all metal tweezers in the Class 100 clean room were replaced with vacuum pick-ups, because industry findings indicated that the metal tweezers scratched the wafer edges and generated particles [2].

```
Least Handling  (1) Cassette-to-cassette
                (2) Dump transfer into cassette
                (3) Wafer hand-loaded onto platen or into boat
Most Handling   (4) Wafer placed on microscope stage
                    and repositioned for inspection
```

FIG. 5--Methods of wafer transfer.

3. VLSI processing at Westinghouse is currently being done in two types of cleanrooms: Class 100,000 open air/Class 100 work areas and Class 100 open air/Class 10 work areas. The Class 100,000 lab is used for fabricating devices with geometries greater than or equal to 3 microns and the Class 100 clean room for geometries less than 3 microns. These cleanroom classes are maintained through rigorous cleaning and daily particle monitoring.

4. Laboratory personnel are required to wear protective clothing in the cleanrooms. Clothing for a Class 100,000 room includes hood, coat, gloves, and mask (only if facial hair is present). The dress code for a Class 100 room includes hood, coveralls, shoe covers, gloves and mask. These dress codes are enforced.

5. Wafer processing chemicals include both inert and reactive liquids and gases (see Fig. 6). Analysis of a wafer-cleaning machine which used unfiltered N_2, DI H_2O, and several acids showed particle counts of 480 for the N_2. Filters were installed on the N_2, and as a precaution, they were also installed on the DI H_2O. This brought the particle levels down to less than 10. Since the N_2 and DI H_2O are supplied in-house, filters were subsequently installed on all equipment used for VLSI processing.

	INERT	REACTIVE
GAS	N_2	CCl_4
	He	Cl_2
	Ar	CF_4
LIQUID	H_2O	HCl
		HF

FIG. 6--Sample processing chemicals.

6. The processing equipment was characterized in two ways. Table 2 lists the eight major categories of equipment with potential particle sources. The equipment had already been classified according to the method of wafer transfer (see Fig. 5). Table 3 presents the typical wafer processing operations. Several different films (e.g. aluminum, polysilicon, thermal oxide) were deposited, patterned, and etched using this equipment, and the wafers were electrically tested for yield.

TABLE 2--Processing equipment particle sources [3].

TYPE	PROBLEM
1. Gas Distribution and Control Systems	Internal welds, seals, solenoid valves, gas filter media, connectors.
2. Spin Dryers	Drive mechanisms, feedthroughs, static charge, exhaust, door seals, loading and handling.
3. Resist Spinners	Exhaust control, splash back edge lip, plumbing lines, resist bottle or package, dispensing pump.
4. Ion Implantation	Mechanical transport, resist chips and residues, gas ports, vacuum components.
5. Low Pressure Deposition	Chemical gas contamination, quartz tube particles, reaction chamber spalling, loading stations.
6. Metal Deposition	Loading mechanisms, internal coating and spalling, source control, internal moving parts.
7. Dry Etching Systems	Autoloaders, pedestals, ionized contamination, sidewall accumulations, internal mechanisms, backside wafer residues.
8. Annealing Systems	Metallic contamination, loading and transfer systems, internal stress spalling, wafer residue transfer, annealing gas inlets.

TABLE 3--Typical wafer process operations.

PROCESS	TYPE OF PROCESSING EQUIPMENT	TYPE OF HANDLING	NUMBER OF TIMES USED
Batch-load RIE	7	3	3
Single-wafer RIE	7	1	3
Barrel etcher	7	3	15
Wet Chemical	--	2	37
Photolithography	3	1	13
Nitride deposition	5	3	2
Dielectric deposition	5	3	4
Polysilicon deposition	5	3	1
Metal deposition	6	3	2
Furnace operation	1	3	11
Ion implantation	4	3	7

Since there were more than 15 pieces of equipment used in the process sequence and limited resources for characterizing this equipment with respect to particles, some priorities had to be established. The machines suspected of being the major VLSI yield detractors were chosen for further study based on frequency of use, type of handling, and electrical yield. The data overwhelmingly pointed toward the equipment associated with metal deposition, photolithography, and dry etch. Therefore, our efforts were concentrated on characterizing the wafer cleaning equipment, the metal deposition system, the photoresist deposition system, and the metal plasma etcher.

EQUIPMENT CHARACTERIZATION

We found that the wafer cleaning equipment and photoresist system were easy to characterize. We have not finished characterizing the metal deposition system. The plasma etcher turned out to be a complex machine to analyze for particle contamination. Therefore, a significant amount of effort went into characterizing that system. We have chosen to present details of the plasma etcher study as an example of particle characterization and control for a specific wafer processing machine.

Wafer Cleaning Equipment

Three pieces of equipment were used to clean the wafers: two chemical sprayers and a water sprayer (hereafter referred to as C1, C2, and W1, respectively). C1, C2, and W1 were characterized both individually and in different combinations. The cleanest combination [(C1+W1) or (C2+W1)] had particle counts as low as the manufacturer's virgin wafers (less than 10). This combination was better than the cleaning procedure currently being used, so a process change was made.

Metal Deposition System

The metal deposition system currently is being characterized. No particle data is available at this time. We did, however, conclude that the exact same piece of equipment in two different cleanrooms with different combinations of targets could in fact produce quite different electrical yields.

Photoresist Deposition System

The photoresist system was characterized two ways: bare silicon wafers were run through the system without depositing resist and were checked for particles by the laser scanning system; wafers were run that did get photoresist and were checked optically for particles. There was no significant increase in particle count in either case.

Plasma Etcher

The etcher in this study is a manually-loaded, batch RIE system. It is normally run one shift per day. Carbon tetrachloride (CCl_4) and helium are employed for both aluminum and titanium etching; carbon tetrafluoride (CF_4) is employed for etching a titanium-tungsten alloy. Anisotropic metal etching is achieved in part by sidewall polymer formation.

A series of preliminary tests were performed in order to isolate the operation in which the particles appear. These experiments included: 1) loading wafers into the etcher (no pumping); 2) loading wafers into the etcher and pumping down to base pressure (1.0×10^{-5} torr); 3) repeat 1) and 2) after removal of top electrode; and 4) rinsing wafers in deionized water (normal post-etch surface passivation) followed by automatic spin drying. These tests showed that pumping the chamber down and subsequent venting had more than an order of magnitude greater increase in particles than any of the other steps.

It was suggested that slowing the pumping and/or venting steps might reduce the number of particles by lessening the turbulence in the chamber. Although there was no easy way to slow the pumping step, the venting speed was readily adjustable. The venting process could be slowed by reducing the pressure on the N_2 that vented the chamber. Although particles deposited in the venting step would not affect the metal etch, they could cause problems in later processing steps. At the time of this work, point-of-use filters (fluoropolymer stacked disc cartridges) were being installed on house gas lines. A filter of this type was put on the N_2 line that vents the chamber. Between slowing down the vent and installing the filter, some reduction in particle counts was observed.

A contamination source was isolated, and some of the counts were reduced, but the particle count was still much higher than desired. The likelihood of the etching product $AlCl_3$ being at least part of the contamination problem pointed the experimental focus towards the chamber cleaning procedure.

A brief discussion of the nature of $AlCl_3$ is necessary at this time. At room temperature, $AlCl_3$ is a white powder that is hygroscopic. Every time the etching chamber is opened, the $AlCl_3$ residue combines with water in the air. The presence of moisture is known to adversely affect the quality of metal etching. In this way, as a particle source and a moisture getter, $AlCl_3$ is doubly undesirable.

It was found that although a chamber cleaning procedure had been specified, the frequency of this clean varied greatly. Over a period of about two months, the type of and interval between chamber cleans were studied extensively. Although decreasing the particle count was the prime objective, certain other factors could not be ignored. The best example of these is that during the cleaning routine the machine is down for production. The goal was to reduce particle counts with little or no change in throughput.

The two types of cleans were plasma and manual. The plasma clean consisted of a high power density oxygen (O_2) burn. The O_2 plasma chemically eliminates organics and moisture in the etching chamber. A 30-minute O_2 plasma was determined to be sufficient for removal of nearly all chamber contaminants on a typical processing day.

The manual clean consisted of wiping the chamber with lint-free cloths dampened with propanol (Note: the manual clean is always followed by a plasma clean). Certain residues are not removed by the plasma clean and therefore accumulate on the chamber walls, lid, and platen. The manual clean removes debris in the chamber, but it also introduces other particles into the chamber. By the end of the clean, propanol is present at a high concentration. Also, moisture-containing cleanroom air freely enters the chamber during the 10-15 minute manual clean. Based on this premise, data was taken that showed that frequent (i.e., daily) manual cleaning actually raised the particle counts.

The last stage of this study was to determine the frequency of manual cleaning. As mentioned previously, daily cleaning was undesirable with respect to both particulate contamination and throughput. Weekly manual cleaning of the machine along with daily plasma cleaning greatly reduced the particle counts. A two-week interval between cleans was also tried. By the middle of the second week, particle counts had risen to an unacceptable level. After repeating this test, the one-week cleaning interval was reinstated. Particle counts decreased as expected. For the desired throughput, the weekly manual clean was found to be optimal, based on the normal loading of 8 hours per day.

The cleaning procedure was implemented as part of the routine maintenance schedule of the machine. The machine operators and technicians have been made aware of the particle problem and are to keep watch as to the cleanliness of the chamber. Particle monitor wafers are run at least once a week, and the results are kept on a control chart. A particle monitor is now also part of machine requalification after any unscheduled down time.

The investigation of the particle problem of this plasma etcher proved worthwhile. Based on this work, particles greater than 2 microns were reduced by more than an order of magnitude (see Figure 7). A particle count threshold was set for this etcher. Based on six months of data, the limit was set at 200 particles greater than 2 microns for a 4-inch wafer. If routine monitors exceed this value, production runs will not be made.

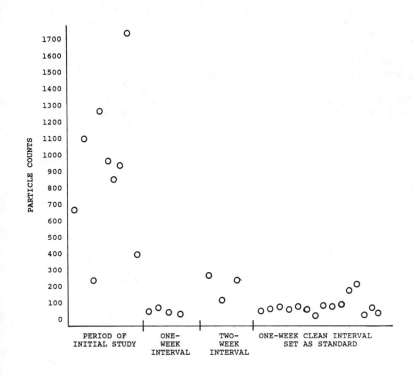

FIG. 7--Selected data from etcher study.

CONCLUSION

Several useful techniques for reducing particles and improving device yields were developed during the course of our studies. Regular particle monitoring should be performed on all critical processing equipment. Maximum particle counts should be established for all monitored equipment. Since process steps ar not equally critical in a wafer fabrication cycle, particle limits for the equipment have to be determined individually. In order to keep the particle counts in an acceptable range, cleaning procedures for the machines must be set up and followed regularly. A control point must be chosen that would be a signal for equipment shutdown and maintenance. This maximum particle count would be based on the acceptable VLSI yield for a given level. Particle counts for each machine should be plotted on a chart that is kept at the machine. A key person should be identified to monitor all charts and evaluate them for any negative trends that might indicate a general contamination problem.

Particle monitoring should be made a part of each operator's regular routine, just like making periodic etch rate measurements. Figures 8 and 9 illustrate the positive impact that our contamination control effort has had on product yield. Over a period of 4 months, optical defect density decreased by a factor of 2 and functionality increased by a factor of 10.

Our intent is to characterize other processing equipment in as much detail as the metal etcher. This type of process contamination control is sure to enhance device yield in a given production line.

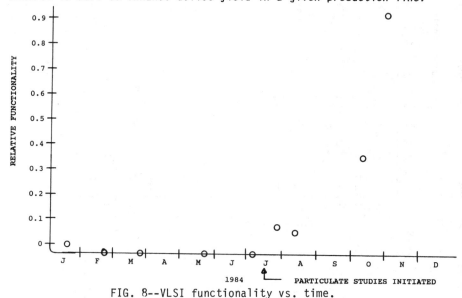

FIG. 8--VLSI functionality vs. time.

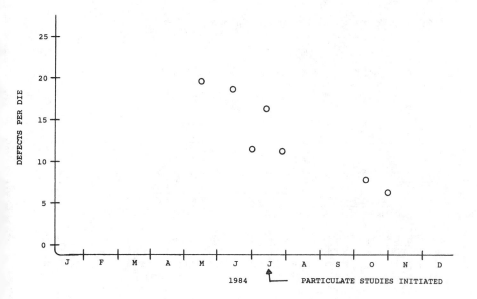

FIG. 9--VLSI optical defect density vs. time.

ACKNOWLEDGEMENTS

We would like to express sincere appreciation to M. Polinsky, C. J. Taylor, and H. Hyman who provided assistance in this investigation. Special thanks go to our management, especially M. Fitzpatrick and A. Turley, who supported both the experimental efforts and the text preparation.

REFERENCES

[1] Burnett, J., "Clean Rooms for ULSI Manufacturing: Class 1 Practice," Solid State Technology, Vol. 28, No. 9, September 1985, pp. 121-123.

[2] Hoenig, S. A., "The Clean Room as a System for Contamination Control," Solid State Technology, Vol. 28, No. 9, September 1985, pp. 129-135.

[3] Tolliver, D. L., "Contamination Control: New Dimensions in VLSI Manufacturing," Solid State Technology, Vol. 27, No. 3, March 1984, pp. 129-137.

Jeffrey M. Davidson and Thomas P. Ruane

PARTICULATE CONTROL IN VLSI GASES

REFERENCE: Davidson, J.M. and Ruane, T.P., "Particulate Control in VLSI Gases", Emerging Semiconductor Technology, ASTM STP 960, D.C. Gupta and P.H. Langer, Eds., American Society for Testing and Materials, 1986.

ABSTRACT: Semiconductor device miniaturization has required continually decreasing levels of particulate contamination in process gases. This paper briefly reviews existing technologies for analyzing and for minimizing particles in VLSI gas streams. Recent advances and research topics in both areas are also addressed. Examples from laboratory and field evaluations of VLSI and non-VLSI systems are discussed.

KEYWORDS: gases, particles, contamination, filtration

As the miniaturization of microelectronic devices continues, increasing emphasis is being placed on particulate contamination control in the semiconductor fabrication environment. Contamination from many sources can be minimized through isolation of the chip from the fabrication environment. In contrast, process gases are now receiving greater scrutiny since they are exposed directly to the chip during fabrication.

The problem of particle control in VLSI grade gases is really twofold: 1) how to analyze particles in gases and 2) how to minimize them. Particle analysis technologies, which have historically been geared mostly towards environmental measurements, must now be improved to meet the more stringent cleanliness requirements of the microelectronics industry. Similarly, ultraclean gas and fluid handling systems originally used for the pharmaceutical and food industries must be further

Drs. Davidson and Ruane are Senior Scientist and Senior Physicist, respectively, at The BOC Group, Inc., Group Technical Center, 100 Mountain Avenue, Murray Hill, NJ 07974. The BOC Group Technical Center performs research and development on behalf of Airco Industrial Gases, BOC Limited, and other Group companies.

developed for the storage and distribution of electronic grade gases. While particle analysis and particle minimization each pose a set of uniquely different problems, they are interdependent. For example, setting goals or specifications for particle minimization is meaningful only if the particles themselves can be measured and characterized. Conversely, as particle analysis capabilities advance, new methods, materials and system designs can be developed for further particle minimization. This paper will briefly review the state-of-the-art in both particle analysis and minimization in VLSI gases, and highlight recent advances in both areas. Discussions will focus on the more heavily used bulk atmospheric gases, but topics of concern to specialty cylinder gases will also be noted.

PARTICLE ANALYSIS

Over the past few years, particle analysis has been used most frequently for gas quality assurance in the field, where specifications have been, and are continuing to be, set for the total number of particles greater than a given particle size per unit volume of gas. Increasingly, particle analysis is also being directed towards quality improvement through identification of particle sources, which can be minimized or eliminated through proper hardware selection or design. Therefore, in addition to particle concentration, the particle-related parameters of size distribution and chemical composition are also of interest.

Particle analysis is most commonly performed with optical particle counters, which operate on the principle of light scattering from particles in a gas stream intercepting a light beam [1,2]. Such instruments provide real-time analyses of particle concentrations and size distributions from measurements of the scattered light. The minimum particle size detectable with counters using white light sources is generally between 0.3 to 0.5 μm. More recently available laser-based instruments can detect particles as small as 0.1 μm diameter, but only in relatively low flow gas samples, typically 5 X 10^{-6} $m^3/s*$. Higher flow sampling is provided in some laser-based instruments, but usually with the trade-off of slightly larger minimum particle size detectable.

To achieve detectability of particles much smaller than 0.1 μm, particles must be pre-conditioned to enhance their light scattering capability. For example, in the condensation nucleus counter (CNC) the sampled gas passes through a saturated alcohol vapor, allowing nucleation and growth of liquid droplets on all particles larger than 0.01 μm. The droplets are subsequently counted by conventional light scattering. As CNC's only count all particles greater than a minimum size, they are generally used in conjunction with particle size classifiers, such as diffusion batteries or electrostatic classifiers [3].

* 1 m^3/s = 2.1 x 10^3 CFM (cubic foot per minute).

One drawback with all of the instruments discussed so far is that they sample gases only at atmospheric pressure and, as mentioned, usually at low flow rates. Since VLSI gas lines generally operate at higher pressures and flow rates, careful attention must be paid to the methods of pressure reduction and sampling in order to ensure that a representative sample is obtained [4]. The use of an aligned isokinetic sampling probe is also frequently recommended, although theoretical studies, which we have verified experimentally, predict that use of such probes is not necessary for measuring sub-micron particles, which are the major particles of interest in high purity gases. Regardless of the sampling technique, the sampling system must be designed and fabricated with great care, so as to minimize the introduction of particles from the sampling hardware itself.

It is not obvious which designs and materials will minimize contamination in the sampling system. Our experience with different sampling systems has yielded some guidelines. We try not to use valves or regulators to achieve pressure reduction, since those components could act as particle generators. It is likely that the use of an expansion chamber or a critical orifice will introduce less contamination, but our work has shown that these methods can still introduce particles, albeit at very low concentrations, depending on the flow conditions. Although the phenomena are not well understood, it seems that conditions of high flow can be effective in removing particles from surfaces. We follow several practices to minimize this. The sampling system is fabricated from stainless steel, and cleaned with an ultrasonic vapor degreaser. We design the samples to minimize the gas velocity, consistent with the other pressure and flow requirements. It is necessary to pay close attention to these details for sampling systems which are to be used with high purity gas, for which even a small contamination may be significant compared to the particles originally present in the gas itself.

The problems of atmospheric pressure sampling are also overcome in various high pressure instruments such as the PMS High Pressure Laser Aerosol Spectrometer (HPLAS), which allows direct sampling of gas lines without pressure reduction [2]. Such high pressure instruments are also finding applications for specialty gases which, due to their toxicity or reactivity, are not amenable to atmospheric pressure sampling. In order to use the HPLAS, window modules must be installed in the gas line to be measured. Then the sensor is mounted on the window, and can detect particles as small as $0.3\ \mu m$ as they pass along the center of the gas line. This eliminates the need to withdraw a sample from the gas system.

Our experience with particle counters has shown that it is essential to know the background level of the measurement system. The background level can contain false particle counts due to electronic noise in the counter, or counts due to particles which have been introduced by an improper sampling system. There is no way to know what this background is without measuring it. In our laboratory and field studies, we determine the background level of the measurement system, including the particle counter and its sampling system, both before and after making any other measurement.

Clearly, each instrument has its own strengths and weaknesses. We have found most success in our dynamic analyses by using multiple instruments on a single studied gas line, as shown in Figure 1. Integrating all instruments to microcomputers allows similarly formatted data to be obtained from all instruments. Also, with computer interfacing, dynamic phenomena can be observed on a real-time basis, and the data can be transferred to a mainframe computer for more sophisticated analysis. The unique feature of optical particle counters is that they determine the times when particles are present in a gas system. Using computers to collect the data takes full advantages of this benefit.

One problem with all dynamic counting is that it reveals very little about the physical characteristics of particles, such as their composition. Therefore, one alternative method of particle analysis finding increasing use is that of capturing particles on flat membrane filters, and then analyzing those filters with a scanning electron microscope (SEM) with attached Energy and/or Wavelength Dispersive X-Ray Spectrometers (EDS and WDS) for chemical analysis. However, conventional SEM evaluations of filters can be extremely tedious and time consuming, which can be appreciated from the fact that a standard 47 mm sampling filter contains approximately 5 million fields of view at 5000X. Furthermore, data accumulated manually is generally not amenable to

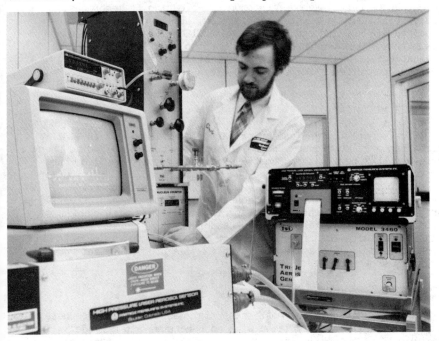

Figure 1: Particle counting experiments being run at BOC Group Technical Center.

statistical analysis. To overcome these limitations of manual SEM evaluation, we have computer interfaced our electron beam microprobe, which is basically an SEM with enhanced chemical analysis capabilities. The computer controls the stage automation, so that any number of pre-selected frames can be sequentially scanned without user intervention. The computer also analyzes the electron beam output for size and shape information, and the spectrometers for quantitative chemical analysis. Figure 2 illustrates the image analysis process of this system. Not only does this system allow for unattended counting, sizing, and compositional analysis of particles collected on sampling filters, it also summarizes and reanalyzes this information for a statistically large particle population. A typical data output includes the time-averaged particle concentration, determined from the number of particles observed and the amount of gas sampled, a compositional breakdown of the entire particle population, and the size distribution of all the particles sampled, and of each compositional type. In a sense, the data output provides a signature of a gas line through which particle origins might possibly be traced.

To accurately analyze particles on membrane filters, particles derived from ambient contamination must be minimized. Membrane filters in the as-received condition will contain some particles, which can be partly removed through pre-cleaning of membranes. Also, to minimize ambient contamination, all of our sample filter housings are cleaned and handled under Class 100 hoods within a clean room. Also, carbon or metal coating of filters prior to evaluation, as well as the microprobe evaluation itself, are likewise performed under Class 100 clean conditions.

PARTICLE MINIMIZATION

The underlying philosophy of particle minimization in VLSI grade gases is to produce the gas as pure as possible, and to then maintain this purity throughout the gas delivery route, from the air separation unit to the final vaporization and gas distribution at a chip fabrication facility. Many of the principles for materials and components selection and operating procedures were originally based on intuitive reasoning. However, with the improvements in particle analysis capabilities discussed above, we can now evaluate the effectiveness of these principles, or make further improvements as necessary in the gas production and delivery schemes.

Production

The raw material for most industrial gases is the atmosphere, which is usually heavily laden with both natural and manmade particles. Atmospheric particle concentrations (all particles greater than 0.02 μm diameter) can range from several million

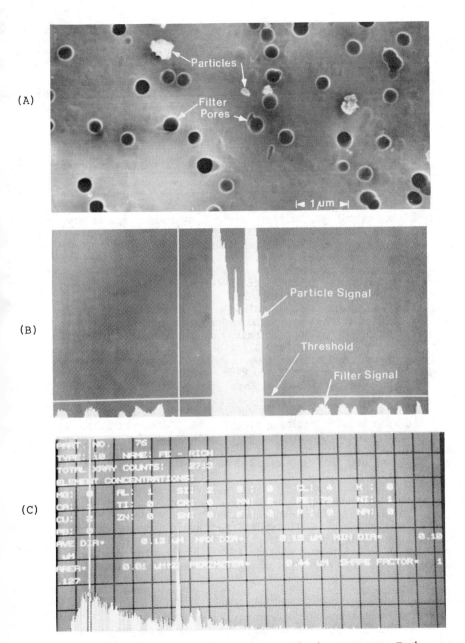

Figure 2: Illustration of particle image analysis process. Each particle collected on the filter, as shown in the secondary electron image (A), also produces a digitized contrast signal (B) exceeding a threshold limit set above the signal from the filter. When a particle is recognized, it is counted and sized by intersecting chords. Also, the particle center is located and an EDS spectrum is obtained, (C).

particles per cubic meter of air in cleaner environments to several billion particles per cubic meter in urban pollution areas. During the air separation process, many particles are removed by filtration, distillation or nucleation of heavy liquid droplets and solid crystals on the particles, which are then swept out of the gas stream. While particles might also be added to the process stream by mechanical moving parts, e.g., valves and compressors, and various adsorbers, these gases will contain orders of magnitude lower particle concentration than the parent atmosphere. Due to elaborate distillation and purification processes, specialty gases also contain far lower particle concentrations than ambient environment.

Further minimization of particles in high purity gases is obtained through high efficiency filtration. Although filters are mostly placed within gas delivery systems, and most importantly at the final point-of-use within a chip fabrication facility, the entire filtration scheme can be considered an extension of the high purity gas production route. For VLSI grade gases, membrane filter elements are preferred over depth fiber filters due to concern over possible release, or shedding, of fibers into the gas stream. The membrane filters are usually stacked or pleated so that the gas passes through multiple filter elements within a single filter cartridge, providing particle removal efficiencies of greater than 99% for most operating conditions [5]. However, even the most advanced membrane filters cannot be relied upon solely to guarantee gas cleanliness. Upset conditions such as pressure surges or mechanical shock can cause sudden release of previously trapped particles, or even loss of filter integrity. Particles in the range of 0.1 to 0.5μm, which are presently the most damaging to VLSI chips, are unfortunately also the most capable of penetrating membrane filters [5]. Also, particle shedding from even well constructed filters cannot be discounted, especially after prolonged aging. For these reasons, particulate contamination must also be minimized in the gas delivery system.

Delivery

Following production, the delivery of gases requires multiple transfers between vessels, e.g., the transfer of liquid nitrogen from a truck to a customer storage tank, and flow through a wide variety of materials and components. The underlying consideration for these high purity gas delivery systems is that they be passive, i.e., not contribute any additional particles to the high purity gas stream. Consequently, materials, components, fabrication techniques and operating parameters are all selected or designed to limit particle contamination of gases during their storage and delivery.

Stainless steels, especially AISI type 304L and 316L, are increasingly becoming the materials of choice for high purity gas handling equipment over the more traditionally used copper, brass, or low alloy steels. Figure 3 shows a typical array of stainless steel components for delivering gases from cryogenic storage tanks

Figure 3: Typical array of stainless steel hardware used to vaporize and distribute VLSI grade atmospheric gases.

to a chip fabrication facility. One advantage of stainless steels lies in their ability to passivate, i.e., achieve a chemical inertness in oxidizing environments such as air. Most theories attribute this passivation to the formation of an integral and tenaciously adhering thin layer of amorphous oxides and oxyhydroxides of primarily chromium [6]. Such passivation should limit the formation of corrosion product particles, e.g., iron oxide rust particles, and also limit any gaseous contamination of the high purity gases. In contrast, copper surfaces rapidly form porous and loosely adhering hydrated oxides and sulfides during atmospheric exposure [7], which can be released into the high purity gas as particles.

In addition to materials specifications, maximum surface roughness, usually expressed in terms of micro-inches* R_A (arithmetic mean), is also frequently specified for most gas-wetted surfaces. For example, tubing for high purity nitrogen delivery systems is now generally provided with a surface roughness of 15 micro-inches or less. In contrast, most commercially available tubing and piping will have surface roughness exceeding 30 micro-inches, and as high as 100 micro-inches. These surface roughness specifications are derived from the belief that rough surfaces may provide numerous trapping

1 micro-inch = 0.025μ m.

sites for particles, moisture, or chemicals used in the finishing operation, which can later be released into the gas stream. Figure 4 illustrates the particle trapping phenomenon on surface striations of a stainless steel tube. In many cases, however, particle adhesion appears to be insensitive to surface features, and may be affected more by molecular interactions and electrostatic attractions [8]. In either case, several operating conditions encountered in typical gas delivery systems can cause these surface adhering particles to become dislodged. Figure 5 illustrates one example of particles in a non-VLSI system being released into the gas stream by mechanical vibration of the gas delivery pipes. Clearly, surface adhesion of particles warrants further research, not only with regard to high purity gas handling equipment, but also in relation to particle adhesion to silicon wafer surfaces [9].

To achieve the mirror-like surface finishes required, electrolytic or chemical polishing is used in addition to possible prior mechanical polishing. Besides reducing roughness, both electropolishing and chempolishing also provide enhancement in corrosion resistance over mechanically polished surfaces. Part of this improvement is due to removal of a surface layer that is highly deformed and possibly contaminated with abrasives, oxide scale and lubricants. As illustrated in the Auger spectra of Figure 6, electropolishing or chempolishing also enhance surface passivity by promoting the development of a chromium-rich passive layer beyond that formed during mechanical polishing.

The final step in surface treatment of gas handling materials is a cleaning operation to remove any residual contaminants. The cleaning procedure will usually consist of several steps that selectively remove specific contaminants, and are sequenced to avoid reintroducing other types of contamination. The final cleaning steps are aimed at removing micron and submicron sized particles, and are preferably performed within a clean room. Prior to transport to the fabrication site, the components are pressurized with filtered, inert gas and capped.

The engineering design of gas handling components, such as valves and regulators, can also affect their particle generating characteristics. Even if these components are fabricated predominantly from electropolished stainless steel, particle generation is still possible due to wear of the seals or other moving parts. Figure 7 illustrates the dynamic particulate emission behavior of two different stainless steel valves, a commonly used ball valve and a diaphragm valve frequently specified for VLSI applications. Both valves were repeatedly throttled between the fully open, two-thirds open, and one-third open positions. All valves produce a burst of particles when they are turned to a new position, most likely due to mechanical wear within the valve. This particle generation is minimized to the greatest extent with the VLSI grade diaphragm valve.

Fabrication of the gas delivery system proceeds by joining individual components into subassemblies, which in turn are joined

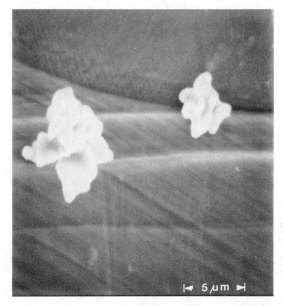

Figure 4: Scanning electron micrograph of particle trapped on surface striations on a stainless steel tube. (4,000 X Magnification)

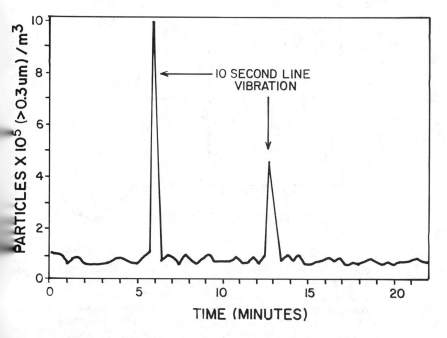

Figure 5: Change in particle concentrations in a non-VLSI line due to 10 second line vibration.

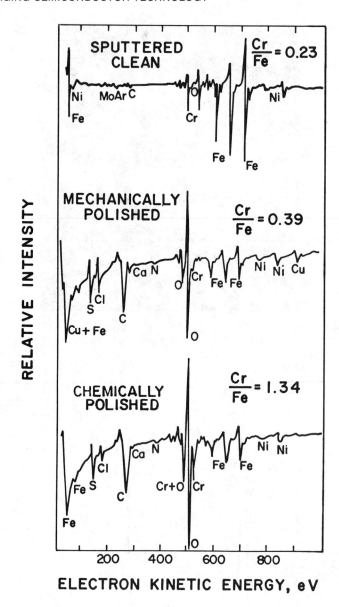

Figure 6: Auger spectra obtained for AISI 316L stainless steel in (A) the sputtered clean condition, (B) as mechanically polished, and (C) as chemically polished.
Progressively greater Cr and O surface enrichments are observed in this sequence. Also, metallic (copper) contamination imparted to the surface during mechanical polishing is removed by the chemical polishing.

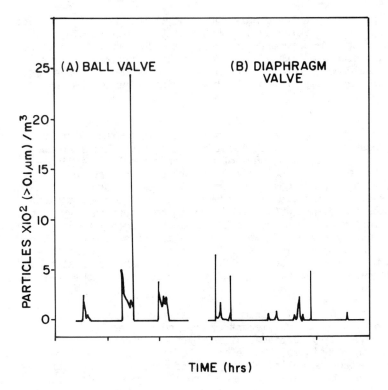

Figure 7: The concentration of particles (> 0.1μm diameter) in nitrogen immediately downstream of (A) a stainless steel ball valve and (B) a stainless steel diaphragm valve. The valves were repeatedly throttled between the fully open, two-thirds open, and one-third open positions.

to form progressively larger modules and ultimately the entire integrated system. As much of this assembling as possible is performed within a clean room and cleaned prior to attachment to the larger system. Because cleaning after the final fabrication steps is difficult, the joining should be performed with the upmost attention paid to contamination minimization. Gas tungsten arc welding (GTAW), in which components are joined together under an inert environment, is preferred for high purity applications due to the cleanliness, lack of fume generation, and control inherent in the process. Here, again, stainless steel has an advantage over other materials in that it is compatible with ultraclean GTAW processes. Usually low carbon grades of steel, such as AISI type 304L or 316L, are preferred for enhanced corrosion resistance of the welds. Furthermore, GTAW of tubular components can be performed with automatic orbital welding machines, which provide consistently smooth welds lacking pockets that can trap contaminants. In contrast, ordinary welding processes produce unacceptable levels of weld surface oxides and spatter, which are likely particle sources. Even more severe contamination problems would occur with copper delivery systems, which are generally brazed with low melting point metals and fluxes.

Following final fabrication, the integrated gas storage and delivery system is purged, and sometimes pulsed with bursts of high pressure, to remove residual contaminants. At this point, the integrity and cleanliness of the system should be evaluated. Integrity testing can be achieved by pressurizing the entire system with an inert gas, and monitoring pressure drops with time, or leak testing individual components. Cleanliness evaluations are necessarily indirect, if they are to be nondestructive to the system integrity. The most simple evaluation technique is to use dynamic particle counters and trace gas monitors before and after flow in the gas handling system to indicate any contaminants picked up from the system itself. Multiple instruments also can be integrated into computer-based systems for continuous on-line monitoring of gas streams.

As has been shown, particle concentrations in a gas can be affected by numerous parameters related to the gas handling system, such as materials, surface finishes, component design, and fabrication techniques. Therefore, the cleanliness level of a gas passing through an entire VLSI system is defined by that parameter which causes the greatest amount of particle generation. From this systems approach, the concept of VLSI gas handling system design can be easily verified. Figure 8 illustrates this through comparison of particle concentrations, as a function of particle size, for non-VLSI and VLSI systems, both without final point-of-use filtration. As shown, the VLSI system reduces particle concentrations by two to three orders of magnitude below the non-VLSI system. Point-of-use filtration further extends this particle minimization to consistently yield gases of maximum cleanliness.

Particle analysis also helps determine the origin of particles, and again points to the benefits of VLSI design. Table 1 details the compositional analyses of particles collected in a non-VLSI system. The majority of particles are iron-rich. The additional presence of nickel and absence of chromium in these iron-rich particles suggests origination in a 9-Ni steel liquid storage tank. The presence of copper-rich and aluminum-rich particles might be traced to the copper lines and aluminum vaporizer. None of these non-stainless steel materials would be found in a VLSI system.

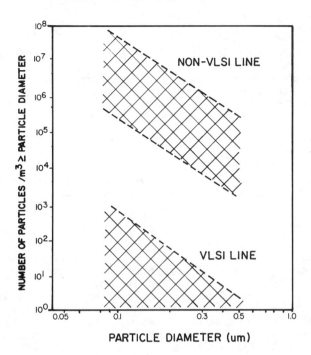

Figure 8: Particle concentration vs. size for non-VLSI and VLSI nitrogen lines without point-of-use filtration.

TABLE 1 -- Non-VLSI Line Particle Compositions

Particle Type [a]	% of Particle Population
Fe-rich[b]	65.7
Cu-rich	4.2
Al-rich	3.4
Mixtures (Fe-Cu-Al rich)	23.7
Other	3.0

[a] Typed according to element present at level greater than 50%, or as mixture if all elements are less than 50%.

[b] contains 5% Ni (average)

CONCLUSIONS

To meet the emerging needs of the microelectronics industry, a multi-faceted research and engineering effort is required for particulate control in gases. Effective solutions depend, in part, on in-depth knowledge of the gas processing and handling materials parameters which can lead to gas contamination. Also required are advanced analytical capabilities to evaluate these parameters and to identify particle sources.

ACKNOWLEDGEMENTS

The technical assistance of W. Gerristead, R. Sherman, and P. Irion, all from the BOC Group Technical Center, is gratefully acknowledged.

REFERENCES

[1] Willeke, K. and Liu, B.Y.H., "Single Particle Optical Counters: Principles and Applications," in Fine Particles, B.Y.H. Liu, Ed., Academic Press, New York, 1976, pp. 698-729.
[2] Knollenberg, R.G., "The Measurement of Particle Sizes Below 0.1 Micrometers," presented at the 30th Annual Meeting of the Institute of Environmental Sciences, Orlando, Florida, 1984, and published in the Proceedings, Institute of Environmental Sciences, Chicago, Illinois.
[3] Keady, P.B., Quant, F.R. and Sem, G., "Differential Mobility Particle Sizer," TSI Quarterly, Vol. 9, No. 2, April-June 1983, pp. 3-11.
[4] Fuchs, N.A., "Sampling of Aerosols," in Atmospheric Environment, Vol. 9, Permagon Press, London, 1975, pp. 697-707.
[5] Liu, B.Y.H., Pui, D.Y.H. and Rubow, K.L., "Characteristics of Air Sampling Filter Media," in Aerosols in the Mining and Industrial Work Environment, V.A. Marple and B.Y.H. Liu, Eds., Ann Arbor Science Publishers, Ann Arbor, 1983, pp. 989-1083.
[6] Kruger, J. and Frankenthal, R.P., Eds., Passivity of Metals, John Wiley & Sons, New York, 1978.
[7] Leidheiser, Jr., H., The Corrosion of Copper, Tin and Their Alloys, John Wiley & Sons, New York, 1971.
[8] Hinds, W.C., Aerosol Technology, John Wiley & Sons, New York, 1982.
[9] Bowling, R.A., "An Analysis of Particle Adhesion on Semiconductor Surfaces," Journal of the Electrochemical Society, Vol. 132, No. 9, September 1985, pp. 2208-2214.

Dopant Profiling Techniques and In-Process Measurements

James R. Ehrstein

SPREADING RESISTANCE MEASUREMENTS - AN OVERVIEW

REFERENCE: Ehrstein, J. R., "Spreading Resistance Measurements - An Overview," Emerging Semiconductor Technology, ASTM STP 960, D. C. Gupta and P. H. Langer, Eds., American Society for Testing and Materials, 1986.

ABSTRACT: Spreading Resistance is the most versatile electrical technique for characterizing depth profiles in silicon. However, it is being increasingly challenged as an analytical method by shrinking device geometries. Consequently, refinement of such aspects as probe conditioning, sample preparation, and bevel angle measurement is needed, and traditional practice regarding calibration, algorithms, and profile interpretation must be reexamined. Based on examples drawn from the author's work, multilaboratory experiments, and recent literature to illustrate and discuss these topics, this paper attempts to summarize the current status of the measurement and its interpretation showing both strong points and apparent limitations.

KEYWORDS: depth profiling, dopant profiling, ion implantation, silicon, spreading resistance

INTRODUCTION

More than 20 years have passed since the first published work [1] on the use of the spreading resistance phenomenon for determining the resistivity depth profiles of semiconductor specimens. The first ten years saw the development of a detailed model for the measurement of graded structures [2], the commercial availability of a refined spreading resistance instrument, and the adoption of the measurement in many semiconductor companies worldwide, both for process control and research applications. These years culminated in a Spreading Resistance Symposium featuring nearly two dozen papers on the state of the art in

Dr. James Ehrstein is a Physicist in the Semiconductor Electronics Division of the National Bureau of Standards, Gaithersburg, MD 20899. Contribution of the National Bureau of Standards. Not subject to copyright.

theory, practice, and application of the measurement [3]. While there were some demonstrations of profile measurements on 1-μm thick layers included in that symposium, the prevalent device structures in semiconductor electronics of that day were generally several to ten or more microns deep. The limited comparisons which could be made were generally with C-V profiles and showed quite satisfactory agreement. The most significant obstacles to effective utilization of spreading resistance profiling were the lack of a real-time data analysis algorithm for a small computer, difficulty of conditioning a set of probes, and lack of probe calibration specimens.

Since that time, many papers have been written on various aspects of the measurement. However, semiconductor device technology has advanced at a more rapid rate than has understanding of the measurement. It has been possible to obtain spreading resistance profile data for some of the shallowest and most abrupt structures currently being fabricated. However, questions of the accuracy of the profiles obtained from such data - never completely answered for less demanding structures - must be raised anew. (In fairness, it is worth noting that similar questions of accuracy of other available profiling techniques have not been fully answered either.)

DEVELOPMENTS IN MEASUREMENT THEORY

The earliest models used to explain the measured spreading resistance in terms of the resistivity variations under the probes were the two-layer image solution of Dickey [1] and the multilayer Laplace equation model of the electrostatic boundary value problem due to Schumann and Gardner [2]. The first suffered from being too simple a model to cope with truly graded structures such as diffusions and ion implants. The second, at least in its original numerical implementation, proved too slow, even on main frame computers of that day, to provide true real-time analysis.

The procedure of Dickey, based on a two-layer image solution was greatly extended by the multilayer image solution due to Nagasawa et al. [4], which appears not to have been used (or cited) in other than the original manuscript. Further work by Dickey [5] led to the "Local Slope" model which was based on the limiting cases of a thin layer over insulating or conducting substrates. While derived from simple limiting-case physical models, it was shown by Albers [6] to give profile analyses in very good agreement with those resulting from the multilayer Laplace solution for a number of micron-depth scale model-data structures. A construction similar to the junction-isolated starting point of the Local Slope method was developed by Kudoh et al. [7] and was shown to give results in good agreement with profiles obtained from incremental sheet resistance measurements for several sub-micron structures. Kudoh's procedure, like the junction-isolated limit of Local Slope, was based on direct analogy with incremental sheet resistance. Because these two profile-analysis procedures were algebraic in nature, they were the first to run in real time on a desktop computer.

A significantly larger amount of work has been reported on the enhancement and understanding of the multilayer Laplace equation methodology. As originally developed by Schumann and Gardner, a structure with graded resistivity is partitioned into lamellae or layers parallel to the surface, each assumed to have a constant resistivity. Laplace's equation is assumed to be applicable for describing the potential distribution in each of the layers. Solution of Laplace's equation is used to obtain a relation between measurement voltage and current and the resistivity, ρ_n, of the top layer, to which the probes make contact, of the following form (restated here for two probes instead of the three-probe configuration assumed by Schumann and Gardner):

$$R_{sp} = \frac{\Delta V}{I} = \frac{2\rho_n}{\pi a} \int_0^\infty (1+2\theta) \left(\frac{J_1(\lambda a)}{\lambda a} - \frac{J_0(\lambda s)}{2}\right) \frac{\sin(\lambda a)}{\lambda} d\lambda \qquad (1)$$

where a is the contact radius of either of the probes, s is the spacing between probes, and J_0 and J_1 are Bessel functions of the first kind. The term $(1 + 2\theta)$ is a structure factor containing information about the resistivity of the top layer and each of the underlying layers. Conceptually, this equation is solved to find the resistivity, ρ_n, of an added top layer so that the calculated resistance agrees with the measured resistance at the corresponding point. The process is repeated in turn for each new top layer (each new measurement point) needed to comprise the structure. (Since the resistivity of the new layer occurs both outside the integral and inside the structure factor, a transcendental relation results which must be solved self-consistently.) The application of the equation is generally looked at in a rather different fashion, however. Once the best value for ρ_n, the resistivity of the new layer, is obtained by solving the transcendental relation, comparison with the relation for measurements on a thick structure of uniform resistivity, ρ_n

$$R_{sp} = \frac{\rho_n}{2a} \qquad (2)$$

results in the integral on the right-hand side of eq (1), (times $4/\pi$) being identified as a "correction factor." The correction factor "removes" the effect of the resistivity variations of the deeper lying portion of the structure sampled by the current distribution so the measured resistance is as though resulting from a thick uniform layer of resistivity, ρ_n.

Unfortunately, early numerical implementation of this formalism was time consuming and required a mainframe computer. Yeh [8] reported that solution for a 17-layer profile required 10 min on an IBM 7090.

More than half a dozen authors have contributed to enhancing the computational speed of the multilayer Laplace algorithm until at least one version [9] can process profile data in real time on the same scale desktop computer as might be used for the Local Slope algorithm. These improvements were made by simplifying the evaluation of the structure factor and examining a number of available numerical integration techniques. These improvements are discussed in detail by Albers [10].

At least as significant have been the contributions of a number of authors to the investigation of several aspects of the "accuracy" of the multilayer Laplace model. These include consideration of the current distribution under the probe necessary to satisfy the assumption of constant potential across the contact [11-13], other formulations of the potential difference between the two probes [14-16], comparisons of various algorithm formulations on model data simulating real profiles [6,17], as well as the effects of sparse data [17,18], and nonconstant radius of contact [19]. Among the most important considerations raised by these works are the following:

1) Lee [14] pointed out that care must be taken when iterating the transcendental equation to find the value for the new top layer resistivity since it is possible to have a high degree of convergence between successive trial solutions yet to have converged to the wrong resistivity value. (This observation is independent of particular assumptions of current distribution at the contacts or methods for performing the numerical integration of the correction factor.)

2) Choo and Leong [17] showed that a paucity of data for any profile will cause errors in final resistivities which range from a few percent to truly significant errors on a relative basis, being largest in the tail of a highly graded structure and significantly less at the surface. Choo suggested a procedure in which the "stair-step" set of discrete resistivities implicit in the usual multilayer model be replaced by a piecewise series of exponentially varying resistivities and demonstrated significant improvement in results for sparse data. However, perhaps because of a predicted factor of 3 increase in computation time, this technique does not seem to have been adopted.

3) Berkowitz and Lux [12] developed a procedure for determining the current distribution through the contact as a function of underlying structure which best satisfies the assumption of constant potential across the contact. His calculations showed that the "best" current distribution is strongly dependent on the relative resistivities of the top layer and the substrate. He also showed that the current density used by Schumann and Gardner, and also that of uniform flux density, to be reasonably good general purpose choices (i.e., usable for all types of layer-substrate combinations). Either of these choices caused a worst-case error in the correction factor of about 10%. The Schumann-Gardner distribution fared worst for the case of heavily conducting substrates, while uniform current flux fared worst for uniform layers and for rapidly changing portions of diffusion and ion-implant-type structures. Separately, Berkowitz noted that an error in the calculated resistivity value for any one point has a rapidly diminishing effect on the evaluation of resistivities for subsequent points.

4) Vandervorst and Maes [16] noted that small integration inaccuracy in evaluating the correction factor may result in inaccuracies a number of times larger for the resistivities eventually obtained from solving the basic transcendental equation. An example of a 2% change in the correction factor resulting in a 15% change in the resultant resistivity was given. This source of error is

independent of the particular form of the integrand in the correction factor and of the numerical procedure used to solve the transcendental equation. It is therefore a source of potential error which is in some manner additive to those others.

5) Choo et al. [13] presented a variational procedure to obtain the current distribution which best satisfies constant potential at the contact. Choo's procedure involves using a mixture of the Schumann-Gardner and uniform flux distributions in varying proportions. Using this procedure for a wide variety of simple structures, he shows that when applied to the case of a single contact (for which exact solutions of the constant contact potential problem are available), the correction factor obtained from the mixed current distribution formulation is within 1% of the exact value.

The net effect of all this theoretical work is that a correction factor can be calculated in real time with a reasonably solid understanding of the errors that may be encountered within the framework of the Laplace equation model. Large errors predicted to arise from sparse data should be minimizable by the exponential approximation of Choo et al., and errors from choosing the wrong current distribution can be significantly reduced with Choo's current admixture procedure. Other observations on the need to evaluate the correction factor integral accurately and to solve the transcendental relation accurately serve as simple precautions which can be adequately addressed in any data reduction algorithm, albeit at the expense of computational time or the size of CPU required. As noted by Choo et al. in their work on the exponential approximation, the biggest practical limitation to effective implementation of the procedure is noise in real measurement data since the multilayer algorithms tend to be noise amplifiers. It is likely that measurement noise will act as even more of an impediment to obtaining numerical stability with any of the more elaborate analysis procedures having adjustable parameters such as the current mixture procedure of Choo or the resistivity-dependent contact procedure of Piessins et al. [19].

The most significant failure of the modeling of spreading resistance measurements has only been fully recognizable as device geometries have dramatically shrunk, and doped layers with thickness below a nominal 1 μm have become the norm. It must be recognized in general that the resistivity or free carrier density profile of a graded semiconductor layer is not commensurate with the doping density profile which gives rise to it. Free carriers will diffuse or spill from high concentration regions until limited by the electric field of the space charge region. In earlier generation device structures with relatively deep and slowly graded structures, such effects were relatively minor and Laplace's equation was perhaps not a bad model of the spreading resistance measurement. The inherent limitation of the Laplace model was noted in the early work of Schumann and Gardner, but considered not to be a serious problem at the time. The compounding effect of electric field line distortion and additional redistribution of carriers due to beveling was noted by Severin [11].

More recently, Hu [20] published the results of an extensive series of calculations of the solution of Poisson's equation for a variety of profile shapes and degrees of unbalance between doping level in the substrate and peak doping density in the graded layer. His results, stated in terms of electrical junction (electron density equals hole density) versus metallurgical junction (donor atom density equals acceptor atom density), show that on beveled specimens, the free carrier profile may be shallower than, deeper than, or commensurate with the dopant profile. Many factors, including shallow- and deep-level surface state densities were shown to affect the specific results obtained. The important point, however, is that the use of Laplace's equation (based on the assumption of space charge neutrality) to model the electrical conduction problem which underlies spreading resistance measurements, may lead to misinterpretation of the graded dopant layer which gives rise to that conduction.

It is worth emphasizing Hu's additional observation that similar effects due to carrier spilling will distort profiles obtained from incremental sheet resistance or Hall effect profiling, and in fact with each successive layer removal, the effects of carrier spilling on these measurements are expected to change. (Additional sources of error in incremental layer removal techniques used for profiling which are due to space-charge and to surface effects were discussed by Kramer and van Ruyven [21] and by Huang and Ladbrooke [22].) Finally, Hu also noted that carrier spilling on a beveled structure is likely to affect the results of junction staining measurements as well.

The question might be asked - "Why continue to do spreading resistance measurements if such problems are known?" Perhaps the best answer is found in Hu's work: there are a number of classes of profiles for which the difference between the electrical and metallurgical junctions is predicted to be small (on the order of 5% - basically about the size of accumulated uncertainty when comparing results of both types of measurements). The principal classes of profiles having the largest predicted differences, Gaussian and erfc profiles and very thin lightly doped epitaxy over a buried layer, certainly have significant application in current semiconductor processing. However, even here, unless the layer is lightly doped, the principal discrepancy is expected to be in the profile tail; evaluation of the more heavily doped portions of the layer (which are the dominant contributions to layer sheet resistance and electrically active fraction of implant dose) is likely to be undistorted. Moreover, there are circumstances when electrical profile data is simply beneficial, ranging from studying specialized annealing cycles, where a complement to secondary ion mass spectrometry (SIMS) analysis is highly desirable for interpretation, to process control applications where accuracy may be less important than precision and turn-around time. In such circumstances, SRP profiling has clear advantages of speed and allowed layer combinations, and perhaps is no less accurate than incremental sheet resistance or Hall effect profiling for junction-isolated layers.

SPREADING RESISTANCE - THE EMPIRICAL SIDE

The foregoing was a summary of the progress that has been made in modeling spreading resistance measurements as an electrical conduction problem. In the world of real measurements with real probes, a number of other considerations must be addressed. In the following, what is known about the behavior of real probes and how it affects calibration of spreading resistance measurements as well as what is known about the resolution, precision, and "accuracy" of spreading resistance profiling is considered.

THE PROBES AND THEIR CALIBRATION

Despite the limitations and caveats regarding the analysis and interpretation of spreading resistance measurements which have been set forth in the many papers on theory, nothing so controls the results of spreading resistance measurements as the quality of the probes themselves. Originally used at a static load of about 45 g to maintain reproducible low noise contacts, probes now commonly are used at loads of 5 to 10 g. The need for lighter working loads arises from the need to reduce physical penetration of the probes into the silicon and to improve spatial resolution of the profile data by reducing contact size. Early benefits of reduced probe load were shown by Mazur and Gruber [23] in conjunction with an improved technique for preparing the contacting area of the probe to obtain stable data.

A set of probes, whether operated at small loads or large, is known to have a nonlinear response; i.e., if a set of bulk silicon specimens with a range of known resistivity values is measured (as is done for traditional probe calibration), the spreading resistance values will not be exactly obtainable from eq (2) for any single value of probe radius, a. This variability, which depends on conductivity type, crystallographic orientation of the silicon surface, silicon surface preparation, probe load, material, and conditioning, was explored by Ehrstein [24]. This same nonlinearity is often found when analyzing depth profiles of epitaxial layers over heavily conducting substrates with a correction factor algorithm which employs a constant radius of contact. For such structures, it has been observed that a value of the contact radius which gives good agreement with a known value of substrate resistivity will not give good agreement with values of epitaxial resistivity obtained from another method such as C-V profiling. In the same fashion, a value of the contact radius which gives good agreement for the epitaxial layer resistivity has been found to cause discrepancies in the substrate resistivity value. This is likely the reason why Vandervorst and Maes [25] find different values for the epitaxial layer concentration versus probe load when all results are normalized to give the same result for the substrate.

Mazur and Dickey [26] suggested this nonlinearity was due to a resistivity-dependent interfacial barrier resistance, $K(\rho)$, which operated in series with the spreading resistance phenomenon. The use of such a series barrier resistance term on the analysis of a graded layer is described by Gruber [27]. Support for the $K(\rho)$ barrier resistance model was provided by Kramer and van Ruyven [28] based on their modeling of the current across the contact interface in terms of tunneling through an interface barrier and thermal emission over the barrier. They also noted evidence for a two-region contact where the central region causes plastic deformation of the silicon and the outer, annular region, causes only elastic deformation. They suggested that this outer region may contribute to the conduction only for very low and very high resistivities.

In contrast to this, Ehrstein et al. [29] demonstrated better agreement with independently known profile parameters if a resistivity-dependent contact radius rather than a fixed contact radius or a barrier resistance was used in the analysis of depth profiles. These results were based on a series of boron implants and a local slope analysis of the measurements. This is not to assert that a resistivity-dependent contact radius is inherently a more correct description of the physics of the contact. Rather, it simply shows that the variable radius treatment was more effective in evaluating some key parameters of the boron implants being tested. Similar tests with correction factors based on the Laplace equation have not been performed - such an algorithm which is compatible with a variable contact radius has only recently been reported [19] and it is not yet clear that it will operate stably with real specimen, i.e., noisy, data.

In point of fact, a full description of probe contact physics which can readily be applied to individual probes is not likely to be achieved. In addition to the current transport mechanisms across the central region of the interface discussed by Kramer and van Ruyven and the possibility that the conduction through the peripheral region is controlled by the resistivity structure under the particular layer being measured, the effects of piezoresistance [30] and the interaction of multiple microcontacts [11,14,25] would need to be taken into account. Moreover, experience with microscopic examination of probe tracks on silicon shows that a wide variety of microcontact distributions can result even from a reasonably well-controlled procedure for preparing the probes for use. This suggests that a unique description of the contact might have to be evaluated for every probe.

For these reasons, it appears that a simple lumped parameter approach to accounting for nonlinearity of contact response as a function of resistivity is likely to be both more expedient and tractable. Significantly more experimental testing needs to be done, particularly in conjunction with a Laplace equation-based correction factor, to determine the relative effectiveness of the barrier resistance and effective radius approaches on a wide variety of structures.

A somewhat different approach to calibration of probe response for graded layer analysis has been suggested by Pawlik [31]. He uses the measured sheet resistance of the layer to be profiled as a point of reference and then analyzes SRP profile data using a fixed contact radius. A value of sheet resistance obtained from integrating the resistivity profile is compared with the directly measured value. If the two values do not agree within acceptable limits, the profile analysis is repeated with a somewhat different radius value, and the process repeated as necessary until acceptable self-consistency of sheet resistance values is obtained.

A more elaborate procedure based on the same principle of self-consistency of electrical measurement values was reported by Albers and Berkowitz [32]. It is applicable to a somewhat wider class of structures than the junction- or insulator-isolated layers to which Pawlik's procedure applies. It requires taking spreading resistance measurements at several values of probe spacing on the top surface of the specimen to emulate the results obtained by a four-point probe, as well as performing several analyses of the depth profile using different trial contact radius values and several auxiliary calculations. The result again is a tightly self-consistent analysis of the depth profile relative to a four-probe determination of the layer's resistance.

While both these procedures have the advantages of enforced self-consistency, of circumventing the requirement of preparing a set of calibration specimens with a surface identical to that of the test specimen, and of applicability to materials such as SOS where an innately higher level of defects may make equivalence with standard calibration material unattainable, they are not without disadvantages. Because both procedures use integrated layer resistance as a checkpoint, they optimize profile accuracy for the most conductive parts of the layer (which are the dominant contributions to the layer resistance). However, their use of a fixed value of radius for each profile analysis risks misinterpreting somewhat the more lightly conducting portions due to lack of accounting for probe calibration nonlinearity. More significantly, these two procedures are inapplicable to any but the top layer of a multilayer structure, e.g., to the emitter of a bipolar transistor, and they are also inapplicable to structures with conducting substrates, e.g., epitaxy over a buried layer.

RESOLUTION OF SRP MEASUREMENTS

The resolution of spreading resistance measurements has an important bearing on whether it has any continuing role to play in the quantitative or qualitative evaluation of profiles in an environment of continually shrinking device dimensions. The question of resolution can be broken into two aspects, geometric resolution and functional electrical resolution.

Geometric resolution can be considered to be the smallest increment in structure depth which can be measured. It is primarily controlled by the beveling angles which can be produced on silicon chips since the bevel angle provides the geometric magnification between the specimen stage translation and the effective increment in depth below the original specimen surface. It has been found that with the use of 0.1-μm diamond and a frosted glass plate, bevels of less than 4 min of arc (tangent $\leqslant 0.001$) can be obtained with good bevel edge acuity and flatness of beveled surface. Figure 1 shows a photograph of the chip having the shallowest bevel yet attained in this laboratory together with the superimposed interferometer fringes used to assess the quality of the bevel as well as to measure its angle. This chip, which has a geometric magnification of nearly 3000:1 would give depth resolution increments of less than 1 nm if used with stage translation increments of 2.5 μm. Such extremely shallow angles are not readily attainable, however. Factors such as nonuniform wear on the base of the beveling fixture, inability to get a perfectly uniform wax film between the silicon chip and the beveling fixture and irregularities in the silicon surface (fine-scale "orange peel") make it difficult to achieve the intended bevel angle for angles below about 4 min of arc. The same factors also make it likely that the bevel will develop at some skew angle to the edges of the chip, thus making it impractical to attempt to work with small patterned test areas at very small angles. Texture of the beveled surface also has a bearing on attainable resolution.

BEVEL EDGE

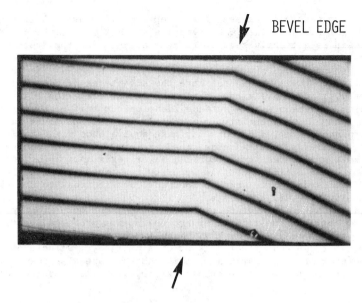

Figure 1. Photomicrograph of a 2.05 × 4.2 mm beveled silicon chip with superimposed interferometer fringes. Beveled surface is at right. Bevel angle is approximately 1.1 minutes of arc.

Measurement analysis is based on the assumption that the layer has uniform lateral properties between the probes with all the points having the same distance above the substrate; this is violated to a degree depending on the roughness of the surface finish which results from the beveling process.

The functional electrical resolution can be considered to be the smallest feature, or change in resistivity, which can be resolved as a change in measured resistance. The feature can be a very small change of resistivity irrespective of physical dimension, or more appropriately, it will be a somewhat larger change of resistivity which occurs in a very small increment of structure depth. While good geometric resolution is a prerequisite, functional electrical resolution is controlled primarily by probe preparation (conditioning) in a way that is not yet completely understood. Low measurement noise is clearly one requirement. Preparing the probes so they operate with minimal penetration into the silicon is another requirement as was demonstrated by Mazur and Gruber [23]. However, reducing physical contact area is also important since it reduces the sampling depth of the probes as was pointed out by Vandervorst and Maes [25]. At present, it is necessary to test for resolution; it cannot be predicted adequately a priori. Figures 2 to 7 illustrate the types of structural and spatial resolution possible with carefully prepared probes.

Figure 2 serves as a point of reference for a bipolar transistor structure that was part of a recently completed ASTM round robin on SRP depth profiling and shows raw data for the emitter, base, and part of the collector. The data came from the lab which shows the best overall

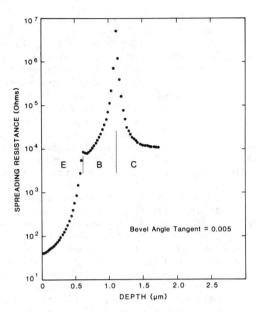

Figure 2. Spreading resistance data for bipolar transistor structure showing good dynamic range but poor resolution of emitter-base junction.

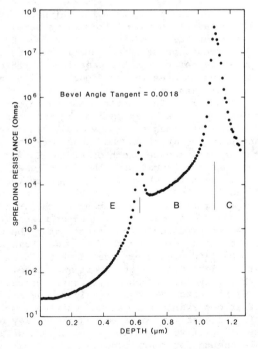

Figure 3. Spreading resistance data from another chip of the same bipolar structure showing improved resolution of emitter-base junction due to improved probe conditioning.

precision and nearly the highest overall dynamic range in raw measurement data - which indicate both well-prepared, stable probes and probes with good resolution of the resistivity changes in the structure. However, note the lack of resolution of a clear blocking junction signature, i.e., a cusp, at the emitter-base transition at about 600 nm of depth; of all the other round-robin data, only one run by one laboratory showed even a weak suggestion of a normal blocking character at this junction. Figure 3 shows measurements taken at NBS on another piece of the same bipolar structure after work to condition a pair of probes to leave as light a probe track as possible. In this figure, a very strong emitter-base junction character is displayed in the form of a resistance peak, or cusp. Subsequent tests were run on other pieces of the same structure at bevel angles from 6 to 72 min and with horizontal measurement steps adjusted to give approximately the same geometric depth resolution between successive measurements. These tests showed that the emitter-base junction cusp decreased somewhat with increasing bevel angle, but it was always sufficiently strong to indicate negligible penetration of the probes through the junction. Further measurements, using a second set of probes which showed only a very weak blocking character at the emitter-base junction, were run on the same set of bipolar chips. Comparison of the two sets of data showed that within the precision limit which is determined by the number of measurement points, there was no dependence of the width of the base layer as a function of which set of probes were used or as a function of bevel angle. It appears that preparation of a set of probes which elicit a strong junction character may have little practical effect on the interpretation of a modestly wide layer (400 nm in this case).

The same does not appear to be true for thinner structures. Figure 4 shows results obtained in this laboratory several years ago on an advanced bipolar structure received from another laboratory. After considerable work to reduce probe damage and improve beveled surface texture, it was possible to profile the base-layer with a geometric resolution of about 1 nm as seen in the figure. While the dynamic range of the base-layer data was thought to be good at the time, no sign of blocking character of the emitter-base junction was found. Further, the width of the base-layer determined from the data shown, as well as from repeated measurements, was typically less than 45 nm, whereas device and process data from the laboratory which provided the material predicted an active base width of 80 to 90 nm. More recently, this same structure was remeasured with the probes used for the measurements in Fig. 3. Results, shown in Fig. 5, show that an improved dynamic range and delineation of the emitter-base junction results from enhanced probe preparation; more significantly, the apparent base width has noticeably increased compared to the earlier results. It must be noted, however, that as used for these latest data, the probes were prepared and operated at a load such that no probe tracks could be detected on polished silicon, either under dark-field illumination or Nomarski contrast. Under this condition, it was found that repeated measurements on this very narrow base layer gave indicated base thicknesses as low as 67 nm, with the smaller values generally being accompanied by lack of an observable cusp at the emitter-base junction. Such nonpenetrating or nearly nonpenetrating probes may be necessary for ultimate resolution of very shallow structures, but at present they are not repeatable enough for routine use.

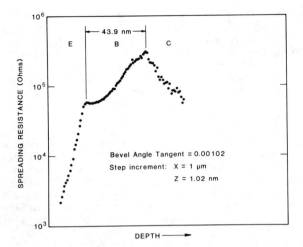

Figure 4. Early spreading resistance data of very narrow-base bipolar structure showing extreme geometric resolution but poor electrical resolution of emitter-base junction.

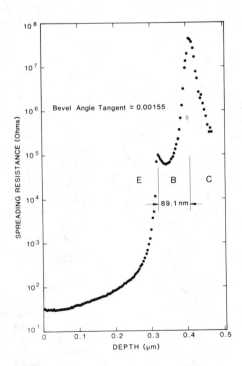

Figure 5. Recent spreading resistance data of the same narrow-base structure showing electrical resolution of emitter-base junction and increased dynamic range due to improved probe conditioning.

Another example of the improved resolution obtainable with very low penetration probes are shown in Figs. 6 and 7. The structure being measured is a $p^+/p/n/n^+$ multiple epitaxial layer impatt diode structure. Both sets of data were taken with probe separations of about 17 μm and probe loads of about 5 g. The two pairs of probes used here were the same ones described above in the round-robin bipolar specimen base-width test, with the pair showing the very strong emitter-base junction character (Fig. 3) being used for the impatt diode data in Fig. 7. The data of interest for the impatt structure are that of the n-epitaxial layer, for which the resistivity is supposed to be constant with an abrupt transition to the n^+ substrate for proper device operation. As can be seen in Fig. 6, the n-epitaxial layer appears graded throughout, indicating that the probes were sensing the shunting effect of the substrate all the way through the epitaxial layer. In contrast to this, the data in Fig. 7 show that it is possible to prepare a set of probes whose response is extremely surface-weighted so that the effect of underlying layers becomes almost negligible.

Recently, Pawlik [33] has demonstrated the ability of spreading resistance measurements to profile boron-modulation-doped MBE silicon where the boron transitions are of about two orders of magnitude and the transitions take place over a distance on the order of 15 nm per decade. He shows quite clearly, however, that meaningful analysis of such data is highly dependent on noise in the data, on numerical accuracy of the integration procedure in the correction factor algorithm, and on the value used for probe contact radius in the algorithm.

As a related matter, several articles have been published which noted significant spatial resolution limitations on spreading resistance measurements due to sensitivity to lateral boundaries on the structure being measured. These results were based on use of a three-probe spreading resistance system with probe separations as large as 1 mm. Since it is possible with care to prepare the probes of a two-probe system to operate with separations in the 10- to 20-μm range (5 μm has been obtained in this laboratory), such observed resolution limitations do not apply. Two precautions are in order for operating at very small separations, however. The first relates to the assumption in all correction factor theories that neither probe modifies the current distribution at the other probe. This suggests that a probe separation of about 10 times the probe radius is a reasonable lower limit. The second is a limitation in conjunction with use of the Berkowitz and Lux correction factor algorithm: a lower limit of 10 times the contact radius is set for the probe separation in order to maintain accuracy of the correction factor integration procedure he developed.

It is seen to be possible to obtain geometric and electrical resolution with spreading resistance probes which is basically compatible with some of the most challenging structures of emerging device technology. The price to be paid to attain this resolution is even more careful attention to probe and specimen preparation than is the norm for most present work and, at least in the case of modulation-doped MBE structures, the likelihood that more exacting numerical procedures for evaluating the correction factor will have to be adopted to obtain adequate analysis.

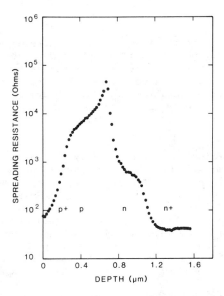

Figure 6. Spreading resistance data for $p^+/p/n/n^+$ impatt diode at probe load of 5 g showing shunting effect of n^+ substrate on n-epitaxy data.

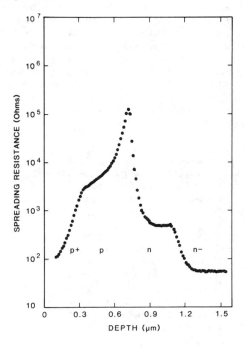

Figure 7. Spreading resistance data for the same impatt structure at a probe load of 5 g. showing decreased sensitivity to the n^+ substrate due to improved probe conditioning.

PRECISION

Satisfactory precision is a primary requirement for the acceptability of any measurement procedure. The only precision figures reported so far for spreading resistance measurements have been for bulk specimens such as those derived from an ASTM round robin on the effects of specimen surface preparation [34]. Recently, an ASTM round robin on spreading resistance depth profiling has been completed. The following is a description of the round-robin procedure and a summary of the analysis.

Fifteen laboratories participated in the round robin which consisted of three parts. The first part was the use of model spreading resistance data generated by Albers [6] to test for variations among the various algorithms being used. The second part was a qualification of probes using four test specimens: two bulk, one thin epitaxial, and one ion-implanted. Measurements on these specimens were used to screen the probes for noise, measured value, equivalence of response, and penetration. One additional specimen was to be beveled and probe impressions made; this was sent to the coordinating laboratory for physical inspection of probe damage tracks and beveled surface quality.

Three chips of the actual test specimen, an unpatterned bipolar transistor structure on (111) silicon, were sent to each laboratory. Each chip was to be beveled at a shallow angle and profile measurements taken of the emitter, base, and part of the collector. The chip was then to be remounted so that the other end could be beveled at a steeper angle allowing measurements of the entire collector, the buried layer, and part of the substrate. This measurement process was to be performed on each of three days using a separate chip each time; the three sets of data gave a measure of each lab's repeatability. Measurements were required to be at a load of no more than 20 g. All profile data were to be analyzed using the customary algorithm at the particular lab and probe calibration was to be established using the customary material and procedure at that laboratory; a description of the analysis algorithm and calibration procedure was required to be reported.

Six laboratories were dropped from final round-robin analysis for a variety of reasons ranging from poor quality of probe track or beveled surface to insufficiency of information returned. All remaining laboratories made measurements at loads of 10 g or less, used 0.1- or 0.25-μm diamond for specimen beveling, and had sufficiently comparable data on the probe qualification specimens that none could be rejected on those grounds. A number of laboratories that used the same or a similar correction factor algorithm reported unusual interpretations of the model spreading resistance data. Since their analyses of the model data gave the same results, they were put into one group, "Lab Group A," for analysis of the bipolar specimen. The remaining laboratories, which used several different algorithms, faithfully reproduced the resistivity profiles used for the model data; these laboratories were put in a second group, "Lab Group B."

It should be noted that among these nine remaining laboratories, there was still a rather wide range of values for the measurement conditions: step size, bevel angle, and probe separation. There were also a variety of methods of analyzing data: two laboratories used the "local slope" algorithm, the remainder used some algorithm based on the multilayer Laplace equation model (at least three different versions were identified). Probe calibration for analysis was accomplished using both variable contact radius and barrier resistance representations. For conversion of resistivity values to dopant density, most laboratories used ASTM Method F 723 [35], but two used other conversion relations and one converted to carrier density. Finally, since NBS calibration standards were not available at the beginning of the round robin, all but the final lab to report used in-house-generated calibration sets, the comparability of which is not known.

A summary of the results for the bipolar structure is given in Table 1 for Lab Group A and in Table 2 for Lab Group B. Additional results which are based on all nine laboratories are given in Table 3. Several comments need to be made regarding the compilation of these data: 1) localized parameters such as maximum density (N_{max}) and minimum resistivity (ρ_{min}) for each of the layers were obtained by first-curve fitting the data around these points - emitter, base, and buried layer data were fitted to a parabola, the central region of the collector was fitted to a straight line - this curve fitting improves the meaningfulness of the comparison by suppressing the effects of measurement noise. 2) The collector-buried layer "junction depth" was arbitrarily chosen as the mid-point of the plotted spreading resistance data between the average value for the collector-layer and the minimum value measured in the buried layer. 3) The standard deviation values listed in the last column of each of the three tables are based on pooled results. At best, they represent estimates of precision (repeatability) only for the case of junction depth values which are directly taken from raw data. For all other parameters, the variety of analysis and calibration procedures which were used to derive the data that were pooled mixes a significant element of the "accuracy" of the individual analysis procedures in the results. 4) The seemingly inconsistent cut-off values of precision seen in the third data column of Table 3 are based on histogram plots of precision values for each parameter. It was desired to represent the precision obtained by at least two-thirds of the labs; beyond this, the cutoff was determined by the point at which a significant increase in reported precision would be required to include one more lab.

It is seen from these results that most laboratories are able to obtain internal precisions that are quite acceptable. Even where the precision figure for the worst lab, Table 3, exceeded 20%, it was usually found to be due to one highly discrepant run - generally recognizable in the raw data. The precision across all labs for the several junction depth values is surprisingly good. The specific cause of the differences among labs for values of various parameters other than junction depth cannot be determined. In general, the disagreement within Lab Group B is no worse than that within Lab Group A, despite a wider variety of correction factor algorithms being used in Group B.

ACCURACY

Any discussion of the accuracy of spreading resistance profiling requires a point of reference with respect to which the accuracy can be measured. Despite the understanding that spreading resistance profiling responds to the resistivity profile as it exists on the

Table 1 - Results from Lab Group A

Parameter	Range of Lab Averages for 3 Runs	Best Single Lab Precision	Average for Group	1σ%
Emitter-Nmax (cm^{-3})	$6.76 \times 10^{19} - 1.14 \times 10^{20}$	2.8%	9.34×10^{19}	19%
Emitter-ρmin ($\Omega \cdot cm$)	0.000688-0.00108	1.9%	0.00834	21%
Emitter-Dose (cm^{-2})	$2.29 \times 10^{15} - 4.13 \times 10^{15}$	1.0%	3.02×10^{15}	25%
Emitter Rs (Ω)	18.2-31.8	1.8%	25.2	22%
Emitter/Base Xj (μm)	0.607-0.660	0.4%	0.632	4.0%
Base-Nmax (cm^{-3})	$1.99 \times 10^{17} - 3.46 \times 10^{17}$	2.0%	2.49×10^{17}	27%
Base-ρmin ($\Omega \cdot cm$)	0.086-0.125	1.6%	0.110	16%
Base-Dose (cm^{-2})	$3.37 \times 10^{12} - 8.79 \times 10^{12}$	0.4%	5.70×10^{12}	34%
Base-Rs (Ω)	3068-5847	2.5%	4357	23%
Base/Coll Xj (μm)	1.021-1.124	1.4%	1.078	3.5%
Base width (μm)	0.414-0.480	1.3%	0.446	6.8%
Coll. Navg (cm^{-3})	$3.78 \times 10^{15} - 7.29 \times 10^{15}$	5.6%	5.74×10^{15}	22%
Coll. ρavg ($\Omega \cdot cm$)	0.692-1.23	1.4%	0.899	23%
Coll.-BL "Xj" (μm)	2.94-3.14	1.4%	3.04	2.4%
B.L. Nmax (cm^{-3})	$1.39 \times 10^{19} - 2.07 \times 10^{19}$	1.7%	1.72×10^{19}	16%
B.L. ρmin ($\Omega \cdot cm$)	0.00322-0.00453	1.5%	0.00398	14%
B.L./subst Xj (μm)	8.08-8.53	1.2%	8.25	2.1%
Subst Navg (cm^{-3})	$7.26 \times 10^{14} - 1.32 \times 10^{15}$	1.2%	9.29×10^{14}	25%
Subst ρavg ($\Omega \cdot cm$)	10.2-18.6	1.0%	15.1	21%

Table 2 - Results from Lab Group B

Parameter	Range of Lab Averages for 3 Runs	Best Single Lab Precision	Average for Group	1σ%
Emitter-Nmax (cm^{-3})	$5.07 \times 10^{19} - 7.14 \times 10^{19}$	4.9%	6.13×10^{19}	16%
Emitter-ρmin ($\Omega \cdot cm$)	0.00108-0.00142	3.4%	0.00127	13%
Emitter-Dose (cm^{-2})	$2.04 \times 10^{15} - 2.99 \times 10^{15}$	2.5%	2.57×10^{15}	16%
Emitter Rs (Ω)	25.3-35.3	1.6%	29.6	15%
Emitter/Base Xj (μm)	0.594-0.639	0.8%	0.608	3.4%
Base-Nmax (cm^{-3})	$8.87 \times 10^{16} - 1.24 \times 10^{17}$	3.4%	1.04×10^{17}	17%
Base-ρmin ($\Omega \cdot cm$)	0.159-0.289	2.7%	0.214	28%
Base-Dose (cm^{-2})	$2.18 \times 10^{12} - 3.30 \times 10^{12}$	7.9%	2.65×10^{12}	18%
Base-Rs (Ω)	6011-9663	6.7%	7823	21%
Base/Coll Xj (μm)	0.992-1.083	1.4%	1.033	4.1%
Base width (μm)	0.397-0.489	4.5%	0.425	10.2%
Coll. Navg (cm^{-3})	$4.56 \times 10^{15} - 9.27 \times 10^{15}$	2.3%	6.20×10^{15}	34%
Coll. ρavg ($\Omega \cdot cm$)	0.629-1.037	2.2%	0.882	20%
Coll.-BL "Xj" (μm)	2.87-3.12	1.6%	3.03	3.8%
B.L. Nmax (cm^{-3})	$1.03 \times 10^{19} - 1.68 \times 10^{19}$	1.5%	1.42×10^{19}	20%
B.L. ρmin ($\Omega \cdot cm$)	0.00424-0.00592	1.0%	0.0048	16%
B.L./subst Xj (μm)	7.75-8.35	0.5%	8.14	3.3%
Subst Navg (cm^{-3})*	$9.26 \times 10^{14} - 9.34 \times 10^{14}$	2.1%	9.30×10^{14}	-
Subst ρavg ($\Omega \cdot cm$)*	14.2-14.5	2.1%	14.35	-

*Two labs did not provide data.

Table 3 Pooled Results - Nine Laboratories

Parameter	Best Single Lab Precision - 3 Runs	Worst Single Lab Precision - 3 Runs	No. Labs With Precision <X%	Nine-Lab Average	1σ%
Emitter-Nmax (cm^{-3})	2.8%	18%	6 < 10%	7.9×10^{19}	28%
Emitter-ρmin ($\Omega\cdot cm$)	1.9%	20%	6 < 10%	0.00105	27%
Emitter-Dose (cm^{-2})	1.0%	20%	6 < 5%	2.82×10^{15}	23%
Emitter Rs (Ω)	1.6%	21%	6 < 5%	27.2	20%
Emitter/Base Xj (μm)	0.4%	4.5%	9 < 4.5%	0.621	4.1%
Base-Nmax (cm^{-3})	2.0%	29%	7 < 8%	1.72×10^{17}	50%
Base-ρmin ($\Omega\cdot cm$)	1.6%	26%	6 < 6%	0.162	42%
Base-Dose (cm^{-2})	0.4%	22%	6 < 14%	4.35×10^{12}	49%
Base-Rs (Ω)	2.5%	19%	6 < 10%	5898	37%
Base/Coll. Xj (μm)	1.4%	5%	9 < 5%	1.058	4.2%
Base width (μm)	1.3%	9.7%	8 < 10%	0.437	8.2%
Coll. Navg (cm^{-3})	2.3%	16%	8 < 11%	5.91×10^{15}	27%
Coll. ρavg ($\Omega\cdot cm$)	2.2%	13%	8 < 11%	0.891	20%
Coll.-BL "Xj" (μm)	1.4%	5.6%	8 < 3.5%	3.04	2.9%
B.L. Nmax (cm^{-3})	1.5%	30%	7 < 10%	1.58×10^{19}	20%
B.L. ρmin ($\Omega\cdot cm$)	1.0%	25%	7 < 9%	0.00439	17%
B.L./subst Xj (μm)	0.5%	3.6%	9 < 4%	8.21	2.6%
Subst Navg (cm^{-3})*	1.2%	6.2%	7 < 7%	9.30×10^{14}	21%
Subst ρavg ($\Omega\cdot cm$)*	1.0%	6.6%	7 < 7%	14.5	18%

* Two labs did not provide data.

specimen bevel, it is common practice to convert the resistivity
profile to a dopant density profile; therefore the actual dopant
density profile will be taken here as the point of reference.
Nevertheless, assessing the accuracy of spreading resistance profiles
is not straightforward for a variety of reasons, and with available
information only a partial assessment can be made.

Before proceeding, three general points regarding profile accuracy
should be made. 1) Different levels of accuracy are expected to apply
to different structures or even to different portions of the same
structure - this follows from a number of the works on the mathematical
accuracy of various correction factor models and from Hu's observation
that the extent of carrier spilling effects depends both on profile
gradient and on differences in dopant density between a layer and its
substrate. 2) Not all errors are equally important - errors which
misrepresent the shape, the maximum layer density or some other feature
of the profile that is important for process modeling or predicting
device performance are generally serious, yet an error which has a
large numerical value, e.g., 25 or 50%, may have negligible effect on
profile interpretation if it occurs on a rapidly changing portion of
the profile. 3) Comparison of spreading resistance profiles with those
of other techniques to test for accuracy must be done with care since
each responds to a different property of the graded layer. The
measurement errors of these other techniques, as well as effects due to
sample preparation, e.g., incomplete annealing of an ion implant, must
be adequately taken into account.

Assuming negligible error of bevel angle measurement, negligible
probe penetration, and adequate preparation and characterization of
calibration specimens, errors in doping density profiles obtained from
spreading resistance measurements can be associated with one of three
sources: 1) numerical accuracy of present tractable profile analysis
models, 2) conversion of the resistivity profile to a dopant density
profile, 3) limited appropriateness of present analysis models for the
description of the real semiconductor problem. The first two relate to
the "relative" accuracy attainable if the semiconductor effects related
to space-charge regions and carrier spilling are ignored, while the
third relates to the additional error due to ignoring the violation of
local space charge neutrality and using the wrong (Laplace) equation in
the model.

Within the framework of correction factor models based on the
Laplace equation, guidelines have been developed for the form of the
correction factor integral and the ensuing numerical analysis to enable
obtaining resistivity profiles with errors limited to a few percent
even for sparse spreading resistance data. The question of probe
radius value to be used in the correction factor can be reduced to a
self-consistent determination in a number of applications; this allows
the effect of nonideal probe physics to be circumvented. Unfortunately, the full embellishment of all the correction factor recommendations
apparently has not been tried anywhere, and if tried, may be numerical-
ly unstable in the presence of measurement noise. Nevertheless, it
appears possible in principle to obtain highly accurate resistivity
profiles (as they exist on the beveled specimen) particularly for the
high concentration portion of a graded layer.

In the case of the data obtained from the previously mentioned round robin, the above comments would lead to expectations of better interlaboratory agreement than was found. However, none of the labs in the round robin used anything but a simple, time-efficient correction factor, and all obtained contact radius estimates from bulk specimen calibration. Only three parameters of the round-robin bipolar structure are known at this time from independent measurements: substrate resistivity - 16 $\Omega\cdot$cm, collector carrier density - 5.9×10^{15} cm^{-3}, emitter sheet resistance - 26 Ω. In fact, one lab reported average values which were within 5% of each of those values and with repeatability of about 5%, even using such simplified procedures. The cause of the variations reported by the other labs is not understood at present, and is being investigated. However, the overall average value from the Group A labs was written 6% of the known value for each of the parameters while for the Group B labs the overall average values differed from the known values by amounts ranging from 5% to 14%.

The nearly universal custom of using a transfer relation such as ASTM F 723 [35] to interpret the resistivity profiles as dopant density profiles will cause errors in at least two cases apart from the effects due to carrier spilling. The first relates to the fact that such a transfer relation is based on empirical data for specific dopants, boron and phosphorus in the case of F-723. Errors will result if these transfer relations are applied to layers doped with species with noticeably different activation energies, or to layers very heavily doped with species having different solid solubility limit properties than do boron or phosphorus. Errors will also result when the dopant in a layer does not have full electrical activation or does not have the mobility implicit in the transfer relation; the effect is that the doping density obtained from the resistivity profile will understate the true value. It is not yet possible to determine unambiguously the extent to which these latter effects are present in the profile of a given sample, but it appears that incomplete activation may be compounding the error due to carrier spilling observed in the tails of implanted boron profiles.

Additional errors caused by carrier spilling in profiles derived from spreading resistance are expected to show up predominantly as layer thickness errors, regardless of maximum layer doping, and to be particularly noticeable for gaussian and erfc profiles. As a result, for diffused layers and for most implants, spreading resistance profiles are expected to understate the depth of the profile tail. However, for moderately or lightly doped layers, carrier spilling may cause errors extending back into the peak density region of the structure. Extreme errors are predicted for thin, lightly doped epitaxy over a heavily doped buried layer.

There is only a limited amount of data in the recent literature which can be used to judge the effectiveness of carrier spilling calculations in explaining the profile tails determined from spreading resistance measurements. The work of Ehrstein et al. [36] on a series of reasonably high fluence, furnace-annealed boron implants at two energies shows very good agreement between spreading resistance and SIMS profiles in the near surface and peak density regions, but a noticeable divergence of the two types of profiles in the tail region, starting at a density of about 10^{18} cm^{-3}. While the integrated differences between the tails of the two types of profiles was found to amount to less than 0.5% of the total implanted fluences, the difference in depth scales was as large as 0.3 μm at 10^{15} cm^{-3}. These differences are about a factor of two larger than are predicted by Hu for Gaussian profiles and surface concentrations appropriate to the measured implants. It is possible that incomplete activation of boron as observed by Hofker [37] and by Fink et al. [38] causes the additional difference observed in the implant tails.

A larger range of differences between spreading resistance and SIMS profiles was reported by Godfrey et al. [39] for a series of boron implants at 25 keV with fluences from 2×10^{12} cm^{-3} to 2×10^{15} cm^{-3}. The differences they report for the three higher fluences are also noticeably larger than appears to be predicted by Hu. For the lightest fluence implant, the differences they report are truly dramatic, extending to the peak of the implant and amounting to 0.6 μm at 10^{15} cm^{-3}. This latter result is in strong contrast to the earlier result of Ehrstein et al. [29] for a boron fluence of 3×10^{12} cm^{-2} and implant energies of 150 keV and 300 keV; here the difference in tail depths at a density of 10^{15} cm^{-3} was approximately 0.25 μm. The peak densities of both these low-fluence implants are below the values for which Hu provides results. However, it is noted that the larger profile difference reported by Godfrey occurs for an anneal of 16 h at 950°C compared to an anneal of 30 min at 800°C used by Ehrstein. This results in a noticeably lower peak density and more slowly graded tail (by SIMS) in Godfrey's specimen; both of these conditions are expected to enhance any effects due to charge spilling.

Rather mixed results were reported by Cohen and coworkers [40] for a large matrix of boron implants which were annealed by thermal, electron beam, or combined thermal-electron beam processes. Their SIMS and spreading resistance profiles were summarized as junction depth determinations. Results showed junction depths by spreading resistance that are smaller than, comparable to, or larger than those determined by SIMS with no apparent pattern of dependence on the implant or anneal parameters.

In a study of heavy dose arsenic implants with junction depths from 0.4 to 1.8 μm, Wagner and coworkers [41] found junction depths from SIMS to be larger than those from spreading resistance by a reasonably constant amount, about 0.09 μm. This better agreement than was found for boron, above, is expected from Hu's work to be due in part to the somewhat steeper gradient in the arsenic profile tail. However, the relatively constant fixed difference in junction depths is inconsistent with the carrier spilling calculations which predict relatively constant percentage differences. Part of this discrepancy may be caused by the relatively limited sensitivity of SIMS to arsenic which necessitated defining the junction depth at a density of 1×10^{17} cm^{-3}.

In the case of an electron-beam annealed As double implant, Pawlik [42] found the spreading resistance profile to track that from SIMS to within about 0.025 μm down to the limits of SIMS sensitivity.

Both of the above works on arsenic implants, which show a closer agreement between spreading resistance and SIMS determination of profile tails than is generally found for boron, are conceptually consistent with carrier spilling calculations which predict that the agreement will improve as the tails become more steeply graded. However, they are also consistent with the possibility that arsenic is more fully activated than boron during annealing.

Available data in the literature appear to be inadequate to test the predicted effects of carrier spilling for the case of an epitaxial layer over a heavily doped substrate. Reported comparisons are found to be limited to epitaxial structures with doping or layer thickness values which place them outside the range for which Hu predicts any meaningful effect due to carrier spilling. Such comparisons as are available show very favorable agreement with respect to epitaxial carrier concentration (C-V measurements) and epitaxial layer thickness (SIMS measurements), however.

SUMMARY

Relative ease, applicability to a wide range of sample types and turn-around time continue to make spreading resistance a desirable profiling technique. It has been shown that it appears possible to obtain the resolution needed for advanced device structures. A multi-laboratory test showed that it is possible for most labs to obtain acceptably good precision for a wide range of profile parameters. The most significant remaining problem is the question of accuracy. Numerous improvements in numerical analysis procedures have recently been suggested which should result in improved computational accuracy, particularly for the peak regions of junction-isolated structures. Most troubling, however, is the uncertainty of the extent to which carrier spilling modifies spreading resistance profiles, particularly for fractional micron layers. The challenge is to better understand and quantify this effect so the analyst can judge whether an intelligent interpretation of the structure of the layer can be obtained from spreading resistance measurements or whether an alternate technique is more suited to the task.

ACKNOWLEDGMENT

The author would like to acknowledge many interesting discussions with Dr. J. Albers on a variety of aspects of modeling spreading resistance measurements and computing correction factors.

REFERENCES

[1] Dickey, D. H., "Diffusion Profile Studies Using a Spreading Resistance Probe," Abstract No. 57, <u>Extended Abstracts of the Electronics Division</u>, Electrochemical Society, Vol. 12, No. 1, 1963.
[2] Schumann, P. A., and Gardner, E. E., "Application of Multilayer Potential Distribution to Spreading Resistance Correction Factors," <u>Journal of the Electrochemical Society</u>, 116, 1969, pp. 87-91.
[3] "Semiconductor Measurement Technology: Spreading Resistance Symposium," J. R. Ehrstein, ed., NBS Special Publication 400-10, National Bureau of Standards, Washington, DC, 1974.
[4] Nagasawa, E., and Matsumura, M., "An Image Method Application to

Multilayer Spreading Resistance Analysis," *Solid-State Electronics*, 20, 1977, pp. 507-513.

[5] Dickey, D. H., and Ehrstein, J. R., "Semiconductor Measurement Technology: Spreading Resistance Analysis for Silicon Layers with Nonuniform Resistivity," NBS Special Publication 400-48, National Bureau of Standards, Gaithersburg, MD, 1979.

[6] Albers, J., "Comparison of Spreading Resistance Correction Factor Algorithms Using Model Data," *Solid-State Electronics*, 23, 1980, pp. 1197-1205.

[7] Kudoh, O., Uda, K., and Ikushima, Y., "Impurity Profiles Within a Shallow p-n Junction by a New Differential Spreading Resistance Method," *Journal of the Electrochemical Society*, 123, 1976, pp. 1751-1754.

[8] Yeh, T. H., and Khokhani, K. H., "Multilayer Theory of Correction Factors for Spreading Resistance Measurements," *Journal of the Electrochemical Society*, 116, 1969, pp. 1461-1464.

[9] Berkowitz, H. L., and Lux, R. A., "An Efficient Integration Technique for Use in the Multilayer Analysis of Spreading Resistance Profiles," *Journal of the Electrochemical Society*, 128, 1981, pp. 1137-1141.

[10] Albers, J. H., "Some Aspects of Spreading Resistance Profile Analysis," *Emerging Semiconductor Technology*, ASTM STP 960, Gupta, D. C. and Langer, P. H., Eds., American Society for Testing and Materials, 1986.

[11] Severin, P. J., "Formal Comparison of Correction Formulae for Spreading Resistance Measurements on Layered Structures," Semiconductor Measurement Technology: Spreading Resistance Symposium, NBS Special Publication 400-10, National Bureau of Standards, Gaithersburg, MD, 1974, pp. 27-44.

[12] Berkowitz, H. L., and Lux, R. A., "Errors in Resistivities Calculated by Multilayer Analysis of Spreading Resistance," *Journal of the Electrochemical Society*, 126, 1979, pp. 1479-1482.

[13] Choo, S. C., Leong, M. S., and Sim, J. H., "An Efficient Numerical Scheme for Spreading Resistance Calculations Based on the Variational Method," *Solid-State Electronics*, 26, 1983, pp. 723-730.

[14] Lee, G. A., "Multilayer Analysis of Spreading Resistance Measurements," NBS Special Publication 400-10, op. cit., pp. 75-94.

[15] Leong, M. S., Choo, S. C., and Tan, L. S., "The Role of Source Boundary Conditions in Spreading Resistance Calculations," *Solid-State Electronics*, 21, 1978, pp. 933-941.

[16] Vandervorst, W. B., and Maes, H. E., "Spreading Resistance Correction Formula More Suited for the Gauss-Laguerre Quadrature," *Solid-State Electronics*, 24, 1981, pp. 851-856.

[17] Choo, S. C., and Leong, M. S., "A Multilayer Exponential Model for Spreading Resistance Calculations," *Solid-State Electronics*, 22, 1979, pp. 405-415.

[18] Albers, J., "Continuum Formulation of Spreading Resistance Correction Factors," *Journal of the Electrochemical Society*, 127, 1980, pp. 2259-2263.

[19] Piessens, R., Vandervorst, W. B., and Maes, H. E., "Incorporation of a Resistivity-Dependent Contact Radius in an Accurate Integration Algorithm for Spreading Resistance Calculations," *Journal of the Electrochemical Society*, 130, 1983, pp. 468-474.

[20] Hu, S. M., "Between Carrier Distributions and Dopant Atomic Distributions in Beveled Silicon Substrates," *Journal of Applied Physics*, 53, 1982, pp. 1499-1510.

[21] Kramer, P., and van Ruyven, L. J., "Space Charge Influence on

Resistivity Measurements," *Solid-State Electronics*, 20, 1977, pp. 1011-1019.

[22] Huang, R. S., and Ladbrooke, P. H., "The Use of a Four-Point Probe for Profiling Sub-Micron Layers," *Solid-State Electronics*, 21, 1978, pp. 1123-1128.

[23] Mazur, R. G., and Gruber, G. A., "Dopant Profiles on Thin Layer Silicon Substrates with the Spreading Resistance Technique," *Solid State Technology*, 24, No. 11, 1981, pp. 64-70.

[24] Ehrstein, J. R., "Effect of Specimen Preparation on the Calibration and Interpretation of Spreading Resistance Measurements," Semiconductor Silicon 1977, Proceedings Volume 77-2, The Electrochemical Society, Princeton, NJ, 1977, pp. 377-389.

[25] Vandervorst, W. B., and Maes, H. E., "Probe Penetration in Spreading Resistance Measurements," *Journal of Applied Physics*, 56, 1984, pp. 1583-1589.

[26] Mazur, R. G., and Dickey, D. H., "A Spreading Resistance Technique for Resistivity Measurements on Silicon," *Journal of the Electrochemical Society*, 113, 1966, pp. 255-259.

[27] Gruber, G. and Pfeiffer, R., "The Evaluation of Thin Silicon Films by Spreading Resistance Measurements," NBS Special Publication 400-10, op. cit., pp. 209-216.

[28] Kramer, P. and van Ruyven, L. J., "The Influence of Temperature on Spreading Resistance Measurement," *Solid State Electronics*, 15, 1972, pp. 757-766.

[29] Ehrstein, J. R., Albers, J. H., Wilson, R. G., and Comas, J., "Comparison of Spreading Resistance with C-V and SIMS Profiles for Submicron Layers in Silicon," *Extended Abstracts*, The Electrochemical Society, Vol. 80-1, Princeton, NJ, 1980, pp. 496-498.

[30] Fonash, S. J., "The Physics of Spreading Resistance Measurements," NBS Special Publication 400-10, op. cit., pp. 17-26.

[31] Pawlik, M., "Dopant Profiling in Silicon," *Semiconductor Processing, ASTM STP 850*, Dinesh C. Gupta, Ed., American Society for Testing and Materials, 1984, pp. 390-408.

[32] Albers, J. H., and Berkowitz, H. L., "The Relation Between Two-Probe and Four-Probe Resistances on Nonuniform Structures," *Journal of the Electrochemical Society*, 131, 1984, pp. 392-398.

[33] Pawlik, M., "On the Determination of Abrupt Doping Profiles in MBE Silicon by Spreading Resistance," *Proceedings of the First International Symposium on Silicon Molecular Beam Epitaxy*, The Electrochemical Society, Princeton, NJ, 1985.

[34] "Method for Measuring Resistivity of Silicon Wafers Using a Spreading Resistance Probe," ASTM Designation F 525-84, *Annual Book of ASTM Standards*, The American Society for Testing and Materials, Philadelphia, PA, 1985, Part 10.05.

[35] "Standard Practice for Conversion Between Resistivity and Dopant Density for Boron-Doped and Phosphorus-Doped Silicon," Designation F 723-82, *Annual Book of ASTM Standards*, American Society for Testing and Materials, Philadelphia, PA, 1985, Part 10.05.

[36] Ehrstein, J., Downing, R., Stallard, B., Simons, D., and Fleming, R., "Comparison of Depth Profiling ^{10}B in Silicon Using Spreading Resistance Profiling, Secondary Ion Mass Spectrometry, and Neutron Depth Profiling," *Semiconductor Processing*, ASTM STP 850, Dinesh C. Gupta, Ed., American Society for Testing and Materials, Philadelphia, PA, 1984.

[37] Hofker, W. K., "Implantation of Boron in Silicon," *Philips Research Reports*, Suppl. No. 8, 1975.

[38] Fink, D., Biersack, J., Carstanjen, H., Jahnel, F., Muller, K.,

Ryssel, H., and Osei, A., "Studies of the Lattice Position of Boron in Silicon," *Radiation Effects*, 77, 1983, pp. 11-33.

[39] Godfrey, D., Groves, R., Dowsett, M., and Willoughby, A., "A Comparison between SIMS and Spreading Resistance Profiles for Ion Implanted Arsenic and Boron After Heat Treatments in an Inert Ambient," *Physica*, 129B, 1985, pp. 181-186.

[40] Cohen, S., Norton, J., Koch, E., and Weisel, G., "Shallow Boron-Doped Junctions in Silicon," *Journal of Applied Physics*, 57, 1985, pp. 1200-1214.

[41] Wagner, H., Schaefer, R., and Kempf, J., "Comparison of Characterization Methods for As-Doped Silicon," *Journal of Applied Physics*, 52, 1981, pp. 6173-6177.

[42] Pawlik, M., "Dopant Profiling in Silicon," *Semiconductor Processing, ASTM STP 850*, Dinesh C. Gupta, Ed., American Society for Testing and Materials, Philadelphia, PA, 1984.

John Albers

SOME ASPECTS OF SPREADING RESISTANCE PROFILE ANALYSIS

REFERENCE: Albers, J., "Some Aspects of Spreading Resistance Profile Analysis," Emerging Semiconductor Technology, ASTM STP 960, Gupta, D. C. and Langer, P. H., Eds., American Society for Testing and Materials, 1986.

ABSTRACT: The calculation of resistivity profiles (and carrier density profiles) from spreading resistance requires the use of a correction factor. The present status of the calculation of the correction factor based upon the Schumann and Gardner multilayer solution of Laplace's equation is reviewed and discussed. Recent calculations of carrier densities from atomic densities are also discussed. In particular, the numerical solutions of the semiconductor equations are reviewed and their implications in the interpretation of spreading resistance measurements for profiling shallow layers are presented. The limitations of the multilayer Laplace equation analysis of spreading resistance in VLSI profiling are also discussed.

KEYWORDS: atomic distributions, carrier distributions, conductivity, correction factor, finite element method, Laplace's equation, Poisson's equation, resistivity, semiconductor equations, spreading resistance analysis.

INTRODUCTION

The present trend in semiconductor technology is towards submicron devices. The push made by VLSI technology into the submicron regime places increased demands upon the ability to adequately measure and interpret dopant profiles. The last two decades have witnessed the growth and development of a number of atomic and electrical profiling techniques. The Secondary Ion Mass Spectrometry

Dr. John Albers is a physicist in the Semiconductor Electronics Division at the National Bureau of Standards, Gaithersburg, Maryland 20899. Contribution of the National Bureau of Standards. Not subject to copyright.

technique has become an important tool in the atomic profiling of semiconductor structures. On the other hand, two-probe spreading resistance profiling and, to a lesser extent, capacitance-voltage profiling are probably the most widely used electrical profiling techniques. The use of the spreading resistance technique for micron and submicron structures necessitates an appreciation for the models which have been developed over the past two decades for the interpretation of the data. This is especially important as the technique does not yield the profile directly. Rather, the resistivity is indirectly obtained from the calculation of the spreading resistance correction factor. This requires a physical model of the measurement in order to connect the data and the underlying resistivity structure. Also, the carrier density which is obtained from the resistivity does not have to be equal to the atomic density. Consequently, it is important to understand the basic model of spreading resistance insofar as its assumptions and possible limitations are concerned. This is especially the case with the number of algorithms which have appeared on the scene over the past ten years. At first, they may appear to be quite different. However, to a large extent, they are all related to the same basic model. Their differences are mostly due to ways of handling probe current densities, calibration data, and the evaluation of integrals which appear in the relation between the spreading resistance and the resistivity.

This paper is divided into two parts. The first contains a review, or rather a history, of the calculations of the spreading resistance correction factor. The history is not an exhaustive bibliography but rather is intended to provide the reader with a genealogy indicating the changes and improvements which have taken place over the years since the original calculations of Schumann and Gardner. In addition, the incorporation of several forms of the probe current density, the possible determination of the probe radius and its relation to calibration data, and the effects of the assumption of finite thickness, uniform resistivity 'layers' will also be considered. The second part of the paper contains a discussion of the results of some recent calculations of carrier densities arising from atomic densities with particular emphasis on beveled geometries and micron and submicron structure sizes. This relates directly to the validity of the use of a multilayer Laplace equation in the interpretation of spreading resistance data. In particular, it will be shown that there is a fundamental problem with the basic assumption used in all spreading resistance algorithms which rely upon the multilayer Laplace equation. The failure of this assumption gets progressively worse as the structure size moves from the micron to the submicron region. It is hoped that this will help to clearly define the problems which must be addressed so that the spreading resistance technique might have some hope of being a truly quantitative submicron electrical profiling technique.

REVIEW OF CORRECTION FACTOR CALCULATIONS

The primary purpose of the theory of spreading resistance is to provide for a simple and yet comprehensive model which relates the spreading resistance to the details of the structure being investigated. As will be seen in this section, the present theory provides for a well-defined procedure for obtaining the spreading resistance from the resistivity. The reverse or inverse process of obtaining the resistivity from the spreading resistance requires the solution of what is known mathematically as the inverse problem. Simply stated, it is necessary to determine part of an integrand of an integral from the numerical value of the integral. In general, there is no mathematical method for uniquely determining the solution to this problem. The solution to this problem in the case of spreading resistance analysis has tested the ingenuity of numerical analysis.

Over the past two decades, there have been a number of advances which have taken place in the theoretical analysis of spreading resistance. The starting point of this history is the multilayer Laplace equation analysis of Schumann and Gardner [1,2].* The multilayer solution of the Laplace equation provides for the calculation of the two-probe spreading resistance on the top surface of a nonuniform resistivity structure. The structure is viewed as being planar and made up of a number of 'layers,' each of uniform resistivity. For the case of measurements made on a beveled sample, the results of the planar calculations are used under the assumption that the bevel does not alter the results. As the probes move down the bevel, the definition of the surface changes to that which is presently in contact with the probes. The material which is uphill, so to speak, is neglected. The fundamental assumption of the multilayer analysis is that Laplace's equation is satisfied in each of the 'layers' in the material. There are a number of assumptions which are clearly spelled out and discussed in [1]. It is important to emphasize that Schumann and Gardner were very careful to point out what they thought to be limitations of the model.

First, consider the case of a single uniform layer. The solution to this problem provides a solution which is assumed to describe the potential in each of the 'layers'. The problem is set up in cylindrical coordinates to emulate the symmetry of a single circular contact on the top surface of the material. The potential is assumed to be independent of the angular variable in this coordinate system. The Laplace equation may then be written as

$$\nabla^2 V(r,z) = \frac{\partial^2}{\partial r^2} V(r,z) + \frac{1}{r}\frac{\partial}{\partial r} V(r,z) + \frac{\partial^2}{\partial z^2} V(r,z) = 0, \qquad (1)$$

where $V(r,z)$ is the potential, r is the radial coordinate, and z is the depth coordinate. This equation may be solved by means of separation of variables with the result that a particular solution is

$$V(r,z) = \exp(-\lambda z) J_0(\lambda r) + \exp(+\lambda z) J_0(\lambda r), \qquad (2)$$

where $J_0(\lambda r)$ is the Bessel function and λ^2 is the separation of variables constant. The boundary condition on the r part of the solution is that $V(r,z)$ approaches

---

* The reader is referred to the book by Koefoed [3] for a comprehensive discussion of Laplace's equation, its solution in cylindrical coordinates, and the multilayer approach applied to geoelectric resistance measurements. The derivation of the recursion relation (to be discussed later in this paper) is also presented in considerable detail in this reference. It should be noted that the mathematical description of spreading resistance and geoelectric resistance are exactly the same even though the length scales are vastly different.

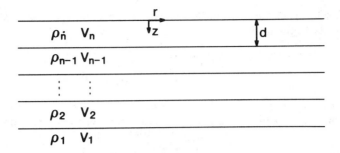

Figure 1. Schematic representation of the multilayer geometry used in the Schumann and Gardner multilayer analysis.

zero as r tends to infinity. This is satisfied by the above for all values of λ. Then, the general solution is an integral of the particular solution with a weighting factor and is of the form

$$V(r,z) = \int_0^\infty \left\{ (1 + \theta(\lambda)) \exp(-\lambda z) + \psi(\lambda) \exp(+\lambda z) \right\} J_0(\lambda r) d\lambda, \qquad (3)$$

where the weighting functions, $\theta(\lambda)$ and $\psi(\lambda)$, are determined from the z dependent boundary conditions. The above is the general solution of Laplace's equation in cylindrical coordinates for a single layer.

For the case of an n-layer structure (pictured in Figure 1), Laplace's equation is assumed to be valid in each of the layers. The single-layer solution as given by Eq 3 then provides the basis for the solution in each of the layers. The solution in the i-th layer may then be written as

$$V_i(r,z) = \int_0^\infty \left\{ (1 + \theta_i(\lambda)) \exp(-\lambda z) + \psi_i(\lambda) \exp(+\lambda z) \right\} J_0(\lambda r) d\lambda. \qquad (4)$$

The boundary conditions used to solve the system of equations (determine $\{\theta_i(\lambda)\}$ and $\{\psi_i(\lambda)\}$, $i = 1, ..., n$) are provided by conditions on the top surface, the intermediate interfaces, and the bottom surface. On the top surface, current flow takes place only through the probe which is modeled as a circular plate of radius a. Then the top surface boundary condition is expressed as

$$-\frac{1}{\rho_n} \frac{\partial V_n(r,z)}{\partial z} = J(r), \qquad (5)$$

where $J(r)$ is the current density. The original Schumann and Gardner calculation makes use of the current density for the case of a semi-infinite slab in the form

$$J(r) = \begin{cases} I/2\pi a(a^2 - r^2)^{1/2} & \text{if } r \leq a \text{ and } z = 0 \\ 0 & \text{if } r > a \text{ and } z = 0 \end{cases}. \qquad (6)$$

Several other forms of the current density can be used. In particular, the uniform current density in the form

$$J(r) = \begin{cases} I/\pi a^2 & \text{if } r \leq a \text{ and } z = 0 \\ 0 & \text{if } r > a \text{ and } z = 0 \end{cases}, \tag{7}$$

would describe the current flow for the case of a layer over a highly conductive layer. On the other hand, the ring delta function form of the current density,

$$J(r) = (I/2\pi r)\delta(r - a), \tag{8}$$

where $\delta(r - a)$ is the Dirac delta function, would describe the current flow for a conducting layer over a highly resistive layer. Berkowitz and Lux [4] have discussed how the above forms of the probe current density enter into the final correction factor equations. It should be noted that the above current densities have different patterns but all have the same total current, i.e.,

$$\int_0^{2\pi} \int_0^{\infty} J(r) d\theta r dr = I. \tag{9}$$

It might be argued that the 'real' current density would be some linear combination of those given by Eqs 6, 7, and 8. This is the basis of the variational calculation which will be discussed later in this section. For the present time, the probe current density given by Eq 6 will be used and the changes introduced by the use of other current densities will be presented.

Returning to the boundary conditions, on the bottom surface, the potential is assumed to be well behaved and, more specifically, is assumed to approach zero, i.e.,

$$\lim_{z \to \infty} V_1(r, z) = 0. \tag{10}$$

At the interfaces between the layers, the potentials and the current densities are assumed to be continuous. These are expressed as

$$V_i(r, z) = V_{i-1}(r, z), \tag{11}$$

and

$$\frac{1}{\rho_i} \frac{\partial V_i(r, z)}{\partial z} = \frac{1}{\rho_{i-1}} \frac{\partial V_{i-1}(r, z)}{\partial z}, \tag{12}$$

where the functions and their derivatives are to be evaluated at the interfacial boundaries.

For the case of an n-layer structure, the substitution of Eq 4 into the boundary conditions given by Eqs 5, 10, 11, and 12 gives rise to a set of $2n$ equations in $2n$ unknowns ($\{\theta_i(\lambda)\}, \{\psi_i(\lambda)\}, i = 1, ..., n$). This simplifies by one equation as $\theta_n(\lambda) = \psi_n(\lambda)$. The analytic solution of this system of equations requires the use of the Cramer's rule of linear algebra. Clearly, this can become rather tedious especially since the expansion coefficients are functions of the continuous variable, λ. Schumann and Gardner worked out the system of equations for the cases up to

$i = 3$. However, it is possible to show that the potential on the top surface of the n-layer structure may be written as

$$V_n(r,0) = \frac{I\rho_n}{2\pi a} \int_0^\infty \frac{A_n(\lambda)\sin(\lambda a)J_0(\lambda r)}{\lambda} d\lambda, \qquad (13)$$

where the kernel function, $A_n(\lambda) = 1 + 2\theta_n(\lambda)$, depends upon the resistivities and thicknesses of the 'layers' in the multilayer structure (through the solution of the above system of simultaneous equations). It is this potential which is of prime importance as the measurement is made on the top surface of the material. Equation 13 represents the potential at a distance r from a single probe. The spreading resistance for a two-probe configuration is defined as the voltage difference between the probes divided by the current, $\Delta V/I$. In order to obtain an expression for the spreading resistance, two things must be done. First, the potential must be averaged over the area of the probe since the probe is not sensitive to the details of the potential but rather the average potential. Second, the voltage difference between the two probes is calculated assuming that the second probe does not perturb the potential due to the first probe. This is known as superposition and leads to the result that the two-probe spreading resistance on the top surface of an n-layer structure may be written as

$$R_n = \frac{\Delta V}{I} = \frac{\rho_n}{2a} C_n, \qquad (14)$$

where C_n is the correction factor given by

$$C_n = \frac{4}{\pi} \int_0^\infty A_n(\lambda) \left\{ \frac{J_1(\lambda a)}{\lambda a} - \frac{J_0(\lambda S)}{2} \right\} \sin(\lambda a) \frac{d\lambda}{\lambda}, \qquad (15)$$

where S is the separation between the probes. The Bessel function term, $J_1(\lambda a)$, arises from an averaging of the potential over the area of the probe.

There are three distinct terms in the integrand of the correction factor integral with each representing different parts of the problem. First, there is the kernel function, $A_n(\lambda)$, which depends upon the resistivities and thicknesses of the layers in the multilayer structure. The second part of the integral, $(J_1(\lambda a)/\lambda a - J_0(\lambda S)/2)$ relates to the two-probe configuration with the probes separated by a distance S. Finally, the term $\sin(\lambda a)/\lambda$ arises from the probe current density (from Eq 6) passing through the probes. In terms of the uniform current density (Eq 7) and the ring delta function current density (Eq 8), the corresponding correction factors would be given by

$$C_n^{un} = \frac{4}{\pi} \int_0^\infty A_n(\lambda) \left\{ \frac{J_1(\lambda a)}{\lambda a} - \frac{J_0(\lambda S)}{2} \right\} 2J_1(\lambda a) \frac{d\lambda}{\lambda}, \qquad (16)$$

and

$$C_n^{rd} = \frac{4}{\pi} \int_0^\infty A_n(\lambda) \left\{ \frac{J_1(\lambda a)}{\lambda a} - \frac{J_0(\lambda S)}{2} \right\} a J_0(\lambda a) d\lambda, \qquad (17)$$

respectively. These forms of the correction factor integral have been discussed by Berkowitz and Lux [4].

Within the confines of the assumptions presented by Schumann and Gardner, Eqs 14 and 15 represent a relation between the spreading resistance and the resistivity. This relation is not a simple one as the determination of the resistivity from the spreading resistance, $(\{R_i\} \to \{\rho_i\}, i = 1, ..., n)$, requires the inversion of the correction factor integral. * There were, in addition, two difficulties associated with the implementation of these equations. First, the determination of the kernel function for a large number of layers required the use of a mainframe computer [5] to carry out the numerical matrix algebra (determination of the set of functions $\{\theta_i(\lambda)\}, \{\psi_i(\lambda)\}, i = 1, ..., n$ by numerical evaluation of Cramer's rule). Power-law interpolation [6] techniques were of limited utility due to the large requirement of CPU (still on a mainframe). Also, the partitioning of the $(2n \times 2n)$ matrix problem into a smaller set of (2×2) matrices had been presented but has not been carried beyond this point [7]. The second problem in the implementation of the correction factor equation has to do with the evaluation of the integral. The last two parts of the integral described above are oscillatory and a direct evaluation of the integral requires the use of at least a minicomputer.

The first breakthrough in the evaluation of Eq 15 effectively removed the numerical difficulty associated with matrix inversion. This was hinted at in the geophysical literature [8] and worked out in detail using the theory of determinants in Koefoed's book [3]. The application of these techniques as applied to the spreading resistance problem culminated with the introduction of a recursion relation by Choo et al. [9]. The utility of a recursion relation is that the kernel for an n-layer structure can be easily generated from the kernel of an $(n - 1)$ layer structure by means of an algebraic relation. In particular, if the kernel is known for the $(n - 1)$ layer case, then the kernel for the n layer case is given by

$$A_n(\lambda) = \frac{\omega(\lambda)\rho_n + \rho_{n-1}A_{n-1}(\lambda)}{\rho_n + \omega(\lambda)\rho_{n-1}A_{n-1}(\lambda)}, \tag{18}$$

where

$$\omega(\lambda) = \frac{1 - \exp(-2\lambda d)}{1 + \exp(-2\lambda d)}, \tag{19}$$

and d is the layer thickness. In practice, the recursion relation is begun with the one-layer case from which the two-layer kernel is generated. This sequence is repeated until the n-layer kernel is determined. This method was used in reference 9 with a great reduction in computation time.

At this particular time, the one major problem which remained was the evaluation of the integral in the correction factor equation. D'Avanzo et al. [10] made use of a pre-stored partial integral technique which usually required the use of a

---

* For the sake of simplicity and brevity, the notation $(\{\rho_i\} \to \{R_i\}, i = 1, ..., n)$ will be used to indicate the calculation of the n spreading resistance values from the n values of the resistivity, while the notation $(\{R_i\} \to \{\rho_i\}, i = 1, ..., n)$ will be used to indicate the inverse process, i.e., the determination of the n values of the resistivity from the n spreading resistance values. The former requires single evaluations of the correction factor integral, whereas the latter require inversions of the correction factor integral.

minicomputer in its implementation. The focal point of this technique was to use a cubic spline approximation for the kernel of the correction factor integral, i.e.,

$$A(\lambda) = \alpha_i \lambda^3 + \beta_i \lambda^2 + \gamma_i \lambda + A(\lambda_i). \tag{20}$$

Then the correction factor integral with the Schumann and Gardner infinite slab current density is given by

$$C = \frac{4}{\pi} \sum_{i=1}^{m-1} \left\{ \alpha_i \int_{\lambda_i}^{\lambda_{i+1}} \lambda^3 f(\lambda) d\lambda + \beta_i \int_{\lambda_i}^{\lambda_{i+1}} \lambda^2 f(\lambda) d\lambda \right.$$
$$\left. + \gamma_i \int_{\lambda_i}^{\lambda_{i+1}} \lambda f(\lambda) d\lambda + \delta_i \int_{\lambda_i}^{\lambda_{i+1}} f(\lambda) d\lambda \right\}, \tag{21}$$

where the upper integration limit, λ_m, is chosen to ensure the convergence of the integrals, and the function, $f(\lambda)$, is the remaining part of the correction factor integral, i.e.,

$$f(\lambda) = \left\{ \frac{J_1(\lambda a)}{\lambda a} - \frac{J_0(\lambda S)}{2} \right\} \frac{\sin(\lambda a)}{\lambda}. \tag{22}$$

In practice, the integrand of the correction factor goes to zero when λ is at most on the order of 100. Then, the $\{\lambda_i\}$ values are chosen as logarithmically equispaced on the interval, $0 \leq \lambda \leq 100$. The integrals in Eq 21 are evaluated for specific choices of the probe radius and probe spacing by means of a program called SPINT and the results of the partial integrals are stored in array form. These arrays need be calculated only once for each value of a and S. The evaluation of the correction factor integral and the extraction of the resistivity from the spreading resistance required the evaluation of the set of 280 (4 by 70) spline coefficients ($\{\alpha_i\}, \{\beta_i\}, \{\gamma_i\}, \{\delta_i\}$). This is carried out in the program called SPRED. Both of these programs require the use of a minicomputer to carry out the calculation.

At about the same time, Choo et al. [11] introduced an integration scheme based upon the Gauss-Laguerre quadrature. They use the uniform current density form of the correction factor integral as given by Eq 16. The general philosophy behind quadrature techniques of evaluating integrals is the search for a set of integration points and weights, $\{x_k\}, \{w_k\}, (k = 1, ..., N)$ such that an integral may be evaluated as a finite sum in the form

$$\int_a^b f(x) dx = \sum_{k=1}^{N} w_k f(x_k). \tag{23}$$

It should be clear that knowing the set of integration points and weights vastly simplifies the evaluation of the integral. This simplification may render the integration amenable to evaluation on a microcomputer system. For the case at hand, the Gauss-Laguerre quadrature makes use of the first 33 roots and weights obtained from the 128th order Laguerre polynomial. For the case of the uniform current density, the correction factor may be evaluated as the weighted finite sum,

$$C_n = \frac{8}{\pi} \sum_{k=1}^{33} w_k^{gl} \left\{ \frac{J_1(\lambda_k a)}{\lambda_k a} - \frac{J_0(\lambda_k S)}{2} \right\} \frac{J_1(\lambda_k a)}{\lambda_k} A_n(\lambda_k), \qquad (24)$$

where the weights and integration points $(a\lambda_k)$ are listed in reference 11.

While the above is formulated for the case of the uniform current density, the Schumann and Gardner current density as well as the ring delta function current densities may also be used. It is important to note that the Gauss-Laguerre technique can be used to carry out the calculation $(\{\rho_i\} \to \{R_i\}, i = 1, ..., n)$. The inverse calculation, $(\{R_i\} \to \{\rho_i\}, i = 1, ..., n)$, requires a convergence or bounding criteria.

From a completely different point of view, Dickey [12] presented a technique known as the local slope method for the calculation $(\{R_i\} \to \{\rho_i\}, i = 1, ..., n)$ This method does not require the evaluation of the correction factor integral but rather makes use of a heuristically derived algebraic relation between the correction factor and the local depth derivative of the spreading resistance.

Starting with the trapezoidal Rhomberg integration technique employed by D'Avanzo et al. [10], Albers [13] generated model spreading resistance data, $(\{\rho_i\} \to \{R_i\}, i = 1, ..., n)$, for a number of resistivity structures. In these calculations, the Schumann and Gardner current density was used with given values of the probe radius and the probe spacing. This technique may be used to carry out this calculation for all three forms of the probe current density. These model spreading resistance data were then used in the D'Avanzo et al. SPINT-SPRED codes as well as the Dickey local slope method to determine the $(\{Ri\} \to \{\rho_i\}, i = 1, ..., n)$ results for each algorithm. It was found that the SPINT-SPRED codes solved the inversion problem to within 1 percent. The local slope method gave very reasonable semiquantitative results. While it has been shown that the local slope equations cannot be derived from the multilayer equations [14], the speed and quality of the results make it a reasonable program to use.

Berkowitz and Lux [15] have carried out an investigation which has shown that it is possible to evaluate the spreading resistance correction factor using a 22-point scheme. The central point of this analysis is the replacement of the correction factor integral by an approximate form as

$$C_n = \frac{8}{\pi} \int_0^\infty A_n(\lambda) \left\{ \frac{J_1(\lambda a)}{\lambda a} - \frac{J_0(\lambda S)}{2} \right\} J_1(\lambda a) \frac{d\lambda}{\lambda}$$

$$= \frac{8}{\pi} \int_L^\infty A_n(\lambda) \left(\frac{J_1(\lambda a)}{\lambda a} \right)^2 d\lambda, \qquad (25)$$

where $L = 1.12292/S$. The choice of the truncated lower limit of integration is dictated by the form of the functions for the case of both insulating and conducting boundaries. The key point is that the approximate integral can be divided into four domains in each of which the Newton-Cotes integration method may be used. The evaluation of the approximate integral in Eq 25 is then performed by means of a finite sum as

$$C_n = \sum_{k=1}^{22} w_k^{bl} A_n(\lambda_k), \tag{26}$$

where the weights, w_k^{bl}, are different from those used in the Gauss-Laguerre technique. Comparison of the correction factor calculated by means of Eq 26 and the more extensive technique discussed by Albers [13] shows that the two agree to within better than 1 percent. Equation 26 provides a quick way of carrying out the ($\{\rho_i\} \to \{R_i\}, i = 1, ..., n$) calculation. In an unpublished (but publicly available program *) Berkowitz and Lux have used the above Newton-Cotes technique to carry out the ($\{R_i\} \to \{\rho_i\}, i = 1, ..., n$) calculation. This makes use of a bounding procedure on the kernel of the integral. This bounding procedure is based upon their realization that the recursion relation for the kernel function could be expressed as

$$A_n(\lambda) = \frac{\tanh(\lambda d)\rho_n + \rho_{n-1} A_{n-1}(\lambda)}{\rho_n + \rho_{n-1} A_{n-1}(\lambda) \tanh(\lambda d)}. \tag{27}$$

For the cases of insulating or conducting boundaries, the kernel function is bounded by the hyperbolic functions, coth and tanh, respectively. This provides for two bounding envelopes between which the kernel must lie. Calculation of ($\{\rho_i\} \to \{R_i\}, i = 1, ..., n$) using Eq 26 followed by the ($\{R_i\} \to \{\rho_i\}, i = 1, ..., n$) calculation using Eq 26 in conjunction with the bounding procedure shows that the original resistivities can be obtained to within less than 1 percent. This bounding procedure is extremely general. It may be used in the Gauss-Laguerre technique for the ($\{R_i\} \to \{\rho_i\}, i = 1, ..., n$) calculation.

All of the above techniques make use of a single form of the probe current density. Choo et al. [16] have presented a calculation of the spreading resistance correction factor based on a variational technique. This makes use of the uniform current density and the Schumann and Gardner semi-infinite slab current densities as a "basis set," and the amount of each current is varied according to a variational principle. The spreading resistance obtained from this variational method is

$$R_n^v = \frac{\rho_n}{2} C_n^v, \tag{28}$$

where

$$C_n^v = \frac{4}{\pi} \frac{K_1 I_{3s} + K_2 I_{2s}}{K_1 + K_2/2}, \tag{29}$$

where K_1 and K_2 are given by

$$K_1 = \frac{I_2 - I_3/2}{I_1 I_2 - I_3^2}, \tag{30}$$

$$K_2 = \frac{I_1/2 - I_3}{I_1 I_2 - I_3^2}. \tag{31}$$

---

* A FORTRAN version of this code may be obtained from Dr. H. L. Berkowitz, U.S. Army Electronics Technology and Devices Laboratory, ERADCOM, Fort Monmouth, New Jersey 07703

The integrals I_1, I_2, I_3 are

$$I_1 = \int_0^\infty A_n(\lambda) \left\{ \frac{\sin(\lambda a)}{\lambda a} \right\}^2 a d\lambda, \tag{32}$$

$$I_2 = \int_0^\infty A_n(\lambda) \left\{ \frac{J_1(\lambda a)}{\lambda a} \right\}^2 a d\lambda, \tag{33}$$

$$I_3 = \int_0^\infty A_n(\lambda) \left\{ \frac{\sin(\lambda a) J_1(\lambda a)}{(\lambda a)^2} \right\} a d\lambda. \tag{34}$$

The integrals I_{3s} and I_{2s} are

$$I_{3s} = \int_0^\infty A_n(\lambda) \left(\frac{\sin(\lambda a)}{\lambda a} \right) \left(\frac{J_1(\lambda a)}{\lambda a} - \frac{J_0(\lambda S)}{2} \right) d\lambda, \tag{35}$$

and

$$I_{2s} = \int_0^\infty A_n(\lambda) \left(\frac{J_1(\lambda a)}{\lambda a} \right) \left(\frac{J_1(\lambda a)}{\lambda a} - \frac{J_0(\lambda S)}{2} \right) d\lambda. \tag{36}$$

Choo et al. [17] have made use of the Berkowitz-Lux approximation [15] to simplify the integrals in Eqs 35 and 36 as

$$I_{3s} = \int_L^\infty A_n(\lambda) \left(\frac{\sin(\lambda a)}{\lambda a} \right) \left(\frac{J_1(\lambda a)}{\lambda a} \right) d\lambda, \tag{37}$$

$$I_{2s} = \int_L^\infty A_n(\lambda) \left(\frac{J_1(\lambda a)}{\lambda a} \right)^2 d\lambda. \tag{38}$$

It should be noted that the integrals given by Eqs 32, 33, and 34 contain all of the information which is needed to evaluate Eqs 28 and 29. This follows from the fact that the approximate integrals given by Eqs 37 and 38 are contained in the corresponding I_3 and I_2 integrals. The various integrals (I_1, I_2, and I_3) may be evaluated by means of a segmented Newton-Cotes method similar to that used by Berkowitz and Lux. The coordinates and weights are listed in reference 17. These equations can be implemented in order to carry out the $(\{\rho_i\} \rightarrow \{R_i\}, i = 1, ..., n)$ calculation. The inverse procedure, $(\{R_i\} \rightarrow \{\rho_i\}, i = 1, ..., n)$, can be carried out by means of a generalization of the Berkowitz-Lux bounding technique where the bounding by the hyperbolic functions is needed for all three of the integrals involved. This analysis shows that the original resistivities can be recovered to within 1 percent.

EFFECTS OF PROBE CURRENT DENSITY

As can be seen from the above discussion, there are three primary current densities which can be used in the calculation of the spreading resistance correction factor. The Schumann and Gardner form seems appropriate to the case of a semi-infinite slab, the uniform current density should be adequate for the case of a layer over a highly conductive layer, while the ring delta function seems appropriate for the case of a layer over a highly resistive layer. The model data calculations presented by Albers [13] (using the trapezoidal Romberg integration scheme) can

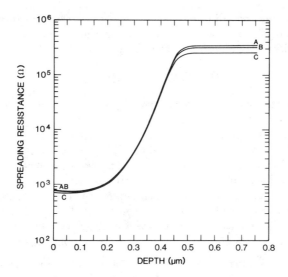

Figure 2. The spreading resistances calculated from a Gaussian implant type structure for the three forms of the probe current density are presented in this figure. The trapezoidal-Romberg technique of reference 13 is used in the calculation. The current densities used are the uniform current density (A), the Schumann and Gardner current density (B), and the ring delta function current density (C).

easily be generalized in order to use any one of the three current densities. The Gauss-Laguerre quadrature [11] may also be used with any one of the above current densities. The Berkowitz-Lux calculations [15] (originally specialized to the uniform current density but, in principle, extendable to the other two current densities) as well as the variational technique employing the Berkowitz-Lux approximate integral [17] can also be used to calculate the spreading resistance from the resistivity. For several structures, these various techniques were used to carry out the $(\{\rho_i\} \to \{R_i\}, i = 1, ..., n)$ calculation. The calculated spreading resistance obtained from a Gaussian implant structure are contained in Figure 2. These data were calculated using the three forms of the probe current density in conjunction with the trapezoidal Romberg integration technique. These data are more or less typical of diffusions and implants. Other examples of resistivity profiles and calculated spreading resistance data may be found in reference 13.

CALIBRATION DATA AND THE PROBE RADIUS

Having looked at the effects of the probe current density on the results of spreading resistance calculations, it is important to understand that there is a parameter in Eqs 14 and 15 which must be measured or at least estimated. This is the probe radius, a. The results of the $(\{\rho_i\} \to \{R_i\}, i = 1, ..., n)$ calculation and the inverse $(\{R_i\} \to \{\rho_i\}, i = 1, ..., n)$ calculation are sensitive to the choice of a. One way of measuring the probe radius was originally proposed by Dickey [12] as part of a conjecture relating spreading resistance and sheet resistance. In particular, an image calculation of the spreading resistance on the top of an isolated layer of thickness, t, and resistivity, ρ, yields the result

$$R = (\rho/\pi t)\ln(S/a) = (\rho/\pi t)\{\ln(S) - \ln(a)\}. \tag{39}$$

As the sheet resistance of this layer is given by $\mathcal{R} = \rho/t$, this would suggest that the spreading resistance may be linear in the logarithm of the probe spacing with a slope related to the sheet resistance and an intercept related to the probe radius. For junction-isolated boron diffusions with junction depths up to 4.3 μm, Dickey showed that there was good agreement between the sheet resistance measured by the standard four-probe technique and determined from the R vs $\ln(S)$ analysis. This agreement provided justification for the proposed relation between the spreading resistance and the sheet resistance but did not give any indication of the validity of the relation of the intercept of the R vs $\ln(S)$ analysis with the probe radius. The values obtained by Dickey were certainly of the expected size.

In a series of investigations to study the validity of the proposed relation between the spreading resistance and the sheet resistance (as well as the possibility of determining the probe radius), Albers [18,19] used the $(\{\rho_i\} \to \{R_i\}, i = 1, ..., n)$ calculation for a number of resistivity structures, probe current densities, several fixed values of the probe radius, and a series of values of the probe separation. The resistivity structures could be used to calculate the sheet resistance directly and the probe radius was a known input quantity. The model spreading resistance data were used in a linear regression analysis in $\ln(S)$ and the slope and intercept correlated with the sheet resistance and the probe radius. The regression analysis of the model spreading resistance data indicated that the relation between the spreading resistance and the sheet resistance was of general validity but that the probe radius was not generally obtainable from the regression analysis. Hence, some other technique seemed to be required to determine this quantity.

One possibility was to adopt the point of view that the probe radius is a function of the resistivity. This is hinted at by the form of the calibration data obtained on various resistivity bulk samples. This point of view was incorporated into the Dickey local slope analysis of boron implant data used in the comparison of spreading resistance, SIMS, and CV profiles by Ehrstein et al. [20]. The local slope analysis is ideally suited to this as the updating of the probe radius takes place in the algebraic equation relating the correction factor and the local slope. All of the other algorithms and integration schemes would be hampered by the updating of the probe radius which would necessitate the reevaluation of the sampling points and the integration weights.

Piessens et al. [21] have discussed a technique based upon Chebyshev polynomial methods for the evaluation of the Schumann and Gardner equations in the case of a resistivity-dependent probe radius. This technique requires a minicomputer system in the evaluation of the resistivity from the spreading resistance. The basis of the assumption of a resistivity dependent probe radius is the level of agreement obtained by Ehrstein et al. [20] in comparing profiles obtained from spreading resistance, SIMS, and CV. As will be discussed in the second part of this paper, there are a number of reasons why carrier and atomic profiles do not have to agree. Indeed, the use of a resistivity-dependent probe radius appears to allow for one more degree of freedom for every data point taken.

The point of view adopted by Albers and Berkowitz [22,23] is that the probe radius is a single parameter which can best be evaluated by optimizing the agreement between the four-probe resistance obtained from the resistivity profile and measured on the top surface of the structure. The multilayer expression for the in-line four-probe resistance, $Z_n(S)$, has been derived in the form [22]

$$Z_n(S) = 2\rho_n \int_0^\infty A_n(\lambda)\{J_0(\lambda S) - J_0(2\lambda S)\} I_\nu(\lambda a) d\lambda, \qquad (40)$$

where $I_\nu(\lambda a)$ is the Hankel transform of the generalized probe-current density normalized to unit current [4], and ν is an index which refers to the specific form of the current density. It was shown that the four-probe resistance could always be obtained from the spreading resistance

$$Z_n(S) = R_n(2S) - R_n(S). \qquad (41)$$

The four-probe resistance was shown to be: (i) independent of the probe radius and the probe-current density, and (ii) simply related to the sheet resistance, $\mathcal{R}_n$, when the distance to an insulating boundary is small compared with the probe spacing. The former observation provided the basis for a calibration procedure for determining the value of the probe radius to be used in spreading resistance profile analysis. This is discussed is detail in reference 22. The evaluation of the four-probe resistance integral was found to be simplified by the alternative form given by [23]

$$Z_n(S) = \frac{\rho_n}{\pi} \int_{1/2S}^{1/S} A_n(\lambda) d\lambda. \qquad (42)$$

In the above form, the four-probe resistance can easily be calculated using the Newton-Cotes method of evaluation,

$$Z_n(S) = \frac{\rho_n}{\pi} \ln(2) \sum_{i=0}^{8} w_{9,i} \lambda_i A_n(\lambda_i), \qquad (43)$$

where $\lambda_i = 2^{i/8}/2S (0 \leq i \leq 8), w_{9,i} = C_i/8$, and where C_i are the Newton-Cotes weighting factors. This procedure allows for the variation of the resistivity profile under the requirement that the measured and calculated four-probe resistances agree within given limits.

EFFECTS OF FINITE LAYER THICKNESS

The original Schumann and Gardner derivation assumed that the material may be viewed as a series of uniform resistivity layers. These authors realized the possible limitation of this assumption as they showed how the correction factor depended upon the number of layers used in the calculation. This same observation was studied in more detail by Choo et al. [9] by means of the recursion relation presented in Eq 18. In order to offset some of the possible difficulties associated with the finite thickness of the 'layers', Choo et al. [24] introduced an exponential model in which each of the layers was assumed to have an exponentially varying resistivity. This piecewise exponential was to be a more realistic representation of the continuously varying resistivity. The differences between the correction factor calculated by means of a finite layer picture and a continuum picture were also considered by Albers [25]. Beginning with the recursion relation, he showed how a nonlinear, inhomogeneous differential equation could be rigorously derived for the kernel of the spreading resistance correction factor integral in the limit as the layer thicknesses approach zero. This equation may be transformed to a Riccatti equation which can be solved analytically for an exponentially varying resistivity. When the correction factor calculated from this technique was compared with that obtained from the finite layer picture, it was shown that finite-layer analysis will underestimate the correction factor for the case of a low resistivity layer over a high resistivity layer. Over the past ten years, the beveling techniques which have evolved have allowed for the preparation of very shallow bevels. These shallow bevels have pushed the measurements into the region where the finite layer thickness approximation is reasonably good.

Before turning to the calculation of carrier densities from atomic densities, it is important to note that Albers and Berkowitz [26] are currently in the process of preparing a detailed report containing background material, computer programs and sample input/output data related to most of the work discussed to this point in this paper.

SIMULATIONS OF CARRIER DENSITIES FROM ATOMIC DENSITIES

In this, the second part of the paper, attention is focused on the calculation of carrier densities from atomic densities for planar and beveled structures. The purpose of this section is to provide insight into the validity of the use of the multilayer Laplace equation which assumes charge neutrality in each of the 'layers' in the material. In terms of the difference between carrier densities and atomic densities along the bevel cut into the material for spreading resistance profile analysis, Hu [27] has presented a calculation based upon the solution of Poisson's equation for a number of profile types into substrates of the opposite conductivity type, i.e., junction-isolated structures. Using a finite difference solution method, he finds that electrical junctions may be much shallower than metallurgical junctions for structures of the Gaussian and exponential type. As the deviation between carrier densities and atomic densities has also been observed for same conductivity-type substrates, Albers et al. [28] have approached the problem from the point of view of the solution of the semiconductor equations using an adaptive finite-element method previously discussed by Blue and Wilson [29,30]. They find that the deviation of the carrier densities from the atomic densities occurs for both same- and opposite-conductivity type substrates.

Consider the structure presented in Figure 3. This consists of an n-type region and a p-type region. A portion of the material is removed producing the beveled surface. The variables of interest to the problem are the electrostatic potential and the electron and hole densities which arise from the atomic donor and acceptor

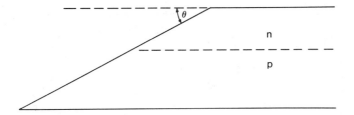

Figure 3. Geometry of the beveled material in which the semiconductor equations are solved by means of a finite element method.

concentrations. From a computational point of view, it is more convenient to formulate the problem in terms of the electrostatic potential and the electron and hole quasi-Fermi levels which are related to the corresponding densities. The model which shall be used to describe the system is based on the basic semiconductor equations [31-32] (steady-state, time-independent):

$$\nabla \cdot (\kappa \nabla V) = -\frac{q}{\epsilon_0}(p - n + N), \qquad (44)$$

$$\nabla \cdot (\mu_n \, n \, \nabla \phi_n) = R, \qquad (45)$$

$$\nabla \cdot (\mu_p \, p \, \nabla \phi_p) = -R, \qquad (46)$$

where κ is the dielectric constant, V is the electrostatic potential, q is the electronic charge, ϵ_0 is the permittivity of free space, N is the net ionized impurity density, $N = |N_D - N_A|$, N_D is the atomic donor density, N_A is the atomic acceptor density, μ_n and μ_p are the electron and hole mobilities, respectively, and ϕ_n and ϕ_p are the electron and hole quasi-Fermi levels, respectively. The Shockley-Read-Hall recombination is given by:

$$R = \frac{(pn - n_i^2)}{\tau_{no}(n + n_1) + \tau_{po}(p + p_1)}, \qquad (47)$$

where n_i is the intrinsic carrier density. The electron density, n, and the hole density, p, are related to the electrostatic potential and the corresponding quasi-Fermi levels (for Boltzmann statistics) by:

$$n = n_i \exp(q(V - \phi_n)/kT), \qquad (48)$$

$$p = n_i \exp(q(\phi_p - V)/kT). \qquad (49)$$

The solutions of Eqs 44, 45, and 46 are obtained by means of an adaptive finite-element technique. The reader is referred to references 28 to 30 for a more detailed discussion of the solution method and strategies. For the purpose of investigating the validity of the use of the multilayer Laplace equation, Figures 4 through 9 contain the results of the calculations for the cases of Gaussian diffusion and Gaussian implantation type structures. These figures contain the net ionized im-

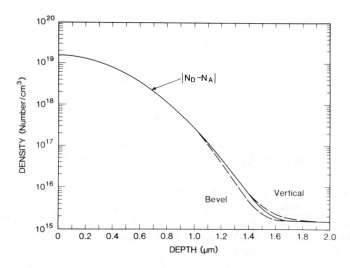

Figure 4. Carrier and atomic distributions for a 1.5-μm Gaussian structure with same conductivity type substrate. The surface concentration is $1.5 \times 10^{19} \text{cm}^{-3}$ and the background concentration is $1.5 \times 10^{15} \text{cm}^{-3}$. The net impurity concentration is denoted by the solid curve while the long dashed curve represents the electron density.

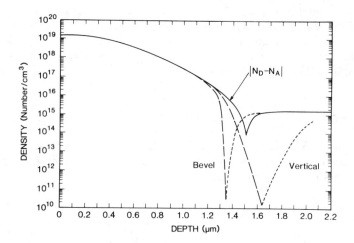

Figure 5. Carrier and atomic distributions for a 1.5-μm Gaussian structure with an opposite conductivity type substrate. The surface concentration is $1.5 \times 10^{19} \text{cm}^{-3}$ and the background concentration is $1.5 \times 10^{15} \text{cm}^{-3}$. The net impurity concentration is denoted by the solid curve, the long dashed curve represents the electron density, and the short dashed curve represents the hole density.

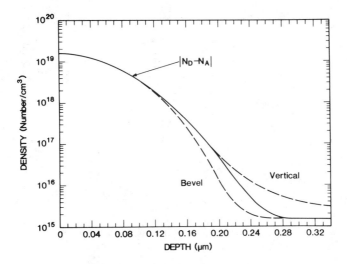

Figure 6. Carrier and atomic distributions for a 0.25-μm Gaussian structure with same conductivity type substrate. The surface concentration is 1.5×10^{19}cm^{-3} and the background concentration is 1.5×10^{15}cm^{-3}.

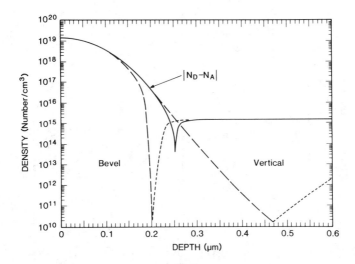

Figure 7. Carrier and atomic distributions for a 0.25-μm Gaussian structure with an opposite conductivity type substrate. The surface concentration is 1.5×10^{19}cm^{-3} and the background concentration is 1.5×10^{15}cm^{-3}.

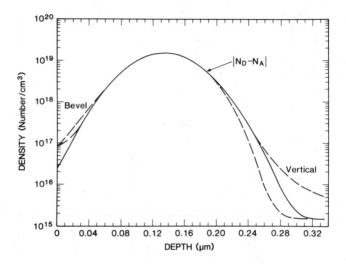

Figure 8. Buried peak Gaussian implant for same conductivity type substrate. The peak concentration is $1.5 \times 10^{19} \text{cm}^{-3}$ and the background concentration is $1.5 \times 10^{15} \text{cm}^{-3}$. The net impurity concentration is denoted by the solid curve while the long dashed curve represents the electron density.

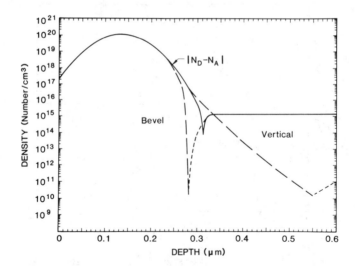

Figure 9. Buried peak Gaussian implant for opposite conductivity type substrate. The peak concentration is $1.5 \times 10^{20} \text{cm}^{-3}$ and the background concentration is $1.5 \times 10^{15} \text{cm}^{-3}$. The net impurity concentration is denoted by the solid curve, the long dashed curve represents the electron density, and the short dashed curve represents the hole density.

purity (atomic) density, the electron and hole densities calculated for a planar material, and the electron and hole densities calculated along the bevel. Figures 4 and 5 contain the results for same- and opposite-conductivity type substrates for a 1.5-μm Gaussian with a surface concentration of 1.5×10^{19} cm^{-3}. Figures 6 and 7 contain the results for same- and opposite-conductivity type substrates for a 0.25-μm Gaussian with a surface concentration of 1.5×10^{19} cm^{-3}. Finally, Figures 8 and 9 contain the results for same- and opposite-conductivity type substrates for Gaussian implant structures with peak concentrations of 1.5×10^{19} cm^{-3} and 1.5×10^{20} cm^{-3}, respectively. The deviations between the carrier densities and the atomic densities become larger as the peak concentrations are reduced beyond the level presented in the figures. These deviations represent regions where there is a nonzero charge. This charge separation is due to the repulsion of the mobile carriers in the restraining field due to the immobile ionized impurities. In the regions where there is no deviation, the net local charge is zero.

PRESENT STATE OF SPREADING RESISTANCE ANALYSIS

The calculation of the spreading resistance correction factor using the Schumann and Gardner multilayer Laplace equation has progressed to the point where the calculation of the spreading resistance from the resistivity and the inverse calculation can be performed on microcomputer systems. The interpretation of the probe radius can be done by using a resistivity-dependent probe radius model or by comparing the interpreted profiles in terms of the four-probe resistance. However, there is a problem with the use of the multilayer equations for the case of shallow layers where the net local charge is nonzero. There is no doubt that the spreading resistance can be measured in submicron regions, but the interpretation of the data in terms of the multilayer analysis Laplace is severly in question. From a conceptual point of view, it is safe to assume that the measured data should be related to the physical description provided by the semiconductor equations. However, the use of the Laplace equation description in the regions where local charge neutrality is violated leaves open the question of the correctness of the resistivity (and carrier density) profiles in these regions. As can be seen from Figures 4-9, this region may begin at concentrations at about 10^{18} cm^{-3} for submicron structures. It is extremely important to keep in mind that these represent a best case scenario where damage, defects, incomplete activation, etc. have not been brought into play. The resolution of this problem for the interpretation of spreading resistance data on shallow VLSI structures requires the investigation of the solution of the Laplace and Poisson equations. The solutions of these two equations should provide an estimate of the differences between the actual carrier distribution and that determined from the Schumann and Gardner analysis. The answer to this problem is unknown at the present time and is under consideration. The resolution of this question is of utmost importance if the spreading resistance technique is to have any hope of being successfully interpreted in the profiling of submicron structures.

REFERENCES

[1] Schumann, P. A. and Gardner, E. E., "Application of Multilayer Potential Distribution to Spreading Resistance Correction Factors," *Journal of the Electrochemical Society*, Vol. **116**, 1969, pp. 87-91.

[2] Gardner, E. E. and Schumann, P. A., "Spreading Resistance Correction Factors," *Solid-State Electronics*, Vol. **12**, 1969, pp. 371-375.

[3] Koefoed, O., *The Application of the Kernel Function in Interpretation of Geoelectrical Resistivity Measurements*, Gebrüder Borntraeger, Berlin-Stuttgart, 1968.

[4] Berkowitz, H. L. and Lux, R. A., "Errors in Resistivities Calculated by Multilayer Analysis of Spreading Resistance," *Journal of the Electrochemical Society*, Vol. **126**, 1979, pp. 1479-1482.

[5] Yeh, T. H. and Kokhani, K. H., "Multilayer Theory of Correction Factors for Spreading-Resistance Measurements," *Journal of the Electrochemical Society*, Vol. **116**, 1969, pp. 1461-1464.

[6] Hu, S. M., "Calculation of Spreading Resistance Correction Factors," *Solid-State Electronics*, Vol. **15**, 1972, pp. 809-817.

[7] Iida, Y., Abe, H. and Kondo, M., "Impurity Profile Measurements of Thin Epitaxial Silicon Wafer by Multilayer Spreading Resistance Analysis," *Journal of the Electrochemical Society*, Vol. **124**, 1977, pp. 1118-1122.

[8] Sunde, E. D., *Earth Conduction Effects in Transmission Systems*, Dover, New York, 1968.

[9] Choo, S. C., Leong, M. S. and Kuan, K. L., "On the Calculation of Spreading Resistance Correction Factors," *Solid-State Electronics*, Vol. **19**, 1976, pp. 561-565.

[10] D'Avanzo, D. C., Rung, R. D., Gat, A. and Dutton, R. W., "High Speed Implementation and Experimental Evaluation of Multilayer Spreading Resistance Analysis," *Journal of the Electrochemical Society*, Vol. **125**, 1978, pp. 1170-1176.

[11] Choo, S. C., Leong, M. S., Hong, H. L., Li, L. and Tan, L. S., "Spreading Resistance Calculations by the Use of Gauss-Laguerre Quadrature," *Solid-State Electronics*, Vol. **21**, 1978, pp. 769-774.

[12] Dickey, D. H. and Ehrstein, J. R., "*Semiconductor Measurement Technology: Spreading Resistance Analysis for Silicon Layers with Nonuniform Resistivity*," NBS Special Publication **400-48**, National Bureau of Standards, Washington, D. C., 1979.

[13] Albers, J., "Comparison of Spreading Resistance Correction Factor Algorithms Using Model Data," *Solid-State Electronics*, Vol. **23**, 1980, pp. 1197-1205.

[14] Albers, J., "The Relation Between the Correction Factor and the Local Slope in Spreading Resistance," *Journal of the Electrochemical Society*, Vol. **130**, 1983, pp. 2076-2080.

[15] Berkowitz, H. L. and Lux, R. A., "An Efficient Integration Technique for Use in the Multilayer Analysis of Spreading Resistance Profiles," *Journal of the Electrochemical Society*, Vol. **128**, 1981, pp. 1137-1141.

[16] Choo, S. C., Leong, M. S. and Tan, L. S., "Spreading Resistance Calculations by the Variational Method," *Solid-State Electronics*, Vol. **24**, 1981, pp. 557-

562.
[17] Choo, S. C., Leong, M. S. and Sim, J. H., "An Efficient Numerical Scheme for Spreading Resistance Calculations Based on the Variational Method," *Solid-State Electronics*, Vol. **26**, 1983, pp. 723-730.
[18] Ehrstein, J. R., Albers, J., Wilson, R. G. and Comas, J., "Comparison of Spreading Resistance with C-V and SIMS Profiles for Submicron Layers in Silicon," *Extended Abstracts*, The Electrochemical Society, Vol. **80-1**, Princeton, N. J., 1980, pp. 496-498.
[19] Albers, J., "Probe Spacing Experiment Simulation and the Relation Between Spreading Resistance and Sheet Resistance," *Journal of the Electrochemical Society*, Vol. **129**, 1982, pp. 599-605.
[20] Albers, J., "Spreading Resistance Probe Spacing Experiment Simulations: Effects of Probe Current Density and Layer Thickness," *Journal of the Electrochemical Society*, Vol. **129**, 1982, pp. 2788-2795.
[21] Piessens, R., Vandervorst, W. B. and Maes, H. E., "Incorporation of a Resistivity Dependent Contact Radius in an Accurate Integration Algorithm for Spreading Resistance Calculations," *Journal of the Electrochemical Society*, Vol. **130**, 1983, pp. 468-474.
[22] Albers, J. and Berkowitz, H. L., "The Relation Between Two-Probe and Four-Probe Resistances on Nonuniform Structures," *Journal of the Electrochemical Society*, Vol. **131**, 1984, pp. 392-398.
[23] Albers, J. and Berkowitz, H. L., "An Alternative Approach to the Calculation of Four-Probe Resistances on Nonuniform Structures," *Journal of the Electrochemical Society*, Vol. **132**, 1985, pp. 2453-2456.
[24] Choo, S. C. and Leong, M. S., "A Multilayer Exponential Model for Spreading Resistance Calculations," *Solid-State Electronics*, Vol. **22**, 1979, pp. 405-415.
[25] Albers, J., "Continuum Formulation of Spreading Resistance Correction Factors," *Journal of the Electrochemical Society*, Vol. **127**, 1980, pp. 2259-2263.
[26] Albers, J. and Berkowitz, H. L., report to appear in the *"Semiconductor Measurement Technology: ,"* NBS Special Publication **400** series, National Bureau of Standards, Gaithersburg, Maryland. Interested individuals may contact the first author to receive a copy of the report once it is published.
[27] Hu, S. M., "Between Carrier Distributions and Dopant Atomic Distribution in Beveled Silicon Substrates," *Journal of Applied Physics*, Vol. **53**, 1982, pp. 1499-1510.
[28] Albers, J., Wilson, C. L. and Blue, J. L., "Two-Dimensional Atomic and Carrier Distributions in Silicon Structures with Planar and Beveled Surfaces," submitted to *Journal of Applied Physics*.
[29] Blue, J. L. and Wilson, C. L., "Two-Dimensional Analysis of Semiconductor Devices Using General-Purpose Interactive PDE Software," *IEEE Transactions on Electron Devices*, Vol. **ED-30**, 1983, pp. 1056-1070.
[30] Blue, J. L. and Wilson, C. L., "Two-Dimensional Analysis of Semiconductor Devices Using General-Purpose Interactive PDE Software," *SIAM Journal Scientific Statistical Computing*, Vol. **4**, 1983, pp. 462-484.
[31] Kurata, M., *Numerical Analysis for Semiconductor Devices*, Lexington Books, Lexington, Mass., 1982.

Marek Pawlik

SPREADING RESISTANCE : A COMPARISON OF SAMPLING VOLUME CORRECTION FACTORS IN HIGH RESOLUTION QUANTITATIVE SPREADING RESISTANCE

REFERENCE: Pawlik, M., 'A comparison of sampling volume correction factors in high resolution quantitative spreading resistance', Emerging Semiconductor Technology, ASTM STP 960, D.C. Gupta and P.H. Langer, Eds., American Society for Testing and Materials, 1986

ABSTRACT: The development of fine geometry devices requires a greater depth of understanding of dopant incorporation mechanisms and the subsequent location of carriers. Spreading resistance is an accepted method for determining carrier concentration profiles. The deconvolution of these profiles from the raw spreading resistance data is a crucial step which uses so called sampling volume correction factors. These are complex and difficult to evaluate but a number of algorithms are available which attempt the necessary integrations and iterations. Five of these algorithms are reviewed and their accuracy and performance are evaluated on real (as opposed to model) data. The need for new methods of data smoothing is highlighted and it is concluded that the so called Berkowitz-Lux algorithm is currently the most successful available.

KEYWORDS: Profiling, spreading resistance, sampling volume correction factors, correction factors, carrier concentration profiles.

INTRODUCTION

The development of technologies for the manufacture of fine geometry devices has precipitated a need to understand (and hence implicitly a need to measure and quantify) many characteristics of devices and materials. One important area receiving considerable practical and theoretical attention is that of understanding the basic processes of dopant incorporation mechanisms. The need to fabricate shallower dopant distributions (using novel fabrication processes such as electron beam or flash lamp annealing), to provide data for the validation of process and device models and to assess novel materials growth techniques (reduced and low pressure chemical vapour deposition and molecular beam epitaxy) has led to a renewed interest in dopant and carrier concentration profiling techniques. Developments of these techniques

M Pawlik, Department Head - Silicon Materials and Characterisation, GEC Research Limited, Hirst Research Centre, East Lane, Wembley, Middlesex HA9 7PP United Kingdom

techniques have been undertaken with a view to establishing higher spatial resolution and greater accuracy in both depth and absolute carrier and dopant concentrations. For dopant profiling, Secondary Ion Mass Spectrometry (SIMS) and for carrier concentration profiling Spreading Resistance (SR) have emerged as the preferred techniques during the last few years. A discussion of the relative merits and implementation of both these techniques can be found in reference [1].

To achieve high resolution and, it is hoped, accuracy in using the SR technique four major issues must be addressed:

1. Sample preparation - capping (when necessary), bevelling, cleaning, and bevel angle measurement

2. Probe preparation - conditioning the probes to give good electrical characteristics

3. Data acquisition - correct choice of operating conditions

4. Data processing - converting the raw spreading resistance data to a carrier concentration profile.

The objective of this paper is to consider the last of these issues. Experimental procedures which have been developed to address the first three have been recently discussed in reference [1]. It must however be stressed that the best possible experimental data must be obtained if accuracy is to be achieved. No amount of data processing will turn poor data into acceptable carrier concentration profiles.

In the next section the origin and need for the sampling volume correction factors are reviewed and expressions are derived for them using varying assumptions in the Models section.

The algorithms which have been used to calculate the factors are then described and results are presented which use the algorithms on real data. Finally the Discussion and Conclusions section considers the efficiency of the algorithms which together with the results obtained on many types of samples, indicates that one algorithm is currently superior. The need for new methods of data smoothing is highlighted in view of the influence of noise in the raw data which is shown in the Results section.

SAMPLING VOLUME CORRECTION FACTORS

The need for Sampling Volume Correction Factors is best understood by considering the fundamental principles of the spreading resistance technique: i.e. where two probes of equal radius 'a' are placed on the surface of an homogeneous, semi-infinite silicon sample, of resistivity ρ, then the spreading resistance is given by:

$$R_S = \rho/2a \qquad (1)$$

It is readily shown that for perfect, non-indenting and circular contacts the equipotentials are hemispherical and the direction of current flow is normal to these equipotentials. About 70% of the potential drop occurs within a distance of four times the probe radius.

Since typically probe radii of 1-3 micrometres are used in SR, the volume that the measurement samples is very small in relation to the total volume of a typical bulk specimen. However, even for homogeneous samples the actual measured resistance R_M is given by:

$$R_M = R_S + R_B \qquad (2)$$

where R_B is a resistivity dependent barrier resistance term. The relative magnitudes of R_S and R_B are strong functions of the resistivity with barrier resistance becoming dominant at high resistivities. The effect of barrier resistance is often incorporated into an effective contact resistance which is resistivity dependent i.e.

$$R_M = \rho/2a(\rho) \qquad (3)$$

The effective contact radius, can be determined from calibration data on bulk specimens and is generally found to be significantly different to the geometric radius used in equation (1).

If the semiconductor is inhomogeneous the current and potential distributions beneath the probes are no longer uniformly radial and hemispherical respectively and their exact nature depends on the form of the inhomogeneity of the material. The overall effect of this is to modify the volume that the probes sample. This is best illustrated by considering the case of a junction isolated layer whose thickness is less than a few probe radii. The current flows in the entire layer being profiled and now has a large lateral component. In the case of a high resistivity layer on a low resistivity layer the current will again flow throughout the top layer but lateral current spreading will be less than in the homogeneous case, as current will attempt to flow into the low resistivity substrate.

Thus, for a thin layer, as the probes are stepped down the bevel, the measured resistance is due to the barrier resistance and the spreading resistance due to the entire layer below the point of measurement. (Indeed, unpublished computer modelling results have shown that in certain circumstances the sampling volume actually extends into regions <u>above</u> the point of measurement).

Thus, equation (3) can be modified to:

$$R_M = CF \times \rho/2a(\rho) \qquad (4)$$

where CF is the sampling volume correction factor and contains all the information relevant to current flows, system boundaries etc. CF is itself a function of the resistivity ρ and (4) therefore represents a highly non-linear equation which must be solved to determine ρ self-consistently.

Having seen how the need for CF arises a model must now be developed which will enable it to be calculated. Two such models are currently in use and these are discussed in the next section.

SAMPLING VOLUME CORRECTION FACTORS - MODELS

The Schumann-Gardner Theory

The model most commonly used for determining the correction factors is that due to Schumann and Gardner [2]. In essence, the inhomogeneous profile is modelled as a multilayer structure, consisting of slabs of constant resistivity. The thickness of these slabs is taken to be the depth increment of the sampling points as the probes are moved down a bevel. The following assumptions are then made:

- The contact area of the probes is uniform and can be represented by an effective radius 'a'.

- The probe-semiconductor contact is ohmic.

- Laplace's equation is valid and there are no accumulation or depletion layers at the multilayer interfaces.

- The current distribution at the contact is determined by the solution of the Laplace equation for a flat circular contact on a semi-infinite substrate.

- Full cylindrical symmetry is assumed.

With these assumptions, Laplace's equation can be solved to give the following expression for the correction factor at any point in a profile:

$$CF = \frac{4}{\pi} \int_0^\infty F(\sigma,\lambda) \left\{ \frac{J_1(\lambda a)}{\lambda a} - \frac{J_0(\lambda s)}{2} \right\} \sin(\lambda a) \frac{d\lambda}{\lambda} \qquad (5)$$

where J_0, J_1 are the zero and first order Bessel functions, a is the probe radius, s is the probe separation, λ is a dummy integration variable and $F(\sigma,\lambda)$ is a term which contains all the information about the profile <u>beneath and including</u> the layer for which the correction factor is being determined.

Subsequently it was shown by Leong et al [3] that the fourth assumption above can be modified to use a uniform current density instead of a variable current density to give:

$$CF = \frac{4}{\pi} \int_0^\infty F(\sigma,\lambda) \left\{ \frac{J_1(\lambda a)}{\lambda a} - \frac{J_0(\lambda s)}{2} \right\} 2J_1(\lambda a) \, d\lambda \qquad (6)$$

A further choice of current distribution was made by Berkowitz-Lux [4] and is the so called ring current distribution which gives the following form for the correction factor.

$$CF = \frac{4}{\pi} \int_0^\infty F(\sigma,\lambda) \left\{ \frac{J_1(\lambda a)}{\lambda a} - \frac{J_0(\lambda s)}{2} \right\} aJ_0(\lambda a) \, d\lambda \qquad (7)$$

The process for actually calculating a sequence of correction factors begins with the deepest measured point where some assumption is made about the nature of the underlying material. Normally, the material is assumed to be uniform in the case of non-junction isolated structures or to have a very high resistivity in the case of junction isolation. The term $F(\sigma,\lambda)$ is then calculated for the next layer and the correction factor integral can be computed. This process is then repeated layer by layer until the top surface is reached.

The practical application of the correction factors given by equations (5-7) is in reality much more complex than the simple procedure outlined above and has led to the development of a number of algorithms. The following points must be considered:

(i) Calculation of the multilayer integration factor $F(\sigma,\lambda)$.

The calculation of $F(\sigma,\lambda)$ was originally performed in a laborious manner using matrix inversion methods which led to excessive computer memory requirements and long computation times. Choo et al [5] made a significant contribution by devising a recurrence relationship for $F(\sigma,\lambda)$ which is straightforward to calculate i.e.

$$F_{i+1} = \frac{F_i \sigma_{i+1} + \sigma_i \tanh(\lambda d)}{\sigma_i + F_i \sigma_{i+1} \tanh(\lambda d)} \qquad (8)$$

where σ_i is the conductivity of the ith layer and d is the layer thickness (normalised to the probe radius). Although subsequently D'Avanzo et al [6] used interpolation techniques to calculate F (and thereby increased efficiency of computation) this extension is rarely used and equation (8) is used in its full form.

(ii) Calculation of the correction factor integral CF.

When the integration factor $F(\sigma,\lambda)$ is calculated, the infinite integral (equations (5,6,7)) must be performed. This integral is particularly complex, since the integrand is a function of the zero and first order Bessel functions. The integrand is highly oscillatory in nature and may contain several thousand maxima and minima. The challenge is therefore to perform this integration to a predetermined degree of accuracy whilst minimising computational speed and memory requirements. The degree of accuracy required still remains an open question. Elegant arguments by Vandervorst et al [6] suggest that an accuracy of much better than 1% is necessary but as will be shown later, for most practical purposes it is sufficient to use a 5% accuracy figure, which has great implications on the efficiency of algorithms.

(iii) Choice of current distributions.

As was shown earlier, three forms of the assumed current distribution are available and a choice must be made as to whether equations 5,6 or 7 should be used. Some preference may be expressed for this choice on purely physical grounds e.g. that the uniform current distribution is more realistic on thick layers in isotype structures and that the variable current density is appropriate to thin layers with insulating boundaries. However, the work of Albers [7] and Berkowitz-Lux [4] has shown that there is little significant difference between correction factors calculated using any of the three current distributions. However, some of the integration schemes proposed dictate this choice since they rely on the integrand having a particular form.

(iv) Choice of the effective probe radius 'a'.

The parametric dependence of CF is on the probe spacing 's' the depth increment 'd', and the effective probe radius 'a'. The first two of these are readily obtained by direct measurement whereas the last cannot be easily measured as it represents an effective radius and not a geometric radius. Indeed, it is undoubtedly a function of the resistivity and so no single value is probably appropriate. However, for most practical purposes a single value of 'a' must be chosen (algorithms which have attempted to use variable radii have been found to be unstable). The determination of 'a' by using a single resistivity value on a bulk calibration curve has been found to be grossly in error by the author and a self-consistent scheme for determining 'a' has been proposed [1]. Following the conditioning of new probes, a number of 'standard' ion implanted samples are run. These samples have known sheet resistivities and implanted doses. For each algorithm to be used, the effective radius is chosen on the basis of the best predicted sheet resistivity and dose as compared to the measured values. Thus an iteration procedure is employed. This must be repeated every time the probes require recalibration. This method has been used in all the results discussed in this paper.

(v) Application of the correction factors.

Equation (4) with 'a' constant, i.e. not dependent on resistivity, represents the simplest way in which the correction factors can be applied with the <u>measured</u> resistance used in the left hand side. Thus, the barrier resistance term in equation (2) is ignored. Since CF is a function of ρ, equation (4) must be iteratively solved since it is transcendental in nature. Thus in effect the zeros of the function $H(\rho)$ must be found where

$$H(\rho) = \frac{\rho}{2a\,R_M} - \frac{1}{CF}$$

A number of iterative schemes have been proposed e.g. binary search, rational approximation, secant method etc. which vary in their efficiency (number of iterations required) before the correct root is found. Maes and Clarysse [8] have shown that in fact <u>all</u> these iterative schemes converge to the same root, albeit with different speeds.

The above procedure is commonly used but is incorrect in ignoring the barrier resistance term. The measured resistance at any point in the profile includes the barrier resistance appropriate to the true resistivity of that layer and the spreading resistance corrected for the underlying layers which should only be a function of the true resistivities. Thus, the barrier resistance must be independently determined in order that it can be subtracted from the measured resistance before the correction factor is applied. The iteration scheme must therefore include adjustment of the barrier resistance at each step of the iteration process as the resistivity converges to the true value. The barrier resistance can be determined by using bulk material calibration curves i.e. a radius 'a' is chosen and $R_B(\rho)$ is determined by subtracting $\rho/2a$ at each point of the calibration curve. In view of the added complexity of this process it is rarely used in practice.

The Local Slope Technique

The local slope technique also utilises a multilayer approximation to an inhomogenous structure and a full derivation is given in [9]. The principle of this method is the knowledge of the limiting values of the correction factor for a thin layer over a nonconducting substrate and a conducting substrate. For a thin uniform layer over an insulating substrate the resistance may be written as:

$$R = \frac{\rho}{\pi t} \ln(s/a) \qquad (9)$$

where t is the thickness of the layer, ρ its resistivity, s and 'a' are the probe separation and contact radius. For a graded layer on an isolating substrate it is easier to think in terms of conductance since in the multilayer approximation conduction occurs in parallel through these layers, between the probes. The conductance at the ith layer is written as:

$$G_i = \frac{\pi}{\ln(s/a)} \sum_{j=1}^{i} \sigma_j \Delta t \qquad (10)$$

which can readily be deduced from equation (9).

The difference in the conductance between the ith and (i-1)th layer can be related to the conductivity of the ith layer using equation (10) to give:

$$\sigma_i = \frac{\ln(s/a)}{\pi} \left\{ \frac{G_i - G_{i-1}}{\Delta t} \right\} = \frac{\ln(s/a)}{\pi} \frac{\Delta G}{\Delta t} \qquad (11)$$

Now, for a thin inhomogeneous layer over a conducting substrate, current flow is primarily in the substrate and thus current flow in the layer of interest can be thought of as columnar. Thus the measured resistance is due to a series of cylindrical resistances in the sublayers of the multilayer approximation and the normal spreading resistance of the substrate. Hence the resistance of the ith layer is given by:

$$R_i = 2 \sum_{j=1}^{i} \rho_j \left\{ \frac{t_j - t_{j-1}}{\pi a^2} \right\} + \frac{\rho_s}{2a} \qquad (12)$$

where ρ_s is the substrate resistivity. Proceeding as before the resistivity of the ith layer can be deduced from that of the ith and (i-1)th to give

$$\rho_i = \frac{\pi a^2}{2\Delta t}\left\{R_i - R_{i-1}\right\} = \frac{\pi a^2}{2}\frac{\Delta R}{\Delta t} \qquad (13)$$

Now, using simple mathematical transformations equations (11) and (13) can be combined to give

$$\frac{1}{\rho} = \frac{\ln(s/a)}{\pi}\frac{1}{R}\left\{\frac{-\Delta \ln(R)}{\Delta t}\right\}$$

for a junction isolated layer and

$$\rho = \frac{\pi a^2}{2} R \left\{\frac{\Delta \ln(R)}{\Delta t}\right\}$$

for a layer on a conducting substrate.

By comparison with equation (1) the two limiting cases of correction factor are found to be:

$$[CF]_1 = k_1 M \quad \text{(isolated)}$$
$$[CF]_2 = -k_2/M \quad \text{(conducting)}$$

where

$$k_1 = \frac{(2.3)(2a)\ln(s/a)}{\pi}$$

$$k_2 = \frac{(2.3)}{\pi a} 4$$

and

$$M = -\frac{d\log R}{dt}$$

Thus M is the local slope of the resistance. Dickey [9] now suggests the following form for the correction factor valid in all cases:

$$CF = \frac{k_1 M}{2} + \sqrt{\frac{k_1 M^2}{2} + 1 + (k_1 k_2 - 1)\frac{2}{\pi}\tan^{-1}(\log R_0/R_s)} \qquad (14)$$

where R_0 is the resistance of the layer being measured and R_s is the resistance of the layer one contact radius deeper. This final form has the correct asymptotic forms for large positive and large negative values of local slope and is also correct when M=0.

Although somewhat empirical in nature, the advantage of the local slope approximation is the simplicity of the final result and hence the speed of computation.

ALGORITHMS

Algorithms that are used to calculate correction factors must operate efficiently in order that they do not limit the throughput of a spreading resistance system. As a result of the experimental problems encountered with spreading resistance it is necessary to obtain the final corrected profiles before another sample is analysed. Thus off-line computation on main frame machines is undesirable.

The algorithms discussed here have been written in Fortran 77 and data processing is undertaken on an HP9836S computer running under the HP Pascal operating system. The only exception to this is the SSM algorithm [10], which runs on the PDP8 computer in the ASR100C/2 Spreading Resistance Probe. All the data are acquired by the ASR and are transferred to the HP9836 where a turn key software package has been developed to process the data using the different algorithms. In Table 1 details are given of the number of lines of coding required for each algorithm, normalised running times for a typical profile and the length of compiled code.

The algorithms used were as follows:

Local Slope

The details of this algorithm can be found in reference [9]. The correction factor used is that given by equation (14). The correction factor is applied to the spreading resistance i.e. the measured resistance with the barrier resistance term subtracted. The barrier resistance is calculated from calibration data which is obtained by direct measurement on bulk samples.

Berkowitz-Lux

The Berkowitz-Lux algorithm uses a 22 point integrator [11]. This integrator neglects the J_0 term in equation (6) and replaces the zero lower limit of integration by a finite limit. Comparisons of this approximation in analytically soluble cases show an agreement of better than 1%. The iteration procedure employed is based on a rational polynomial approximation and a convergence criterion of 10^{-3}% in resistivity has been used. No allowance for finite barrier resistances is made.

PROF

The PROF algorithm uses a modified version of equation (5) where the Schumann-Gardner theory is modified by including the effects of a voltage induced in one probe by the current in the other. This leads to a more complex integral which Vandervorst et al [6] find is more amenable to numerical integration. A 150 point Gauss-Laguerre integrator is used which gives an accuracy of around 0.15%. The iteration scheme used combines both the secant method and binary search techniques and convergence to 10^{-3}% is achieved. Barrier resistance effects are ignored.

PROFA

The algorithm PROFA [12] uses the correction factor as shown in equation (5). An integration scheme is used which combines analytical integration, polynomial approximation and Gauss-Kronrod numerical integration. The idea behind using this complex but efficient integration scheme was to incorporate a resistivity-dependent radius in the calculation. The algorithm was however found to be unstable when used in this fashion and so only constant, fixed values of the radius were used. The integration is accurate to better than 0.1% and the iteration schemes used are the same as in the program PROF. Barrier resistance is again neglected. Convergence to 10^{-3}% in resistivity was sought.

SSM

This proprietary algorithm is supplied with the ASR 100C/2 spreading resistance probe and little is known about it. It is believed to utilise a 33 point Gauss-Laguerre integrator which is at best accurate to 5%. The barrier resistance terms are incorporated by a method described in reference [13]. It is not known what level of convergence is achieved using the iteration scheme of successive approximation.

RESULTS

In order to initially validate the algorithms, model data were used. A number of model resistivity profiles were considered and theoretical spreading resistance profiles were generated from these using the algorithm of D'Avanzo et al [14]. These model profiles included shallow and deep ion implants into substrates of similar or opposite polarity, ion implants with and without buried peaks, n on n^+ and p on p^+ epitaxial layers, modulated structures with rapid dopant transitions etc., etc. This algorithm, using a highly accurate Romberg integration scheme, is accurate to better than 0.1%. The prediction of R_S from ρ is straightforward compared to the normal problem of determining ρ from R_S. (An analogy that can be used here is the difference between a scattering problem in physics, when the scattering potential is known and an inverse scattering problem when the scattering potential is to be determined from a knowledge of the wave functions. The former involves a simple solution of the appropriate wave equation whereas the latter involves finding the appropriate term of the wave equation when its solutions are known).

The R_S model data was then converted back to resistivity using the five algorithms. (Care was taken to use the current distribution appropriate to the assumptions of each algorithm). It was found that PROF, PROFA and Berkowitz-Lux all successfully reproduced the model profiles whereas Local Slope and SSM failed. This was very significant on profiles where the substrate was of the same type as the layer and was of higher conductivity. In these cases significant smearing of the interface took place. Examples of this effect can be found in reference [15]. Since no barrier resistance was included in the model data, no problems were found in the deconvoluted resistivity values in absolute terms. Although differences were found between the PROF, PROFA and Berkowitz-Lux generated profiles these were well within the 5% accuracy that could optimistically be claimed for the spreading resistance technique i.e. when compared as a typical graph (8 decade logarithmic) the differences were undetectable.

The second stage of the comparison procedure was to determine the appropriate contact radius to be used with each algorithm. The self consistent method outlined in the Models section was used and a wide variation in 'a' was found e.g. for n-type material 2.8 μm for SSM and 0.9 μm for Berkowitz-Lux were found to be the correct values.

Although a large number of real data sets have now been analysed with the algorithms only three distinct types are shown here. With the exception of the SSM algorithm data smoothing was used before all data processing. The data smoothing used was the smoothed first difference algorithm as described by D'Avanzo et al [14]. Although far from ideal, this data smoothing technique is fast and readily implemented on the type of computer systems used in this work. Sample 1 is a 25 keV Boron implant at a dose of 5×10^{14} cm^{-2} through oxide. The raw spreading resistance data is shown in Figure 1. The resistivity profile determined by the PROF, PROFA and Berkowitz-Lux algorithms is shown in Figure 2. (No significant differences are observed using these algorithms and their results are indistinguishable). Figures 3 and 4 show the SSM algorithm and local slope profiles. It should be noted that both these show significant deviations, particularly at the start of the profiles and at the part of the profile close to the junction. Such differences are significant if the data is to be used in device and process models. However, the only algorithm capable of providing a meaningful result in the underlying substrate is the SSM algorithm.

The remaining algorithms show anomalous behaviour in the substrate. The origin of this anomalous behaviour is not yet understood but cannot be attributed to errors in the implementation of the four algorithms, since they all use a widely different formalism. However, these anomalies serves to demonstrate the stability of the algorithms. The effect of an 'error' in one point of the profile i.e. the deepest point in this case, is quickly damped and does not affect the remainder of the profile. Investigations are currently in process to discover the origin of these anomalies.

Figures 5-8 show results of the use of the algorithms on data obtained on epitaxial material such as would be used for CMOS processing. Figure 5 shows the spreading resistance data which has been smoothed. Figure 6 shows the resistivity profile obtained using the SSM algorithm (on unsmoothed data). Figure 7 shows the profile deduced using either the PROF, PROFA on Berkowitz-Lux algorithms. Again, variations between the results of these algorithms are not visible. Finally, Figure 8 shows the results when using Local Slope. The Local Slope algorithm evidently requires a greater degree of smoothing if the ripples on the profiles are to be eliminated. (It is unlikely that these are 'real' resistivity fluctuations and we believe them to be artefacts). However, when a greater degree of smoothing is used, smearing of the transition layer occurs and the profile becomes distorted. The absolute resistivity values shown in Figure 6 (for the epitaxial layer and substrate) have been shown to be correct using four point probe measurements and SIMS. The values shown in Figure 7 and 8 are incorrect, again due to the lack of contact resistance terms. A peak in resistivity is also evident at the interface. This appears to be an artefact of the algorithms and on physical grounds can be

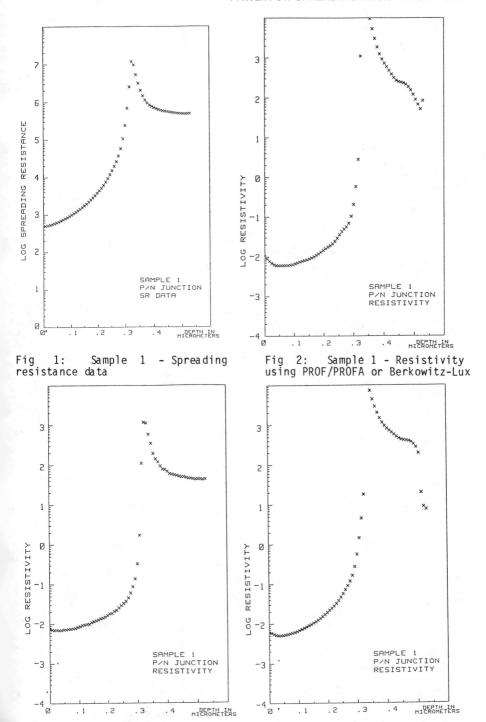

Fig 1: Sample 1 - Spreading resistance data

Fig 2: Sample 1 - Resistivity using PROF/PROFA or Berkowitz-Lux

Fig 3: Sample 1 - Resistivity as calculated using the SSM algorithm

Fig 4: Sample 2 - Resistivity as calculated using Local Slope

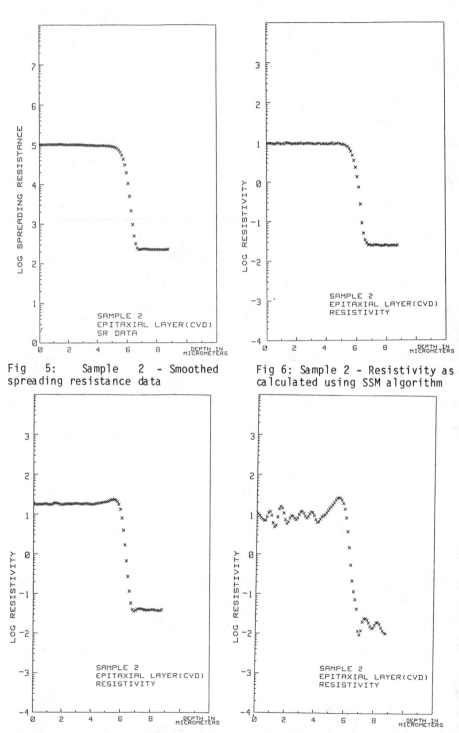

Fig 5: Sample 2 - Smoothed spreading resistance data

Fig 6: Sample 2 - Resistivity as calculated using SSM algorithm

Fig 7: Sample 2 - Resistivity PROF/PROFA or Berkowitz-Lux

Fig 8: Sample 2 - Resistivity as calculated using Local Slope

PAWLIK ON SPREADING RESISTANCE 515

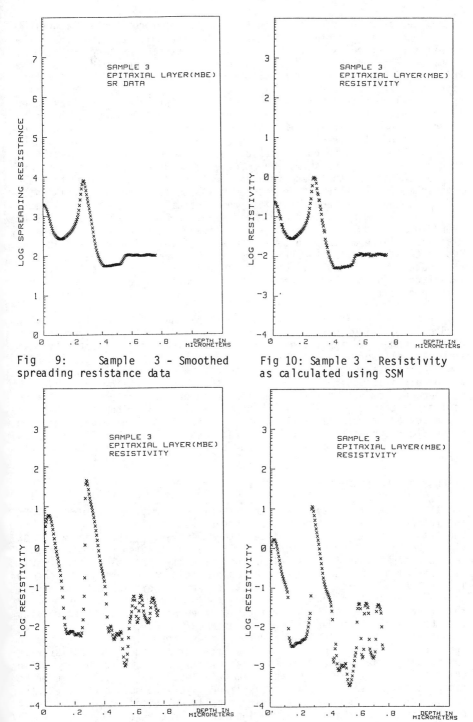

Fig 9: Sample 3 - Smoothed spreading resistance data

Fig 10: Sample 3 - Resistivity as calculated using SSM

Fig 11: Sample 3 - Resistivity using PROF/PROFA or Berkowitz-Lux

Fig 12: Sample 3 - Resistivity as calculated using Local Slope

discounted as a real feature. In this case therefore, the SSM algorithm would appear to perform the best, with the only caution that the transition region is smeared to a slight extent, as can be seen by comparing Figures 6 and 7. The conclusion reached here, about the relative success of the SSM algorithm would however be invalid for thin layers. The SSM algorithm is adequate only on relatively thick epitaxial layers (greater than 1 μm). This point is clearly illustrated in the third example shown in Figures 9-12. This sample, grown by Molecular Beam Epitaxy has two low resistivity regions sandwiched between two high resistivity regions on an n^+ substrate i.e. a high-low-high-low-high doping structure, all within a 0.6 μm total layer thickness. Figure 9 shows the smoothed spreading resistance data and Figure 10 shows the deconvoluted profile obtained using the SSM algorithm. Virtually no correction has been made and the resistivity and spreading resistance profiles are identical in shape. No evidence of the expected structure is seen. Figure 11 shows the PROF/PROFA/Berkowitz-Lux result with the expected structure now clearly visible. There appears to be an excessive level of noise close to the substrate with unexpected structure in the profile. (This has been attributed to a high level of crystalline imperfections at the interface, which are known to disrupt spreading resistance measurements).

The profile determined by the Local Slope algorithm is shown in Figure 12. It follows the same trends as the profile in Figure 11 but with a number of significant differences. The first resistivity peak appears shallower with a significant 'shoulder' just below it. The next low resistivity region shows a gradient whereas Figure 11 shows a reasonably flat bottom. Again, on the next high/low transition a pronounced shoulder appears. The substrate region also shows very large doping fluctuations. Therefore the Local Slope algorithm is unable to reproduce exact details of the profile, details which have been verified using Electrochemical CV and SIMS. The absolute values in Figure 12 are however more exact, again because barrier effects are taken into account.

DISCUSSION AND CONCLUSIONS

It has been found that of the five algorithms, three, PROF, PROFA and Berkowitz-Lux show very similar results and in terms of accuracy there is little to choose between them. However, the Berkowitz-Lux algorithm is considerably faster than the other two (see Table 1) and so can be nominated as the best algorithm currently available. This result is surprising in view of the apparent need for a highly accurate integrator [6]. These three algorithms are at their weakest when used on $n-n^+$ (or $p-p^+$) structures. Since they do not include any barrier resistance corrections, the resistivity predicted on the n^+ substrates can be nearly two orders of magnitude in error. As a direct consequence, the lowly doped layer in these profiles will also be in error since it is corrected on the basis of an incorrect resistivity substrate. The latter becomes particularly problematic on very thin epitaxial layers i.e. those approximately less than an effective probe radius thick. In thin junction isolated layers no such problem exists. Here, barrier resistance effects are totally dominated by the resistance of the layer itself and so may be ignored in practise.

All three algorithms have been found to be stable. Stability is used here in the mathematical sense i.e. if a smooth set of raw data is taken and one point is changed significantly in value, the corrected profile will only show a significant distortion from that calculated from the unperturbed profile within the locality of the perturbed point. Specifically, if the value at the boundary (deepest point) is changed then the profile remains unchanged except in regions close to the boundary.

However, all three algorithms require smoothing of the raw spreading resistance data prior to processing. Even relatively low levels of noise (< 1%) are apparently 'amplified' to give huge fluctuations in resistivity. It should be noted that this is not true amplification of noise but a correct prediction of the algorithms. To understand this phenomenon consider a sample of homogeneous resistivity which is being depth profiled. In the ideal case of no noise in the data, successive measurements of R_s would be identical. With noise in the data they may however be a few percent different. Since in SR, typical depth increments are of the order of a 100 Å the noise appears as an apparent change in resistivity in a very thin layer. Thus, a few percent change in spreading resistance in a very thin layer must represent a very large apparent change in resistivity and an accurate algorithm will interpret this correctly and calculate a large correction factor. This argument does not contradict the earlier statement about stability of the algorithms i.e. if noisy data is considered then perturbing one data point will still lead to identical profiles a few points away from that data point. In this case the 'local' amplification will be even greater if the perturbation increases the difference between adjacent spreading resistance values but within three or four points, the effect of this will be undetectable. The need for data smoothing now poses new challenges. Raw spreading resistance data varies over as much as eight orders of magnitude and this data must be smoothed to show a level of noise less than 0.1%. Under normal circumstances this would be difficult, but spreading resistance data often contains rapid swings which make the problem even harder. Thus again, the problem is most acute on n and n^+ structure but is relatively straightforward on ion implanted or diffused layers. Work is in progress on new approaches to data smoothing of these difficult profiles.

The remaining two algorithms, Local Slope and SSM were found to give consistently poor results on most profiles. The SSM algorithm is fundamentally unstable i.e. changing the value of spreading resistance at the layer boundary (particularly the highest measured value at a junction) is found to affect the entire profile. This algorithm also fails in thin layers where there are deeper values of higher conductivity e.g. an implant with a peak well below the surface. In this case the algorithm fails to correct the profile sufficiently above the peak and hence masks the existence of the peak. However for thin layers where the peak occurs at the surface, a valid correction is made. The SSM algorithm is robust in the sense that data need not be smoothed prior to processing. This desirable feature is achieved by some damping of the size of the correction factor and hence results in a loss of accuracy. Since the SSM algorithm does take barrier resistance terms into acount it is able to predict correct values of resistivity in n on n^+ structures in both n and n^+ regions. But, the corrections in the transition region are in error and so parameters such as the n layer thickness may be wrong. In the case of thick epitaxial layers (greater than approximately 3 μm), the SSM algorithm is found to be adequate. However in very thin layers, involving large correction factors, it has already been shown [15] that this algorithm fails catastrophically.

In view of these results it is considered that the Berkowitz-Lux algorithm provides a fast and sufficiently accurate method for calculating the sampling volume correction factors within the limits and assumptions of the original Schumann-Gardner theory. Further improvements can be made by including barrier resistance corrections and by investigating new data smoothing methods. When modifications have been implemented and their effects analysed a fundamental re-evaluation of the assumptions of the Schumann-Gardner theory must be undertaken. Only then, will the effects of dimensionality (e.g. bevel geometry), true contact models (e.g. multiple contacts) etc on the accuracy of resistivity profiles be determined. Such studies are necessary in view of the constant need for more accurate data on carrier concention profiles, in the semiconductor industry.

ACKNOWLEDGEMENTS

The author would like to thank R D Groves for his assistance in providing the experimental data used in this paper, H Berkowitz-Lux, H Maes and W Vandervorst for their assistance in implementing their algorithms, J Ehrstein and R G Mazur for many helpful discussions and E J Townsend for help in analysing the data. He would also like to thank A Casel for supplying one of the samples used in this work.

The author would like to thank the European Economic Community for partial financial support under the ESPRIT programm.

REFERENCES

[1] Pawlik, M., 'Dopant profiling in silicon', Semiconductor Processing, ASTM STP850, D.C. Gupta, Ed., American Society for Testing and Materials, 1984

[2] Schumann, P.A. and Gardner, E.E., 'Application of multilayer potential distribution to spreading resistance correction factors', J. Electrochemical Society, Vol 116, No 1, 1969, pp 87-91

[3] Leong, M.A., Choo, S.C. and Wang, C.C., 'Spreading resistance calculations for graded structures based on the uniform flux source boundary condition', Solid State Electronics, Vol 20, 1977, pp 255-264

[4] Berkowitz, H.L. and Lux, R.A., 'Errors in resistivities calculated by multilayer analysis of spreading resistances', J. Electrochem. Soc., Vol 126, 1979, pp 1479-1484

[5] Choo, S.C., Leong, M.S. and Kuan, K.L., 'On the calculation of spreading resistance correction factors', Solid State Electronics, Vol 19, No 7, 1976, pp 561-565

[6] Vandervorst, W.B. and Maes, H.E., 'Spreading resistance correction formula more suited for the Gauss Laguerre Quadrature', Solid State Electronics, Vol 24, No 9, 1981, pp 851-856

[7] Albers, J., 'Spreading resistance probe spacing experiment simulations : effects of probe-current density and layer thickness', J. Electrochem. Soc., Vol 25, No 12, 1982, pp 2788

[8] Maes, H.E. and Clarysse, T. Private Communication

[9] Dickey, D.H. and Ehrstein, J.R., 'Semiconductor measurement technology : spreading resistance correction factors for silicon layers with non-uniform resistivity', NBS Special Publication 400-48, 1979, pp 400-448

[10] Solid State Measurements Inc., 110 Technology Drive, Pittsburgh, PA 15275, USA

[11] Berkowitz, H.L. and Lux, R.A., 'An efficient integration technique for use in the multilayer analysis of spreading resistance profiles', J. Electrochem. Soc., Vol 128, No 5, 1981, pp 1137-1141

[12] Piessens, R., Vandervorst, W.B. and Maes, H.E., 'Incorporation of a resistivity dependent contact radius in an accurate integration algorithm for spreading resistance calculations', J. Electrochem. Soc., Vol 130, No 2, 1983, pp 468-474

[13] Solid State Measurements Inc., Users manual

[14] D'Avanzo, D.C., Rung, R.D., Gat, A. and Dutton, R.W., 'High speed implementation and experimental evaluation of multilayer spreading resistance analysis', J. Electrochem. Soc., Vol 125, No 7, 1978, pp 1170-1176

[15] Pawlik, M., 'On the determination of abrupt doping profiles in MBE silicon by spreading resistance', Proc. of the 1st Int. Symp. Silicon Molecular Beam Epitaxy, Vol 85-7, The Electrochemical Society, 1985

TABLE 1: Comparison of algorithms - coding and speed.

Algorithm (reference)	Length of Program (lines)	Length of compiled code (bytes)	Running time per 100 points (secs)
PROF [6]	790	35.8 K	200
PROFA [12]	1170	42.4 K	280
LOCAL SLOPE [9]	135	7.7 K	8
BERKOWITZ-LUX	243	10.7 K	22
SSM [10]	?	?	340

George G. Sweeney and Tony R. Alvarez

COMPARISON OF IMPURITY PROFILES GENERATED BY SPREADING RESISTANCE PROBE AND SECONDARY ION MASS SPECTROMETRY

REFERENCE: Sweeney, G. G., and Alvarez, T. R. "Comparison of Impurity Profiles Measured by Spreading Resistance Probe and Secondary Ion Mass Spectrometry," Emerging Semiconductor Technology, ASTM STP 960, D. C. Gupta and P. H. Langer, Eds., American Society for Testing and Materials, 1986.

ABSTRACT: Silicon wafers, implanted with As or BF_2, were subjected to isochronal anneals and the resulting impurity profiles were measured with both spreading resistance probe (SRP) and secondary ion mass spectrometry (SIMS). The two measurements converged as the anneal temperature was increased. This is to be expected as SRP measures the electrically active impurity concentration, while SIMS measures the atomic concentration of the impurity present. An anomalous kink in the BF_2 profile observed in SIMS measurements was not present in the SRP measurement. Junction depths measured by SRP were consistently shallower than measurements made by SIMS. The SRP and SIMS profiles were compared to those profiles simulated with SUPREM, a process modeling computer program.

KEYWORDS: silicon, arsenic, boron, secondary ion mass spectrometry, spreading resistance, depth profiling

INTRODUCTION

As design rules are scaled down in integrated circuit fabrication, vertical scaling of impurity profiles must also be pursued. It follows that the ability to measure, understand and control impurity profiles becomes more important. This is true both in the vertical and lateral

G. Sweeney and T. Alvarez are senior scientists at Motorola Inc., 2200 West Broadway, Mesa, AZ 85202.

dimensions. Because of the difficulty inherent in lateral profile measurement, this paper is restricted to a discussion of vertical impurity profiles, in particular, shallow junction profiles, typical of advanced Bipolar and MOS processes.

Three techniques are commonly used to measure impurity profiles in silicon: MOS or Schottky CV, SRP, and SIMS (1-3). All three techniques have been automated in recent years and in most cases, measurements can be made by skilled technicians. Because of its simplicity, the CV technique has historically been very popular. However, sample preparation and the inability to profile heavily doped silicon restricts the CV technique to a certain range of impurity levels and applications. The technique therefore requires frequent calibration. SRP is a two-point measurement of the spreading resistance under the point contacts. From the raw spreading resistance data, resistivity, and therefore, impurity concentration is obtained. SRP measures the electrically active impurity concentration; i.e., concentration of impurity atoms on substitutional sites in the silicon lattice. The SIMS measurement generates an impurity profile by detecting secondary ion emission from a host substrate. If suitable standards exist, the ion count is transformed into a quantitative measure of the impurity profile in the host. Note, that while all three measurements profess to measure impurity profile, each measures a slightly different property of the material.

In this investigation, As and BF_2 ions implanted into silicon were subjected to isochronal anneals and the resulting impurity profiles were measured using SRP and SIMS. The profiles obtained were compared to each other as well as to computer simulations using SUPREM III. In the next two sections, an overview of SRP and SIMS measurements is provided.

Spreading Resistance Probe

In the SRP measurements, the spreading resistance of a point contact between the silicon surface and two metal probes is measured. By stepping the probes along a beveled silicon surface, a measurement of spreading resistance versus depth is obtained. The raw resistance data is "corrected", typically using a multilayer analysis, producing a curve of resistivity versus depth. The resistivity profile is then converted to dopant density using empirical curves of resistivity versus doping concentration. The SRP technique is widely used because of its large dynamic range, its ability to measure the electrically active impurity concentration, and the commercial availability of instrumentation that performs the measurement and calculations.

If the surface barrier resistance of a metal-silicon contact is negligible, it has been shown that for a two point probe the spreading resistance is

$$Rs = \rho/2a$$

where ρ is the resistivity of the silicon and "a" is the radius of the point contact (2). In the derivation of Rs, it is assumed that the region surrounding the contact is homogeneous. This assumption is almost always violated in shallow structures. The volume sampled by the measurement extends several micrometers into the silicon, approximately 1.5 times the effective radius of the electrical contact (4,5). If a junction is present, the volume sampled is constrained by the junction depth. Any nonuniformity in the doping profile distorts the potential distribution beneath the probes, resulting in a spreading resistance measurement which depends on the doping throughout the structure. In order to obtain the correct resistivity, this effect must be taken into account when calculating resistivity from the raw spreading resistance data. This is typically done by approximating the continuous concentration profile with multiple layers. Each layer is assumed to be homogeneous, with thickness equal to the spacing of adjacent measurement points. The multilayer approximation allows Laplace's equation to be solved, yielding the potential distribution in each layer (6).

Various techniques have been used to reduce the computational complexity involved in obtaining a solution to the multi-layer problem. Recent efforts by D'Avanzo, et al, reduced the execution time (for an analysis of an epi-buried layer structure) to 2.5 minutes on an HP 2100 (7). Once the resistivity profile is obtained, it is converted into a doping profile using empirical curves of impurity dopant concentration versus resistivity (8).

A number of parameters affect the accuracy of SRP measurements (9). These include SRP load and spacing, sample preparation, angle measurement, and probe preparation. Probe condition appears to be extremely critical. In general, a light probe load (10 - 15 grams) is preferred as it minimizes probe penetration. Too light a load makes it difficult to obtain repeatable measurements. Small probe spacing (20 - 25 µm) has also been found to be preferable. Chemical-mechanical polishing is typically used in order to obtain a consistent finish. Since bevel angles range from 0.4 to 5 degrees, accurate determination of the bevel angle is of considerable importance. A laser interferometer has recently been used to obtain very precise measurement of the bevel angle and surface edge rounding. Probe tip technology is constantly being improved. Conditioning the tip and use of metal alloys are common techniques used to improve measurement repeatability.

Secondary Ion Mass Spectrometry

SIMS employs a primary ion beam to bombard the silicon surface. The ion bombardment causes sputtering of the sample to occur, resulting in secondary ion emission from the silicon surface. The emitted secondary ions are collected and mass analyzed in order to obtain the surface elemental composition. It follows that SIMS measures the atomic concentration of the impurity of interest in silicon. This is in contrast to SRP which measures the net electrically active impurity concentration. Since sputtering is inherent in the measurement, depth profiling is built-in. SIMS is one of the more sensitive of the physical/chemical measurement techniques, being able to detect impurity concentrations as low as $1 \times 10^{14}/cm^3$.

In the simplest case, a SIMS system consists of a primary ion gun, a sample manipulator, a secondary ion energy filter and a quadrapole mass spectrometer, all mounted in an ultrahigh vacuum chamber (10). Typically, more sophisticated instrumentation is used in order to extend the capability of the analysis. In this experiment, a Cameca IMS-3f, a direct-imaging analyzer, was employed. The primary ion beam, generated in a duoplasmatron, is focused by a series of lenses into a target spot of (in this work) 100 μm diameter. Cs is used as the primary beam for creating negative ions (P, As, Sb) and O_2 is used to create positive ions (B, Al). In order to prevent crater artifacts, two things are done: the primary beam is rastered across the sample creating a flat bottom crater (in this work) 250 μm x 250 μm, and a 60 μm diameter field (physical) aperture is used to insure that only secondary ions emitted from the center of the crater are collected. The transfer optics allow analysis of areas of different dimensions, and optimization of the instrument sensitivity for the different areas. The secondary ions are analyzed by a combination spherical electrostatic sector and magnetic prism. The detected secondary ions are recorded using a pulse counting system. If a standard is available, the ion count can be converted into an impurity concentration versus time. The sputter etch rate, assumed constant for each sample, is calculated by dividing the crater depth as measured by a profilometer, by the time spent sputtering. Knowing the sputter etch rate, and impurity concentration versus time, the impurity profile can be generated.

The capabilities of SIMS measurements are characterized by several benchmark operating characteristics: sensitivity, speed, spatial resolution, depth resolution, and the ability to perform quantitative analysis. Unfortunately, sensitivity, data acquisition time, and spatial resolution are not independently adjustable. The output signal of the measurement, ion counts, is given by (11):

ion count = (sputter rate) (data acquisition time per point) (analysis area) (volume concentration) (ion yield) (instrument efficiency).

Volume concentration is the parameter of interest. The ion yield and instrument efficiency are set by use of a particular instrument and the ion being studied. Only sputter rate, data acquisition time, and analysis area are under the investigator's control. While it is possible to obtain sensitivities as low as 1×10^{14} cm^{-3}, sensitivities of 1×10^{15} cm^{-3} to 1×10^{16} cm^{-3} are more realistic. The spatial resolution is limited to 0.5 - 1.0 μm. Depth resolution is limited by ion knock-on and sputter induced nonuniformities to approximately 3 nm.

Accurate quantitative analysis using SIMS is in general a difficult problem. Since use of SIMS is predicated on knowledge of the ion yield, a standard of composition similar to the samples of interest must be available for accurate results. This is especially significant in light of the fact that sensitivity varies by orders of magnitude depending on the element analyzed and the matrix. Silicon standards are readily prepared, and matrix effects are minimal. Consequently, accurate quantitative analysis of common impurities in silicon is possible (12).

EXPERIMENTAL PROCEDURE

Both phosphorus and boron doped (100) wafers with nominal impurity concentrations of 1×10^{15} cm^{-3} were implanted through 10 nm of SiO_2 with As and BF_2. The screen oxide was used in order to eliminate potential surface dopant depletion during subsequent anneals, minimize possible contamination during implantation, and minimize initial ion channeling. Table 1 provides details of the implants and anneals.

TABLE 1 -- Summary of Implants and Anneals

Ion	Dose, cm^{-2}	Energy keV	Temp°C	Time/Min
BF_2	5×10^{15}	60	800	30
BF_2	5×10^{15}	60	900	30
BF_2	5×10^{15}	60	1000	30
As	5×10^{15}	80	800	30
As	5×10^{15}	80	900	30
As	5×10^{15}	80	1000	30

Implantation was carried out using a Varian-Extrion implanter at 7° off-angle. Four-point probe sheet resistance measurements were made on all wafers. Wafers were scribed in half and SRP and SIMS measurements were made on pieces from each wafer. The SRP wafer halves received an additional 1 µm plasma oxide (350°C) deposition to improve measurement accuracy. In addition, three samples were prepared from one wafer for SRP analysis. These three samples were analyzed on three consecutive days; they were repolished between measurements. Junction depth measurements were repeatable to within 25 nm, and peak doping concentration to within 15%.

The SRP measurements were performed on a Solid State Measurements, Inc. 100-C. The measurement conditions were: probe load = 7 grams, bevel angle = .4°, X-step = 1 µm (effective probe step of 17.5 nm), and probe separation = 25 µm. The angle measurement was made using a small angle measurement ring (SAM) that adapts to the instrument for in situ angle measurements. Calibration was performed using NBS standard reference material SRM2526 - 2529. The SIMS measurements were performed on a Cameca IMS-3f. Measurement conditions varied depending on whether As or B was being analyzed. A 10 keV (14.5 keV effective) primary Cs beam at 200 nA was used to analyze the As doped wafers. A 15 keV (10.5 keV effective) primary O_2 beam at 3 microamps was used to analyze the B doped wafers. Data reduction consists of converting the ion intensity or count rate obtained as a function of sputtering time into an elemental concentration scale as a function of depth. The concentration scale was established by normalizing the data to that of a known implant analyzed made under the same analytical conditions. The depth scale was determined by measuring the depth of each sample crater with a Dektak (Sloan Technology Corp.) or an Alpha-step 200 (Tencor Instruments). The silicon sputter rate was approximately 2 nm/second.

SUPREM III is the third version of the Stanford University integrated circuit process simulator program, SUPREM. Several modifications and refinements relative to earlier versions have been implemented. The program is capable of simulating most fabricating steps and is designed to simulate these steps either individually or sequentially, just as they would occur in the actual process. Given an input file that describes the process, SUPREM solves the diffusion equation described by Ficks' law, moving boundary epitaxial growth, and moving boundary oxidation in one dimension and outputs both chemical and electrical concentration profiles. SUPREM version 3-2 distributed by Technology Modeling Associates, Inc. (TMA) was used to simulate the implant and anneal cycles. Simulation is based on several process models for example ion implantation, diffusion, etc., see for example Ho, et al (13). Physical parameters used in these models are stored in the program and constitute the default values. No changes were made to these default model parameters.

Results and Discussion

Composite depth profiles of boron generated from SRP and SIMS measurements and the simulated profiles obtained from SUPREM III are given in Figures 1-3. Table 2 is a summary of the data giving peak concentrations Np, junction depths Xj, and sheet resistance Rs. It should be noted that the SIMS data was extrapolated to 1×10^{15} cm^{-3} to allow comparison to SRP data. Figure 1 shows the boron profiles obtained after an anneal at 800°C for 30 minutes. The SIMS peak ion intensity for boron corresponds to a concentration of 1×10^{21} cm^{-3}, whereas, the SRP data has a peak at 5×10^{19} located at a depth of 0.1 μm, which corresponds to the small kink observed in the boron profile. Although implant conditions (7° tilt, screen oxide) were chosen to minimize channeling, the channeling effect of the implanted boron can be seen as a tail in the SIMS profile. This channeling is caused by preferential penetration of ions which are scattered along paths parallel to the crystal lattice. Since SUPREM does not have a model that takes into account the extra damage generated by the BF_2 implant (compared to B^{11}), the default boron model was used and the implant energy was scaled by 11/49. Recent modifications to the default parameters as suggested by Simard-Narmandin and Slaby (14) will be incorporated into future experiments.

Figure 2 shows the profiles obtained for the 900°C anneal. The SRP data shows a rounder profile especially in the near surface region indicating an increase in the electrical activity of the dopant. The slight kink in the SIMS profile of Figure 1 now shows as a small peak. A number of different explanations for the kink were considered. An implanter with a post acceleration analysis tube was used, minimizing

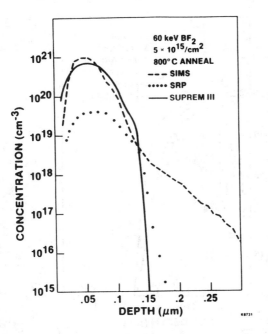

Figure 1. BF$_2$ implant, annealed at 800°C for 30 minutes.

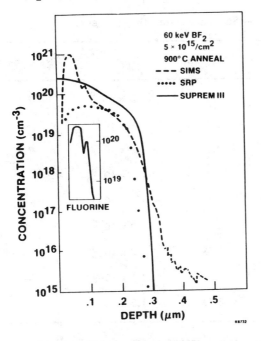

Figure 2. Profiles of BF$_2$ implant annealed at 900°C

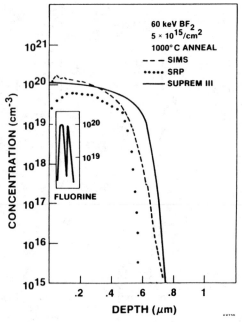

Figure 3. Comparison profiles of BF$_2$ implant annealed at 1000°C

Table 2 -- Summary Data - BF$_2$ Implants

Method	Temp°C	^{a}Xj, m	^{b}Np, cm^{-3}	^{c}Rs ohms/sq.
SRP	800	0.18	7.2 x 10^{19}	230
SIMS	800	0.37	1.0 x 10^{21}	---
SUPREM	800	0.15	6.8 x 10^{20}	216
SRP	900	0.28	7.5 x 10^{19}	118
SIMS	900	0.32	1.0 x 10^{21}	---
SUPREM	900	0.3	3.0 x 10^{20}	59
SRP	1000	0.58	8.2 x 10^{19}	46
SIMS	1000	0.76	1.1 x 10^{20}	---
SUPREM	1000	0.77	1.0 x 10^{20}	26

^{a} Junction Depth; ^{b} Peak Concentration; ^{c} Sheet Resistance

the possibility of B^{11} contamination during BF_2 implant (15). This explanation was rejected because the magnitude of the boron peak in the kink region was greater than 10% of the peak boron doping level. Hofker observed similar kinks in high dose (1.0E15cm^{-2} B^{11}) implants annealed at low temperature (800° - 900°C) (16). He proposed that silicon interstitials created during implantation were mobilized during the anneal process, displacing boron from substitutional to interstitial lattice sites. At high concentrations, the boron interstitials precipitate into cluster formations that are electrically inactive. The kink is apparently due to an electrically inactive precipitated boron phase superimposed on the electrically active profile. At 1000°C, the kink disappears because the maximum boron concentration falls below the solid solubility level, and the precipitates dissolve.

It has been reported that when BF_2 is used as the implant species, it is possible to obtain a near surface amorphous layer (peak damage region) and a highly damaged, but not amorphous subsurface region (17). Wilson has shown that the distribution of both boron and fluorine exhibits different characteristics depending on the implant dose and the magnitude and distribution of the damage that remains after annealing (18). Boron in the amorphous region is activated during low temperature recrystallization. The activation of boron implanted into the damaged (non-amorphous) crystalline region is due to migration of thermally generated vacancies to interstitial B atoms, a process which requires a higher temperature anneal. The subsurface damaged region getters fluorine, resulting in a fluorine peak below the boron peak. While it is not certain that an amorphous surface layer was created in this experiment, the gettering of the fluorine is evident in the SIMS analysis. Thus an alternative explanation for the boron kink is that boron may redistribute preferentially into the damaged, fluorine-rich region during anneal.

Figure 3 shows that as the temperature increases to 1000°C total activation of the electrical impurities is attained. The profiles of SIMS and SUPREM III converge but are deeper than the SRP profile. The SIMS profile has lost its characteristic shape with only a slight indication of a double peak structure. The fluorine distribution is well defined with two peaks, one that correspond to the amorphous region after implantation and the other which is attributed to the plane of damage and acts as a getter site (19).

The composite profiles for arsenic are given in Figures 4 - 6 and the data are summarized in Table 3. Figure 4 shows arsenic profiles obtained after an 800°C anneal for 30 minutes. The peak concentration of SIMS and SUPREM III are in excellent agreement, but as expected the SRP profile is much lower due to lack of activation in the region of the surface.

The bulge in the SIMS profile at approximately 0.15 μm is also reproduced by the SRP data but not simulated by SUPREM III. Junction depths measured by SRP are always less than junction depths measured by SIMS. Hu attributed this difference between the electrical and metallurgical junction depth to carrier spilling (20). He argued that Gauss' law required the potential lines to bend perpendicular to the beveled surface. This is accomplished by a divergence of carrier spilling which becomes two dimensional in the

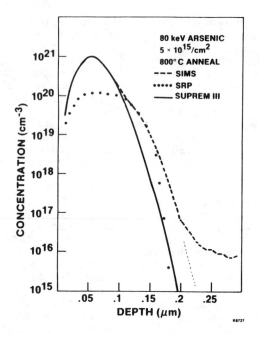

Figure 4. Profiles obtained for an arsenic implant annealed at 800°C

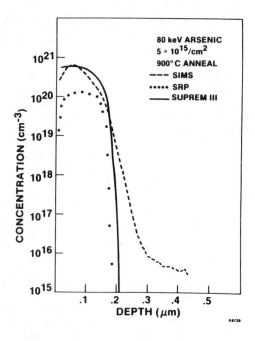

Figure 5. Profiles of arsenic implant, annealed at 900°C

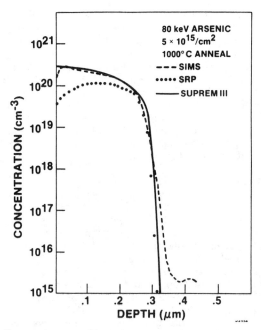

Figure 6. Comparison profiles of arsenic implant, annealed at 1000°C

Table 3 -- Summary Data - Arsenic Implants

Method	Temp°C	[a]X_j, m	[b]N_p, cm^{-3}	[c]R_s ohms/sq.
SRP	800	0.18	1.1 x 10^{20}	115
SIMS	800	0.24	1.0 x 10^{21}	---
SUPREM	800	0.19	1.0 x 10^{21}	80
SRP	900	0.19	1.4 x 10^{20}	88
SIMS	900	0.25	5.8 x 10^{20}	---
SUPREM	900	0.21	5.5 x 10^{20}	53
SRP	1000	0.31	1.1 x 10^{20}	33
SIMS	1000	0.35	2.9 x 10^{20}	---
SUPREM	1000	0.32	2.8 x 10^{20}	29

[a] Junction Depth; [b] Peak Concentration; [c] Sheet Resistance

vicinity of the bevel. This divergence can make the beveled electrical junction depth appear either shallower (for Gaussian, erfc, or exponential profiles) or deeper (superlinear gradient profiles) than the metallurgical junction depth. This is in spite of the fact that (one dimensional) carrier spilling makes the electrical junction depth deeper in the bulk. Using two-dimensional simulations of the beveled surface geometry, Albers, et al confirmed Hu's work (21). They concluded that rapid variation of the ionized atomic profile works against the free carriers remaining in regions of the structure where large gradients exist. Similar to the results observed in this work, their SRP junction depth measurements were always less than the corresponding junction depth measured by SIMS.

The difference between the electrically active and total As concentration observed in Figures 4-6 may be explained by a recently proposed As clustering model (22). The model consists of three As atoms and one electron (23). Above $1.0E20 cm^{-3}$, As atoms form clusters that are electrically inactive at room temperature. The diffusivity of the As clusters is negligible below 1000°C. Above 1000° nearly all the As atoms are electrically active, clusters dissolve and diffuse as separate As species. Therefore, to obtain a high degree of electrical activation, cluster formation should be avoided. These effects are borne out by comparing the SRP and SIMS profiles. By annealing at high temperatures for short periods of time (Dt constant) maximum electrical activation is achieved without a corresponding increase in junction depth. This implies that long temperature ramp downs, following activation anneals, may increase As cluster formation and reduce the conductivity of the implanted layer.

SUMMARY

SRP and SIMS measurements were made on isochronal anneals of BF_2 and As implants. The results indicate a convergence of the profiles as the anneal temperature is increased; indicating, as expected, a greater degree of electrical activation with increasing temperature. BF_2 profiles obtained by SIMS reveal a kink in the diffusion tail that was not clearly evident in the SRP profiles. Hofker's work was used to provide a plausible explanation for the large amount of electrically inactive boron centered around 0.05 µm. The amount of electrically inactive dopant is important because it may play a dual role. The precipitated or clustered impurities may serve as getter sites for heavy metals, or may punch out prismatic dislocation loops that can grow during subsequent temperature cycling resulting in excessive junction leakage. The anomaly around 0.1 µm is only seen in BF_2 implants as indicated by the work of Wilson and may be related to fluorine pile up at the depth of maximum damage. The presence of large amounts of fluorine may in fact increase the ion yield, an artifact often seen in SIMS profiles. SUPREM III results, even without making modifications to default coefficients, are in fair agreement with the data. SUPREM III, however, does not model well the difference between the electrically active and the atomic impurity present. SIMS and SRP measurements are complementary; at times both may be necessary to fully characterize shallow, highly doped implanted layers.

REFERENCE

1. van Gelder, W., Nicollian, E. H., "Silicon Impurity Distribution as Revealed by MOS C-V Measurements," Journal of the Electrochemical Society, 118, 1971, pp 138-141.
2. Mazur, R. G., Dickey, D. H., "A Spreading Resistance Technique for Resistivity Measurements on Silicon," Journal of the Electrochemical Society, 113, 1966, pp 255-259.
3. Helms, C. R., "A Review of Surface Spectroscopies for Semiconductor Characterization," Journal of Vacuum Science Technology, 20, 1982, pp 948-952.
4. Ehrstein, J. R., "Two-Probe (Spreading Resistance) Measurements for Evaluation of Semiconductor Materials and Devices," in Nondestructive Evaluation of Semiconductor Materials and Devices, Plenum., 1979 pp 1-66.
5. Schroen, W. H., "Multilayer Analysis of Spreading Resistance Measurements," in Process and Device Modeling for IC Design, Noorhoff, 1977.
6. Schumann, P. A., Gardner, E. E., "Application of Multilayer Potential Distribution to Spreading Resistance Correction Factors," Journal of the Electrochemical Society, 116, 1969, pp 87-91.
7. D'Avanzo, D. C., Rung, R. D., Gat, A., Dutton, R. W., "High Speed Implementation and Experimental Evaluation of Multilayer Spreading Resistance Analysis," Journal of Electrochemical Society, 125, 1978, pp 1170-1176.
8. "Standard Practice F 723-82," Annual Book of ASTM Standards, American Society for Testing and Materials, Philadelphia, Pa., 1983, Part 10-05.
9. Kulkarni, S. B., "Characterization of Silicon Epitaxial Films," in VLSI Electronics Microstructure Science, Volume 6: Materials and Process Characterization, Academic Press 1983, pp 74-145.
10. Honig, R. E., "Surface and Thin Film Analysis of Semiconductor Materials," Thin Solid Films, 31, 1976, pp 89-122.
11. Helms, C. R., "A Review of Surface Spectroscopies for Semiconductor Characterization," op. cit.
12. Leta, D. P., Morrison, G. H., "Spectrometric Determination of the 1a-7a Group Elements in Semiconductor Mattrices," Analytical Chemistry, 52, 1980, pp 514-519.
13. Ho, C. P., Plummer, J. D., Hansen, S. E., Dutton, R. W., "VLSI Process Modeling - SUPREM III," IEEE Transactions on Electron Devices, Vol. ED-30, No. 11, November 1983, pp 1438-1453.
14. Simard-Narmandin, M., Slaby, C., "Empirical Modeling of Low Energy Boron Implants in Silicon," Journal of the Electrochemical Society, 132, 1985, pp 2218-2223.
15. Glawisching, H., Noack, K., "Ion Implantation System Concepts," in Ion Implantation Science and Technology, Academic Press, 1984, pp 358-361.
16. Hofker, W. K., "Implantation of Boron in Silicon," Thesis, University of Amsterdam, 1975.
17. Kulkarni, S. B., "Characterization of Silicon Epitaxial Films", op. cit. pp 74-145.
18. Wilson, R. G., "Boron, Flourine, and Carrier Profiles for B and BF_2 Implants into Crystalline and Amorphous Si," Journal of Applied Physcis, 54, 1983, pp 6879-6889.

19. Tsai, M. Y., Streetman, B. G., Day, D. S., Williams, P., Evans, C. A., "Recrystallization of Implanted Amorphous Silicon Layers II. Migration of Fluorine in BF_2 Implanted Silicon", Journal of Applied Physics, 59, 1979 pp 188-192.
20. Hu, S. M., "Comparison Between Carrier Distribution and Dopant Atomic Distribution in Beveled Silicon Substrates," Journal of Applied Physics, 53, 1982, pp 1499-1510.
21. Albers, J., Wilson, C. L., Blue, J. L., "Effects of Surface Beveling on Carrier Profiles," Extended Abstracts. The Electrochemical Society, Vol. 831-1, 1983, pp 641-642.
22. Tsai, J. C., "Diffusion" in VLSI Technology, McGraw-Hill, 1983, pp 169-218.
23. Tsai, M. Y., Morehead, F. F., Baglin, J. E. E., "Shallow Junctions by High-Dose As Implants in Si: Experiments and Modeling," Journal of Applied Physics, 51, 1980, pp 3230-3235.

John Albers

MONTE CARLO CALCULATION OF PRIMARY KINEMATIC KNOCK-ON IN SIMS

REFERENCE: Albers, J.,* "Monte Carlo Calculation of Primary Kinematic Knock-On in SIMS," *Emerging Semiconductor Technology, ASTM STP 960*, D. C. Gupta and P. H. Langer, Eds., American Society for Testing and Materials, 1986.

ABSTRACT: Secondary Ion Mass Spectrometry (SIMS) occupies a central position in atomic profiling of semiconductor device structures. One of the possibilities for distortion of the profiles is the phenomenon of knock-on where the incident sputtering ion transfers enough kinetic energy to the impurity atoms to push them deeper into the material before they can be sputtered and counted. The effects of sputtering and primary kinematic knock-on are investigated by means of a Monte Carlo code previously used to study ion implantation processes. In particular, the dependence of the primary kinematic knock-on on the mass and energy of the sputtering ion as well as the mass of the impurity atom are presented.

KEYWORDS: atomic profiling, displacement damage, ion implantation, Monte Carlo calculation, recoil implantation, Secondary Ion Mass Spectrometry, sputtering.

INTRODUCTION

When an energetic ion strikes a solid target, a number of physical processes can take place [1]. In a certain sense, all or most of these processes may occur simultaneously. The cataloging of the processes is intended to focus attention on the principal process in each case.

First, the incident ion can penetrate into the target and come to rest some distance from the surface. As the slowing down is a random process resulting from

---

* Dr. John Albers is a physicist in the Semiconductor Electronics Division at the National Bureau of Standards, Gaithersburg, Maryland 20899. Contribution of the National Bureau of Standards. Not subject to copyright.

collisions of the incident ion with the ions and electrons of the solid target, there will be a depth distribution of incident ions which have come to rest. The incorporation of the stopped ions into the solid (crystalline or amorphous) gives rise to compositional and structural changes in the target material which in turn may dramatically alter its physical and/or electrical properties. Ion implantation of impurities in semiconductors is a prime example of the use of this kind of process. Ion implantation is quite useful not only for the control which can be exercised over incident ion species and energy but also because solid solubility does not limit the compositions which may be attained. That is to say, materials of compositions which are thermodynamically unstable or metastable (but with sufficiently long relaxation times at normal temperatures) and impossible to fabricate using conventional techniques are readily made using implantation. In this respect, implantation is a much more versatile technology when compared to diffusion technology, for example.

The second process which takes place arises from the slowing down of the incident ion by elastic scattering processes with the ion cores of the target. In the elastic scattering process, the total kinetic energy of the colliding pair is conserved. For conceptual and computational purposes, the binding of the target ion to the lattice is not included in the description of the scattering process. After the elastic scattering is completed, the target ion may be viewed as having a certain amount of kinetic energy inside the binding well which can be characterized by a binding or displacement energy, E_d. If the elastic energy transfer is larger than the displacement energy, then target ions can be displaced from lattice sites thus creating damage in the material. The displacement damage which results consists of an ion which has been knocked off a site (creating an interstitial) and leaving behind a vacancy in the site. This damage usually begins as vacancy-interstitial or Frenkel pairs but may undergo recombination or condensation depending upon the physical situation. The resulting damage may be desired for some situations but usually must be removed in semiconductor device fabrication in order to achieve optimal electrical activation and carrier mobility. The processes in which the damage participates form the basis for the study of defects and their interactions. In the above picture, the ion which is knocked off a site is referred to as a primary displacement (or just primary). Of course, the primary may have enough energy to create further damage by extended displacement damage processes. The ions which are displaced by the primary are referred to as secondaries and they themselves may give rise to additional displacement damage. The extended displacement damage network created by the incident ion is referred to as a collision cascade.

The third process is related to the displacement damage. Consider the case of a target consisting of a substrate covered on the top surface with a thin layer of different chemical composition. If the elastic energy transfer is sufficient to displace an ion from a lattice site in the top surface layer, then there is the possibility that the displaced ion may be driven into the underlying substrate portion of the target and thus be considered as an impurity which is being "recoil implanted" into the substrate. One example of this situation is that of a thin SiO_2 layer on top of a Si target. When this target is bombarded with argon ions, both Si and O can be displaced from the SiO$_2$ layer into the underlying Si target [2]. The number and energies of the Si and O recoils will depend upon the scattering process as well as the mass ratios (relative to that of the incident ion). This process can also take place for a nonuniformly doped target.

The fourth process to be discussed is also related to displacement damage effects and usually is the one which arises in the region of low-beam energy. In the normal range of beam energies used in ion implantation (10 to 300 keV), the incident ion moves into the material and away from the surface. Hence, displacement damage is not confined to the region of the target surface. If the beam energy is lowered below a certain value (depending upon the masses of the incident ion and the target ion), displacement damage may become the dominant energy loss mechanism. There may then be sufficient damage confined to the surface region which will overcome the surface binding energy and cause ejection of the surface atoms. The surface atoms are then sputtered from the material. The sputtering may be of importance in and of itself. However, sputtering in combination with mass separation and analysis forms the basis of the Secondary Ion Mass Spectrometry (SIMS) technique of surface analysis. It is important to note that sputtering takes place over a wide range of energies and incident ion masses but is most marked for heavier ions ($Z \geq 8$) at low energies ($E \leq 30$ keV). Sputtering depends upon the mass and energy of the incident ion, the masses of the target atoms as well as the angle of incidence of the ion beam. In addition to sputtered ions, neutral species, electrons, and photons may be emitted from the target during ion bombardment. Further, if the target is composed of several components, the sputtering process may be more rapid in one of the "sublattices" leading to the phenomenon of preferential sputtering.

The remaining processes form the wide class of all other processes that can take place. Most of these are in the categories of thermal, chemical, and/or nonequilibrium thermodynamic processes. Among these are chemical sputtering, radiation-enhanced diffusion, segregation (during preferential sputtering), etc. These processes fall outside of the realm of the present work which focuses upon kinematic phenomena.

The advances in the understanding of these various processes are evident from the frequent extensive conferences concerned with these items [3]. In particular, the SIMS technique has advanced to the point where state-of-the-art instrumentation is commercially available and the results of work in this field are reported on a regular basis [4-7].

Clearly, the above listed physical processes occur simultaneously when an energetic ion beam impinges on a solid target. Consequently, there may be collaboration or competition between these processes. One possibility is the competition between recoil implantation (or collisional mixing) and sputtering during the atomic profiling of a material by the SIMS technique. This recoil implantation in the presence of sputtering, commonly known as knock-on, can present serious difficulties in the interpretation of SIMS data especially if the sputtered ion which is to be monitored and depth profiled is also the same ion which is being driven further into the material. The basis of this situation can be understood as follows. Consider a two-component target which is bombarded by an ion beam which is meant to sputter the target for mass analysis. If the incident ion has high mass and low energy (and the beam tilted away from the target normal), then the incident ion will lose much of its energy by displacement damage in the first 1 to 10 nm (10 to 100 Å). In this picture, the displacement damage in the surface region will contribute to the sputtering process. However, the incident ion will penetrate further into the material than just the surface layer. During its slowing down in the bulk of the material, the incident ion will transfer much of its energy to the two components in displacement processes. The relative contribution to the recoil implantation of each of the two target species will depend upon the scattering processes as well as the masses. The purpose of the present work is to investigate the sputtering and primary kinematic knock-on (recoil implantation) from the vantage point of a Monte Carlo code which has been used previously to investigate ion implantation.

MONTE CARLO CALCULATION

There are two principal techniques which are currently employed for calculating the effects of the interaction of an energetic ion beam with a solid target. The first category consists of the calculations which employ the Boltzmann transport equation [8-12]. The second category makes use of Monte Carlo methods. The work reported in this paper makes use of the TRansport of Ions in Matter (TRIM) Monte Carlo code originally developed by Biersack and Haggmark [13]. This code has been used previously to investigate the various processes which take place during ion implantation of unmasked and masked silicon targets [14-16].

The TRIM code used in the present work makes use of the assumption that the target is random. The randomness of the target is as is determined by the incident ion. That is, the incident ion does not sense any long range order in the target. This condition may be achieved by considering either an amorphous (or heavily damaged) target or a crystalline target which is suitably tilted away from open axes. The computational reason for this choice is that the coordinates of the target atoms do not have to be stored for successive scattering events. This allows for a random number generator to determine the distance between elastic collisions as well as the values of the variables in the geometry of the scattering event. The various processes in which the incident ion participates arise from the elastic collisions with the core ions of the target and the inelastic scattering (viscous drag) with the target electrons.

The elastic two-body collisions are responsible for the displacement damage and are described by means of classical scattering theory [17]. In particular, the calculation focuses upon the asymptotic scattering angle in the center of mass coordinate system which is given by

$$\theta(p, E_r; V(r)) = \pi - 2p \int_{r_0}^{\infty} \frac{dr}{r^2 \left(1 - \frac{p^2}{r^2} - \frac{V(r)}{E_r}\right)^{1/2}}, \qquad (1)$$

where r is the scalar distance between the colliding partners, p is the impact parameter, E_r is the relative energy in the center of mass system, r_0 is the distance of closest approach which is determined from the equation

$$\left(1 - \frac{p^2}{r_0^2} - \frac{V(r_0)}{E_r}\right) = 0, \tag{2}$$

and $V(r)$ is the two-body screened Coulomb potential which is written as

$$V(r) = \frac{Z_1 Z_2 e^2}{ar} \Phi(r), \tag{3}$$

where Z_1 and Z_2 are the atomic numbers of the incident ion and the target ion, respectively, e is the electronic charge, a is the screening length, and $\Phi(r)$ is the screening function. Hundreds or perhaps thousands of two-body elastic collisions may take place before the incident energetic ion comes to rest. In principle, this would require repeated evaluations of the integral in the expression for the asymptotic scattering angle (Eq (1)). The TRIM code circumvents this difficulty by using a fitted algebraic relation between the scattering angle and the impact parameter and the energy for several forms of the screening potential. The krypton-carbon (Kr-C) potential [18] is used in the present calculations as it represents one of the most recent scattering potential functions. The elastic energy transferred to the target atom is determined from the asymptotic scattering angle (which is determined from the impact parameter, the energy, and the specific form of the potential). Specifically, the elastic energy transfer which is responsible for the displacement damage is given by

$$T = \gamma E \sin^2\left(\frac{\theta}{2}\right), \tag{4}$$

where E is the current value of the kinetic energy of the incident ion and the recoil factor, γ, is

$$\gamma = 4 \frac{M_1 M_2}{(M_1 + M_2)^2}, \tag{5}$$

where M_1 and M_2 are the atomic masses of the incident ion and the target ion, respectively. The recoil factor, γ, is plotted in Figure 1 in which it can be seen that the maximum possible elastic energy transfer takes place for equal mass collision partners (where $\gamma=1$) and falls away from unity on either side of this maximum.

Between the two-body elastic collisions, the incident ion undergoes continuous inelastic scattering off the target electrons. This inelastic scattering may be represented as a velocity-dependent viscous drag caused by the electron gas in the target (arising from the outer valence electrons of the target atoms). The inelastic energy loss due to the viscous drag of the electron gas follows from the calculations of Lindhard [19] and Firsov [20] in the low-energy regime (below about 1 MeV for all but the lightest incident ions) and is expressed as

$$\Delta E_e = LNS_e(E), \tag{6}$$

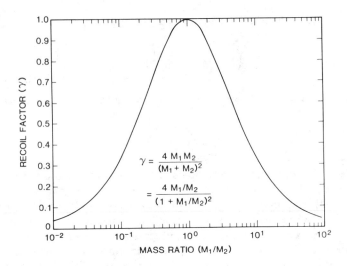

Figure 1 Recoil factor, γ, as a function of the mass ratio, M_2/M_1.

where L is the path length, N is the number density of the target, and $S_e(E)$ is the electronic stopping given by

$$S_e(E) = S_L = kE^{1/2}, \tag{7}$$

where

$$k = k_L = \frac{1.212 Z_1^{7/6} Z_2}{\left(Z_1^{2/3} + Z_2^{2/3}\right)^{3/2} M_1^{1/2}}. \tag{8}$$

The analogy to a viscous drag model is contained in the fact that the electronic stopping is proportional to $E^{1/2}$ or the incident ion velocity. Deviations from the above electron gas model are known to exist [21] and may be accounted for in terms of an energy-independent correction term which multiplies the electronic stopping in Eq (7). The best estimate for the electronic stopping is used in the calculations even though the phenomena of sputtering and knock-on are primarily displacement damage processes (related to the elastic scattering).

In the present calculations, the target consists of a uniform two-component system. The concentration of each of the components is entered in relative atomic fraction (such that the fractions add to unity). The probability that an elastic collision will take place between the incident ion and a particular component is determined by the relative atomic fraction. It is this fraction which enters into the use of a random number generation of a particular colliding pair combination.

The ion is assumed to be incident on the top surface of the target. The trajectory of each ion which is incident on the target is followed through the sequence of binary collisions with the core ions as well as the viscous drag with the electron gas along the path between the collisions with the core ions. If the energy transferred to the lattice ions is below the displacement energy, the collision is considered to excite phonons. If the energy transferred is above the displacement energy (25 eV for silicon), then the energy transfer as well as the target component are recorded. The energy lost to electronic drag is also recorded. These calculations follow the incident ion in the target until its energy falls below a predetermined energy (usually 5 eV). In addition, the number and energy of the incident ions which are backscattered are also calculated. This calculation is repeated for each of the ions incident upon the target until the total number of particles (usually 10^4) in the beam have been considered. Typical calculations of this type take on the order of from about 15 min to about 1 h of CPU on a minicomputer. The times vary directly with energy and inversely with projectile mass. For the total of all of the ion histories, the results of the various energy loss mechanisms and particle processes are stored. In particular, the energy lost to phonon excitation, primary displacement damage (in both the host and impurity "sublattices"), and electronic excitation are stored in one-dimensional arrays. Further, the number of stopped ions and the number of primary displacements (in both the host and impurity "sublattices") are accumulated in another set of one-dimensional arrays. These arrays correspond to the energies and numbers as functions of a discrete depth increment (which is predetermined as input in the code). These energies and numbers depend upon the number of histories and the depth increment used in the calculation. The choice of the number of histories used in the calculation is usually dictated by CPU time constraints. As indicated above, the present calculations make use of 10^4 histories. This has proven to provide adequate statistics with reasonable expenditure of computer time. The depth increment, on the other hand, is chosen based upon a certain amount of experience. If the depth increment is too small, then the calculations will show a fair amount of noise. If the depth increment is too large, some of the important features and structure of the data may be washed out. The important considerations in the choice of the depth increment are the incident ion mass and energy as well as the number density of the target. For the calculations presented here, the depth increment was chosen from 1 to 2 nm (10 to 20 Å) with a major portion of the calculations carried out with the former value. This was found to provide reasonable data. It is important to obtain the information from the calculations in terms of probability distributions (which are independent of the number of histories and the depth increment used). The calculation of these probability distributions is based upon the calculation of the probability distribution for the stopped ions. This distribution is constructed to have the following features. First, it is normalized to unity. Second, when multiplied by the beam dose or fluence, it will yield the number density of stopped ions. If $N_p(i)$ is the number of stopped incident ions which are located in the i-th element in the array, then the normalized (to unity) stopped ion distribution, $F_p(i)$, may be written as

$$F_p(i) = C \times N_p(i). \tag{9}$$

The constant, C, is determined from the unit normalization of $F_p(i)$ as

$$C \sum_{i \geq 1} N_p(i) \Delta x = \sum_{i \geq 1} F_p(i) \Delta x = 1, \tag{10}$$

where Δx is the step increment (in cm). This leads immediately to the following expression for the constant

$$C = \frac{1}{\Delta x \sum_{i \geq 1} N_p(i)}. \tag{11}$$

The constant C is in units of 1/cm such that if the dose, D, (number/cm^2) is multiplied by the stopped ion distribution, $D \times F_p(i)$, then the resulting number density is in units of number/cm^3. If the number and energy loss arrays are multiplied by this factor, then the resulting probability distributions (in 1/cm) and energy loss probability distributions (in eV/cm) are independent of the number of particle histories and the step increment used in the calculation. This is especially important when the number of histories and/or the step increment must be changed in the calculation. The distributions as constructed from the procedure described above provide a useful common ground upon which to study the various processes. In addition, they provide useful statistical information. For example, consider the energy loss probability distributions for phonon, electronic, and displacement damage processes (in both the host and impurity "sublattices") which are denoted by $F_{ephon}(i)$, $F_{eelec}(i)$, $F_{edam1}(i)$, and $F_{edam2}(i)$, respectively. These probability distributions do not integrate to unity. The integrals of these functions represent the energy lost (on the average) by the incident ion to phonon, electronic, and displacement damage processes (in both the host and impurity "sublattices"). Two particle-like probability distributions are also constructed. These are the primary displacement probabilities (or simply "primary probabilities") in the host and impurity "sublattices" which are denoted by $F_{pdd1}(i)$ and $F_{pdd2}(i)$, respectively. The integral of these probabilities yields the average number of primaries created by the incident ion in the respective sublattices. As can be easily seen, these probabilities provide a potential wealth of information about the various processes. The one other piece of information which is of central importance to the analysis is the average energy for the host and impurity primaries. These can be readily calculated as

$$E_{prim1}(i) = \frac{F_{edam1}(i)}{F_{pdd1}(i)}, \tag{12}$$

and

$$E_{prim2}(i) = \frac{F_{edam2}(i)}{F_{pdd2}(i)}, \tag{13}$$

respectively. The primary probabilities and the average energy per primary play a central role in the calculations. The assumption which is implicitly contained in the use of Eqs (12) and (13) is that the average energy provides an adequate description of what takes place in the i-th depth increment. A more complete description would make use of the standard deviation of the energy in the i-th depth increment in order to provide an estimate of the local spread of energies. This is not considered in the present work but will be under future consideration.

The version of TRIM used in these calculations does not follow the trajectory of the primaries created by the incident ion. Consequently, collision cascades which are responsible for the secondary displacement damage created by the primary are not explicitly considered. This would lead to a significant increase in the computation time. Monte Carlo calculations which involve the following of the trajectories of the primaries require several hours of CPU on a supercomputer [22-25].

SPUTTERING

As discussed above, the sputtering process involves the ejection of ions from a solid target by the interaction with the energetic ion beam. The phenomenon has been known for some time and the reader is referred to the two books edited by Behrisch [26,27]. These books contain discussions of physical, chemical, and preferential sputtering as well as sputtering by electrons, photons, and neutrons. Also, the first volume contains an excellent review of the sputtering data for a variety of solid targets. The information pertaining to sputtering is usually reported in terms of the sputtering yield, Y. If n_s is the total number of sputtered ions and n_i is the total number of incident ions, then the total sputtering yield (or just the sputtering yield) is defined as

$$Y = \frac{n_s}{n_i}. \tag{14}$$

The above definition may be used to calculate Y from the number density of the target, the crater depth, the rastered area, and the beam current (or total dose). In addition, differential sputtering yields may be defined in terms of the energy and/or angular dependence of the sputtered ions. Generally, sputtering yields may be between zero and about 10^4. However, typical sputtering yields fall in the range from about 10^{-2} to about 5. The lower end of this range is produced by low mass incident ions (H^+, D^+, and He^+ over a wide energy range from about 0.1 keV to 10 keV) while the upper end is produced by heavier ions ($Y \approx 4$ for Xe^+ at about 100 keV). For oxygen and cesium in the energy range from about 5 keV to 10 keV, the sputtering yields fall in the region of about 0.5 to about 2. In a certain respect, sputtering is a very inefficient method for erosion of a solid since $1/Y$ ions are required to remove a single target ion. However, it is the inefficiency which allows sputtering to be a controlled process and to be used in conjunction with mass analysis. If the sputtering yield were very large, profiling on the nanometer level might be difficult except for low-beam currents.

The assumption which is made in the interpretation of the Monte Carlo calculations is that sputtering is primarily a phenomenon related to the surface displacement damage created by the incident ion beam in the host sublattice. On the other hand, knock-on is envisioned as being determined by the subsurface displacement damage caused by the incident ion in the impurity sublattice. It is important to keep in mind that there are a number of probability distributions for the host which might be used in this study. These are: the energy loss probability distribution for displacement damage, the primary displacement probability distribution, and the average energy per primary.

In order to determine the relation between the surface damage and the sputtering yield, the collection of sputtering yield data as assembled by Blank and Wittmaack [28] was investigated. These data are for the sputtering yield of silicon for the cases of argon and xenon ions incident at energies from 0.1 keV to about 700 keV. The use of argon and xenon provides for sputtering yields for incident ions of about the same mass as silicon and much larger than that of silicon. The energy

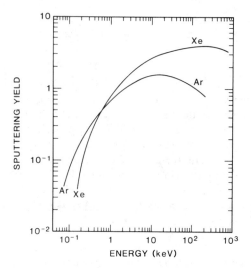

Figure 2 Measured sputtering yield as a function of energy for argon and xenon incident upon silicon (after Blank and Wittmaack [28]).

dependence of the sputtering yield for argon and xenon normally incident on pure silicon targets is presented in Figure 2. There are a number of features which are of interest. At low energies (below about 600 eV), the sputtering yield of argon is larger than that of xenon. At about 600 eV, the sputtering yield curves for these two elements cross. For energies above 600 eV, the sputtering yield for xenon is larger than that of argon. Also the argon sputtering yield goes through a maximum at about 10 to 15 keV and then decreases as the energy increases. For xenon, on the other hand, the sputtering yield increases with beam energy until about 200 to 300 keV where the maximum in the sputtering yield occurs. These maxima may be related to the changes which take place with the relative contribution of the elastic (nuclear) and inelastic (electronic) energy loss mechanisms. As the beam energy increases, elastic processes eventually diminish and move away from the surface and are gradually overcome by inelastic processes which do not contribute to sputtering.

The Monte Carlo calculation was performed for pure silicon targets with argon and xenon ions incident at energies from 100 eV to 100 keV. The surface values of the energy loss probability distribution for displacement damage, the primary displacement probability distribution, and the average energy per primary were then plotted as functions of the beam energy. It was found that the surface values of the displacement damage energy loss probability distribution provided the closest link with the sputtering yield data. For the purposes of brevity and clarity, the surface values of the displacement damage energy loss probability distribution will be referred to as simply the surface displacement damage probability. This notation will be used in the remaining figure where appropriate. Figure 3 contains the calculated surface values of the energy loss displacement damage for argon and xenon ions normally incident upon a silicon target. The interesting thing to note is that these curves reproduce the features of the experimental sputtering yield vs energy curves presented in Figure 2. In the present case, the argon curve lies above that of xenon at low energies (below 300 eV) and crosses at 300 eV. Above 300 eV, the xenon curve lies above that of argon. In addition, the argon curve goes through a maximum over a broad energy range centered at about 10 to 20 keV. These observations lead naturally to a consideration of plotting the measured sputtering yield against the calculated surface value of the displacement damage energy loss probability. This is presented in Figure 4. The general features for both argon and xenon indicate that there is an approximately linear relation between the sputtering yield and the surface damage. The connected curves presented in this figure are presented to show the reasonable nature of the linearity.

In order to establish the general validity of the approximately linear relation presented in Figure 4, calculations were performed with oxygen, cesium, and krypton beams incident upon pure silicon targets. The first two species were chosen as being the ones used in SIMS analysis while the krypton was used as it lies between argon and xenon (in mass) and there were sputtering yield data experimentally available [26]. In all of these cases, the surface displacment damage was calculated and was used to obtain a "predicted" sputtering yield. Using the surface energy displacement damage calculated by the Monte Carlo code and the approximately linear relation in Figure 4 leads to very good agreement with experimentally determined sputtering yields. It is important to emphasize that the results presented in Figure 4 should not be extrapolated beyond the range of surface displacement damages presented. There is the possibility that this relation may break down as more than the surface value may be at play. However, sputtering yields from about 0.5 to about 3.2 cover a rather wide range.

Having established the validity of the approximately linear relation presented in Figure 4, an extensive set of computations were performed with oxygen and cesium beams incident on pure silicon targets. The purpose of these calculations was to investigate the energy and angular dependence of the surface displacement damage as well as the sputtering yield. The beam energies were chosen from 1 keV to 10 keV (in steps of 1 keV) for normal incidence ($\alpha = 0$) in one set of calculations. The results of the energy series for oxygen and cesium at normal incidence are contained in Figures 5 and 6, respectively. The interesting thing to note from Figure 5 is that the surface displacement damage is approximately constant over the range of beam energies considered. This would indicate that the sputtering yield for oxygen incident upon silicon should be fairly constant over this energy interval. This is physically based upon the fact that oxygen being about half the mass of silicon should produce much less surface damage than argon (which is slightly more massive than silicon). On the other hand, the results contained in Figure 6 show that the surface displacement damage for cesium increases markedly with increasing beam energy. In this respect, cesium should show the same kind of sputtering behavior as xenon (which it is next to in the periodic table). In another set of calculations, the beam energies were chosen at 5 keV (for oxygen) and 10 keV (for cesium) and the beam angle was varied. These beam energies are more of less typical of those used in SIMS analysis. The beam angle, α, with the target normal was chosen as 0, 15, 30, 45, 60, and 75 degrees.

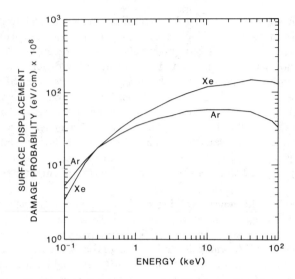

Figure 3 Surface displacement damage probability as a function of energy calculated by the TRIM Monte Carlo code for argon and xenon incident upon silicon at normal incidence.

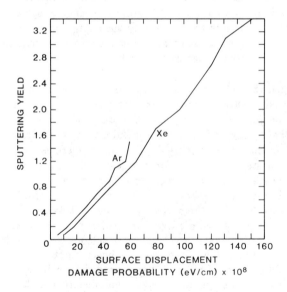

Figure 4 Measured sputtering yield plotted against the calculated surface displacement damage probability.

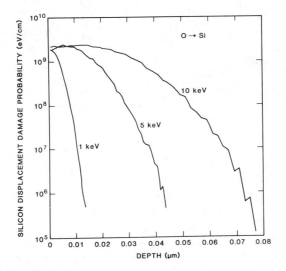

Figure 5 Calculated displacement damage probability for oxygen normally incident upon a silicon target. The beam energies of 1 keV, 5 keV, and 10 keV are marked in the figure. In this and the remaining figures, the data are presented as calculated with no attempt at any smoothing. The lack of smoothness of the results arises from the number of histories used in the calculations.

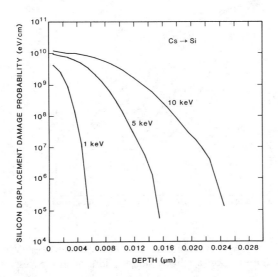

Figure 6 Calculated displacement damage probability for cesium normally incident upon a silicon target. The beam energies of 1 keV, 5 keV, and 10 keV are marked in the figure.

548 EMERGING SEMICONDUCTOR TECHNOLOGY

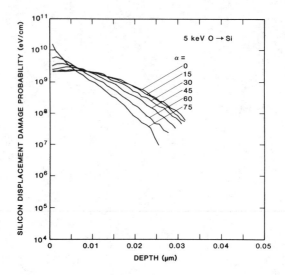

Figure 7 Calculated displacement damage probability for a 5-keV oxygen beam incident at various angles upon a silicon target. The beam angles are marked in the figure.

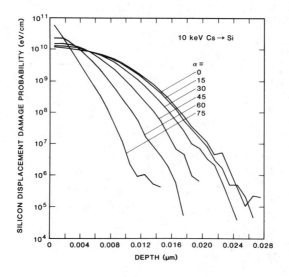

Figure 8 Calculated displacement damage probability for a 10-keV cesium beam incident at various angles upon a silicon target. The beam angles are marked in the figure.

The effects of the angle of the beam from the target normal on the displacement damage probability are depicted in Figures 7 and 8 for 5-keV oxygen and 10-keV cesium, respectively. The changes in the surfaces values are apparent in both cases. The surface values of the displacement damage probability may be used to obtain an estimate of the sputtering yield as a function of the beam angle in these cases. These figures indicate that the sputtering yield increases dramatically with increased beam angle. This is in keeping with experimental observations and may be explained on the grounds that increasing the angle decreases the vertical range of the incident ion and hence leads to higher surface energy deposition.

PRIMARY KNOCK-ON FOR BORON AND ARSENIC IN SILICON

In addition to the above ranges of values of beam energies and beam angles for pure silicon targets, the Monte Carlo calculations were performed on silicon targets with various uniform concentrations of boron and arsenic. As the code is written in terms of the relative atomic concentration (normalized to unity), it is important to keep in mind that the atomic (or number) density of silicon is approximately 5×10^{22} cm^{-3}. The relative atomic concentrations of boron and arsenic were chosen as 0.1, 0.01, 0.001, and 0.0001 which correspond to impurity concentrations of 5×10^{21} cm^{-3}, 5×10^{20} cm^{-3}, 5×10^{19} cm^{-3}, and 5×10^{18} cm^{-3}, respectively. As previously, the number of histories used in the simulation was set at 10^4. The energy loss mechanisms and the primary probability for the host, as well as the atomic distribution for the incident ions, were calculated. In addition the contribution to the impurity was also calculated. In particular, the energy loss into displacement of impurities (creating impurity primaries), the impurity primary probability, and the average energy per impurity primary were calculated. As these were carried out for the 10^4 histories, the best statistics were obtained for the case of a relative impurity concentration of 0.1.

One of the more interesting effects of the presence of the impurity is the effect upon the silicon displacement damage. This is especially pronounced in the case of a 10-keV cesium beam incident upon a silicon target uniformly doped with arsenic. The results of these simulations are contained in Figure 9. The decrease of the silicon displacement damage in the region from the surface down to about 0.01 μm is due to the more favorable energy transfer from the cesium to the arsenic as opposed to the silicon (as can be seen from Figure 1). This is most evident for the relative arsenic concentration of 0.1. For relative arsenic concentrations of 0.01 and 0.001, there is no difference in the silicon displacement damage.

The major thrust of the calculations related to primary kinematic knock-on is the calculation of the impurity primary probability and the average impurity primary energy. When the impurity primary probability is multiplied by the dose of the beam, the result is the number of impurity primaries created by the beam. Within the confines of the present model, this cannot exceed the initial number of impurities. Beyond some critical value of the dose, the description would have to include impurity primaries which are recoil implanted several times. This would be akin to the situation which arises in high fluence ion implantation where the stopped ions may be recoil implanted by the incident beam [29]. The average impurity primary energy determines the average distance which the primary will

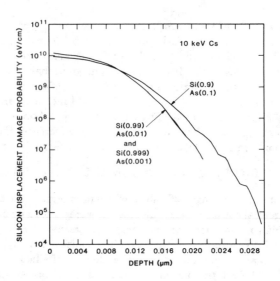

Figure 9 Silicon displacement damage probability for a 10-keV cesium beam normally incident upon a silicon target with several uniform concentrations of arsenic. The relative arsenic concentrations range from 0.1 to 0.001.

travel into the material. In order to provide for the determination of the average distance traveled by the primary, use was made of the Winterbon four-moment code [11,12] to obtain the range of the various impurities as a function of the recoil energy (which is the same as the range as a function of the beam energy if the beam were the impurity). The best known values of the electronic stopping coefficients were used in this code (the value of the electronic stopping correction term was taken as 1.59 for boron recoils and 1.00 for arsenic recoils). A least-squares third order polynomial (in the particle energy) was determined to fit the range vs energy relation for both arsenic and boron (over the range from 50 eV to 10 keV). The average impurity primary energy was used in this polynomial in order to determine the average penetration of the impurity primary recoils. If $F_{pdd2}(x)$ is the primary probability for the impurity at the depth x, then the shift of this function was determined from the average energy to give the displaced primary probability, $F_{pdd2}(x + r)$, where r is the range. The procedure used to determine the displaced primary probability from the primary probability, the average primary energy, and the range-energy curve determined from the Winterbon code is summarized in the plots presented in Figure 10. The depth dependent average primary energy is used to obtain a shift corresponding to the range and the average primary probability curve is shifted accordingly at each depth by the range for that depth.

Typical plots of the displaced impurity primary probabilities for the various situations considered are contained in the remaining figures. The relative concentrations are 0.99 for silicon and 0.01 for the impurity. The beam is normally incident. Figure 11 contains the results of the displaced arsenic primary probability for a

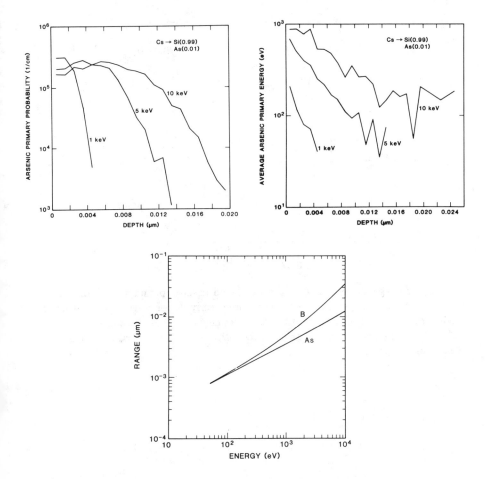

Figure 10 These curves contain the arsenic primary probability, the average arsenic primary energy and the range-energy curves used to construct the displaced arsenic primary probability due to the recoil of the arsenic impurity by the sputtering ion (cesium in this case). This same scheme is used for all displaced impurity primary probabilities presented.

cesium beam at energies of 1 keV, 5 keV, and 10 keV. Similar results were obtained for a boron-doped target as well as arsenic-doped targets bombarded with oxygen beams at the same energies. It is important to note that the displaced primary probabilities have their origins from the slowing down of the incident ion and the energetic recoil of the impurity at each depth. It is the slowing down which determines the magnitude of the function while it is the recoil energy which determines the shifting of the function. The energy dependence of the curves presented in Figure 11 is in keeping with what might be expected from simple kinematic arguments.

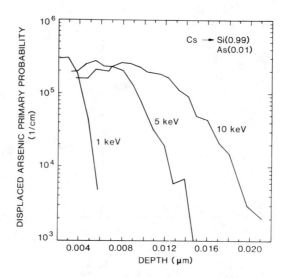

Figure 11 Energy dependence of the displaced arsenic primary probability for a cesium beam incident upon silicon target uniformly doped with 1 percent arsenic. Similar results were obtained for a boron-doped target bombarded with a cesium beam as well as arsenic- and boron-doped targets bombarded with oxygen beams at several energies.

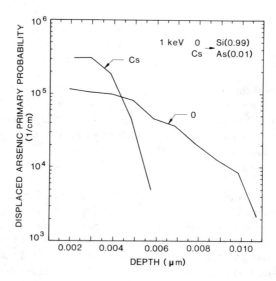

Figure 12 Dependence of the displaced arsenic primary probability on the incident ion beam at fixed energy.

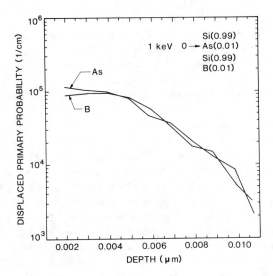

Figure 13 The displaced arsenic and boron primary probability arising from a 1-keV oxygen beam. Similar results were obtained for 5-keV and 10-keV oxygen beams.

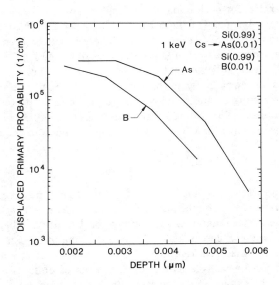

Figure 14 The displaced arsenic and boron primary probability arising from a 1-keV cesium beam. Similar results were obtained for 5-keV and 10-keV cesium beams.

Of considerable interest is the effect of the two different sputtering ion species on the displaced impurity probability. This is illustrated in Figure 12 where the displaced arsenic probability is presented for a 1-keV oxygen beam and a 1-keV cesium beam. There are several features which provide some insight into the dependence of the mass of the sputtering ion on the primary knock-on of the impurity. For small depths, the cesium ion creates approximately three times as many arsenic primaries as does the oxygen ion (for the same beam energy). At some depth, there is a crossing of the curves. Beyond this point, the effect of the cesium drops off rapidly while the oxygen ion is still creating impurity primaries. This can be understood from the following point of view. As the cesium is much more massive than the oxygen, it is expected to create more displacements (in both the host and impurity subsystems). However, because of its larger mass, its penetration into the silicon host will be much smaller than that of oxygen. The same general trend presented in Figure 12 is also found from the calculations for higher energy beams as well as for boron-doped targets. It should be noted that this effect might be accentuated by the fact that the sputtering yield for oxygen is smaller than for cesium which requires a proportionally larger dose in order to sputter to the same depth. It is also important to note that the larger penetration of the primary knock-on due to oxygen was calculated on purely kinematic grounds and that radiation enhanced diffusion and chemical effects have not been invoked.

Another item of interest is the effect of a given sputtering beam and energy on the displacements in arsenic and boron impurity sublattices. Figure 13 contains the results for arsenic and boron impurities with a 1-keV oxygen beam. The feature which is most striking is the fact that the displaced impurity primary probabilities are virtually identical for both impurities. Similar results were obtained for 5-keV and 10-keV oxygen beams. On the other hand, Figure 14 contains the results of the investigation for a 1-keV cesium beam. There it can be seen that the mass effects are quite pronounced. Similar results were obtained for 5-keV and 10-keV cesium beams. These indicate that the kinematic contribution to the primary knock-on should be the same for both arsenic and boron with an incident oxygen beam, while the results should be more marked for arsenic (than boron) for the case of a cesium beam. This is understandable from the fact that arsenic is about seven times more massive than boron and is closer in mass to the cesium ion.

RESULTS AND CONCLUSIONS

The TRIM Monte Carlo code has been used to investigate sputtering and primary kinematic knock-on. In particular, an approximate relation between the sputtering yield and the surface displacement damage probability has been determined. The energy and angular dependence of the sputtering yield can be understood in terms of this approximate relation. Further, the primary kinematic knock-on phenomenon has been investigated for a number of uniformly doped targets bombarded by both oxygen and cesium beams over a range of energies. One of the more interesting effects is the larger knock-on of impurities due to an oxygen beam. This, as discussed above, is easily understood on the grounds of the larger penetration depth of the oxygen ion as compared with the cesium ion.

It is important to emphasize that collision cascades, nonequilibrium thermodynamic processes, and high-fluence effects are not contained in the calculation. However, the model provides a considerable amount of physical insight into the sputtering and knock-on processes. Improvements in the calculation can be achieved by including nonuniform targets and probably an increase in the number of histories (especially in the region of low concentration). Also, the "loading" of the target with the incident ions (changing the composition as well as the inelastic contributions since $\gamma = 1$ would be achieved) is of interest. These items are being considered and may provide insight into a number of experimental observations [30-35].

ACKNOWLEDGMENTS

The author would like to express his appreciation to Dave Simons of the National Bureau of Standards for a number of interesting and enlightening discussions.

REFERENCES

[1] Williams, J. S. and Poate, J. M., Eds., *Ion Implantation and Beam Processing*, Academic Press, New York (1984).
[2] Tsong, I. S. T. and Monkowski, J. R., "Ion-Beam Induced Atomic Mixing at the SiO_2/Si Interface," *Nuclear Instruments and Methods*, Vol. **182/183**, 1981, pp. 237-240.
[3] Manfred Ullrich, B., Ed., *Ion Beam Modification of Materials Parts I and II*, Proceedings of the Fourth International Conference on Ion Beam Modification of Materials held at Cornell University, Ithaca, NY, July 16-20, 1984, North-Holland, Amsterdam, 1985.
[4] Benninghoven, A., Evans, C. A., Powell, R. A., Shimizu, R. and Storm, H. A., Eds., *Secondary Ion Mass Spectrometry SIMS II*, Springer-Verlag, New York, 1979.
[5] Benninghoven, A., Giber, J., Laszio, J., Riedel, M. and Werner, H. W., Eds., *Secondary Ion Mass Spectrometry SIMS III*, Springer-Verlag, New York, 1982.
[6] Benninghoven, A., Okano, J., Shimizu, R. and Werner, H. W., Eds., *Secondary Ion Mass Spectrometry SIMS IV*, Springer-Verlag, New York, 1984.
[7] Benninghoven, A., Colton, R. J., Simons, D. S., and Werner, H. W., Eds., *Secondary Ion Mass Spectrometry SIMS V*, Springer-Verlag, New York, 1986.
[8] Lindhard, J., Scharff, M. and Schiott, H. E., "Range Concepts and Heavy Ion Ranges," *Matematisk-fysiske Meddelelser Det Kongelige Danske Videnskabernes Selskab*, Vol. 33, No. 14, 1963, pp. 1-39.
[9] Sigmund, P., "Mechanism of Ion Beam Induced Mixing of Layered Solids," *Applied Physics*, Vol. A30, 1983, pp. 43-46; "Collision Theory of Displacement Damage, Ion Ranges, and Sputtering," *Reviews of Romanian Physics*, Vol. 17, 1972, pp. 823-870; "Collision Theory of Displacement Damage. IV Ion Range and Sputtering," *Reviews of Romanian Physics*, Vol. 17, 1972, pp. 969-1000.
[10] Gibbons, J. F., Johnson, W. S. and Mylroie, S. W., *Projected Range Statistics*, 2nd Edition, Dowden, Stroudsburg, 1975.
[11] Winterbon, K. B., *Ion Implantation Range and Energy Deposition Distributions Volume 2-Low Incident Ion Energies*, Plenum, New York, 1975.

[12] Winterbon, K. B., "Computing Moments of Implanted-Ion Range and Energy Distributions," Atomic Energy of Canada Limited Report AECL-5536, 1976.

[13] Biersack, J. P. and Haggmark, L. G., "A Monte Carlo Computer Program for the Transport of Energetic Ions in Amorphous Targets," *Nuclear Instruments and Methods*, Vol. **174**, 1980, pp. 257-269.

[14] Albers, J., "Monte Carlo Calculation of One- and Two-Dimensional Particle and Damage Distributions for Ion Implanted Dopants in Silicon," *IEEE Transactions on Electron Devices*, Vol. **ED-32**, 1985, pp. 1930-1939.

[15] Albers, J., "Monte Carlo Calculation of One- and Two-Dimensional Particle and Damage Distributions for Ion Implanted Dopants in Silicon," *IEEE Transactions on Computer-Aided Design of Integrated Circuits and Systems*, Vol. **CAD-4**, 1985, pp. 374-383.

[16] Albers, J., "*Semiconductor Measurement Technology:* Results of the Monte Carlo Calculation of One- and Two-Dimensional Distributions of Particles and Damage: Ion Implanted Dopants in Silicon," NBS Special Publication **400-79**, National Bureau of Standards, Gaithersburg, MD, to be published.

[17] Goldstein, H., *Classical Mechanics*, Addison-Wesley, Reading (1950).

[18] Wilson, W., Haggmark, L., and Biersack, J., "Calculation of Nuclear Stopping, Ranges, and Straggling in the Low-Energy Region," *Physical Review*, Vol. **B15**, 1977, pp. 2458-2468.

[19] Lindhard, J. and Winther, A., "Stopping Power of Electron Gas and Equipartition Rule," *Matematisk-fysiske Meddelelser Det Kongelige Danske Videnskabernes Selskab*, Vol. **34**, No. 4, 1964, pp. 1-21.

[20] Firsov, O. B., "A Quantitative Interpretation of the Mean Electron Excitation Energy in Atomic Collisions," *Soviet Physics JETP*, Vol. **36**, 1959, pp. 1517-1523.

[21] Eisen, F. H., "Channeling of Medium-Mass Ions Through Silicon," *Canadian Journal of Physics*, Vol. **46**, 1968, pp. 561-572.

[22] Biersack, J. P. and Eckstein, W., "Sputtering Studies with the Monte Carlo Program TRIM.SP," *Applied Physics*, Vol. **A 34**, 1984, pp. 73-94.

[23] Möller, W. and Eckstein, W., "TRIDYN - A TRIM Simulation Code Including Dynamic Composition Changes," *Nuclear Instruments and Methods in Physics Research*, Vol. **B2**, 1984, pp. 814-818.

[24] Möller, W. and Eckstein, W., "Ion Mixing and Recoil Implantation Simulations by Means of TRIDYN," *Nuclear Instruments and Methods in Physics Research*, Vol. **B7/8**, 1985, pp. 645-649.

[25] Eckstein, W. and Möller, W., "Computer Simulation of Preferential Sputtering," *Nuclear Instruments and Methods in Physics Research*, Vol. **B7/8**, 1985, pp. 727-734.

[26] Behrisch, R., Ed., *Sputtering by Particle Bombardment I*, Springer-Verlag, New York, 1981.

[27] Behrisch, R., Ed., *Sputtering by Particle Bombardment II*, Springer-Verlag, New York, 1983.

[28] Blank, P. and Wittmaack, K., "Energy and Fluence Dependence of the Sputtering Yield of Silicon Bombarded with Argon and Xenon," *Journal of Applied Physics*, Vol. **50**, 1979, pp. 1519-1528.

[29] Littmark, U. and Hofer, W. O., "Recoil Mixing in High-Fluence Ion Implantation," *Nuclear Instruments and Methods*, Vol. **170**, 1980, pp. 177-181.

[30] Wittmaack, K. and Wach, W., "Profile Distortions and Atomic Mixing In SIMS Analysis Using Oxygen Primary Ions," *Nuclear Instruments and Methods*, Vol. **191**, 1981, pp. 327-334.

[31] Wittmaack, K., "Impact-Energy Dependence of Atomic Mixing and Selective Sputtering of Light Impurities in Cesium-Bombarded Silicon," *Nuclear Instruments and Methods*, Vol. **209/210**, 1983, pp. 191-195.

[32] Boudewijn, P. R., Akerboom, H. W. P., and Kempeners, M. N. C., "Profile Distortion in SIMS," *Spectrochemica Acta*, Vol. **39B**, 1984, pp. 1567-1571.

[33] Vandervorst, W., Maes, H. E., and De Keersmaecker, R. F., "Secondary Ion Mass Spectrometry: Depth Profiling of Shallow As Implants in Silicon and Silicon Dioxide," *Journal of Applied Physics*, Vol. **56**, 1984, pp. 1425-1433.

[34] Wittmaack, K., "Beam-Induced Broadening Effects in Sputter Depth Profiling," *Vacuum*, Vol. **34**, 1984, pp. 119-137.

[35] Albers, J., Roitman, P., and Wilson, C. L., "Verification of Models for Fabrication of Arsenic Source-Drains in VLSI MOSFET's," *IEEE Transactions on Electron Devices*, Vol. **ED-30**, 1983, pp. 1453-1462.

M Pawlik, R D Groves   R A Kubiak   W Y Leong and E H C Parker

A COMPARATIVE STUDY OF CARRIER CONCENTRATION PROFILING TECHNIQUES
IN SILICON: SPREADING RESISTANCE AND ELECTROCHEMICAL CV

REFERENCE: Pawlik, M., Groves, R.D., Kubiak, R.A., Leong, W.Y. and Parker, E.H.C., 'A comparative study of carrier concentration profiling techniques in silicon: Spreading Resistance and Electrochemical CV', Emerging Semiconductor Technology, ASTM STP 960, D.C. Gupta and P H Langer, Eds., American Society for Testing and Materials, 1986.

ABSTRACT: The need to be able to measure carrier concentration profiles in silicon has led to the improvements in well known techniques e.g. spreading resistance (SR) and the development of new techniques e.g. electrochemical CV profiling (ECV). Both of these techniques have demonstrated features such as depth resolution, sensitivity and dynamic range but each show some limitations and problems in practical implementation. This paper demonstrates the use of both techniques on a number of samples, in particular silicon grown by MBE showing rapid dopant transitions. The profiles obtained by both techniques are compared and reasons for differences are discussed.

It is concluded that both techniques offer certain advantages in carrier concentration profiling e.g. dynamic range with SR and speed of data acquisition with ECV but both have limitations e.g. the need for correction factors with SR and limited dynamic range in ECV. As a consequence, both techniques should be viewed as being complementary and not competing.

KEYWORDS: Profiling, spreading resistance, CV, electrochemical CV, carrier concentration profiles, Molecular Beam Epitaxy.

INTRODUCTION

In the semiconductor industry there is a need to be able to measure dopant and carrier concentration profiles in silicon and in III-V materials. Such an outwardly simple task does however pose immense challenges to those developing measurement techniques. The needs of the

M Pawlik and R D Groves are Head and Principal Research Scientist of the Silicon Materials and Characterisation Department, GEC Research Limited, Hirst Research Centre, Wembley, Middlesex HA9 7PP, and R A Kubiak, W Y Leong and E H C Parker are Reader, Head of Silicon MBE and Research Fellow at City of London Polytechnic, 31 Jewry St, London EC3N 2EY.

industry now encompass a wide range of differing sample types and each may pose a different set of problems. Any individual technique may possess characteristics which make it applicable to only a subset of these sample types. Thus when measurement techniques are compared they should be compared with reference to the type of structure under investigation. The 'ideal' measurement technique would have the following characteristics:

(a) Applicability - all sample types including: multiple layers, junction and nonjunction isolated, rapid doping transitions, high and low doped regions, thick and thin layers etc.

(b) Dynamic range - all doping levels of current significance i.e. 10^{21}-10^{11} carriers cm^{-3}.

(c) Dynamic sensitivity - a sensitivity of better than 5% in carrier concentration is required and this must be preserved over the entire carrier concentration range of interest.

(d) Depth resolution - a depth resolution capability is required which is adjustable to suit the circumstances i.e. in very thin structures a resolution of 5-10 Å per point which can be changed to some micrometres per point in thick structures. The depth resolution capability should be maintained irrespective of the carrier concentration levels being measured.

(e) Interferences - No interferences in measurements due e.g. to sample geometry or factors not related to profiles e.g. crystal perfection of sample.

(f) Interpretation - No detailed interpretation or complex data processing should be required.

(g) Practicality - Push button, turnkey operation without the need for skilled personnel. Rapid turnaround with minimal sample preparation.

Naturally, no such 'ideal' technique exists. It is the objective of the paper to compare the results of two techniques which are currently being used to determine carrier concentration profiles in silicon. The first, spreading resistance (SR) has been used for the last twenty years for this purpose. Although development of SR still continues, there are a number of reasons which preclude it from being considered ideal. These fall mainly in categories (e), (f) and (g) above. The second technique, electrochemical CV, (ECV) has been used during the last ten years with great success on III-V semiconductors. During the last few years developments in ECV have been made which have extended the range of applicability of the technique to silicon. Such developments have been rewarded with a profiling capability on some types of sample but limitations in categories (a),(b) and (d) above have been found. It is worth noting that SR has been applied to III-V materials with only very limited success. Due to high contact resistance in III-V materials only a small spreading resistance component is measured which greatly limits the carrier concentration range that can be profiled. The ECV technique is therefore more universal in its application to semiconductor profiling in general.

This paper does not intend to be a tutorial in the applications of either technique. The details of both methods have already or will be published. Initially we discuss the SR technique and critically evaluate it within the context of points (a) to (g) above. A similar discussion of ECV is then given. A number of profiles determined by both techniques are then compared with particular attention being paid to samples grown by Molecular Beam Epitaxy (MBE) having rapid dopant transitions. It should however be emphasised that both techniques discussed here measure carrier concentration profiles and care must be taken in inferring dopant profiles from carrier concentration profiles and vice versa.

Finally, conclusions are drawn about the relative strengths and weaknesses of both techniques.

SPREADING RESISTANCE

The spreading resistance (SR) technique was pioneered by R G Mazur in 1966, and the first practical applications were reported by Mazur and Dickey in reference [1]. During the succeeding 20 years many developments have taken place in the technique and it is now widely used in the semiconductor industry. The fundamentals of the technique are described in reference [2] and for the purposes of this paper it is necessary only to outline the basic ideas behind the method.

In SR the resistance of a metal-semiconductor point contact is measured and from this resistance a 'spreading resistance' component can be deduced which is a function of the resistivity of the underlying semiconductor. If the material is homogeneous then the relationship is:

$$R_M = \rho/2a(\rho) + R_B(\rho) \qquad (1)$$

where R_M is the total measured resistance, ρ is the resistivity, R_B is a resistivity dependent barrier term and a is the effective probe radius. If the material is inhomogeneous then equation (1) must be modified to give:

$$R_M = CF \times \rho/2a(\rho) + R_B(\rho) \qquad (2)$$

where CF is the sampling volume correction factor whose significance is discussed in reference [3]. In order to utilise the method to perform depth resolved measurements of carrier concentrations, a magnification factor is achieved in the vertical scale by shallow angle bevelling. The probes are then stepped down the bevel and the measurement is repeated at each point to give a spreading resistance profile of the semiconductor. This profile is then deconvoluted using the correction factors, calibration data and published carrier concentration - resistivity data.

In order to realise this in practice, the metal-semiconductor contact must be made reproducible over many thousands of measurement points. The preparation of the mechanical probes which attempt to do this remains very much an art, and procedures have been developed which are largely based on experience. Considerations such as depth of probe penetration, area of the contact and ohmic behaviour of the contact over many orders of magnitude of resistivity mean that a very high level of skill must be developed and maintained to keep a SR facility running. The sample preparation by shallow angle bevelling using a rotating glass plate and diamond paste as described in reference [4] has alleviated one of the historical problems of SR i.e. the preparation of electrically stable bevelled surfaces. Although it is not claimed that these surfaces are the best that can be achieved they have proven to be reproducible (over many years) and stable over a period considerably longer than the measurement time required to complete a profile. Finally, angle measurements must be made on every bevel and again, with the advent of small angle measurement systems and stylus profilometers, an accuracy of 2% is readily achieved. Although the collection of data is now achieved automatically, the time taken to prepare a sample and acquire a profile, the need to have skilled operators and frequent 'down-times' due to probe failure limit the throughput of a SR system. Thus in terms of (g) above, SR is far from ideal.

In complete contrast, SR may be used on all types of samples irrespective of the presence of junctions, rapid doping transitions etc. Thus, raw spreading resistance data may be acquired on deep structures (>100 μm) characteristic of power devices or thin (<500 Å) layers such as the bases of state-of-the-art bipolar devices. It is necessary to emphasise that although raw data is readily acquired, the understanding of this data and the conversion to a carrier concentration profile poses a severe problem in many cases, e.g. low dose implants, thin multiple layer samples etc. However, this raw data can often still be of great value in the detection of layers, determination of junction depths etc. Also there are certain classes of sample, e.g. $n-n^+$ epitaxial layers where interferences such as noise in the data preclude any detailed profile structure to be detected in the top n layer. Such a structure is discussed in the Results Section. Therefore in terms of (a) above SR, although not ideal, again has a lot to commend it.

There is virtually no limit to the dynamic range that can be measured by SR. Providing the system hardware is capable of measuring resistance values from $1 - 10^9$ ohms then the carrier concentration range quoted in (b) above can be comfortably covered. The dynamic sensitivity that can be achieved over this range is governed by the quality of the probe-semiconductor contact and the accuracy of the measurement electronics. Thus the dynamic sensitivity is limited by the total noise in the system. Ironically, with the probes conditioned as described in reference [4], the worst sensitivity is achieved in the resistivity range of 0.1 - 10 Ω cm (p-type). Above and below this resistivity range the noise is around 5% and hence a comparable sensitivity in carrier concentration is achieved. In the middle range noise levels of 10% are not uncommon and these limit the sensitivity. The very nature of the technique precludes normal noise reduction techniques (e.g. averaging) and data smoothing is very difficult.

In principle there is no limit to the depth resolution that can be achieved by SR. By the correct choice of bevelling angles (<5'), the use of hard capping layers to aid bevel edge definition, minimising probe contact areas etc an effective depth increment of less than 50 Å can be achieved. It is important to note that such depth increments can be obtained irrespective of doping levels. Further increases in depth resolution will be achieved with only minimal effects on other parameters such as dynamic sensitivity. Therefore in terms of (c) and (d) above, SR has many desirable characteristics.

Unfortunately, there are a number of interferences in SR which limit the accuracy of the technique. As discussed earlier, noise is a limiting factor. It has been found that if the material being profiled contains large numbers of crystalline imperfections, noise levels increase significantly. This is a consequence of the small sampling volume of the SR technique, i.e. the maximum lateral dimension of this volume is less than the probe spacing (~20 µm) and the depth dimension is (in junction isolated structures) limited by the depth of the junction. Therefore, in the presence of a significant number of defects the probes will be sampling regions that lie within the strain fields of individual extended defects such as stacking faults and dislocations. The level of noise will increase as the various barrier resistance and piezoresistive components of the measured resistance change. On bulk samples such effects are troublesome, but if the material contains both rapid dopant transitions and crystalline defects then the difficulties associated with both features can combine to give virtually meaningless data. Thus in terms of both (e) and (f) above SR may have significant problems.

Two other interferences are also of major concern. One is due to Debye length smearing, also called 'carrier spilling' and the other is distortion in the profiles that occurs because of the presence of the bevel. The former limits the resolution of most carrier concentration techniques (ECV and SR) since the Debye length is the minimum distance over which the electric field in a semiconductor may change (by a factor of e). In the presence of a sharp dopant transition this implies that the carriers have to 'spill' and lead to a smeared distribution in response to the local field conditions. Corrections for this effect can be applied to convert carrier concentration profiles into dopant distributions and are sometimes used in CV analysis. Although users of carrier concentration profiling techniques must be aware of this phenomenon it is only when dopant profiles are inferred from carrier concentration profiles, that this effect is relevant. A discussion of these constraints is given by Iyer in reference [5].

The latter interference arises from the effect a shallow angle bevel has on the carrier distribution. Hu in reference [6] has shown, using model calculations, that a large displacement of the carrier distribution can occur in the presence of a bevel. This can either push junctions deeper or shallower depending on the exact nature of the dopant distribution. Although this effect can be very significant, its magnitude is a strong function of the boundary conditions, e.g. surface charge. The surface charges present on a bevelled surface are not known and so it is hard to estimate the magnitude of the effect in practice. It is however anticipated that in lowly doped samples this effect could be severe.

SR data requires detailed interpretation due to the need for sampling volume correction factors. A number of models have been proposed and lengthy computer codes are necessary to implement these models. A discussion of these algorithms is given in reference [3]. The successful application of these algorithms depends on the level of noise in the data and/or the ability to smooth the data. Thus smoothing of data has become as great a problem as choosing the correct algorithm. It should be noted that the sampling volume correction factors can be very large, particularly in thin layers and so cannot be ignored. Further corrections of an even more complex nature will also be needed if bevel edge effects and carrier spilling are to be taken into account. The severity of these interpretation problems depends on the type of structure being profiled and on probe condition and other interferences.

A good spreading resistance facility will be aware of these problems and constraints and will know what degree of confidence to place in the results.

ELECTROCHEMICAL CV

It has been observed that the space charge region within a semiconductor adjacent to an electrolyte, under certain bias potentials approximates to a Schottky barrier [7]. In 1975 Ambridge and Faktor [8] demonstrated the feasibility of utilising the CV characteristic of such a barrier, in conjunction with a simultaneous anodic dissolution to yield carrier concentration profiles in GaAs. These ideas were later extended to silicon by Sharpe and Lilley [9].

The fundamental principle behind the electrochemical CV (ECV) profiling technique is a series of repeated differential capacitance measurements combined with in-situ anodic dissolution. The Schottky barrier at the interface between the electrolyte and semiconductor has a capacitance given by:

$$C = A \sqrt{\varepsilon_r \varepsilon_0 \, nq/2V}$$

where A is the area, n is the carrier concentration at the edge of the depletion layer, C is the capacitance, V is the applied bias, q is the electronic charge and $\varepsilon_r \varepsilon_0$ is the permittivity of the material. The depletion width W_D is related to C by:

$$W_D = \frac{\varepsilon_r \varepsilon_0 A}{C}$$

so the depth at which the carrier concentration measurement is made is given by the sum of this depletion width and the amount of material removed by the dissolution process W_R. This latter parameter can be obtained by integrating the dissolution current I using Faraday's law, i.e.:

$$W_R = \frac{M}{DANF} \int I \, dt$$

where F is the Faraday constant, M and D are the molecular weight and density of the semiconductor and N is effective valence number. The effective valence number represents the number of electronic charges required to remove one molecule of the semiconductor. The inherent advantage of electrochemical CV profiling over conventional CV profiling is that in the latter the maximum depth that can be profiled is limited by the junction breakdown under high bias. In ECV the profile depth is advanced by the anodic dissolution and hence the local carrier concentration can be measured at the optimum bias potential for that concentration level. (The choice of optimum bias level is discussed more fully at the end of this section). Thus ECV can profile through highly doped structures of unlimited depth.

The practical implementation of ECV depends on fewer variables than SR. An electrochemical cell must be constructed which allows the anodic dissolution to be performed over a fixed area of material. The electrolyte must be chosen to give favourable dissolution rates and good Schottky characteristics. Finally, the capacitance measurement system must be optimised. Again, in comparison to SR sample preparation is simple and minimal and involves a surface clean and preparation of non-critical contacts.

We have implemented ECV using a commercially available instrument [10]. The accuracy and reproducibility depends critically on the ideality of the electrochemical characteristics of the electrolyte-semiconductor interface. The accuracy of the carrier concentration value depends on the ideality of the Schottky barrier. The accuracy in depth and depth resolution that can be obtained depend on how well the anodic dissolution process can be controlled as a function of time. Finally, the reproducibility depends on the characteristics of the cell used in the dissolution process.

ECV was originally developed for III-V materials, as near ideal electrolytes were readily available [8,11]. Similar attempts in Si led to usable electrolytes but also revealed a number of problems [9,12,13]. One of these problems was that during the anodic dissolution process, hydrogen bubbles are evolved at the surface which hinder the uniformity of the dissolution process and lead to a degraded depth resolution. This has been overcome by using a pulsed jet of electrolyte in the cell. The electrolyte used is NaF/H_2SO_4 with the ratio and concentration varied to produce an optimum dissolution rate. Rates above 3 $\mu m\ hr^{-1}$ can be achieved using concentrated electrolytes. However for uniformity of dissolution and greater depth resolution, diluted electrolytes are used giving dissolution rates of around 1 $\mu m\ hr^{-1}$. Indeed, in principle the layer removal can be reduced to virtually monolayer level, but in practice the best that can be achieved is about 40 Å per measurement point in depth. Such high resolution profiles can be seen in reference [13] where regularly spaced dopant spikes of 150 Å width and peak carrier concentration of 10^{19} cm^{-3} (in a total layer width of 1.5 μm) have been successfully profiled.

The effective valence number N shows some variability with the anodic stripping conditions and electrolyte strength used. With careful choice of these parameters it is possible to control N to within 5%. Such variability can be compensated for by assuming that N remains constant during the profiling sequence. Using digital data logging, the depth scale can be corrected after all the data has been acquired by measuring the final etch pit depth using a surface profilometer. The final accuracy in the depth scale is therefore only limited by the accuracy of the stylus instrument which can easily be calibrated.

Finally, the effective area A must be determined, which poses particular problems in n type material where illumination must be used to provide a source of holes for the dissolution to take place. A discussion of the problems associated with this part of the process can be found in reference [14]. The actual measurement of the capacitance is made using a modulation technique where a small modulation of the bias potential is added to produce a corresponding modulation of the depletion width and hence the capacitance. To obtain good measurements some flexibility exists in the choice of modulation frequencies etc and the optimisation of this is currently in progress [15].

With the choice of operating conditions outlined above an automated carrier concentration profile is readily generated. In contrast to SR no further data processing is necessary with the carrier concentration profile appearing directly on the chosen output medium.

The problems that arise in ECV are mostly related to sample types. With high doping levels (i.e. above 10^{18} cm^{-3}), the leakage current must be kept low for reliable measurements. At high concentration levels the maximum carrier concentration level that can be realistically measured is 2-3 x 10^{19} cm^{-3} before avalanche breakdown or defects cause sufficient leakage to make the measurement unreliable. At low doping levels (below 10^{15} cm^{-3}) high reverse biasing of the electrolyte -Si interface is required. The need for a high bias potential arises from the requirement that profiling should be undertaken in an 'ideal' part of the CV characteristic. The electrolyte-Si interface deviates from being a perfect Schottky barrier and the extent of the 'ideal' region of the CV characteristic is dependent on the doping level of the sample. Thus for a more highly doped sample the bias potential can be lower and still lie in the ideal part of the CV characteristic. For these reasons, the CV, GV and IV characteristics of each sample are measured initially for each sample to assist in the selection of the bias potential for subsequent measurements. The optimum bias potential is chosen such that the conductance is low and steady with voltage change (as discussed by Sharpe et al in reference [12]) and the CV characteristic lies in the ideal region. Therefore bias conditions must be adjusted during a profile if large swings in carrier concentration are to be measured and optimum conditions are to be maintained. This requires operator input on such samples. If relatively flat profiles are being measured, biasing levels can remain constant. In addition, profiling in surface regions of lowly doped samples is impossible due to surface depletion effects. The need for higher bias potential in lowly doped material does imply a loss in depth resolution in such samples. In practice we have chosen to use high bias potentials and hence measure more accurate carrier concentration levels and hence to accept lower depth resolution.

Significant problems have arisen when profiling through p-n junctions is attempted, but progress has recently been reported in reference [16]. The procedure involves the inclusion of a sample software modification to detect the abrupt change in the phase angle (dC/dV) as the depletion region approaches the p-n junction. When this change is detected the dissolution and measurement parameters are changed to more appropriate values. This has been shown to work particularly well in profiling from p to n. However, with certain combinations of doping levels some difficulties are experienced in profiling from n to p. Although some progress has been made (as will be shown in the results section) interpretation of the profile at the junction must be made with caution.

ECV has been shown to be a technique capable of high depth resolution, good dynamic sensitivity and relatively practical and simple to use. It suffers from few interferences but is limited in terms of applicability and dynamic range. In general samples with very low or very high doping levels and those with multiple-type layers are less suited to ECV. But ECV has distinct advantages as will be seen in the Results Section where isotype structures grown on heavily doped layers are profiled. ECV offers a reasonably rapid measurement with minimum operator intervention and for certain classes of sample offers an attractive alternative to SR.

RESULTS

Three samples have been chosen to demonstrate some of the advantages and disadvantages of both techniques. Two of these (1 and 3) were grown by Molecular Beam Epitaxy (MBE) and the third (2) is a CVD epitaxial layer on a buried layer from a bipolar process. Details of these samples are as follows:

(1) Sample 1

This sample represents an attempt to grow an $n-n^+$ structure by MBE. During deposition the dopant flux had not reached thermal equilibrium and the temperature was oscillating around the final steady state. As a consequence, the dopant flux was also oscillating and the epitaxial layer was not uniformly doped.

Figure 1 shows the carrier concentration as determined by spreading resistance. Although some structure appears in the first 0.6 μm of the layer, this is not the periodic variation that was expected. In fact, such structure in an artifact of the smoothing that was used. The smoothing alogrithm that was used (moving weighted averages) cannot cope with true random noise and produce 'pseudo' features such as those observed. Periodic variations were however detected using chemical staining techniques. Figure 2 shows a number of ECV profiles. As can be seen there appears to be a variation in doping across the wafer. It is therefore not surprising that the epitaxial thickness is different in Figures 1 and 2. Near the centre however, periodic doping variations can clearly be seen whose 'period' matched the variations seen by staining. The differences between the SR and ECV profiles could not be explained in terms of cross wafer variations since the SR data was acquired close to the same spot where the ECV measurement was made. It is concluded that SR is unable to resolve small variations in doping in $n-n^+$ structures. This occurs because the majority of the current flows in the underlying n^+ substrate and very little of the resistance measured is due to the n layer (for reasons of parallel superposition). The magnitude of the doping variations found in this sample thus leads to a spreading resistance variation well below the noise level of the measurement. It should however be noted that the noise level of this sample is much higher than is normally found in n type epitaxial material. During subsequent defect evaluation this sample was found to be heavily defected, which explains the high noise level. In contrast, the ECV profile is unaffected by the presence of these defects. The advantages of having a large sampling area with ECV can be seen since the profile is unaffected by noise. Although

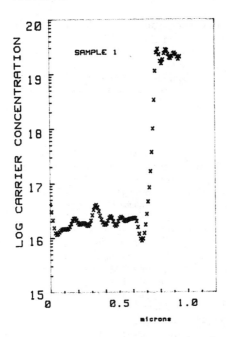

Fig 1: Carrier concentration profile of n-n+ epitaxial layer as determined by SR

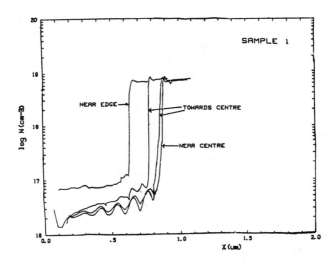

Fig 2: Carrier concentration profiles of the same sample as Fig 1 as determined by ECV. The profiles were measured on different parts of the sample

leakage currents may increase, the capacitor still sees a majority of undefected material and so is able to make a valid measurement. It should also be noted that the ECV profile is unable to correctly predict the carrier concentration of the substrate since it lies above 10^{19} cm^{-3}. Subsequent instrumental improvements have been made in the capacitance measurement part of the system which has resolved this problem.

(2) Sample 2

This sample was grown by conventional CVD epitaxy, the target being a 10^{16} cm^{-3} p type layer over an antimony buried layer and p type substrate. Figure 3 shows the ECV profile which is able to detect both layers and to provide some information on junction regions as discussed in Section 3. There is good agreement between both techniques in ascertaining the buried layer - substrate junction depth as can be seen in Figures 3 and 4. There is however considerable disagreement in the depth of the epitaxial layer - buried layer junction depth. We have no satisfactory explanation of this discrepancy at the present time. The top epitaxial layer in the SR profile is again disturbed by the presence of noise in the data, also caused by crystalline imperfections in the material. This sample effectively shows the capability of ECV in multilayer structures, including junctions. For SR this is a relatively easy sample and hence in any comparisons the SR profiles can be considered definitive.

(3) Sample 3

This sample, again grown by MBE was intended to assess the ability of MBE to produce rapid dopant transitions. A number of doping 'spikes' were introduced into an otherwise intrinsic layer. As such, this structure poses a challenge to any profiling technique as it contains features less than 0.25 µm wide with doping transitions of 3½ orders of magnitude. Figure 5 shows the carrier concentration profile as determined by SR. Four doping spikes are clearly seen and a fifth small feature is visible at the layer/substrate interface. (This last feature has been attributed to boron contamination in the UHV system). The four peaks are separated by the intrinsic layers with doping levels around 5 x 10^{15} cm^{-3}. The spikes are seen to be symmetric with no obvious differences between upward and downward transitions. The doping levels at the peaks of the first three spikes are however higher than expected and the peaks show anomalous structure. Again, noise in the data was correlated with crystalline defects and great care had to be taken in choosing a good smoothing algorithm which did not smear out excessively the features in the profile. The ECV profile is able to show the four peaks clearly as well as the fifth contamination peak as can be seen in Figure 6. The intentional doping peaks occur at different levels which disagree with the SR profile. The levels predicted by ECV are however more reliable at the peaks than the SR result shows. There is believed to be a degree of overcorrection applied, due to difficulties encountered in applying the sampling volume correction factors in such structures, a point further discussed in reference [3]. The data available on the downward transitions is lacking in spatial

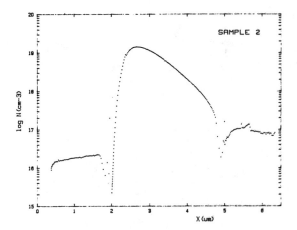

Fig 3: Carrier concentration profile of p-n-p structure as measured by ECV

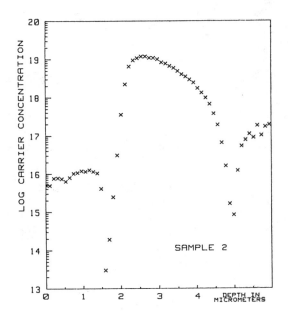

Figure 4: Carrier concentration profile of the same sample as in Figure 3, determined by SR

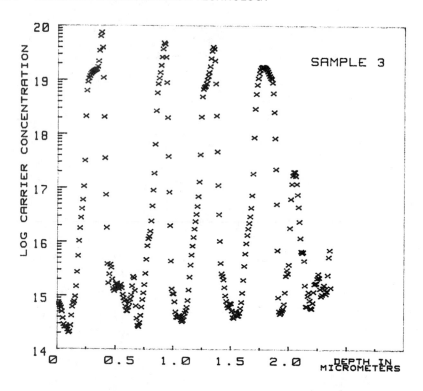

Figure 5: Carrier concentration profile of MBE layer as determined by SR

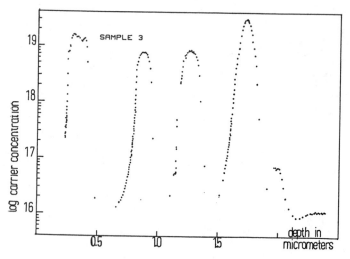

Figure 6: Carrier concentration profile of the same sample as Figure 5 determined by ECV

resolution. Again, there is a lack of data at the contamination spike in contrast to the SR result which is able to track this peak with a constant spatial resolution. Nevertheless, the agreement between Figures 5 and 6 can be considered to be good in particular, with respect to the location of the doping spikes.

CONCLUSIONS

The strengths and weaknesses of SR and ECV have been discussed and compared with three fairly difficult profiles being used for illustrative purposes. It can be concluded that each technique possesses different advantages and disadvantages which means that conclusions about the validity of either technique must be made with reference to the type of sample being analysed. SR is undoubtedly applicable to a wider range of samples but has drawbacks in terms of interferences, interpretation and ease of use. ECV can be used on a more limited range of samples but when properly implementedrsuffers from fewer interferences, requires virtually no interpretation and is much easier to use. ECV can therefore be considered as a superior technique if it is used to assess, e.g. epitaxy from a CVD or MBE deposition system where fast turn round is required. Conversely if an assessment facility is required which is capable of profiling virtually any kind of sample and where turnaround time is less important then SR would be the primary choice.

Comparative studies of results from both techniques must be continued to gain further insight into the limitations of both techniques.

ACKNOWLEDGEMENTS

One of us (MP) would like to acknowledge the assistance of Miss E J Townsend for her help in preparing the data used in this paper.

REFERENCES

[1] Mazur, R. G., and Dickey, D. H., 'A spreading resistance technique for resistivity measurements on silicon', J. Electrochem. Soc, Vol. 113, No. 3, 1966, pp 255-259.
[2] Ehrstein, J. R., 'Two probe (Spreading Resistance) Measurements for Evaluation of Semiconductor Materials and Devices', in Non destructive Evaluation of Semiconductor Materials and Devices, J. N. Zemel, Ed. Plenum Press, New York, 1979, pp 1-66.
[3] Pawlik, M., 'A comparison of sampling volume correction factors on high resolution quantitative spreading resistance', Emerging Semiconductor Technology, ASTM STP 960, D. C. Gupta and P. H. Langer, Eds., American Society for Testing and Materials, 1986.
[4] Pawlik, M., and Groves, R. D., 'Development of Spreading Resistance Techniques for Shallow Dopant Profiling', Recent News Paper 862 163rd Meeting of the Electrochemical Society, May 1983, San Francisco.
[5] Iyer, S. S and Allen, F.G., 'On the determination of sharp doping profiles', Proceedings of the 1st International Symposium on Silicon Molecular Beam Epitaxy, Toronto, Canada, May 7-10th, Vol. 85-7, pp 208-229, The Electrochemical Society, 1985

[6] Hu, S. M., 'Between carrier distributions and dopant atomic distribution in beveled silicon substrates', J. Appl. Phys, Vol. 53, No 3, 1982, pp 1499-1510
[7] Holmes, P. J., 'Electrochemistry of Semiconductors', Academic Press, London, 1972
[8] Ambridge, T., and Faktor, M. M., 'An automatic carrier concentration profile plotter using an electrochemical technique', J. Applied Electrochemistry, Vol. 5, pp 319-328, 1975
[9] Sharpe, C. D., and Lilley, P., 'The Electrolyte-silicon interface: Anodic dissolution and carrier concentration profiling', J. Electrochem. Soc, Vol. 127, No 9, pp 1918-1922, 1980
[10] Bio-Rad Polaron Equipment Ltd, 53-63, Greenhill Crescent, Watford Business Park, Watford, Hertfordshire, UK.
[11] Ambridge T, and Asham, D. J. 'Automatic Electrochemical Profiling of Carrier Concentrations in Indium Phosphide', Electronic Letters, Vol. 15, No 20, pp 647-648, 1979
[12] Sharpe, C. D., Lilley, P., Elliott C. R., and Amridge, T., 'Electrochemical Carrier Concentration Profiling in Silicon', Electronics Letters, Vol. 15, No 20, pp 622-24, 1979
[13] Leong, W. Y., Kubiak, R. A. and Parker, E. H. C, 'Dopant profiling of Si-MBE Material using the Electrochemical CV Technique' Proceedings of the 1st International Symposium on Silicon MBE, Toronto, Canada, May 7-10th, Vol. 85-7, pp 140-148, The Electrochemical Society, 1985
[14] Ambridge, T., Stevenson, J. L., and Redstall, R. M., 'Applications of Electrochemical Methods for Semiconductor Characterisation', J.Electrochem. Soc., Vol. 127, No 1, pp 222-228, 1980
[15] Briggs, A., Paper presented at Polaron Profiler Users Meeting, October 1985, Watford U.K. (unpublished)
[16] Leong, W. Y., Kubiak, R. A., and Parker, E. H. C., Paper presented at Polaron Profiler Users Meeting, October 1985, Watford U.K. (unpublished)

George W. Banke, Jr., Khodadad Varahramyan, and George J. Slusser

ANALYSIS OF BORON PROFILES AS DETERMINED BY SECONDARY ION MASS SPECTROMETRY, SPREADING RESISTANCE, AND PROCESS MODELING

REFERENCE: Banke, Jr., G.W., Varahramyan, K., and Slusser, G.J., "Analysis of Boron Profiles as Determined by Secondary Ion Mass Spectrometry, Spreading Resistance, and Process Modeling," Emerging Semiconductor Technology, ASTM STP 960, D.C. Gupta and P.H. Langer, Eds., American Society for Testing and Materials, 1986.

ABSTRACT: A controlled experiment was performed for a comparative investigation of dopant profiles by secondary ion mass spectrometry (SIMS), spreading resistance (SR), and process modeling. Boron profiles of varying junction depth were considered. These profiles were generated by boron ion implantation in n-type silicon and by varying the subsequent drive-in times. There has been emphasis on the quantification of the day-to-day repeatability of both SIMS and SR techniques and their comparison to process modeling. SUPREM III process simulator was used to generate the modeling profiles. Statistical regression analysis techniques were utilized to characterize the measured dopant profiles. Although there are distinct differences between the profiles determined by SIMS, SR, and Process Modeling, some consistent relationships, as for junction depth, are shown.

KEYWORDS: secondary ion mass spectrometry, spreading resistance, process modeling, boron, junction depth, dopant profiles

Emerging submicron technologies have brought the need for more accurate and reliable dopant profiling techniques. Presently, SIMS and SR are two of the most commonly used techniques for dopant profiling. A wide range of papers have dealt with various aspects of each method and their comparison [1-5]. However, there has not been much focus on the quantitative aspects of measurement quality, including the repeatability of these methods. As the requirements for achieving more shallow dopant profiles increases, the requirement for higher quality measurements also increases. For determination of

---

Mr. Banke, Dr. Varahramyan, and Dr. Slusser are engineers/scientists at IBM Corporation, Essex Junction, VT 05452.

shallow profiles, accuracy and precision become of utmost importance if the measurements are to have meaning.

The purpose of this paper is to investigate differences and similarities between characterization of dopant profiles by SIMS, SR, and process modeling. The emphasis is towards quantification of the quality of both measurements, and the comparison to process modeling. The only designed variable in this experiment was the controlled variation of the drive-in diffusion time. This served to generate profiles of varying junction depth for comparison of SIMS and SR measurements. Boron was the impurity investigated due to its importance as a p-type dopant in silicon. Dopant profiles generated by process modeling have also been examined. It is important to realize that it is not only desirable to have the capability of determining dopant profiles accurately but also in a rapid and economical manner. SIMS and to some extent SR are relatively time consuming and expensive, and thus may not be feasible to use by themselves. An advanced alternative is one based on a complementary use of process modeling to enhance the understanding of the experimental data as well as reducing measurement needs.

EXPERIMENTAL PROCEDURE

Sample Preparation

N-type silicon wafers of <100> orientation and 0.8-2 ohm-cm resistivity were used as starting material. The wafers were oxidized in a dry O_2 ambient at 1000°C to grow 50 nm of screen oxide. Singly-charged boron ions at mass 11 were implanted into the wafers at an energy of 20 keV to a dose of $4 \times 10^{14} cm^{-2}$. The implantations were performed at room temperature and at 7 degrees off normal incidence. The samples were preannealed at 600°C for 1 hour in a nitrogen ambient. The final boron profiles were achieved by drive-in in a nitrogen ambient. Table 1 provides the drive-in conditions used for the samples considered in this study. The substrate dopant concentration values given in Table 1 are average values from spreading resistance measurements conducted in two different labs.

TABLE 1--List of samples and corresponding drive-in conditions considered in this study.

Sample	Time (min)	Temp (°C)	Substrate Doping (cm^{-3})
A	15	950	3.9×10^{15}
B	30	950	3.9×10^{15}
C	60	950	5.1×10^{15}
D	120	950	5.8×10^{15}
E	240	950	3.3×10^{15}

SIMS and SR Measurements

SIMS analysis was performed utilizing a Cameca IMS-300 unit. An O_2^+ ion beam at an energy of 11 keV and current density of 90 μA/cm² impinged the surface of the sample which was biased at positive 4.5 kV. Secondary ions were collected from a 400 μm circle centered in the 750 μm square raster. An oxygen bleed was supplied at the sample surface in order to minimize SIMS artifacts. Mass 11 was monitored for boron and ratioed to the silicon signal at mass 30. This boron/silicon ratio was compared to a boron implant standard in order to interpret the signal to an appropriate boron concentration. The depth scale was calculated by measurement of the resultant sputter crater using a surface profilometer.

The SR measurements were made using a Solid State Measurements, Inc., ASR-100 system. The samples were beveled using a Leco Vari/Pol VP-50 polishing table with 0.1 μm diamond paste. The bevel angles were measured using a goniometer. The angles for the samples ranged from 15 to 40 minutes of arc, as required by junction depth. A nominal probe step size of 1 μm and a nominal probe spacing of 5 μm were used. Probes were fabricated from Fidelitone instrument pivots having tungsten-ruthenium alloy tips. Internally generated <100> standards were used to calibrate the system. The Schumann-Gardner [6] algorithm was chosen to correct the data. Irvin's [7] data were used to convert the resistivity values to dopant concentration.

SIMS and SR measurements were both conducted on samples taken from the center (within a 10 mm radius) of each wafer. The center of the wafers was shown to have 3-sigma variations in sheet resistance of less than 2.5% as determined by four-point probe. The profile measurements were taken on three different days, to estimate the day-to-day variation of each technique. This resulted in multiple profiles for each sample. Statistical regression modeling is well-suited for this type of experimental design.

The statistical model used is that model determined to best represent the relationship between measured dopant concentration and sample depth. Regression analysis techniques [8] were chosen because this type of mathematical analysis allows the estimation of confidence bounds of the best-fit dopant profile. The confidence bounds estimate the expected limits of the mean concentration at a given depth. The statistical modeling was done as the logarithm of concentration to better represent the entire range of concentration values. The following regression model was used:

$$\log_{10} C = a_0 + a_1 x + a_2 x^2 + \cdots + a_n x^n + \varepsilon \qquad (1)$$

where

C = concentration values,
x = depth into the sample,
ε = random variable, normally distributed with a mean of 0,
$a_0, a_1, a_2, \cdots, a_n$ = coefficients of the linear regression.

Process Modeling

Over the past several years improved understanding of process physics has allowed the development of increasingly accurate models for semiconductor fabrication processes. The implementation of these models in computer programs has made possible simulation of entire sequences of process steps. In this work, process modeling was used as a complementary method for determination of the dopant profiles and their analyses. This has been accomplished using the SUPREM III process simulator [9]. In particular, the profiles have been generated using what here will be referred to as models 1, 2, and 3. Model 1 specifies the case where the simulations have been conducted using the default boron diffusivity values [9,10] in SUPREM III, as opposed to model 2 where a recently suggested set of boron diffusivity values [11] have been adopted. Model 3 will be defined in the Results and Discussion section.

Junction Depth Determination

Junction depth is one of the most important parameters associated with dopant profiles in semiconductors, and its accurate determination is of great importance. However, due to differences in the nature of various measurement techniques, junction depths may be obtained differently. For SR, junction depth is determined from the point where maximum resistance occurs. This may also be defined as an electrical junction. The SIMS technique is less suitable for direct measurement of junction depth. In this case, junction depth can be realized from the point where the dopant concentration becomes equal to the opposite-type substrate concentration. This is also known as the metallurgical junction. For the present study, the wafers used are phosphorus-doped at relatively low concentrations where SIMS measurements become too noisy. Thus, the SR-determined substrate doping values shown in Table 1 have been used. The process modeling junction depths were determined in the same manner as SIMS.

RESULTS AND DISCUSSION

Each sample was measured by SIMS and SR techniques on three different days in order to estimate the measurement repeatability and the bounds of the best-fit statistical model. Figures 1 and 2 show, respectively, the three SIMS and SR profile measurements for sample D along with the best-fit regression line and the corresponding 95% confidence bounds. If future measurements were made under the same conditions, then 95% of the future confidence bounds would contain the true mean of the concentration at a given depth. As illustrated in Fig. 1, generally, the 95% confidence bounds for SIMS measurements came very close to the best-fit regression line and are barely visible. Also note that the discrete SIMS data in the 0.45 μm region appear as horizontal lines due to the close proximity of the data points. Due to the more complex shape of the SR profiles, the SR data typically required more terms in Eq. 1 than the SIMS data. Note that the SR profiles exhibit some rather interesting characteristics that show a distinct transition region from 0.2 μm to 0.3 μm between the high concentration and the tail regions. It is not known whether these regions are real, or part of a measurement artifact. In addition, it is clearly shown from Figs. 1 and 2 that the predominant measurement error is the day-to-day variation.

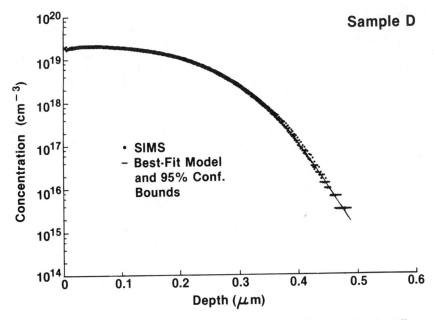

FIG. 1--SIMS profiles of boron taken on three different days, and the best-fit regression curve with corresponding 95% confidence bounds.

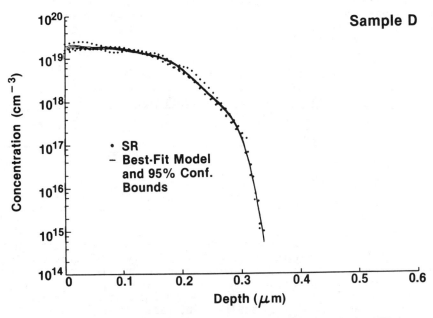

FIG. 2--SR profiles of boron taken on three different days, and the best-fit regression curve with corresponding 95% confidence bounds.

Precision estimates of dopant concentration at a given depth is another feature obtained from the regression analysis. This estimates the day-to-day repeatability, which is derived from the relative standard error of the estimate. These estimates are shown in Table 2 for the SIMS and SR measurements of samples A through E. Note that the SIMS precision is from 25% to 50% better than the SR precision, except for sample A. It is unlikely that the error in the SIMS measurement of sample A would significantly increase solely due to random variation in concentration or depth. As in the case of the SIMS measurements, the SR error increases for samples with more shallow profiles. This demonstrates the increased difficulty in measuring shallow concentration profiles.

TABLE 2--Relative 1-sigma standard error (±1s) of the estimate in the concentration

SAMPLE	SR ±1s(%)	SIMS ±1s(%)
A	47	75
B	28	21
C	34	24
D	22	12
E	34	15

The junction depth precision estimates were obtained directly from the measured data. These estimates are shown in Table 3, where the precision is defined as the (3-sigma) standard deviation normalized with respect to the junction depth average. It is observed that while the SR precisions are more consistent from sample to sample and worsen for samples with more shallow profiles, the SIMS precisions are noticeably more variable. As in the case of the concentration precision, the SIMS junction depth precision decreases substantially for sample A.

TABLE 3--Relative 3-sigma precision (±3s) in the junction depth and the average junction depth ($\overline{X_j}$)

SAMPLE	SR		SIMS	
	$\overline{X_j}$ (μm)	±3s(%)	$\overline{X_j}$ (μm)	±3s(%)
A	0.24	9	0.35	42
B	0.27	8	0.39	1
C	0.28	9	0.40	9
D	0.33	4	0.45	1
E	0.42	4	0.54	4

Dopant profiles for samples A through E have also been determined by the SUPREM III modeling program. Figure 3 shows SIMS and modeling profiles for sample D. Model-generated boron profiles tend to give lower concentration values than SIMS in the region of maximum concentration. For sample D, using models 1 and 2, this difference in concentration is as much as 20% and 16%, respectively. In model 1 there is a more rapid increase in diffusivity due to doping as compared to model 2. It is well known [12] that at high doping levels the diffusivity becomes concentration dependent. It appears that, at higher doping levels, the concentration-dependent diffusivity values used in models 1 and 2 are larger than indicated by measured results.

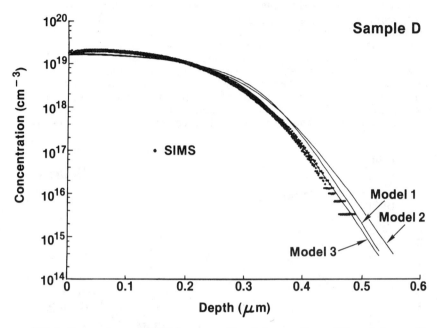

FIG. 3-- Comparison of SIMS boron profiles, taken on three different days, with those generated by process modeling.

Considering the junction depths as determined by SIMS and process modeling, the deviations in the junction depth become larger for profiles with longer drive-in times. For sample D, using model 1 and 2, the junction depths are found to be, respectively, 6% and 10% deeper than the average value by SIMS. This may be attributed to the magnitude of the intrinsic diffusivity used in modeling. In model 1, the intrinsic diffusivity is about 16% less than the one in model 2. At doping levels below the intrinsic carrier concentration, diffusivity becomes its intrinsic value. Thus, deeper junction depths could be predicted by modeling when the intrinsic diffusivity is larger than expected.

The above results indicate that while profiles by model 1 are in better agreement with the corresponding SIMS measurements near the junction region, model 2 profiles compare better with the SIMS results in the region where diffusivity becomes concentration dependent. It appears that a combination of these two models would give a better choice for boron diffusivity. Thus, model 3 is defined as one where, at doping levels below the intrinsic carrier concentration the diffusivity approaches the intrinsic value as in model 1, and at higher concentrations it reaches values as in model 2. As shown in Fig. 3, the comparison with SIMS is better for the profile generated by model 3. In the remaining parts of this paper, all of the modeling results have been obtained by model 3. Work is underway to formulate an improved model for the diffusion of boron at high concentrations for the two-dimensional process simulator FEDSS [13].

Figure 4 illustrates the statistical estimates of the SIMS and SR dopant profiles and the process modeling profile for sample D. It is seen that SIMS and modeling profiles, which are in fairly good agreement with each other, do not compare well with the SR profile. At high concentrations, the three set of plots agree reasonably. In the case of the other samples, the high concentration SR values were about 10% to 20% less than the SIMS. The differences become greater in the junction region. All of the samples measured consistently showed about 1×10^{19} cm^{-3} as the concentration value corresponding to the point at which the SIMS and SR-determined profiles begin to

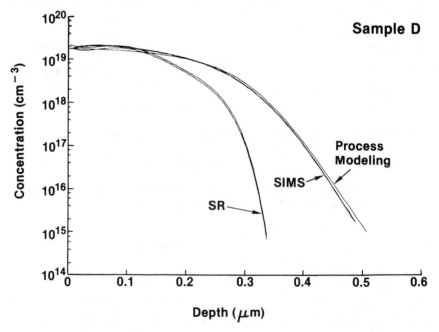

FIG. 4--95% confidence boundaries of the best-fit curves for SIMS and SR boron profiles and the process modeling-generated profile.

diverge. As shown in recent work [14], for a variety of boron profiles, the onset of divergence between SIMS and SR profiles was observed to occur at high concentrations. The sharp decay in the tail region of the SR profile relative to the one by SIMS has been observed for all the samples measured and is consistent with previous experimental results [14,15]. This, in part, is attributed to the fact that while SIMS and modeling directly determine the total atomic distribution, SR measures the carrier distribution associated with the atomic profile. Furthermore, a beveled structure is used by SR to measure the vertical carrier profile within the sample. To consider that the vertical carrier profile is simply linearly imaged on the bevel with a magnified spatial scale is not a valid assumption [16]. The carrier profile measured along the bevel is influenced by several factors, including surface charge and the bevel structure itself. Profiles that gradually decay in their tail region tend to display an electrical junction shorter than the metallurgical junction on the bevel. This is consistent with the findings reported here.

The SIMS, SR, and process modeling determined junction depth values versus the drive-in time are plotted in Fig. 5. The respective best-fit linear regression lines are also shown. For the given range of data, all three junction depth determinations were approximated by a linear function of drive-in time. The linear approximation gave a better fit than a power-law function. It is interesting to note that the junction depths determined by SIMS and SR have almost the same functional relationship with drive-in time, while the process modeling has a slightly steeper slope with time. The model does not appear to entirely explain the variation of the junction depth as a function of drive-in time, as determined by SIMS. There is an observed offset of about 0.1 μm between the SIMS and SR regression lines. This offset is nearly constant for the range of measured junction depths.

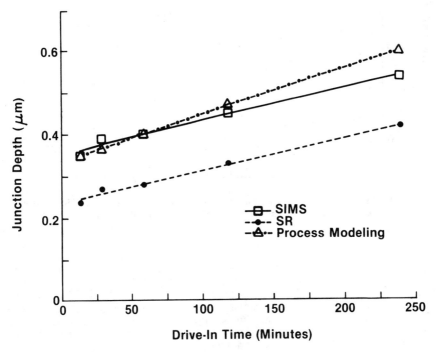

FIG. 5--SIMS, SR, and process modeling junction depths versus drive-in time, with the respective best-fit linear regression lines.

Figure 6 shows the junction depths and the best-fit linear correlation lines of SR and process modeling versus SIMS data. For reference, the SIMS versus SIMS data (x=y line) is also illustrated.

As also noted in Fig. 5, the functional relationship between SR and SIMS is very good. The offset is nearly constant for the range of measured junction depth values because the slope of the correlation line is close to unity (0.96). Extrapolation of these correlations are not valid outside the experimental range considered.

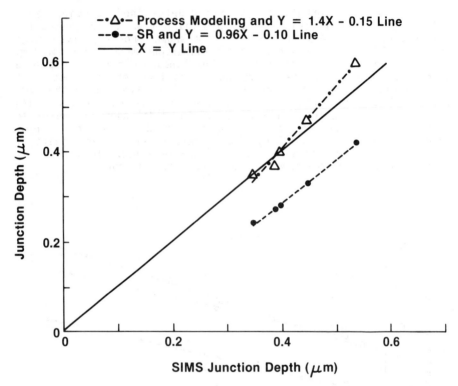

FIG. 6.-- SIMS, SR, and process modeling junction depths versus SIMS junction depth, with the respective best-fit linear correlation lines.

The ratio of the SR- to SIMS-determined junction depth versus the SIMS-determined junction depth has also been plotted as illustrated in Fig. 7. This is a similar plot to that presented by Hu [16], assuming that the SIMS- and SR-determined junction depths represent, respectively, the metallurgical and the on-bevel electrical junctions. From Fig. 11 of Hu's paper it can be observed for a Gaussian type of profile, there is a linear region in the range less than 0.5 μm where the slope of the ratio of the on-bevel electrical to metallurgical junction depth versus metallurgical junction depth becomes constant. The profiles obtained in this investigation are roughly Gaussian, and the results tend to compare favorably with the analysis by Hu. A more detailed comparison would not be appropriate here without considering all of the experimental conditions of this study. Such an analysis is beyond the scope of the present paper.

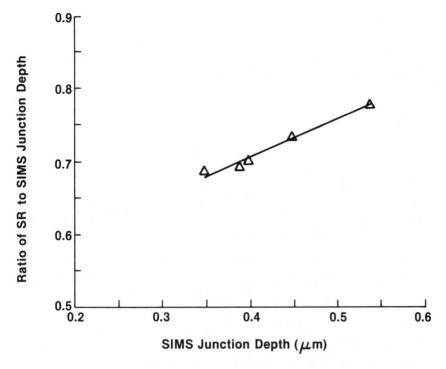

FIG. 7--Ratio of SR to SIMS junction depths versus SIMS junction depth, and the best-fit linear regression line.

SUMMARY AND CONCLUSIONS

In this work, similarities and differences between characterization of boron profiles by SIMS, SR, and process modeling have been investigated. In particular, the quantitative aspects of measurement quality have been realized. It has been found that SIMS concentration precision is overall better than SR. The day-to-day concentration repeatability (1-sigma) for SIMS and SR, respectively, range from 12% to 75% and 22% to 47% for the various samples considered. SIMS showed difficulty in repeatably measuring profiles in the range of 0.3 μm junction depth. For the given range of data, the SIMS, SR, and process model determined junction depths were best fitted by a linear function of drive-in time, which varied from 15 to 240 minutes, at 950°C. Overall, process modeling junction depths compared fairly well with SIMS, except it showed a slightly steeper slope with time. The SIMS junction-depth repeatability is not as consistent from sample to sample as SR. A consistent 0.1 μm offset between SIMS and SR has been observed over the entire range of measured junction depths. A qualitative comparison to Hu's work was also made, which showed reasonable support of the given theoretical relationship between the on-bevel electrical and metallurgical junction.

ACKNOWLEDGMENTS

The authors would like to thank G. W. Blanchard for the SIMS measurements and R. H. Carter for the SR and four-point probe measurements.

REFERENCES

[1] Wagner, H. H., Schaefer, R. R., and Kempf, J. E., "Comparison of Characterization Methods for As-doped Silicon," Journal of Applied Physics, Vol. 52, No.10, October 1981, pp. 6173-6177.
[2] Goldbach, G., Roesch, E., and Wallis, D., "Dopant Profiling Techniques in Semiconductor Technology," Research Report BMFT-FB-T 83-126, Federal Ministry for Research and Technology, Bonn, W. Germany, June 1983.
[3] Albers, J., Roitman, P., and Wilson, C. L., "Verification of Models for Fabrication of Arsenic Source - Drains in VLSI MOSFET's," IEEE Transactions on Electron Devices, Vol. ED-30, No. 11, November 1983, pp. 1453-1462.
[4] Cohen, S. S., Norton, J. F., and Koch, E. F., "Shallow Boron-Doped Junctions in Silicon," Journal of Applied Physics, Vol. 57, No. 4, February 1985, pp. 1200-1213.
[5] McMahon, R. A., Hasko, D. G., and Ahmed, H., "Electron Beam Processing for MOS Devices with Shallow Junctions," Solid State Technology, Vol. 28, No. 6, June 1985, pp. 208-215.
[6] Schumann, P. A. and Gardner, E. E., "Application of Multilayer Potential Distribution to Spreading Resistance Correction Factors," Journal of Electrochemical Society, Vol. 116, No. 1, January 1969, pp. 87-91.
[7] Irvin, J. C., "Resistivity of Bulk Silicon and of Diffused Layers in Silicon," The Bell System Technical Journal, Vol. XLI, No. 2, March 1962, pp. 387-410.
[8] Draper, N. R. and Smith, H., Applied Regression Analysis, John Wiley and Sons, Inc., New York, 1966.
[9] Ho, C. P., Plummer, J. D., Hansen, S. E., and Dutton, R. W., "VLSI Process Modeling-SUPREM III," IEEE Transactions on Electron Devices, Vol. ED-30, No. 11, November 1983, pp. 1438-1453.
[10] Fair, R. B., "Concentration Profiles of Diffused Dopants in Silicon," in Impurity Doping Processes in Semiconductors, Wang, F. F. Y., Ed., North-Holland Publishing Co., Amsterdam, 1981, pp. 317-436.
[11] Hansen, S. E., "Current Status of SUPREM III," in Computer-Aided Design of Integrated Circuit Fabrication Processes for VLSI Devices, Technical Report DXG501-85, Stanford University, Stanford, CA, July 1985, pp. 217-224.
[12] Tsai, J. C. C., "Diffusion," in VLSI Technology, Sze, S. M., Ed., McGraw-Hill Book Co., New York, 1983, pp. 169-217.
[13] Borucki, L., Hansen, H. H., and Varahramyan, K., "FEDSS - A 2D Semiconductor Fabrication Process Simulator," IBM Journal of Research and Development, Vol. 29, No. 3, May 1985, pp. 263-276.

[14] Godfrey, D. J., Groves, R. D., Dowsett, M. G., and Willoughby, A. F. W., "A Comparison Between SIMS and Spreading Resistance Profiles for Ion Implanted Arsenic and Boron After Heat Treatments in an Inert Ambient," Physica B & C, Vol. 129 B-C, No. 1-3, March 1985, pp. 181-186.

[15] Ehrstein, J. R., Albers, J. H., Wilson, R. G., and Comas, J., "Comparison of Spreading Resistance with C-V and SIMS Profiles for Submicron Layers in Silicon," Journal of Electrochemical Society Extended Abstracts, Vol. 80-1, Abstract No. 188, 1980, pp. 496-498.

[16] Hu, S. M., "Between Carrier Distributions and Dopant Atomic Distribution in Beveled Silicon Substrates," Journal of Applied Physics, Vol. 53, No. 3, March 1982, pp. 1499-1510.

Robert G. Mazur

MAPPING SILICON WAFERS BY SPREADING RESISTANCE

REFERENCE: Mazur, R. G., "Mapping Silicon Wafers by Spreading Resistance," Emerging Semiconductor Technology, ASTM STP 960, D. C. Gupta and P. H. Langer, Eds., American Society for Testing and Materials, 1986.

ABSTRACT: Whole wafer mapping is a useful method for studying inhomogeneities in as-grown or processed silicon. However, its usefulness has been limited to structures on which four-point-probe measurements can be made. This paper shows that the spreading resistance technique can be used for surface mapping in those situations where four-point-probe measurements are difficult or impossible. The strengths and weaknesses of the two mapping techniques are compared for the following types of samples: P on P^+ or N on N^+ epi wafers, low-dose ion implants, ion implants into same conductivity-type substrates, heavily-doped as-grown materials, and very thin films.

KEYWORDS: resistivity mapping, spreading resistance, four-point probe, silicon process control, epitaxial layers, ion implants.

Control of resistivity variations across the surfaces of silicon wafers has been necessary for many years. In early days, the need was more or less satisfied with radial profiling by the four-point-probe technique. This allowed both those who grew crystals and those involved in wafer processing to determine the radial resistivity variation on a given substrate or processed wafer. As wafers grew larger, it became important to get a more complete picture of surface resistivity variations. The demand for these measurements was further increased by the advent of ion-implantation as a standard doping method, because it was found that the scanning processes used during implantation could easily produce non-uniform implants.

D.S. Perloff was the first person who not only realized the need for whole-wafer resistivity mapping but also generated a practical way to

Mr. Mazur is President of Solid State Measurements, Inc., 110 Technology Drive, Pittsburgh, PA 15275.

meet it. In the 1970's, he began working with a conventional four-point-probe head mounted on a wafer-prober. He made measurements as a function of position on ion-implanted wafer surfaces and experimented with various ways to present the resulting uniformity data, finally settling on the iso-contour map as the best type of display [1,2].

In the last several years, economic factors have continued to dictate the use of larger and larger wafers. At the same time, it is obvious that there is no economic benefit to converting entire wafer-fab lines to 125 or 150 mm wafers if only the center four inches can be used to make devices. Clearly, it is essential to ensure that all processes are uniform over the entire wafer surface. As a result, whole-wafer resistivity mapping has become a standard technique for controlling device processes.

Until recently, whole-wafer resistivity mapping was completely dependent on four-point-probe measurements. However, some silicon structures of interest cannot be mapped sucessfully by a four-point probe. These structures include N on N^+ and P on P^+ epi wafers, some very thin epi layers or ion implants, some very low-dose ion-implants, and ion-implants into same conductivity-type substrates. This paper will demonstrate that the spreading resistance technique can be sucessfully used to map any structure that cannot be mapped by a four-point probe. The differences between the conventional four-point probe technique and techniques based on the use of spreading resistance probes are discussed and illustrated with examples of whole-wafer resistivity maps.

COMPARISON OF THE FOUR-POINT-PROBE AND SPREADING RESISTANCE TECHNIQUES

Both the four-point-probe technique and the spreading resistance technique are widely used for characterizing semiconductor materials and processes. Each method is covered by highly-developed written procedures from the ASTM, and, through the use of standard resistivity samples, each can be traced back to the National Bureau of Standards [3,4]. However, there are also some fundamental differences between the two techniques. For example, the four-point-probe method is an absolute measurement technique; by contrast, the spreading resistance method is a comparison technique. The spreading resistance method measures the contact resistance of specially-prepared point-contact diodes set down on the surface of a sample. The raw data are converted to resistivity values through a calibration curve generated by making a series of measurements with the same probes on samples of known resistivity.

Spatial Resolution

The two techniques also differ significantly in spatial resolution. Four-point probes are typically used in an in-line array, with a probe spacing in the range of .025 inches (.635 mm) to 1/16-inch (0.0625 inches or 1.6 mm). The spatial resolution is limited to a few probe spacings, i.e., several millimeters. In a spreading resistance measurement, the probes are much closer together, typically in the range of

.025 mm to 0.1 mm. In addition, the four-point probe method averages over the depth of the layer on which the probe is resting, whereas a spreading resistance measurement is sensitive primarily to the material directly beneath the probes. Taken together, these factors account for the very significant difference in spatial resolution between the two methods.

Probe Wear and Penetration

Other differences between conventional four-point-probe and spreading resistance measurements result from the type of contact used. In conventional four-point-probe heads, the probes move laterally or "scrub" as they are set down on a sample surface. This leads to two problems. First, mechanical scrubbing causes probe wear, leading to a change in the mechanical and electrical characteristics of the contacts, which in turn leads to poor long-term reproducibility. Second, mechanical scrubbing causes the probes to rip or plow through thin layers. This often leads to uncertainty about what is actually being measured: is it the top layer of the sample, the substrate, or a combination of the two? It is true that under laboratory conditions, some expert users are able to get good four-point-probe measurements on thin layers by minimizing scrubbing and by shaping the probe tips to a relatively large size to reduce the local pressure. However, under normal operating conditions, scrubbing generally occurs. Unfortunately, it is almost impossible to know when probes are scrubbing or to quantify the effects except by observing the degradation of measured data.

In contrast, spreading resistance probes are mounted on a kinematic bearing system such that there are no degrees of freedom when the probes touch a sample surface. This means that in the absence of vibration, there is no scrubbing, i.e., no motion of the probe relative to the sample [5,6]. This fact has three significant results. First, the probes do not wear in use; therefore, they are highly reproducible both mechanically and electrically. Many spreading resistance users have reported running 20,000 to 30,000 measurement points before the probes must be reconditioned; recently, one user reported that a set of probes has been in use for over 370,000 measurement points. Second, since the probes do not scrub, they do not rip or plow through thin layers. Third, spreading resistance probes are conditioned by controlling the micro-roughness of the probe tips to control the degree of penetration into a silicon surface. Thus, the probes will not penetrate even very thin layers. In addition, it is possible to check the degree of probe penetration quantitatively, by measuring certain test samples [7,8].

RESISTIVITY MAPPING BY SPREADING RESISTANCE

Some time ago, these comparative features of the spreading resistance and four-point probe techniques suggested that useful whole-wafer resistivity maps could be generated by making spreading resistance measurements over the surfaces of as-grown or processed silicon wafers. This would allow mapping structures that cannot be mapped with a four-point probe, such as P on P⁺ or N on N⁺ epi layers and ion-implants into the same conductivity-type substrate. These considerations led to the development of a whole-wafer resistivity mapping system using spreading resistance probes.

As this system was developed, it was also obvious that even where four-point-probe measurements were applicable, they could be improved by using the type of contacts generated by spreading resistance probes instead of the conventional, scrubbing four-point-probe head. Therefore, the system was designed to operate either in a four-point-probe mode or in a spreading resistance mode, as needed. The examples that follow illustrate the utility of the spreading resistance technique for generating whole-wafer resistivity maps. In some cases these maps are compared with four-point-probe maps of the same structure. In all cases, the four-point-probe maps were made with the spreading resistance-type four-point probe, i.e., where all four probes are non-scrubbing, low-penetration probes. The probes were arranged in an in-line array, with a probe spacing of 1.0 mm. Where data are shown as sheet resistance (R_{sh}), configuration switching was used [2]. Configuration switching was used to eliminate the need for geometric edge corrections; it was not needed to eliminate the effects of probe wander, which does not occur when non-scrubbing probes are used.

Figure 1 shows a sample mapped first with the non-scrubbing four-point probe and then by the spreading resistance technique. The maps look almost identical, demonstrating that in some cases the spreading resistance technique can give essentially the same results as a four-point probe in whole-wafer resistivity mapping.

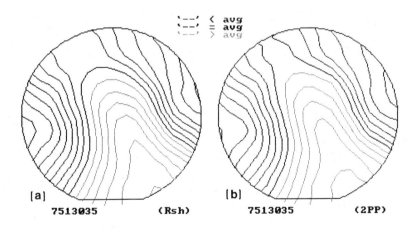

FIG. 1a--Sheet resistance contour map of an N on P epitaxial layer, thickness 1.5 μm, wafer diameter 150 mm, four-point probe in-line array, probe spacing 1.0 mm, average value of sheet resistance 9348 ohms, contour interval 1.0%.

FIG. 1b--Spreading resistance contour map of the same structure shown in Fig. 1a, average measured resistance is 18,407 ohms, contour interval 1.0%.

Figures 2a and 2b demonstrate a situation in which the four-point-probe map and the spreading resistance map do not look exactly alike. This difference may be due to the fact that the two techniques measure different parameters. The four-point probe measures the average resistivity throughout the entire thickness of the top layer, while spreading resistance values depend primarily on the resistivity at the surface of the layer. Spreading resistance measurements are also subject to a boundary correction; the correction factor depends on the depth of an underlying P/N or low-high junction. Therefore, if the junction depth varies, the spreading resistance may vary, even if the resistivity throughout the layer is uniform. In addition, if the epi layer is not uniform in resistivity throughout its depth, the spreading resistance values on the surface will also be affected by variations in this depth profile. In one sense, the sensitivity of the spreading resistance data to these parameters has a positive result: the combination of spreading resistance and four-point-probe techniques can provide more information than may be obtained from either technique on its own.

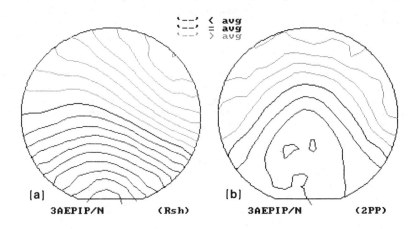

FIG. 2a--Sheet resistance contour map of a P on N epitaxial layer, thickness 2.9 µm, wafer diameter 75 mm, four-point probe in-line array, probe spacing 1.0 mm, average value of sheet resistance 3503 ohms, contour interval 0.5%.

FIG. 2b--Spreading resistance contour map of the same structure shown in Fig. 2a, average measured resistance value is 7544 ohms, contour interval 1.0%.

N on N^+ or P on P^+ Epitaxial Wafers

Among the most important structures that cannot be mapped by a four-point probe are N on N^+ and P on P^+ epi wafers. In the early days of epitaxial growth, the standard method for evaluating the then-commonly-used N on N^+ material was to include a P-type control wafer in each run. The resistivity of the layer grown on the control wafer was then measured with a standard four-point probe. However, the record shows that this procedure generally failed to provide the necessary degree of resistivity control because local auto-doping effects caused the resistivity of the material grown on the control wafer to differ from the resistivity of the epitaxial layer on the N^+ substrates. In recent years, epitaxial growth of both N on N^+ and P on P^+ material has been controlled primarily by using spreading resistance in a depth profiling mode or by capacitance voltage (CV) measurements [9]. However, CV is not currently available for large-area whole-wafer mapping, whereas this paper shows that spreading resistance can do such mapping very effectively. In addition, because the spreading resistance technique provides very high spatial resolution, N on N^+ and P on P^+ layers can be mapped regardless of thickness of the grown layer (see Fig. 3).

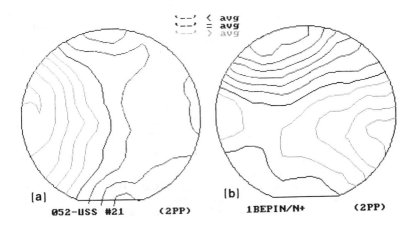

FIG. 3a--Spreading resistance contour map of a P on P^+ epitaxial layer, thickness 7 μm, wafer diameter 150 mm, contour interval 2.0%.

FIG. 3b--Spreading resistance contour map of an N on N^+ epitaxial layer, thickness 2.9 μm, wafer diameter 75 mm, contour interval 1.0%.

It is important to remember that thickness variations or changes in depth profile across the surface of a wafer also affect measured spreading resistance. In order to gain complete control of a structure of this type, spreading resistance measurements should be combined with other measurements, e.g., IR measurements of the epi layer thickness and possibly spreading resistance depth profiles taken at various positions on a selected sample basis. In order to get well-reproduced contour maps, it is also important that the spreading resistance probes have well-reproduced characteristics, especially with respect to the degree of probe penetration. This control can be easily achieved by making spreading resistance measurements on standard test wafers.

Another example of epi material where conventional four-point-probe mapping may not be generally applicable is sub-micron epi, for which the demand is increasing as we progress from VLSI to ULSI. These very thin structures (as thin as 2,000-3,000Å) impose requirements that the conventional, scrubbing four-point-probe technique may not be able to fulfill. However, with non-scrubbing probes, either four-point-probe or spreading resistance maps can be made easily, as shown in Figures 4a, 4b, and 5. The contour maps in Figures 4a and 4b are similar but not identical; this may be due to the reasons cited in the discussion of Figures 3a and 3b.

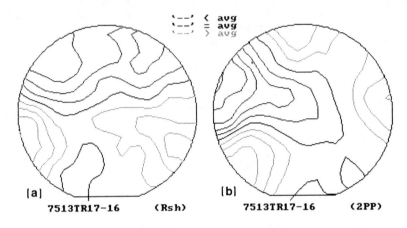

FIG. 4a--Sheet resistance contour map of a thin N on P epitaxial layer, thickness 0.2 μm, wafer diameter 100 mm, four-point probe in-line array, probe spacing 1.0 mm, average value of sheet resistance 297 ohms, contour interval 0.5%.

FIG. 4b--Spreading resistance contour map of the same structure shown in Fig. 4a, average measured resistance value is 16,555 ohms, contour interval 1.0%.

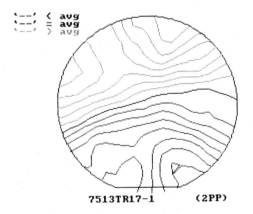

FIG. 5--Spreading resistance contour map of a thin N on N^+ epitaxial layer, thickness 0.2 μm, wafer diameter 100mm, contour interval 1.0%.

Ion Implants

Figure 6 illustrates that the spreading resistance technique can also be utilized to generate contour maps in ion-implants. In the case of ion-implants where there is a high surface concentration, spreading resistance mapping will produce almost exactly the same results as mapping by four-point probe. This occurs because both techniques really are measuring the same thing: the actual spreading resistance part of total resistance resistance is small, compared with the sheet resistance of the layer. Under these conditions, a spreading resistance probe and a four-point probe can measure the same thing. This type of result is indicated clearly in Figure 6.

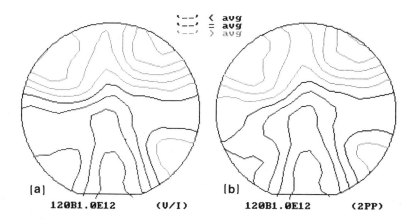

FIG. 6a--Four-point probe V/I contour map of a phosphorous ion implant into a P-type substrate, thickness approximately 1.0 μm, diameter 100 mm, four-point probe in-line array, average V/I value is 5643 ohms, contour interval 1.0%.

FIG. 6b--Spreading resistance contour map of the same structure shown in Fig. 6a, average measured resistance value is 48,374 ohms, contour interval 1.0%.

Just as in the case of epi layers, there are also ion-implants that cannot be measured with a conventional four-point probe. An example is an implant into a substrate of the same-conductivity-type, such as that used for threshhold-voltage-adjust of MOS devices. Once again, while a four-point probe cannot map such a structure, the spreading resistance technique can be used (see Fig. 7). In addition to the threshhold-voltage-adjust situation, there are other cases where the effects of ion-implants could be better studied using a same-conductivity-type substrate rather than an opposite-conductivity-type; such a choice would be permitted with spreading resistance mapping.

Spreading resistance would also be the method of choice when an ion-implant layer is very thin, because there is basically no lower limit on the thickness of layers that can be profiled. Dopant profiles using the spreading resistance technique have ben done on layers as thin as 200Å, with a resolution of 10Å per point [10]. These results also should allow four-point-probe mapping on thin layers, providing that non-scrubbing spreading resistance-type probes are used.

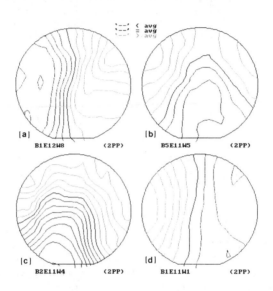

FIG. 7a-d--Spreading resistance contour maps of a series of low-dose boron implants into the same-conductivity-type substrate.

7a--Dosage 1.0×10^{12} cm^{-3}
Contour interval 1.0%

7b--Dosage 5.0×10^{11} cm^{-3}
Contour interval 2.0%

7c--Dosage 2.0×10^{11} cm^{-3}
Contour interval 2.0%

7d--Dosage 1.0×10^{11} cm^{-3}
Contour interval 5.0%

Spreading resistance could also have a role to play with regard to recently-voiced concerns about micro-inhomogeneities that may occur as a result of the ion-implantation process. Conventional four-point-probe mapping cannot reveal these inhomogeneities, because the spatial resolution is not high enough. Even though the four-point probe used in the present work could be adjusted to place all four probes in a 100 μm square region, the resulting resolution might still be insufficient to map ion-implant inhomogeneities effectively. However, the system's spreading resistance capability permits a much higher lateral resolution. The ideal way to study micro-inhomogeneities would be to implant into a same-conductivity-type substrate, then use a conductive lower contact on the wafer and a single spreading resistance probe. In this case, the spatial resolution would be determined by the diameter of the contact, which is about 5 microns.

TABLE 1—Comparison of conventional four-point-probe, spreading resistance, and non-scrubbing, controlled-penetration four-point-probe techniques.

	Conventional Four-Point Probe	Spreading Resistance	Controlled Penetration Four-Point Probe
1)	Absolute technique.	Comparison technique.	Absolute technique.
2)	Averages over entire thickness of layer.	Sensitive primarily to surface layer; high depth resolution.	Averages over entire thickness of layer.
3)	Typical probe spacings (0.635 mm, 1 mm, 1.6 mm = 1/16 inch result in relatively low spatial resolution.	Probe spacing down to 0.1 mm yields much better lateral spatial resolution; single probe mode can give spatial resolution to less than 5 μm.	Better spatial resolution than conventional four-point probe, especially with square array (probe spacing as 0.1 mm).
4)	Only traceable to NBS through ASTM Standard F-84, which requires a probe spacing of 1.6 mm.	Traceable to NBS at any probe spacing through standard resistivity samples.	Only traceable to NBS through ASTM Standard F-84, which requires a probe spacing of 1.6 mm.
5)	Probes scrub on contact and therefore wear, leading to changes in contact size and shape and therefore to a change in contact resistance; affects measurement reproducibility.	Non-scrubbing probes; therefore no probe wear and extremely high reproducibility of probe contact resistance.	Non-scrubbing probes; therefore no probe wear and extremely high reproducibility of contact resistance, both short- and long-term.
6)	Probes scrub on contact and therefore can rip or "plow" through thin surface layer; can invalidate measurements by contacting substrate.	Non-scrubbing probes are conditioned to control probe penetration; degree of probe penetration can be quantified through spreading resistance measurements with individual probes on special test samples.	Non-scrubbing probes; spreading resistance conditioning methods and measurements can be used to control and quantify probe penetration; measurements possible on layers as thin as 200Å or less.
7)	Probes scrub on contact and therefore wander; four-point-probe spacing is variable and therefore configuration switching is required.	Non-scrubbing probes maintain probe positions within ± 1 μm; all mechanical and electrical properties of contacts are highly reproducible.	Non-scrubbing probes based on a kinematic bearing system provide extremely reproducible probe spacing; configuration switching not necessary; more precise measurements possible with calculated correction factors.

CONCLUSION

Table 1 summarizes the major differences between the conventional four-point-probe technique, the spreading resistance technique, and the non-scrubbing, controlled low-penetration four-point-probe technique. This summary and the specific examples illustrated in this paper show that the spreading resistance technique can play a significant role in whole-wafer resistivity mapping. Spreading resistance can be used to map wafers that cannot be measured with a four-point probe. The most important structures in this group are N on N^+ and P on P^+ epi layers and ion-implants into same-conductivity-type substrates. It has also been shown that the four-point-probe measurement technique can be improved by replacing the scrubbing probes in the conventional four-point-probe head with non-scrubbing spreading resistance contacts.

REFERENCES

[1] Perloff, D. S., Wahl, F. E., and Reimer, J. D., "Contour Maps Reveal Non-Uniformity in Semiconductor Processing." Solid State Technology, Volume 13, Feb. 1977, pp. 31-36+.

[2] Perloff, D. S., Gan, J. N., and Wahl, F. E., "Dose Accuracy and Doping Uniformity of Ion Implantation Equipment." Solid State Technology, Volume 17, Feb. 1981, pp. 112-120.

[3] "Standard Method for Measuring Resistivity of Silicon Slices with a Collinear Four-Point Array," in Annual Book of ASTM Standards, Volume 10.05, American Society for Testing and Materials, Philadelphia, annual.

[4] "Standard Method for Measuring Resistivity Profile Perpendicular to the Surface of a Silicon Wafer Using a Spreading Resistance Probe," in Annual Book of ASTM Standards, Volume 10.05, American Society for Testing and Materials, Philadelphia, annual.

[5] Mazur, R. G., "The Effects of RF Electromagnetic Radiation on Spreading Resistance Measurements." In D. C. Gupta, Ed., Semiconductor Processing, ASTM STP 850, American Society for Testing and Materials, Philadelphia, 1984.

[6] Mazur, R. G., "Dopant Profiles by the Spreading Resistance Technique." Presented at the American Chemical Society Symposium on Materials Characterization in Microelectronics Processing, St. Louis, April 9, 1984.

[7] Mazur, R. G. and Gruber, G. A., "Dopant Profiles on Thin Layer Silicon Structures with the Spreading Resistance Technique." Solid State Technology, Volume 24, Nov. 1981, p. 69+.

[8] "Probe Conditioning: Why and How?" SSM CONTACT, Number 1, August 1984, pp. 1-5.

[9] Perloff, D. S., Gan, J. N., and Wahl, F. E., "Dose Accuracy and Doping Uniformity of Ion Implantation Equipment." Solid State Technology, Volume 17, Feb. 1981, pp. 112-120.

[10] Nicollian, E. H. and Brews, J. R., MOS Physics and Technology, Wiley Interscience, New York, 1982.

[11] Ehrstein, J. R., "Spreading Resistance Measurements - An Overview." Emerging Semiconductor Technology (pending publication).

W. Andrew Keenan, Walter H. Johnson, and Alan K. Smith

PRODUCTION MONITORING OF 200MM WAFER PROCESSING

REFERENCE: Keenan, W. A., Johnson, W. H., and Smith, A. K., "Production Monitoring of 200mm Processing", Emerging Semiconductor Technology, ASTM, STD 960, D. C. Gupta and P. H. Langer, Eds., American Society for Testing Materials, 1986.

ABSTRACT: A successful transition to 200mm wafers requires careful monitoring of substrate quality and rapid feedback on process performance. Sheet resistance mapping will play a key role in this transition because it allows precise control of repeatability and uniformity for a wide range of processes. The critical role of sheet resistance mapping in the transition to larger wafers will be reviewed. Advanced, fully automated sheet resistance mapping systems, capable of measuring 200mm wafers, will also be discussed. Data obtained from 200mm wafers will be presented for various process applications, including epitaxial and metal deposition, and ion implantation.

KEYWORDS: resistivity, sheet resistance, uniformity, large diameter wafers, 3-D maps, contour maps, diameter scans, ion implantation, epitaxial layers, trend analysis

INTRODUCTION

Semiconductor processing equipment is now being developed for 200mm wafers. The transition from 150 to 200mm represents a 33% diameter increase and a 77% increase in area. The last increase of this magnitude was in going from 3 inch to 100mm which represented a 31% diameter increase and a 72% area increase. However the amount of real estate involved here is 4 times that of a 100mm wafer! In order to make a viable transition to 200mm wafers and to justify the capital investment required for this increased capacity, the process yield available on 100-150mm wafers must be maintained or improved.

Sheet resistance is used to monitor many semiconductor processes as well as incoming wafers, these processes include ion implantation, metal film sputtering and evaporation, chemical vapor deposition of conducting films, diffusion, epitaxial layer growth and rapid annealing. In order to achieve a comparable process yield on 200mm wafers sheet resistance mapping will be used to carefully monitor the repeatability and uniformity of the process.

The authors are with the Resistivity Product Group, Prometrix Corporation, 3255 Scott Blvd., Bldg. 2 Santa Clara, CA 95054.

A new, fully automated sheet resistance mapping system is needed, capable of measuring wafers up to 200mm in diameter. This system would minimize the opportunity for operator error by providing the engineer with the software to define the tests used for production monitors. Operator involvement would be reduced to loading the wafer and selecting the test. By automatically setting the desired current or voltage for the test a common source of error would be eliminated.

The importance of current configuration switching for wafer mapping will be reviewed. The electronics and probe performance of the new system will be demonstrated. Test site geometry and density will be discussed. Engineering set-up of a monitor test will be demonstrated. Examples of contour and 3-D maps and diameter scans will be given for 200mm wafers.

WAFER ECONOMICS

Table I shows the results of a recent survey of silicon wafer vendors indicating a price per square centimeter of between 13-25 cents for wafers between 2 inch and 150mm in diameter.

TABLE 1

DIAMETER	AREA (CM^2)	PRICE ($)	COST/CM^2
2 inch (50.8mm)	20	3.50-4.50	0.20
3 inch (76.2mm)	46	5.00-7.00	0.13
100mm	79	10.00-12.00	0.14
125mm	123	18.00-25.00	0.17
150mm	177	40.00-50.00	0.25
200mm	314	225.00-275.00	0.80

The price per unit area for 2 inch diameter wafers is more than for 3 inch, perhaps because of reduced production. The big jump in price is for 200mm wafers, which were quoted at between $225.00 and $275.00 per slice, or about 80 cents per square centimeter. This is more than three times the per unit area price of the 150mm wafers. The 200mm wafers available have been mostly mechanical samples, with open electrical or mechanical specifications. The price will of course drop as demand and production increase. Meantime, developing a production tool with 200mm capability will require a significant expense for both dummy and monitor wafers. The only way to capitalize on this expense is to obtain all the information possible from the wafers used. To do this the process parameters (Rs, Tox, Vfb, etc.) must be mapped over the whole wafer in order to fine tune the uniformity of the process and enhance the process yield. In order to justify the expense of tooling up for 200mm the process yield must be at least as good as for the smaller diameter wafers. Wafer mapping will play an important role in this transition.

CONFIGURATION SWITCHING

Ion implantation was the ideal process to focus attention on sheet resistance mapping because every implanter set-up could potentially give drastically different results. Ion implantation also claimed a uniformity and repeatability much better than any other process (originally 2% and presently less than 1%).

Test sites [1] were originally used to measure ion implant sheet resistance repeatability and uniformity because the performance of the four point probe was much worse than the process being monitored. In two round-robins [2] reported in the 1971 Book of ASTM Standards involving six laboratories and five wafers the repeatability (1 standard deviation/mean) for wafers between 500 and 2000ohm-cm was 5%. These results were obtained by measuring the wafer center ten times following ASTM procedure F84-70. Clearly a 5% measurement could not be used to monitor a process claiming 2% repeatability and uniformity. It must be stressed that these 5% results were achieved under ideal conditions, and routine performance in a production line would be much worse.

Significant improvements have been made in both the four point probe itself and in the electronics used in the sheet resistance measurement. However the major enhancement to the measurement was the addition of current configuration switching [3] by D. Perloff. For a homogeneous conducting layer of thickness t and resistivity r the sheet resistance is

$$R_s = \frac{r}{t}$$

In the conventional application of the four point probe the current is forced through the outer two probes and the voltage developed in the semiconductor is sensed across the inner two probes as shown in Figure 1A :

$$R_A = \frac{V_{23}}{I_{14}}.$$

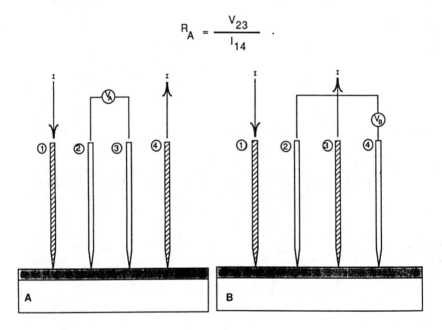

FIGURE 1. Four point probe schematic showing (A) standard current-voltage configuration and (B) second current-voltage configuration which is used to calculate correction factor.

For a thin layer ($t < s/2$, where t is the thickness of the layer and s is the probe spacing), infinite in extent, the sheet resistance (R_s) can be expressed as:

$$R_s = \frac{\pi}{\ln 2} \frac{V_{23}}{I_{14}}$$

$$= \frac{\pi}{\ln 2} R_A$$

assuming the probe spacings are all equal. The fraction $\pi/\ln 2$ is the theoretical geometrical correction factor (C.F.) for the ideal situation where these assumptions apply. The two main sources of error in four point probe measurements are due to, (1) the fact that the probe spacings are not equal (and in fact vary from site-to-site) and (2) the fact that the wafer is not infinite and current crowding results when the probe approaches the edge of the wafer.

A second reading can be made at the same site as shown in Figure 1B, by forcing the current through probes 1 and 3 and measuring the voltage between 2 and 4,

$$R_B = \frac{V_{24}}{I_{13}}$$

the actual geometrical correction factor (C.F.) for the site can be then calculated from the ratio of the two readings

$$\text{C.F.} = f\left(\frac{R_A}{R_B}\right).$$

The calculation of the actual geometrical correction factor for each site compensates for the variations in probe spacing (probe wobble) and for any current crowding due to the proximity of the wafer edge.

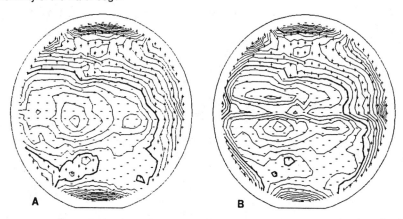

FIGURE 2. Sheet resistance map of broken 200mm epi wafer measured using (A) dual current-voltage configurations and (B) single current-voltage configuration. Dark line represents mean value and light lines represent 1% contours above (+) and below (-) the mean.

602 EMERGING SEMICONDUCTOR TECHNOLOGY

Configuration switching and correction factor calculation became practical with the advent of the mini-computer, and has revolutionized sheet resistance measurements, making accuracy and repeatability of better than 1% possible [4].

Configuration switching provides accurate sheet resistance measurements to within one probe spacing of the edge of the wafer. The two maps in Fig. 2 are for a 200mm epitaxial layer, with a target sheet resistance of 9.5 kohm/square. Map A was measured using configuration switching. This wafer had been broken along an E-W direction. There is no evidence of the break in the map of Figure 2-A. The average sheet resistance is 9.47 kohm/square with a uniformity of 3.79% which is truly excellent for test diameter of 188mm out of a possible 200mm. Figure 2-B is a map of the same wafer, with identical test conditions except only the single current configuration was used. The discontinuities in the contour lines above the equator of the wafer are a clear indication of the location of the break. The higher sheet resistance readings distort the contour lines and are due to current crowding as the probe approaches the break. Current crowding is also obvious in the extra contour lines around the perimeter of the wafer.

Figure 3 shows two scans along a N-S diameter across the break of the same wafer. The diameter scan in Figure 3-A was made with configuration switching, the diameter scan in Figure 3-B was made using the single current configuration. Both diameter scans were done using 121 sites and a 1.2 inch test diameter for step size of 10 mil. At least 12 sites will be lost in moving the probe (40 mil probe spacing) across the break using 10 mil steps. Using configuration switching accurate readings were obtained up to 20 mil from the wafer edge (one-half the probe spacing). Without configuration switching the presence of the break is observed 200 mil from the wafer edge (five probe spacings). The dramatic increase in current crowding as the probe approaches the break is completely compensated by configuration switching.

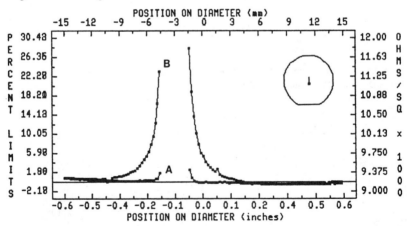

FIGURE 3. A diameter scan across the break in the 200mm epi wafer made with 10 mil steps using (A) dual current-voltage configurations and (B) single current-voltage configurations.

AUTOMATIC RESISTIVITY TESTER (ART)

A completely new sheet resistance mapping system has been developed based on configuration switching and capable of measuring samples ranging in size from 1 cm^2 chips up to 200mm diameter wafers. State of the art electronics provide the capability to measure sheet resistance from a few milliohm/square (e.g. thick aluminum films) up to several megohm/square (e.g. thin, high resistivity epitaxial layers). The Automated Resistivity Tester (ART) is computer controlled and provides automatic set-up of the selected test current or voltage. Operator selection of incorrect current or voltage has always been a problem and often lead to incorrect results [5][6]. A typical test configuration is shown in Fig. 4 where 361 sites on a 180mm test pattern have been chosen for a 200mm wafer. In this case, an automatic test voltage of 7.5mv has been chosen. The ART will start the test by varying the current until a voltage reading of 7.5mv is obtained, this current will then be used for the 361 sites.

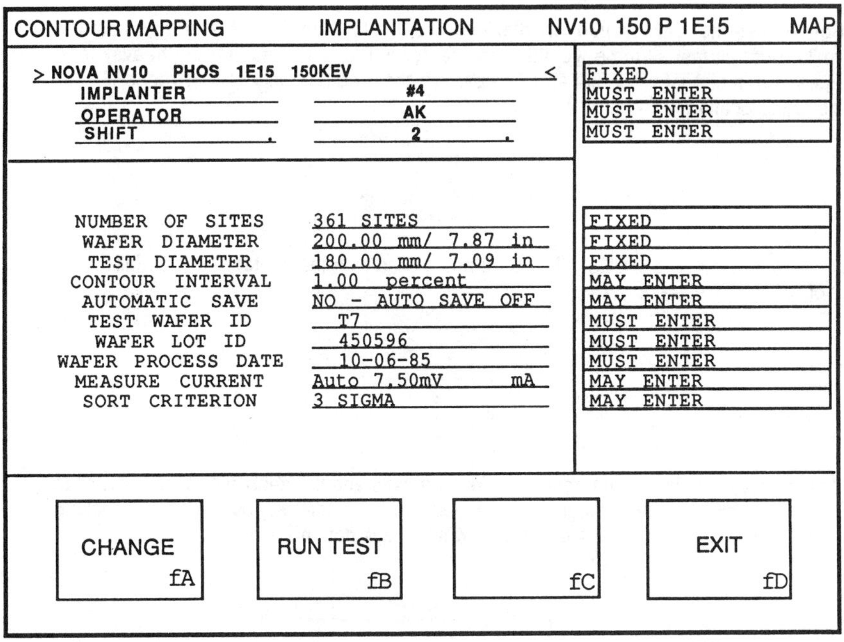

```
CONTOUR MAPPING        IMPLANTATION        NV10  150 P 1E15       MAP

> NOVA NV10   PHOS  1E15  150KEV              <    FIXED
  IMPLANTER                  #4                    MUST ENTER
  OPERATOR                   AK                    MUST ENTER
  SHIFT                  .    2          .         MUST ENTER


  NUMBER OF SITES      361 SITES                   FIXED
  WAFER DIAMETER       200.00 mm/ 7.87 in          FIXED
  TEST DIAMETER        180.00 mm/ 7.09 in          FIXED
  CONTOUR INTERVAL     1.00  percent               MAY ENTER
  AUTOMATIC SAVE       NO - AUTO SAVE OFF          MAY ENTER
  TEST WAFER ID          T7                        MUST ENTER
  WAFER LOT ID         450596                      MUST ENTER
  WAFER PROCESS DATE    10-06-85                   MUST ENTER
  MEASURE CURRENT      Auto 7.50mV        mA       MAY ENTER
  SORT CRITERION       3 SIGMA                     MAY ENTER



  ┌─────────┐   ┌─────────┐   ┌─────────┐   ┌─────────┐
  │ CHANGE  │   │RUN TEST │   │         │   │  EXIT   │
  │      fA │   │      fB │   │      fC │   │      fD │
  └─────────┘   └─────────┘   └─────────┘   └─────────┘
```

FIGURE 4. Automatic Resistivity Tester (ART) test configuration selection screen for 200mm wafer using 180mm test diameter, 361 test sites and automatic set-up of 7.5mv test voltage.

The ART provides a wide choice of test parameters, which include contour map or diameter scan, number of sites, wafer diameter and test diameter, automatic test current or voltage and value of test current or voltage. To help define the test and provide information for later analysis, descriptive information may be entered, such as wafer number, process, tool, date of process, operator identification, shift and run number. For routine testing of production monitors the process engineer will want to specify many of the test parameters and require operator input of certain descriptive information.

TEST SITE PATTERNS

Original sheet resistance measurements involved only 5 or 9 sites on a wafer. For a 1 inch diameter wafer test patterns of 5 and 9 sites corresponded to a test area or sample size about 1 to 0.5 cm^2 per test site. Considering the repeatability and accuracy of early sheet resistance measurements and the problems associated with probe wobble and current crowding, little could be gained by adding more test sites as the wafers increased in size.

The accuracy and repeatability presently available encourage the use of additional test sites to gather data more representative of the sheet resistance variations across the wafer. Care must be exercised in generating any test pattern to insure:

(1) that it preserves the symmetry of the wafer and thus the process,
(2) that each area of the wafer is sampled equally, and
(3) that the density of sites be sufficient to detect any significant variation on the wafer.

Some systems have relied on diameter scans to sample the sheet resistance uniformity, but this test scheme weighted the center of the wafer much more heavily than the edges. For a diameter scan of equal steps, Δ, the area per test site (Ar) varies directly with the radial location of the site, r,

$$A_r = \pi r \Delta$$
$$= \pi n \Delta^2,$$

since $r = n\Delta$, ($n = 1, 2, 3 ...$). The area associated with the center site is defined by a radius of $\Delta/2$ and is equal to:

$$A_o = \frac{\pi \Delta^2}{4}.$$

Clearly the area associated with each test site increases and the density of sites decreases as the distance from the center increases. If additional diameter scans are added (making a total of N diameter scans) the sampling unifromity does not improve

$$A_r(N) = \frac{\pi r \Delta}{N} = \frac{\pi n \Delta^2}{N}.$$

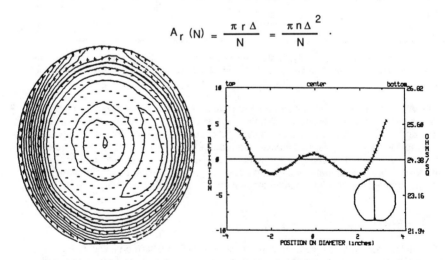

FIGURE 5. Sheet resistance map of 200mm tungsten silicide CVD layer with $<R_s> = 24.77$ ohm/square and uniformity of 3.77% measured using 361 sites.

FIGURE 6. Diameter scan of 200mm tungsten silicide layer results in lower uniformity of 2.26% due to non-uniform distribution of test sites.

Test patterns made up of diameter scans (wagon wheels) weight the center region of the wafer so heavily that the wafer uniformity is misrepresented, often giving a very optimistic indication of the process variability. In order to measure uniformity a uniform test pattern is necessary. The map in Figure 5 is for a 200mm wafer with a tungsten silicide CVD layer showing excellent uniformity in the center area (the so-called "sweet-spot"). The layer becomes thin near the edge of the wafer giving rise to rapid increase in sheet resistance. A diameter scan of the same wafer in Figure 6 shows details of the sheet resistance variation from top to bottom. The calculated uniformity is significantly less for the diameter scan (2.26% versus 3.77%) because the center region is favored with more test sites.

A set of uniform test patterns has been designed to provide equal area for each test site. The test patterns are made up of a series of equally- spaced concentric circles which maintain the symmetries of the wafer. If N is the number of circles and R is radius of the largest test circle, the area per test site is:

$$A_s = \pi \left(\frac{R}{2N}\right)^2,$$

the area associated with the center test site.

The total area sampled can be shown to be a circle of radius $R(1 + 1/2N)$, the area of which is

$$A_T = \pi R^2 \left(1 + \frac{1}{2N}\right)^2$$

$$= \pi R^2 \frac{(2N+1)^2}{4N^2}.$$

The total number of points P_T is given by the total area A_T divided by the area per site A_s:

$$P_T = A_T / A_s$$

$$= (2N+1)^2.$$

The test sites are equally spaced along the circumference of these test circles. The number of sites for any circle can easily be calculated, for the Nth circle

$$PN = (2N+1)^2 - [2(N-1)+1]^2 = 8N.$$

For example, the third circle will have *3 x 8 = 24* test sites and each additional circle picks up 8 more test sites. The first circle will have eight sites at 360/8 degree intervals the second circle will have 16 sites at 360/16 degree intervals and so forth. Table II shows development of test site patterns up to a total of 12 circles and 625 sites.

TABLE 2

CIRCLE NUMBER N	0	1	2	3	4	5	6	7	8	9	10	11	12
SITES PER CIRCLE 8N	1	8	16	24	32	40	48	56	64	72	80	88	96
TOTAL SITES $(2N+1)^2$	1	9	25	49	81	121	169	225	289	361	441	529	625

Figure 7 is the 625 test site pattern with 12 cirlces shown for a 200mm wafer. Each pattern provides an equal test area per site and thus weights each area of the wafer equally in arriving at a mean and standard deviation and in presenting the sheet resistance map.

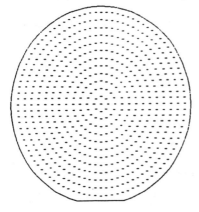

FIGURE 7. Distribution of test points for 625 site pattern made up of 12 equally spaced concentric circles.

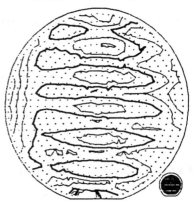

FIGURE 8. Sheet resistance map of ion implanted 100mm wafer tested using 625 sites and a 85mm test diameter. Good uniformity of 1.15% belies a very serious scan lock-up problem.

DENSITY OF TEST SITES

These test patterns sample the wafer uniformly and perserve the wafer symmetry. The question remains as to how many test sites are necessary for mapping the parameter being measured. Should this number be based on the uniformity expected or on the size of the wafer? If the objective were to obtain a mean and standard deviation, the density of sites required would be less than for mapping. The goal here is not only an accurate mean and standard deviation, but a map that adequately displays the sheet resistance variations across the wafer. Maps indicative of very different problems can have the same uniformity. In fact wafers with reasonably good uniformity can have very serious problems. Figure 8 is a map of an ion implanted 100mm wafer with a uniformity of 1.15%. The map shows a classical scan lock-up pattern of high dose stripes across the wafer.

The cost of wafers requires that they be processed and used efficiently. If only 5 sites were to be measured on a 100mm wafer the cost per site is about $2.00. Such a measurement could be very misleading because it could sample a "sweet-spot" in the central region of the wafer where the process is normally uniform, as seen in Figure 5 above. If however, an 81 test site pattern were used, the area/site would be about 1cm^2 and the cost per site would be about 14 cents! By using 1cm^2 per test site or less, sufficient data is collected to recognize and address most problems. Once the wafer is committed, adequate testing will insure a reasonable return on the investment.

The transition to large diameter wafers is a particularly challenging problem because the wafers are so expensive ($225-275/wafer). Since all processes interact with the wafer, two 100mm wafers (or even four) cannot duplicate the results for a 200mm wafer. Several smaller wafers may be substituted in the initial set-up phases, but eventually the 200mm wafers will have to be committed. The cost of the wafer clearly demands that all of the information available be effectively used to tune and refine the process. This will minimize the number of monitors required, speed up the development and reduce the cost of the final equipment.

The ART provides test site patterns ranging from 5 and 9 sites (for a quick check) to 625 for detailed mapping of a 200mm wafer. Table III shows the sample area per test site for the ten patterns versus wafer diameter. To adequately characterize a wafer, variations on a chip scale must be measured. The area of an integrated circuit chip is of the order of 1cm^2 and a sample area of this magnitude should be used to characterize the wafer. For fine diagnostic work and process tuning the area would be reduced to 0.5cm^2/site or less.

TABLE 3

SAMPLE AREA PER TEST SITE (CM^2)

#OF SITES	1" 22mm	2" 51mm	3" 76mm	100mm	125mm	150mm	200mm
1	5.1	20.3	45.6	78.5	122.7	176.7	314.2
5	1.02	4.06	9.12	15.70	24.54	35.34	62.84
9	0.57	2.26	5.007	8.72	13.63	19.63	34.91
25	0.20	0.81	1.82	3.14	4.91	7.07	12.57
49	0.10	0.41	0.93	1.60	2.50	3.61	6.41
81		0.25	0.56	0.97	1.51	2.18	3.88
121		0.17	0.38	0.65	1.01	1.46	2.60
225		0.09	0.20	0.35	0.55	0.79	1.40
361			0.13	0.22	0.34	0.49	0.87
441			0.10	0.18	0.28	0.40	0.71
625				0.13	0.20	0.28	0.50

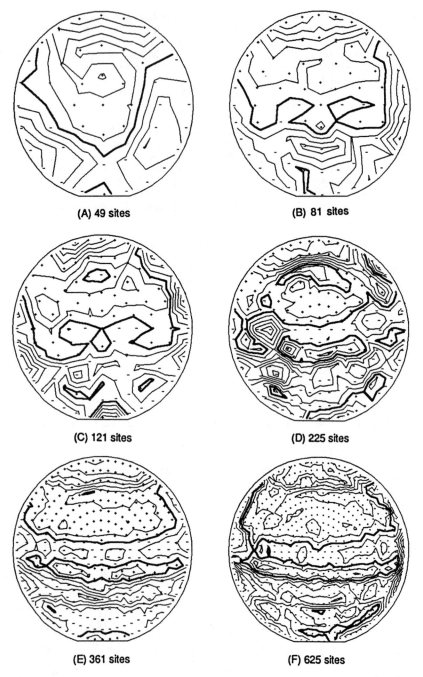

FIGURE 9. Sheet resistance maps of 200mm wafer implanted with 5E15 ions/cm^2 arsensic at 100KeV using mechanical scan. Same test conditions were used for (A) 49 sites, (B) 81 sites, (C) 121 sites, (D) 225 sites, (E) 361 sites and (F) 625 sites.

Table III shows that for a 1 inch diameter wafer 5 and 9 sites corresponded to 1 and 0.5 cm^2 per test site, respectively. The pattern corresponding to approximately 1cm^2 /test site is high lighted for each wafer diameter. For a 100mm wafer this corresponds to an 81 site test and for a 200mm wafer a 361 site test is required. Recommended values are highlighted.

The maps in Figure 9 demonstrate the need for adequate sampling. These maps show a 200mm wafer ion implanted with 5E15 ions/cm^2 arsenic at 100KeV. The wafer was mechanically scanned on a rotating disk that was translated through a fixed ion beam as shown in Figure 10. This was an early attempt at implanting a 200mm wafer and the uniformity achieved (2.4%) was indeed excellent for this first pass. The 49, 81 and 121 site maps in Figure 11 show no sign of the pattern that begins to develop on the 225 site map. The 361 site map, with a sampling area of 1 cm^2 per test site clearly shows a striping across the wafer due to a hesitation in the translation of the rotating disk. It can be determined from looking at Figure 9 that the wafer was loaded with the wafer flat toward the edge of on the disk. The test site density of at least one site per square centimeter is required to uniquely define sheet resistance variations.

Figure 9-F shows a 625 site contour map (2 test sites/cm^2) of the same wafer. The 3-D map of Figure 11 aids in visualizing the variations in sheet resistance across the wafer. The software provides for rotation and tilt of the 3-D maps to highlight particular variations of sheet resistance. Here the 3-D map clearly shows the sheet resistance variations due to the hesitation in the translation of the rotating disk.

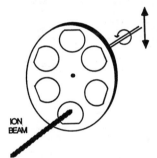

FIGURE 10. Schematic drawing of the mechanical scan system used to ion implant 200mm wafers. Spinning disk is translated through stationary ion beam.

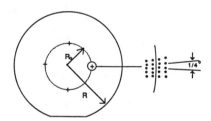

FIGURE 12. Test site pattern used to verify probe repeatability. Five sites are measured, 1/4° apart, at each of four locations 90° apart on test radius Ro.

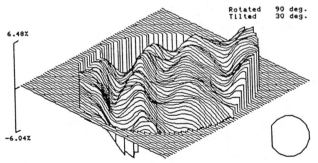

FIGURE 11. A three-dimentional (3-D) map of sheet resistance for same ion implanted wafer using the 361 site data of Figure 11-E. Map shows wafer rotated 90° and tilted 30° for clearer presentation.

PROBE AND ELECTRONICS

In studying of any measurement variability it is always useful to consider the noise introduced by the components. Two main sources of noise in the sheet resistance measurement are the electronics and the four-point probe itself.

The electronic noise can be isolated by testing an appropriate resistor network. The probe performance will be measured using the electronics and thus will include both probe and electronics noise. The ART system is capable of measuring and reporting both. The single most troublesome component in the four-point probe sheet resistance mapping system is the four-point probe itself. It is the probe which must make reliable and repeatable contact to the silicon. The four-point probe used in a mapping system encounters much heavier use than a normal four-point probe; typically 49-625 impressions are made per wafer for a lifetime of thousands of wafers. Clearly one thousand wafers on a mapping system would be equivalent in probe usage to 10 thousand or more wafers, assuming only five sites were measured.

PROBE QUALIFICATION

PROBE ID. : 8216
TEST DIAMETER : 11.43 mm / 0.45 in
PROCESS DATE : - -
LOT ID :
WAFER ID: TCEPI #12
CURRENT : Aut 7.44mV   0.0062mA

	R_s		R_a		R_b		CF
SITE #1	5388.	ohms/sq	1197.	ohms	952.4	ohms	4.503
	5388.	ohms/sq	1196.	ohms	952.3	ohms	4.503
	5385.	ohms/sq	1196.	ohms	952.3	ohms	4.503
	5385.	ohms/sq	1196.	ohms	951.9	ohms	4.503
	5388.	ohms/sq	1195.	ohms	950.8	ohms	4.508
MEAN:	5387.	ohms/sq	1196.	ohms	951.9	ohms	4.504
STDV:	0.030 %		0.059 %		0.069 %		0.049 %
SITE #2	5298.	ohms/sq	1182.	ohms	944.2	ohms	4.481
	5299.	ohms/sq	1183.	ohms	944.4	ohms	4.481
	5300.	ohms/sq	1183.	ohms	944.7	ohms	4.481
	5299.	ohms/sq	1183.	ohms	944.5	ohms	4.481
	5299.	ohms/sq	1183.	ohms	944.7	ohms	4.481
MEAN:	5299.	ohms/sq	1183.	ohms	944.5	ohms	4.481
STDV:	0.013 %		0.037 %		0.022 %		0.000 %
SITE #3	5314.	ohms/sq	1189.	ohms	950.8	ohms	4.470
	5322.	ohms/sq	1189.	ohms	950.8	ohms	4.476
	5324.	ohms/sq	1190.	ohms	951.1	ohms	4.476
	5316.	ohms/sq	1189.	ohms	951.5	ohms	4.470
	5324.	ohms/sq	1190.	ohms	951.1	ohms	4.476
MEAN:	5320.	ohms/sq	1189.	ohms	951.1	ohms	4.474
STDV:	0.088 %		0.046 %		0.030 %		0.073 %
SITE #4	5336.	ohms/sq	1191.	ohms	951.1	ohms	4.481
	5339.	ohms/sq	1191.	ohms	951.9	ohms	4.481
	5337.	ohms/sq	1191.	ohms	951.1	ohms	4.481
	5338.	ohms/sq	1191.	ohms	951.8	ohms	4.481
	5338.	ohms/sq	1191.	ohms	951.6	ohms	4.481
MEAN:	5338.	ohms/sq	1191.	ohms	951.5	ohms	4.481
STDV:	0.021 %		0.000 %		0.040 %		0.000 %

FIGURE 13. Probe qualification test results for probe with excellent repeatability (0.01-0.09%) showing R_S, R_A, R_B and the calculated correction factor CF.

The requirements on a four-point probe in a mapping system are much more stringent than a probe used for 5 or 9 sites. A probe required to measure ion implant monitors with a specified uniformity of 0.75% should not have a repeatability of 0.5% or greater because the net result would be:

$$s = \sqrt{s_{WFR}^2 + s_{PROBE}^2}$$

$$= \sqrt{.75^2 + .5^2}$$

$$= 0.90\%$$

and the monitor wafer would be out-of-spec. But the requirement on probe performance applies not only for very uniform wafers. If the probe were to introduce ±0.5% noise, it would change the value of the reading at each site, giving the map a busy or noisey appearance.

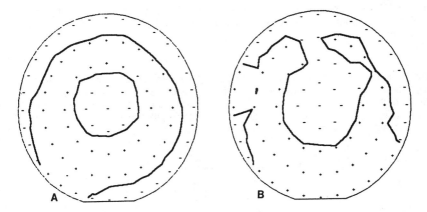

FIGURE 14. Two sheet resistance contour maps of the same wafer using identical test conditions and two probes: (A) probe with good repeatability (0.15-0.26%) gave <R_s> = 15.88 ohm/square and uniformity = 0.35% and (B) probe with poor repeatability on qualification wafer (0.48-0.69%) gave <R_s> =15.72 ohm/square and uniformity = 0.48%. Maps are significantly different.

In order to verify probe repeatability, five sites are tested at each of four locations, the five sites being 1/4° apart, as shown in Figure 12. The repeatability required for an acceptable probe is < 0.2%. Figure 15 shows the output for a probe which has excellent repeatability. The importance of probe repeatability is shown in Figure 14. These two maps are the results of measurments on the same wafer using the same test conditions. In the case of map A the probe repeatability on the four sites ranged from 0.15 to 0.26%, in the case of map B the probe repeatability ranged from 0.48 to 0.69%. The average values for the two maps are within 1% of each other and the uniformities are close but the maps are significantly different. The contour lines in A are smooth and continuous while the contours of B are disjointed, busy and noisey. The noise added to the maps by the probe's lack of repeatability has masked the symmetry of the distribution. These maps clearly indicate the extra requirements on probe performance for mapping.

As the probe repeatability degrades the quality of the maps degrades first, the second parameter to degrade is the standard deviation and the last parameter to degrade is the average sheet resistance. The ability to test and monitor probe repeatability is critical to the performance of any four-point probe sheet resistance measurement, particularly if it is going to be used for mapping. To insure the quality of the maps, as well as the accuracy of the mean and standard deviation it is recommended that the probe repeatability be checked frequently.

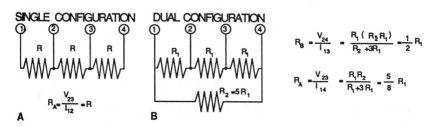

FIGURE 15. Resistor networks for verifying repeatability and accuracy of four point probe electronics (A) for single current voltage configuration and (B) for dual current voltage configuration.

The electronics must be carefully monitored to detect any noise which could effect the maps, or any long term drift which would effect the accuracy and repeatability. The electronics of the four-point probe are usually checked using a standard resistor network as shown in Figure 15-A. The current is forced through terminals 1 and 4, and the voltage measured at terminals 2 and 3. The sheet resistance is given by:

$$R_s = \frac{\pi}{\ln 2} \frac{V_{23}}{I_{14}}$$

$$= 4.532 \, R.$$

In order to test the electronics for configuration switching a new resistor pack was designed and is shown in Figure 15-B. An extra resistor in parallel with the original three resistors is necessary to test the second current configuration. If R_2 is choosen to equal $5 \times R_1$, the equivalent resistances for the two configurations are:

CONFIGURATION A: $R_A = \dfrac{V_{23}}{I_{14}} = \dfrac{5}{8} R_1$, and

CONFIGURATION B: $R_B = \dfrac{V_{24}}{I_{13}} = \dfrac{1}{2} R_1$.

The ratio $R_A / R_B = V_A / V_B = 5/4 = 1.25$ is used to calculate the correction factor (C.F.) with the result

$$R_s = C.F. \cdot (V23 / I14)$$

$$= 4.470 \cdot (5/8) \cdot R1$$

$$= 2.794 \cdot R1.$$

The resistor packs are made with precision 0.1% resistors with R_1 ranging from 1.00 ohm to 10,000 ohm.

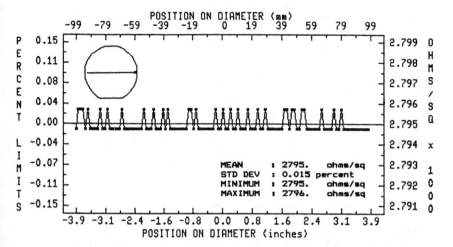

FIGURE 16. Diameter scan presentation of 121 tests of 2794 ohm/square resistor pack, the average value is 2795 ohm/square and repeatability is 0.015%.

To verify the absence of noise the resistor pack is measured in place of a wafer, using 121 site diameter scan. The results for a 2794 ohm/square resistor pack are shown in Figure 16. The diameter scan format is used for plotting the results for the resistor pack. For this short term test the repeatability is better than 0.02% for the 2794 ohm/square resistor pack.

To establish long term repeatability the same resistor pack was tested daily using 49 sites and the data stored for trend analysis. The plot in Figure 17 shows the data gathered for more than a month. For each day the mean R_s is plotted along with the ±0.1% and ±0.2% limits for the trend analysis. The long term stability of the electronics is excellent and will not cause any shift in the measured sheet resistance.

SUMMARY

The cost and area of 200mm wafers require that wafer mapping be used to develop and tune new processing equipment. Mapping will no longer be limited to ion implantation monitors. A new, automated sheet resistance mapping system, capable of wafers up to 200mm in diameter, uses current configuration switching to eliminate problems due to probe wobble and current crowding. This system provides probe repeatability of better than 0.2% and accuracy of better than 1.0%. Tests are provided to verify probe

614 EMERGING SEMICONDUCTOR TECHNOLOGY

performance and the stability and accuracy of the electronics. System software provides for 3-D as well as contour maps, trend analysis and engineering definition of production test. Operator involvement is reduced to test selection and wafer loading. Automatic set-up of the desired test conditions eliminates a serious source of measurement error. Special test patterns preserve the symmetry of the wafer, provide equal area per test site and provide a density of sites to accurately characterize small scale variations on the wafer. The advantages of this new mapping system provides the results essential to effectively characterize 200mm process equipment.

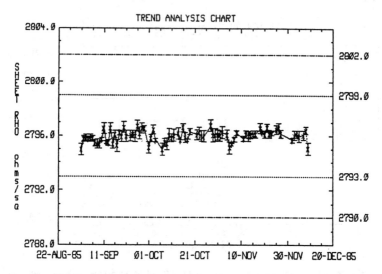

FIGURE 17. Trend analysis plot of test results for 2794 ohm/square dual configuration resistor pack made up of 0.1% precision resistors. Average value for the three month period is 2795.6 ohm/square with a repeatability of 0.015%.

REFERENCES

[1] Seirmarco, J.A. and Keenan, W.A., "Ion Implantation Uniformity Measurements", Abstract No. 152, Electrochemical Society Fall Meeting, Boston, MA (1973).

[2] 1971 Annual book of ASTM Standards, Part 8, Designation F84-70, pp 697-711.

[3] Perloff, D. S., Gan, J. N. and Wahl, F. E., "Dose Accuracy and Doping Uniformtiy of Ion Implantation Equipment", Solid State Technology, Vol. 24, No. 2, Feb. 1981, pp 112-120.

[4] Huang, R. S. and Hadbroake P. H., "The Use of a Four-Point Probe for Profiling Sub-Micron Layers", Solid State Electronics, Vol. 21, No. 9, Sept. 1978, pp. 1123-1128.

[5] Eranna, G. and Kakati, D., "Limitations on the Range of Measurements of Sheet Resistivity of Shallow Diffused Layers for Profiling by the Four-Point Probe Technique", Solid State Electronics, Vol. 25, No. 7, July 1982, pp. 611-614.

[6] Keenan, W. A., Johnson, W. H. and Smith, A. K., "Advances in Sheet Resistance Measurements for Ion Implant Monitoring", Solid State Technology, Vol. 28, No. 6, June 1985, pp. 143-148.

Toshio Shiraiwa, Takashi Ochiai, Mamoru Sano, Yoshifumi Tada, and Tomoya Arai

APPLICATIONS OF X-RAY FLUORESCENCE ANALYSIS TO THE THIN LAYER ON SILICON WAFERS

REFERENCE: Shiraiwa, T., Ochiai T., Sano M., Tada Y., and Arai T., "Applications of X-Ray Fluorescence Analysis to the Thin Layer on Silicon Wafers", Emerging Semiconductor Technology, ASTM STP 960, D. C. Gupta and P. H. Langer, Eds., American Society for Testing and Materials, 1986.

ABSTRACT: Thin layered materials on silicon wafer were analyzed with X-ray fluorescent spectrometer which is specially designed for the analysis of semiconductor devices. The spectrochemical analysis using hard X-rays (22Ti - 92U) can be applied to the thicker layered materials (0.5 - 50 μm). By soft and ultrasoft X-rays (5B - 21Sc), thinner layered materials (0 - 2μm) can be analyzed for the determination of concentration of layered material and film thickness. In analysis of SiO2 Layer, the intensity of O-Kα X-ray was measured and the thickness was calculated by the theoretical approach. The precision of measurement is a few percent and it can distinguish the difference of natural oxide according to the cleaning method. Other experimental results of typical silicon wafer samples are informed; which are the analysis of NSG, PSG, BPSG, metal and alloy films, and As or P implanted elements.

KEYWORDS: X-ray fluorescence, oxide layer analysis, BPSG, PSG, Silicon wafer.

The spectroscopic analysis of medium and heavy elements (hard X-ray group) by fluorescent X-rays has become a useful method for many analytical applications and also the recent developments in the analysis of low atomic number elements or the measurements of soft and ultrasoft X-rays have been carried out which are indebted for high efficient excitation for soft and ultrasoft X-rays and wavelength dispersive devices with high reflectivity (1), (2). As the absorption of soft and ultrasoft X-rays to material are very large, the measurements of thin layered material on the substrate can easily be made on account of the increase of sensitivity.

Shiraiwa is research scientists at Osaka Titanium Co., Hyogo, Japan; Ochiai and Sano are research scientists at Kyushu Electric Metal Co., Saga, Japan; Tada and Arai are research scientists at Rigaku Industrial Co., Takatsuki, Osaka, Japan.

616 EMERGING SEMICONDUCTOR TECHNOLOGY

The electron, proton and ion excitation techniques for thin layer analysis(0-100A) are very popular because of their high sensitivity and easy use, however there are many different kinds of analytical problems with respect to thicker layer measurements (over 100A), reproducibility or precision of intensity and complexed material analysis.

In Table 1, the comparison of transmissions between electrons and X-rays is shown concerning aluminum metal film and air path. The transmitting thickness for X-rays versus electrons is one thousand times in both materials, therefore the X-ray method is more useful to measure thicker layered material than electron method.

In Table 2 transmission of soft and ultrasoft X-ray is shown for various materials. The 50% transmission thickness is the approximate maximum analyzing depth (or measuring thickness) in the case of film thickness measurement by reflection configuration. It is expected from Table 2 that the maximum measurable SiO_2 layer on the silicon wafer is about 0.56 μm when O-Kα X-rays are used.

TABLE 1--Comparison of Electron and X-Ray Absorption

Aluminum (Transmission: 1 %)

Absorbed Beams: Energy, Kev	Electron Thickness[a], μm	X-ray Thickness[a], cm
100	130	9.9
80	80	8.3
50	27	4.5
40	16	2.9
30	8.7	1.5
20	3.2	0.49
10	0.39	0.066

Air(dry) (Transmission: 1%)

Absorbed Beams: Energy, Kev	Electron Thickness[a], cm	X-Ray thickness[a], m
100	28	250
80	18	230
50	6.1	190
40	3.6	160
30	1.9	110
20	0.71	53
10	0.087	8.6

[a]calculated absorption thickness

TABLE 2--Absorption of soft and ultrasoft X-Rays
(Transmission: 1 %)

Absorbing Materials: X-Rays ,nm		Gold Thickness[a], μm	SiO$_2$ Thickness[a], μm	Palypropylene Thichness[a], μm
Al-K	83	0.12	16	82
Zn-L	123	0.48	5.7	28
Cu-l	133	0.41	4.5	22
Ni-L	146	0.35	3.6	17
Fe-L	176	0.27	2.2	10
Cr-L	216	0.23	1.3	6.0
O-K	236	0.20	3.7	4.8
C-K	447	0.16	0.9	24

(Transmission: 50 %)

Absorbing Materials: X-Rays ,nm		Gold Thickness[a], μm	SiO$_2$ Thickness[a], μm	Polypropyline Thickness[a], μm
Al-K	83	0.018	2.4	12
Zn-L	123	0.072	0.86	4.1
Cu-L	133	0.062	0.68	3.3
Ni-L	146	0.053	0.55	2.6
Fe-L	176	0.041	0.34	1.5
Cr-L	216	0.034	0.20	0.90
O-K	236	0.031	0.56	0.72
C-K	447	0.024	0.13	3.6

[a] calculated absorption thickness

PRINCIPLE

The principle of film thickness measurement is based on the Beer's law $(I/Io = \exp(-(\mu/\rho)\rho d))$ and the relations between the intensity (I) and the film thickness (d) are shown in Fig. 1 and Fig. 2.

There are two excitation techniques for fluorescent X-rays which are monochromatic and polychromatic. In the case of monochromatic excitation, the incident X-rays have a single value of the absorption coefficient for an analyzing sample and the simple relation between X-ray intensity and thickness can be obtained subject to the equations in Fig. 1 and Fig. 2. In polychromatic excitation, as incident X-rays consist of various X-rays with intensity distribution that is commonly used, the relation of the intensity against layered material thickness become a bent line which is arising from the change of excitation condition at shallow and deep portion in the sample. In Fig. 3 and 4 the relation of Zn-Kα intensity emitted from the zinc metal film and Fe-Kα intensity from steel substrate against the thickness of zinc metal film are shown respectively using polychromatic excitation techniques.

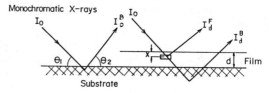

Fluorescent X-ray intensity from a coating film

$dI = k \csc\theta_1 I_0 \exp\{-(\mu_1 \csc\theta_1 + \mu_2 \csc\theta_2)\rho x\} dx$

$a = \mu_1 \csc\theta_1 + \mu_2 \csc\theta_2$

$I_d^F = k \csc\theta_1 I_0 \int_0^d e^{-a\rho x} dx$

$= \dfrac{k \csc\theta_1 I_0 (1 - e^{-a\rho d})}{a\rho}$

$I_\infty^F = \dfrac{k \csc\theta_1 I_0}{a\rho}$

$\dfrac{I_d^F}{I_\infty^F} = 1 - e^{-a\rho d} \longrightarrow 1 - \dfrac{I_d^F}{I_\infty^F} = e^{-a\rho d}$

$\ln\left(1 - \dfrac{I_d^F}{I_\infty^F}\right) = -a\rho d$

μ_1, μ_2 : mass-absorption coefficient
ρ : density

Fig.1--Relation between fluorescent X-ray intensity and thickness

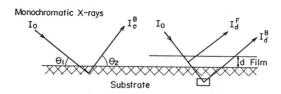

Fluorescent X-ray intensity from a substrate

$I_d^B = K \cdot I_0^B \cdot \exp\{-(\mu_1 \csc\theta_1 + \mu_2 \csc\theta_2)\rho d\}$

$a = \mu_1 \csc\theta_1 + \mu_2 \csc\theta_2$

$\dfrac{I_d^B}{I_0^B} = e^{-a\rho d}$

$\ln\left(\dfrac{I_d^B}{I_0^B}\right) = -a\rho d$

Fig.2--Relation between fluorescent X-ray intensity and thickness

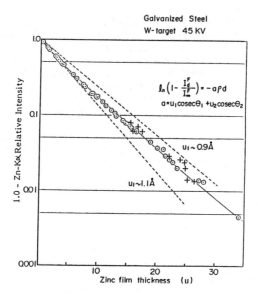

Fig.3--Relation between measured Zn-Kα intensity and zinc coating thickness

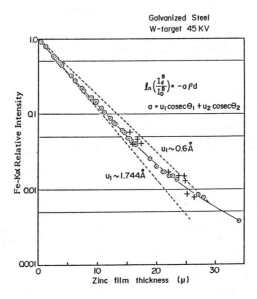

Fig.4--Relation between measured Fe-Kα intensity and zinc coating thickness

In Fig. 3, the experimental results by means of polycromatic excitation techniques are in far agreement with the predicted values which are calculated using the excitation of 1.0A incident X-rays. In Fig.4, as the decrease of Fe-K intensity against zinc metal film thickness becomes small in the case of thicker samples, it is clarified that the soft X-rays of incident X-rays are absorved by zinc metal film and the hard X-rays excite the iron fluorescent X-rays. It is elucidated that the equations illustrated in Fig.3 and 4 are useful in practical applications though they are derived on the base of monochromatic X-ray excitation techniques. For the polychromatic X-ray excitation, the full theoretical approach, namely the fundamental parameter method, was carried though by Shiraiwa and Fujino and the resulting equations were experiementally verified (3), (4).

APPLICATION

Oxide Layer on Silicon Wafer

The SiO_2 layer on silicon wafer is very important in LSI devices and its thickness is usually measured by ellipsometry or contact stylus method. For the measurement of thinner layers (0-1000A) of SiO_2 films, these theoretical and experimental approaches were made using the X-ray fluorescent method.

Experiment : An Oxide layer on a wafer was prepared by thermal oxidization in dry oxygen at 950° and 1000°C. Besides artficial oxide layered samples, naturally oxidized samples were provided after the cleaning treatment of two processes, H_2O_2 + NH_4OH solution and HF solution. The thickness of oxide layers were measured by ellipsometry.

The X-ray fluorescent spectrometer used for this experiment is RIGAKU "WAFER ANALYZER" and measuring condition is shown in Table 3.

TABLE 3--Measuring condition of O-K

X-Ray TubeO-K :	Rh-Target, End Window Type
Tube Voltage :	45KV-60mA
Analyzing Crystal :	RX-40
X-Ray Detector :	Gas Flow Proportional Counter

Calculation : The fluorescent intensity of O-K X-rays from oxide layer consist of three kinds of X-rays ; I_1, the primary fluorescent X-rays directly excited by the irradiating X-rays, I_2, the secondary fluorescent X-rays excited by the fluorescent X-rays (Si-K) from the oxide layer and I_3, the secondary fluorescent X-ray excited by the fluorescent X-rays(Si-K) from silicon substrate. These three kinds of X-rays intensity were computed by the fundamental parameter method.

The irradiating X-rays consist of continuous and charactaristic X-rays emitted from an X-ray tube. Their intensity distribution against X-ray wave length was adopted from reference (5). For the standard of O-K X-ray intensity, the SiO_2 plate was used as an infinit thickness layer sample.

SHIRAIWA ET AL. ON X-RAY FLUORESCENCE ANALYSIS 621

Result : Fig.5 shows the profiles of O-K from oxide layers on a silicon wafer. These thickness values were obtained by the ellipsometry. The estimated detection limit of SiO_2 thickness was 0.5nm. Fig. 6 shows relation between O-K intensity and the thickness measured by the ellipsometry. The linear relation is observed, because the thickness range of measured samples is thin. As the thickness increases, the X-ray intensity approaches a constant value (see Fig. 1) in accordance with the equation in Fig.1.

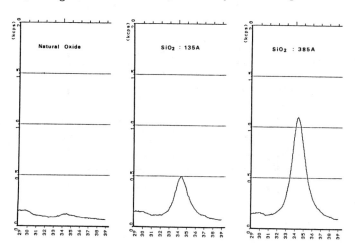

Fig.5--Profile of O-K X-rays

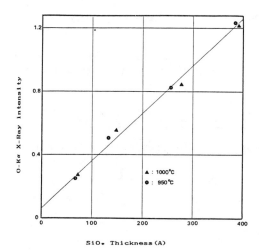

Fig.6--Relation of O-K X-rays against SiO_2 thickness

Fig.7 shows measured O-K intensity versus thorectically calculated one. A good linear correlation is found.

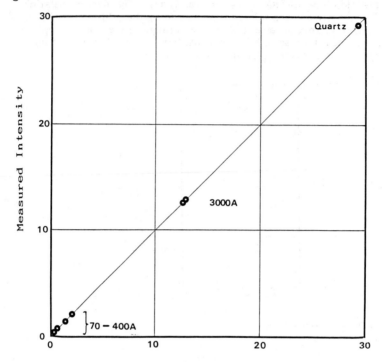

Fig.7--Relation between theoretical and measured intensities

In Table 4, the measurements of natural oxide layer thickness after cleaning are shown and the difference of two cleaning method is clearly exhibited.

TABLE 4--Natural oxide layer on silicon wafer

Surface Cleaning Method	Thickness of Oxide Layer, nm
RCA	1 ± 0.3
HF	0.5 ± 0.2

BPSG and PSG layer on Silicon Wafer

Result : Fig.8 shows relation between B-K intensity and B_2O_3 concentration of BPSG determined chemically. In Fig.1. u1, u2, and are the funcion of matrix element concentration. So it is possible to analyse both concentration and film thickness with X-ray intensity. The measured precision is a few percent.

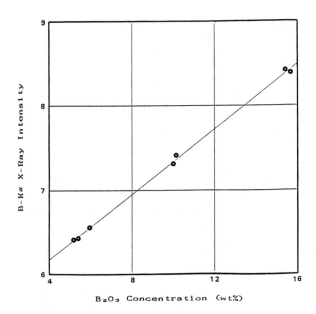

Fig.8--Relation of B-K intensity against B_2O_3 concentration

Fig.9 and Fig.10 show the experimental results of PSG thickness and P_2O_5 concentration versus the chemically determined values respectively. By means of the measurement of Si-K and P-K X-rays, the analytical accuracy of P_2O_5 concentration in PSG film is 0.18 mol% in the range of 7-25 mol%, and the accuracy of PSG film thickness is 0.004 um in the range of 0.07-1.3 um.

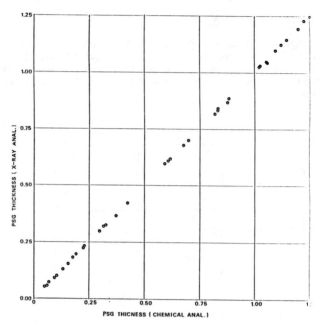

Fig.9--Relation between chemical and X-ray values of PSG thickness

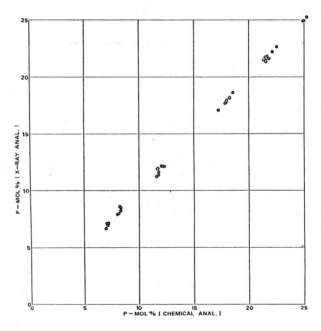

Fig.10--Relation between chemical and X-ray values of P_2O_5 concentration

Data Processing for Non-Standard Sample

The conversion from measured X-ray intensity to film thickness and concentration is carried out based on the fudamental parameter method with mathematical treatments using a computer.

$$X_j = F(P_1\ P_2\ P_3\ ....U_jV) \qquad (1)$$

- X_j : Intensity of j element.
- $P_1\ P_2\ P_3\ ...$: Film thickness, concentration etc.
- U_j : Physical constants (mass absorption coefficients etc.)
- V : Instrumental factors (irradiating angle, take-off angle etc.)

In Equ.(1), the generalized formura (X_j) of the relation between X-ray intensities (X_j) and parameters (P_j) is shown. The film thickness and concentration of film can be determined combining the Equ.(1) and measured X-ray intensities with mathematical treatment and iteration calculation which is shown in Fig. 11.

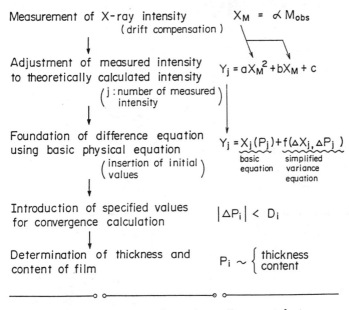

Using standard samples, intensity adjustment factors should be determined before measurement of unknown samples

Fig.11--Calucalating procedures for the determination of thickness and content

EMERGING SEMICONDUCTOR TECHNOLOGY

In Fig. 12, the typical experimental results of nickel and iron alloy film on glass plate and compared results between chemical and X-ray values are shown.

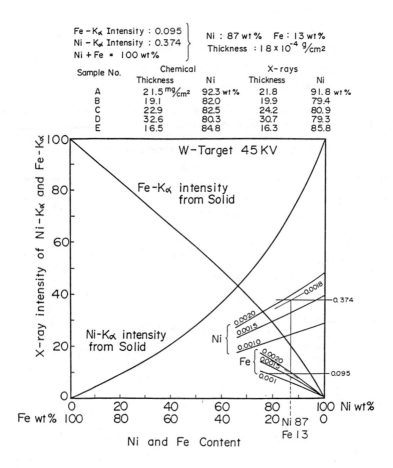

Fig. 12--Determination of thickness and content of Ni-Fe binally alloy film by computation.

SUMMARY

In Table 5, the samples are shown which can be analyzed by X-ray techniques.

TABLE 5--Analysis of layered materials on silicon wafer

Analyzing Sample	Component, wt%	Thickness, um
Oxide film		
BPSG	B_2O_3, 5-10/P_2O_5, 5-15	0.3-1.0
PSG	P_2O_5, 5-25	0.1-1.5
BSG	B_2O_3, 5-20	0.3-1.0
NSG	SiO_2, 100	0.01-0.5
Metal film		
Al	Al, 100	0.2-1.0
Al-Si	Si, 1-2	0.2-1.0
Al-Si-Cu	Si, 1-2/Cu, 1-2	0.2-1.0
Ti-W	Ti, 5-15	0.1-0.5
Silicide film		
Mo-Si	Si, 30-40	0.05-0.5
W-Si	Si, 20-30	0.05-0.5
Ta-Si	Si, 20-30	0.05-0.5
Ni-Si	Si, 40-50	0.05-0.5
Ti-Si	Si, 50-60	0.05-0.5
V/Ni/Au	V, 100/Ni, 100/Au, 100	0.1-0.5
Substrate		
As implanted	(5×10^{14}-2×10^{16} atoms/cm^2)	0.1-0.3[a]
P implanted	(1×10^{14}-1×10^{16} atoms/cm^2)	0.1-0.5[a]
Natural oxide	SiO_2, 100	0.0005-0.003

[a]Implanted depth

REFERENCE

(1) Arai, T., <u>Japanese Journals of Applied Physics,</u> Vol.21, 1982, P1347

(2) Ryon, R.W., Arai, T., and Shoji, T., <u>Advances in X-Ray Analysis,</u> Vol.28, 1985, P137

(3) Shiraiwa, T. and Fujino, N., <u>Japanese Journals of Applied Physics,</u> Vol.5, 1966, P886

(4) Shiraiwa, T. and Fujino, N., <u>Advances in X-Ray Analysis,</u> Vol.12, 1968, P63

(5) Robinson, J.W., <u>Handbook of Spectroscopy,</u> Vol.11, CRC Press, 1974

James F. Black, James M. Berak, and Gregory G. Peterson

QUALIFICATION OF GaAs AND AlGaAs BY OPTICAL AND SURFACE ANALYSIS TECHNIQUES

REFERENCE: Black, J. F., Berak, J. M., and Peterson, G. G., "Qualification of GaAs and AlGaAs by Optical and Surface Analysis Techniques", <u>Emerging Semiconductor Technology, ASTM STP 960</u>, D. C. Gupta and P. H. Langer, Eds., American Society for Testing and Materials, 1986.

ABSTRACT: Optical techniques, including photoluminescence and infrared spectroscopies, have been investigated as a means of evaluating GaAs and AlGaAs wafers. Methods of surface analysis such as Auger Electron Spectroscopy and X-Ray Photoelectron Spectroscopy were used to support the optical measurements, as well as provide additional information, in the following areas: (1) evaluation of semi-insulating GaAs (a) by room temperature photoluminescence that showed strong intensity variations on a microscopic scale, suggestive of corresponding variations in semi-insulating properties, and (b) by low temperature photoluminescence that revealed substantial differences in carbon content in wafers with essentially the same resistivity; (2) self-quenching of the band-gap photoluminescence in GaAs at room temperature that appears to be connected with the formation and thickening of native oxide layers; (3) monitoring the course of post-implant annealing of Be-implanted GaAs and AlGaAs by measurements of photoluminescence at room temperature; (4) qualification of photoluminescence as an accurate means for determining composition in AlGaAs layers with aluminum content up to 0.58 mole fraction.

KEYWORDS: GaAs, AlGaAs, photoluminescence and quenching, infrared spectroscopy, Be-implants, surface analysis

The authors are members of the research staff of United Technologies Research Center, Silver Lane, East Hartford, CT 06108.

INTRODUCTION

Over the past several years, optical techniques have become widely used to evaluate semiconductor wafers. In silicon technology, for example, measurements of infrared absorption bands that are characteristic of the Si-O and Si-C bond energies are routinely used to determine ppm concentrations of dissolved oxygen and dissolved carbon. Before the advent of automated optical testing, only a small sampling of starting wafers could be pre-qualified in a cost effective way. Now, with optical testing, one can not only qualify all of the starting wafers, but in addition, can also monitor the effect of a subsequent process step such as a diffusion.

A main advantage of optical testing is that it is non-contacting and hence, precluding very high power density, inherently nondestructive. In Addition, no special sample preparation is required except for well polished surfaces. Most, if not all, semiconductor wafers are in this condition anyway prior to the start of fabrication. Another advantage is that the surface of a wafer can be optically scanned much faster than it can be electrically scanned where the making and breaking of electrical contacts is required. In addition, the surface or an appreciable volume (thickness) of a sample can be examined by using appropriate optical wavelengths that are either absorbed or transmitted by the sample.

One of the major obstacles to the more widespread use of optical methods for semiconductor wafer evaluation is the lack of data that demonstrate its usefulness. The purpose of this paper is to describe and discuss several optical techniques that are being developed to evaluate GaAs and AlGaAs wafers. In the interests of speed, simplicity and flexibility of apparatus and techniques, measurements at room temperature have been emphasized.

EXPERIMENTAL APPROACH, TECHNIQUES AND EQUIPMENT

Infrared measurements were accomplished on a single beam Fourier Transform Infrared (FTIR) spectrometer that covered the spectral range 4400 to 200 wavenumbers. Reflectance was determined with the help of an attachment which brought the optical beam onto the inverted surface of the sample at a 30 degree angle of incidence.

Beryllium implants were made at voltages up to 200kV with Be+, as well as with Be++ to produce implant depths equivalent to 400kV. Special source heaters built into the implanter allowed the passage of halogen gases over the solid Be charge to produce high implant fluences at moderate source temperatures.

The photoluminescence spectroscopy system employed for the studies described in this paper was similar to many others described in the recent literature. The following specifics deserve further comments. A He-Ne or an argon-ion laser was used to excite photoluminescence (PL) in samples mounted on a rotatable X-Y-Z stage with motorized micrometer drives. The He-Ne and argon-ion lasers could be focused to a e^{-2} spot size of 40 micrometers and 140 micrometers, respectively. The system optical sensitivity was determined to be 1×10^{-13} watts (into the monochromator entrance slit) at a signal/noise ratio of unity at the recorder amplifier output.

Samples to be examined for room temperature photoluminescence were mounted onto aluminum plates by means of a low melting solid detergent (Igepal - available from GAF Corp., Linden, N.J.). This method of mounting was quick and easy, and the Igepal was readily and completely removed from the demounted wafer by multiple rinses in deionized (DI) H_2O. For low temperature PL measurements, hydrocarbon-base greases loaded with high thermal conductivity particulates were used to attach the samples to the cold finger of the cryostat.

In general, crystal damage in semiconductors causes a strong decrease in the intensity of band-edge photoluminescence. Certain point defects, including impurity atoms, also lead to strong reductions in band-edge photoluminescence. This makes PL measurements well suited to the study of implantation and its associated crystal damage, and to the study of post-implant anneals, where recovery of crystal perfection and electrical activation of free carriers and deep levels are involved. Previous use of photoluminescence spectroscopy to follow the course of implantation/activation has been restricted to cryogenic temperatures [1]. Some of the work to be described in this paper was undertaken to see if PL spectroscopy at room temperature could also be useful in tracking the change in GaAs and AlGaAs properties after implantation and activation.

The generation of room temperature band-edge photoluminescence in semiconductors is often difficult because of the strong effects that impurities and crystal defects have in promoting non-radiative or non-band-gap recombination of the photogenerated electrons and holes. One can, to some extent, get around this problem by working at high excitation power density, since the likelihood of producing band-edge recombination increases with increasing electron-hole pair density, approaching a quadratic behavior at sufficiently high power density [2].

The maximum power density possible in the 40 micrometer spot of the He-Ne laser, averaged over the guassian distribution, was 550 watts/cm^2. This power density was employed in obtaining much of the photoluminescence data described and discussed herein, which raises important questions of surface heating. Localized heating was investigated by precise determinations of the peak wavelength of photoluminescence as a function of excitation power density. As the excitation power density was reduced from 550 watts/cm^2 to 5.5

watts/cm^2, a PL peak shift of -2nm was observed, corresponding to an average temperature rise in the laser spot of 5.6°C [3].

Examination of samples by X-Ray Photoelectron Spectroscopy and Auger Electron Spectroscopy was accomplished with a PHI model 610/5100 Surface Analysis System. Auger electron spectroscopy (AES) was used to determine the AlGaAs composition and to verify the sample uniformity. AES is well suited for quantitative analysis of thin films, because Auger electrons have maximum escape depths of 100Å in a solid sample, making the technique good for sensing non-uniformity throughout thin films and at interfaces where interdiffusion and compositional grading are likely to occur. However, sputter etching is usually employed to profile through films greater than 100Å thick, and ion-mixing and surface texturing which accompany this procedure often makes the depth resolution no better than several hundred angstroms.

AlGaAs epitaxial samples selected on the basis of their photoluminescence peaks to cover a wide range of aluminum content, were used for the standardization of Auger elemental sensitivity factors. Quantitative analysis of the AlGaAs standards was performed by electron microprobe analysis (EMPA) using AlCu eta phase as a primary standard [4]. Gallium and arsenic standards were provided by a sample of high purity (100) GaAs. In order to keep the X-ray analysis volume within the thin film, EMPA requires a film thickness greater than one micrometer. Non-uniform samples introduce error in the analysis since the X-ray analysis depth is close to one micrometer. Uniformity of standards was determined with AES by sputter etching through the epitaxial film to the substrate. The above procedure resulted in AES atomic sensitivity factors derived from matrix corrected EMPA data which were used to generate quantitative depth profiles for aluminum, gallium and arsenic in AlGaAs.

EXPERIMENTAL RESULTS AND DISCUSSION

Gallium Arsenide: Be-implantation and annealing

Beryllium is of interest for p-implants in GaAs because it diffuses slowly compared to other p-dopants, and because deep implants can be produced at moderate voltages. A series of undoped semi-insulating Liquid Encapsulated Czochralski (LEC) samples were implanted with Be+ and Be++ using several different combinations of fluence and voltage to produce a profile that, according to ion implant (LSS) calculations, was 2×10^{19}/cm^3 from 0 - 1.2 micrometers, tailing off to 1×10^{18}/cm^3 at 2.0 micrometers [5]. After implantation, the wafers were annealed using the GaAs proximity cap technique, under a high purity nitrogen atmosphere in a rapid thermal anneal system (AGA Heat Pulse, Palo Alto, CA). Each wafer was examined by FTIR spectroscopy and by photoluminescence spectroscopy and was then subjected to Hall effect and resistivity measurements. The results of these examinations are shown in Table 1. Note that for the as-implanted sample there is some electrical activity, suggesting that in-situ annealing has occured during the implant process. The thickness of the electrically active layer deduced from differential resistivity measurements is in better agreement with the LSS range calculations than with the FTIR measurements. The layer thicknesses determined by FTIR measurements assumed a constant refractive index of 3.34. It has been reported, however, that heavy Be-implantation of GaAs increases the refractive index by 10-15% [6]. Recalculation of the implanted layer thickness gives a corrected thickness of 1.44 to 1.39 micrometers, in better agreement with the resistivity measurement and LSS calculations. Notice also, that the electrical measurements show a continual decrease in the implant activation with extended heat treatment, while the PL measurements suggest that crystal perfection and minimum deep-level activity are optimum somewhere between the shortest and the longest anneals. Further work will have to be done to expand on these results but the PL measurements indicate that annealing conditions may have to be tightly controlled to provide acceptable Be activation yields and at the same time produce high crystal quality and minimum trap density.

In a study of masked Be implants into semi-insulating GaAs, wet chemical etching was employed to render the annealed implants visible under the microscope. An alternate way of delineating these regions was desired because chemical etching was time consuming, requiring careful control of etching conditions, and it was essentially destructive. In order to determine if photoluminescence testing could be used to differentiate the implanted from the masked regions, a focused laser spot was scanned across the surface of the sample, which was mounted on an X-Y stage driven back and forth by a motorized micrometer drive. The He-Ne laser was used instead of the argon-ion laser for almost all of the scanning work because it allowed a more tightly

TABLE 1--Measurements of Be-implants in LEC semi-insulating GaAs

PROCESS	FTIR THICKNESS (μm)	PL SIGNAL (A.U)*	ELECTRICAL PROPERTIES
As implanted	1.6	N.D. <0.02	7.2 ohm-cm in 1.3μm LAYER BY DIFFERENTIAL ϱ MEASUREMENT
950°C/10 sec	1.5	50	$p = 1.8 \times 10^{19}$, $\mu_H = 81$
950°C/50 sec	2.0	395	$p = 8.9 \times 10^{18}$, $\mu_H = 95$
880°C/1500 sec	2.9	47	$p = 3.8 \times 10^{18}$, $\mu_H = 115$

*THIS IS AN ABREVIATION FOR ARBITRARY UNITS WHICH IS USED FOR CONVENIENCE IN COMPARING RELATIVE SIGNAL STRENGTHS.

focused spot (40 micrometers vs. 140 micrometers), and thus provided higher spatial resolution. The photoluminescence signal at 867nm, the peak of the room temperature spectral energy distribution, was recorded on a strip chart whose speed was synchronized with the micrometer drive to produce a line scan with a scale of 100 micrometers per cm to as much as 2mm per cm.

In Fig. 1a, a section of a Be-implanted wafer is shown where one set of a group of masked implants has been chemically etched to reveal the implant pattern. The circular Au-based contacts helped to locate the implanted structures and orient the sample so the line scans would cut straight across the Be-implanted "fingers". Most of the finger pairs gave line scan profiles such as the one shown on the right in Fig. 1b, but a few showed no evidence of Be-implant/activation at all as in Fig. 1c. Subsequent analysis of these regions, by secondary ion mass spectroscopy, showed that Be was not present and it was later determined that these areas were near the hold down clamps on the ion implanter target plate, and were probably shadow masked by one of these clamps. It is noteworthy that even though the laser spot diameter was 40 micrometers, we are able to detect the 5 micrometer separation between the two closest Be-implant fingers. The average power density in the focused laser beam that was used to produce these line scans

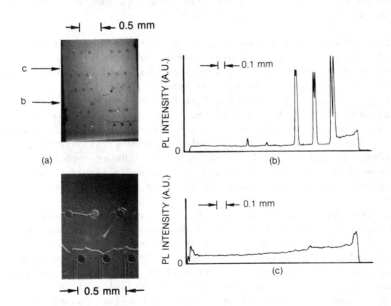

FIG. 1--Masked Be-implants in semi-insulating GaAs.
(a) photomicrographs of implanted samples after post implant annealing and ohmic contact metalization, (b) band-edge photoluminescence line scan at room temperature along the direction indicated by arrow (b) in the photo micrograph, (c) photoluminescence line scan along the direction indicated by arrow (c).

was 550 Watts/cm^2. Similar line scans were seen with a laser spot power density of 55 Watts/cm^2 and below, albeit with appropriate increases in the gain of the photomultiplier output amplifier to compensate for the reduced PL signal.

Evaluation of Substrate Wafers

Because of the large numbers of semi-insulating wafers now being processed by many organizations, a most attractive application of optical testing is the evaluation, or screening, of these wafers prior to their use in device fabrication lines. If a clear correlation could be demonstrated between a wafer's room temperature photoluminescence properties and variations in device properties, during and after fabrication, it would be of great help in advancing the practical technology of gallium arsenide. There have been numerous publications describing the application of near-infrared absorption [7], low temperature band-edge photoluminescence [8], and deep level infrared photoluminescence [9] to the characterization and evaluation of semi-insulating GaAs. It was the intent of the present investigation to see if similar aplications could be demonstrated using band-edge photo-luminescence at room temperature. Room temperature techniques have advantages in speed, simplicity, and flexibility of apparatus over low temperature techniques.

The present work showed room temperature photoluminescence in all semi-insulating material tested except for material made semi-insulating by ion implantation. A search of the open literature failed to uncover any previous reports of measurements of band-edge photoluminescence in semi-insulating GaAs at room temperature. The present results can be attributed to the high laser power density used in our studies and to the high optical sensitivity of our monochromator system. In addition, as will be described in some detail shortly, the photoluminescence was measured immediately after detergent cleaning, which in some cases produced PL signals considerably higher than what they would have been if the sample had not been freshly cleaned.

During the initial series of measurements, strong self-quenching of the band-edge photoluminescence (867nm) was encountered. Since the magnitude of the PL signal was to be used as a measure of material quality, data had to be taken almost immediately after the sample was exposed to the excitation source. More will be said of this quenching phenomenon later. Suffice it to say here, the scanning of the focused laser beam across the surface of the sample at speed of 2mm/minute, reduced the single element (40 micrometer diameter) dwell time to a little more than one second, a time period within which quenching of the photoluminescence was generally less than 10%. The self-quenching of the photoluminescence could also be reduced by working at a lower excitation power density. At 55 watts/cm^2 in the laser spot, PL quenching was usually negligible. However, since the PL signal showed a superlinear dependence on excitation power density (a 10X reduction

in laser power lowered the PL signal by about 80X), it was very
difficult to produce meaningful PL data in some samples at low
excitation power density. As a result, all photoluminescence
measurements of semi-insulating GaAs were made at the maximum power
density to allow straight-forward comparisons in as many samples as
possible.

It was noted that the line scans showed substantial variation in
the PL signal as the scan progressed across the surface of the wafer.
Figure 2 shows a good example of the kind of structure that was
observed in an (100) oriented wafer of GaAs. This structure, as in all
of the other line scans of the photoluminescence, was reproduced
exactly upon repeated scans, the only change being the gradual
diminution of the PL signal strength over a number of scans, due to the
self quenching effect. Another feature observed in these line scans
was a significant decay of the PL signal after the sample had been
exposed to normal room ambient for some time. Again, the PL structure
was found to be exactly preserved, but the contrast was reduced, with
the high signal areas decaying more than the low signal areas. In one
sample studied in detail, the decay in signal strength of adjacent high
and low signal areas was measured to be 4.9X and 2.3X respectively,
after the sample had been allowed to remain on the x-y stage under
normal room illumination for a period of four days. Furthermore, a
brief cleaning of the sample surface with detergent solutions such as
Liquinox, or Decontam, (available from Alconox, Inc., New York, N.Y.
and Electronic Space Products, Inc., Los Angeles, CA respectively)
brought about essentially complete recovery of the original PL signal
strength. In some cases, where samples had not been detergent cleaned
after they were mounted for examination, subsequent cleaning brought
the photoluminescence signal up above that initially observed. As a
result of these observations, only those measurements that had been
made within 15 minutes after cleaning and DI H_2O rinsing, were
compared.

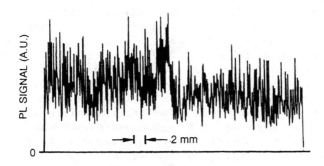

FIG. 2--Typical photoluminescence line scan at room temperature. Sample #73.

One of the obvious structures to look for in the measurement of (100) LEC wafer properties is the well known "W" distribution along a (110) diameter. This feature has been observed in dislocation etch pit counts [10], infrared absorption measurements of the EL2 level [11], sheet resistance measurements [10], bulk resistivity measurements [12], deep level photoluminescence [13], and near-band-edge photoluminescence at liquid nitrogen temperature [14]. Our studies of band-edge photoluminescence at room temperature in LEC wafers of undoped semi-insulating GaAs also yielded evidence of the "W" distribution in many of the wafers examined, a good example of which is seen in Fig. 3.

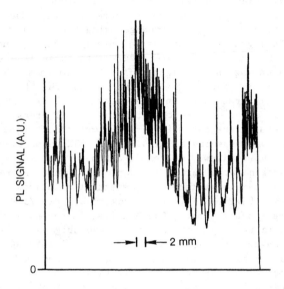

FIG. 3--The "W" distribution of the photoluminescence at room temperature in sample #84.

In the course of screening dozens of different wafers of GaAs according to their (867nm) PL line scan profiles, a variety of structures was observed which are illustrated in Fig. 4. Most semi-insulating LEC wafers had PL profiles of the type shown in Fig. 4a or in Fig. 2, where high and low signals differed by as much as an order of magnitude, and where the high and low signal areas were often separated by no more than 150 micrometers. In these profiles, a base line could sometimes be discerned and the PL structure could be described as microscopically sharp increases in photoluminescence above this baseline. Several horizontal Bridgman wafers doped with chromium, and LEC wafers doped with indium, had profiles as shown in Fig. 4b to 4d, where the PL structure could generally be described as negative signal excursions extending below a maximum base line. In some wafers of In-doped LEC gallium arsenide, large regions of the 2" diameter wafer were observed to have quite uniform PL line scans with little fine structure present (see Fig. 4b). In type (b) to (d) profiles, the extremes in PL signal were usually not more than a factor of three. It should be noted that the scale of some of the photoluminescence structure seen here is close to that recently observed in other studies of LEC gallium arsenide by X-Ray diffraction topography [15], thermally induced acoustic waves [16], cathodoluminescence [17], and chemical etching [10].

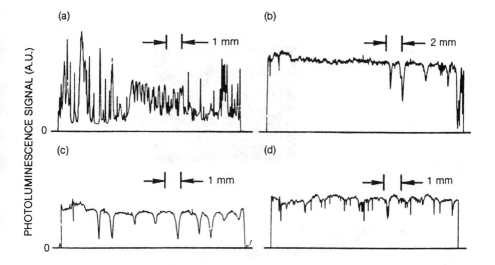

FIG. 4--Line scans of the photoluminescence in various wafers of semi-insulating GaAs, (a) sample #80, undoped LEC material, (b) sample #57 In-doped LEC material, (c) sample #88, In-doped LEC material, (d) sample HB-1, Cr-doped horizontal Bridgman material. The PL signal was amplified by 3X in (c) and (d), and by 3.5X in (b) compared to (a), so that differences in the photoluminescence structure can be more easily seen.

Photoluminescence Quenching

As pointed out above, quenching of the band edge photoluminescence in GaAs causes problems when one wants to study the spectral energy distribution of the photoluminescence or use PL signal amplitude as a measure of material quality, but there are other more interesting aspects to this phenomenon. Quenching of this band edge photoluminescence in (100) oriented GaAs at room temperature has been reported previously and was attributed to photo-enhanced oxidation of the surface [18]. The present investigation, which expands on this earlier work and provides additional useful information, also concludes that the PL quenching is due to oxidation. Two examples of quenching of the photoluminescence are shown in Fig. 5. As reported earlier by others, this quenching was persistent at least over a period of 96 hours. Also for a given excitation power density, quenching was more rapid in those areas of the sample which had the stronger PL signal. A corrolary effect observed was that at higher excitation power density (stronger PL signal) the self quenching was also more rapid. In two samples of undoped semi-insulating LEC GaAs, the quenching was 5.0X faster (sample #90) and 5.5X faster (sample #80) for a doubling of the excitation power density.

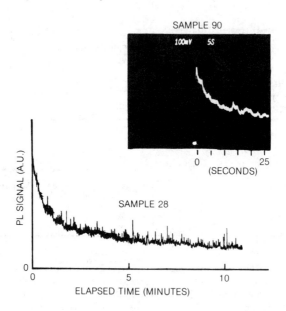

FIG. 5--Examples of self-quenching of the band-edge photoluminescence (867nm) at room temperature in undoped semi-insulating GaAs. The trace shown in the inset was recorded on an oscilloscope.

The effect of temperature on the self-quenching of the band-edge photoluminescence has not yet been investigated in detail, but a few measurements at 77°K. indicate that both the extent and the speed of PL quenching are substantially reduced at low temperature. At this time, it is not known if bringing these samples to temperatures above 300°K. would cause the recovery of the original PL properties. As mentioned before, however, a brief treatment of the surface of the sample with detergent solutions, such as Liquinox or Decontam, followed by thorough rinsing in DI H_2O, would essentially restore the original PL signal.

Besides observing self-quenching of the photoluminescence under Argon-ion and He-Ne laser illumination, quenching of the photoluminescence was also observed in freshly cleaned samples that were exposed to ordinary room illumination (about 60 microwatts/cm^2), and

even in samples that were kept in the dark. The quenching in these
circumstances was quite slow, with time constants (1/e of the original
PL signal) of hours in room light, and days in the dark, compared to
minutes under focused laser illumination. Again, the original PL
properties could be recovered by a one minute detergent treatment.

Self-quenching of band-edge photoluminescence was also observed in
a sample contained in a vacuum of 10^{-4} torr. Although this observation
seems to be at odds with a picture of oxidation induced self-quenching
of photoluminescence, gas (oxygen) adsorption rates at 10^{-4} torr are
still high enough (of the order of 100 monolayers/second) to allow
rapid formation or thickening of thin oxides. Examination of a series
of detergent cleaned wafers, using interference fringe measurements and
a surface profiler, showed that the detergent treatment mentioned above
was removing no more than 35Å of material. Keeping in mind the fact
that the region within which 86% $1-e^{-2}$ of the electron-hole pairs are
generated by 6328A radiation in GaAs exends 0.5 micrometer below the
surface, it is remarkable that such a thin layer can have such a
powerful effect in "killing" the band-gap recombination.

Recently, much work is being done using photoelectron spectroscopy
to investigate the surface chemistry of solids [19]. Depending on how
the surface chemistry of GaAs is changing in concert with the quenching
of the photoluminescence, this surface analysis technique might help us
to understand the quenching phenomenon. X-ray photoelectron
spectroscopy was used to examine a freshly cleaned GaAs wafer that had
been exposed to room ambient under normal illumination (flourescent
lighting, 60 microwatts/cm^2) for four days. Previous exposure of this
wafer under these same conditions was found to reduce the photoluminescence signal by about 5X. The same sample was analysed after
detergent cleaning with a solution of three parts water, one part
Decontam, followed by thorough DI H_2O rinsing, and drying in a stream
of dry nitrogen. The freshly cleaned wafer was transferred into the
analysis chamber within five minutes of cleaning to minimize surface
contamination build-up. The total analysis time was less than 30
minutes for each sample. The X-ray photoelectron spectroscopy was
accomplished in a vacuum of 10^{-9} torr, using 400 watt MgKa X-rays to
generate the photoelectron spectra.

Arsenic, rather than gallium, 3d-photoelectrons were studied
because they show greater energy shifts relative to changes in the
valence state. Shifts of several electron volts which are typical for
As-3d photoelectrons, are readily measured, while the Ga-3d shifts of
one electron volt or so are more difficult to resolve. Photoelectron
spectra were taken at two different electron emergence angles, 55° and
15° to yield information on the depth distribution of different
oxidation states. For a material like gallium arsenide, the mean free
path for As-3d photoelectrons is about 25Å at an angle of 55° to the
GaAs surface, and 8Å at an angle of 15° [20].

Figure 6 shows the results of these measurements. It is apparent that the surface which shows strong quenching of photoluminescence (Fig. 6b and 6d) also shows substantial difference in the X-Ray photoelectron spectrum. The peak at 44 eV is attributable to As-3 (arsenic bound to gallium in the GaAs), while the peak at 48 eV is due to As+3 (arsenic bound to oxygen). It should be noted that these spectra are not corrected for surface charging effects, which would make the true binding energies about three electron volts less than those recorded. However, the energy difference (ΔE) between As(GaAs) and As(oxide) corresponds closely to the reported ΔE for As-3 and As+3 states [21].

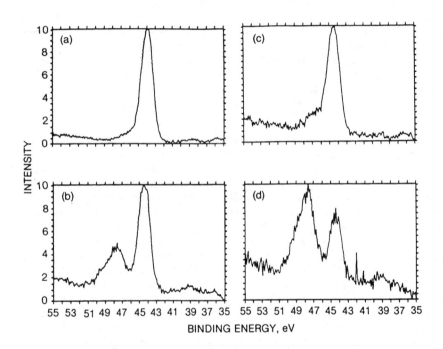

FIG. 6--X-ray photoelectron spectra of the As-3d electrons exiting the GaAs surface at two different angles, (a) freshly cleaned surface at 55°, (b) the same surface after a four day exposure to normal room ambient, (c) clean surface at 15°, (d) the same surface after four days in normal room ambient.

It appears that self-quenching of the band-gap photoluminescence in GaAs at room temperature is indeed associated with the formation or thickening of a native oxide film. According to a recent publication [22], photoelectron spectroscopy measurements at different emergence

angles can be used to deduce the thickness of a native oxide. The
expression relating oxide thickness to photoelectron spectra is:

$$\frac{I_{ox}}{I_{sub}} = \frac{D_{ox}}{D_{sub}} \cdot \frac{\lambda_{ox}}{\lambda_{sub}} \cdot \frac{1 - \exp(-d_{ox}/\lambda_{ox} \sin\theta)}{\exp(-d_{ox}/\lambda_{ox} \sin\theta)} \quad (1)$$

where I_{ox} and I_{sub} are the intensities observed for the 3d peak
corresponding to the As-O, and the As-Ga bond, respectively, D_{ox} and
D_{sub} are the densities of the oxide and the substrate, respectively,
λ_{ox} and λ_{sub} are the mean free paths of the photoelectrons in the oxide
and in the substrate, d_{ox} is the thickness of the oxide, and θ is the
emergence angle to the electron spectrometer, referred to the sample
surface. Using this expression, and assuming that the oxide density
was that of As_2O_3 (4.15 gm/cm^3), and the mean free paths in the oxide
and substrate were the same, oxide layer thicknesses of 10Å (55°
angle), and 13Å (15° angle), were calculated. These native oxide
thicknesses are somewhat less than the values that are generally taken
to apply to GaAs that has been oxidized by exposure to the atmosphere
at room temperature.

Low Temperature Photoluminescence

Measurement of the photoluminescence at low temperture is useful in
detecting and identifying various acceptor impurities in gallium
arsenide. Presently, low temperature spectra of all the common
acceptor impurities in GaAs have been catalogued as an aid in using PL
measurements for impurity analysis [23]. A number of small sections
from a variety of semi-insulating wafers were subjected to PL
measurements at 10°K to reveal the presence of acceptor impurities.
For each sample, a series of spectra were produced covering a range of
excitation power densities from 550 watts/cm^2 to 5.5 watts/cm^2 at
6328Å. In each case, the highest energy peaks increased the most with
increasing power density eventually becoming superlinear with
excitation power density. In each case, the highest energy peaks could
be assigned as donor free-to-bound transitions, and the next highest as
carbon free-to-bound transitions. In several samples, small peaks 33-
38 meV below the carbon peaks were also observed which are believed to
be phonon replicas of the carbon peak. Spectra of two samples of
undoped semi-insulating LEC GaAs that showed large differences in
carbon free-to-bound and donor free-to-bound peaks are shown in Fig. 7.
Because some samples, like sample #116, only showed appreciable donor
free-to-bound peaks at the highest excitation power density, all
comparisons of the different samples were made at a power density of
550 watts/cm^2.

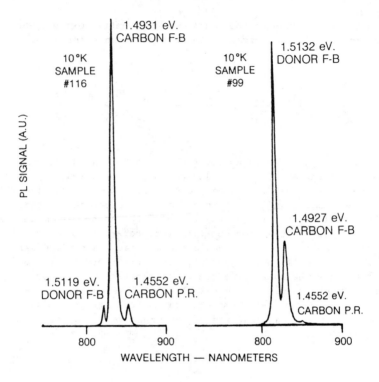

FIG. 7--Low temperature photoluminescence of two wafers of LEC semi-insulating GaAs which showed large differences in the carbon free-to-bound and donor free-to-bound photoluminescence. The PL signal of (b) has been amplified 3.3X compared to (a).

Table 2 lists the dominant PL peaks at 10°K and reveals the rather wide range of PL signal strengths that were encountered. It is to be noted that the extremes in the carbon-related PL signal listed in Table 2 differ by almost 200 for two samples whose room temperature resistivity differed by less than 2 (1.9 x 10^8 ohm-cm for sample #194 and 2.7 x 10^8 ohm-cm for sample #1-2). It should be further noted that the room temperature resistivity reflects materials properties averaged over a sample about 1.5cm square, while the photoluminescence measurement reflects materials properties in a region only 40 micrometers in diameter. A better comparison of room temperature resistivity and low temperature carbon free-to-bound photoluminescence was afforded by averaging the PL signal in line scans made at 10°K similar to previously described line scans of band-edge photoluminescence at room temperature. Comparing averages, the extremes in carbon-related PL peaks differed by 20 x for the two samples mentioned above.

TABLE 2--Comparison of dominant photoluminescence peaks in
semi-insulating GaAs under He-Ne 6328 Å excitation at
10°K.

SAMPLE	TYPE	DOPANT	PHOTOLUMINESCENCE SIGNAL (A.U.)	
			DONOR F-B	CARBON F-B
1-2	LEC	None	11.2	6.9
99	LEC	None	36.6	10.7
116	LEC	None	0.75	11.9
154	LEC	None	77	101
8-2	HB	Cr	820	220
194	LEC	None	1145	1348

<u>Aluminum Gallium Arsenide: Be-Implantation and Annealing</u>

The use of photoluminescence to monitor implantation and annealing in Be-implanted GaAs was discussed in a previous section of this paper. Using the same methods, a similar study was carried out with beryllium implants in AlGaAs. The starting material was a 1.15 micrometer thick n-type ($3 \times 10^{16}/cm^3$) layer of $Al_{0.33}Ga_{0.67}As$, grown on a 1.5cm x 2.2cm Cr-doped GaAs substrate by transient mode liquid phase epitaxy [24]. Beryllium was implanted into the AlGaAs film using four different combinations of fluence and voltage, chosen to create a layer which, according to LSS range theory, had a constant Be concentration of $1 \times 10^{19}/cm^3$ from the surface to a depth of 0.45 micrometer, dropping to $1.3 \times 10^{18}/cm^3$ at 0.65 micrometer, and tailing off gradually to one micrometer. The average Be concentration in the one micrometer thick implant was $5.7 \times 10^{18}/cm^3$.

After implantation, no photoluminescence could be detected. Examination of the as-implanted wafer by FTIR spectrometry yielded clearly discernable interference fringes. Assuming the refractive index was 3.23, that which we would normally use for $Al_{0.33}Ga_{0.67}As$, the thickness of the implanted layer calculated from the interference fringes was 0.82 micrometers. As was pointed out before, however, implantation has been found to cause 10% - 15% increases in the refractive index of semiconductors, including GaAs. Since it is likely that a similar effect occurs in AlGaAs, the correct thickness of the as-implanted layer is probably closer to 0.70 micrometer, which is in reasonable agreement with the LSS calculation.

To protect the material against arsenic loss during rapid thermal annealing, a Si_3N_4 film 0.19 micrometer thick was sputtered onto the implanted surface. The sample was then cleaved into 5mm - 7mm square pieces for separate annealing treatments. The sputter deposition, which had been carried out at 220°C for two hours was found to have produced partial annealing of the implant damage, as evidenced by a weak photoluminescence signal. (This and all subsequent PL measurements were made with the Si_3N_4 cap still in place). There was also a moderate degree of electrical activation, detected by Hall effect measurements of a 5mm square piece on which the Si_3N_4 had been stripped off. Further annealing of other pieces showed a substantial increase in the photoluminescence signal with increasing annealing temperature. Table 3 summarizes the results of this study. Based upon an integration of the LSS range profile, the degree of electrical activation is surprisingly low, about 30% at most. Unlike the Be-implants in GaAs, it is not clear whether the maximum photoluminescence signal has been realized in the annealing treatments of the Be-implanted AlGaAs. Further anneals at higher temperature are required to decide this question.

TABLE 3--Comparison of room temperature photoluminescence and electrical properties of Be-implanted $Al_xGa_{1-x}As$ (x = 0.33).

CONDITION	1.873 eV PL*	ELECTRICAL PROPERTIES
AS IMPLANTED	N.D. <0.1	—
AFTER Si_3N_4 DEPOSITION	0.9	$p = 1.2 \times 10^{17}$, $\mu = 48$, $t = 1.0 \mu m$
600°C/60 sec	237	$p = 1.6 \times 10^{18}$, $\mu = 73$, $t = 1.0 \mu m$
800°C/11 sec	310	$p = 1.1 \times 10^{18}$, $\mu = 85$, $t = 1.0 \mu m$
900°C/5 sec	960	$p = 1.3 \times 10^{18}$, $\mu = 95$, $t = 1.0 \mu m$

*Argon-ion 4880 A excitation power density was 45 Watts/cm^2

Determination of Composition

In the case of AlGaAs, one of the most common applications of photoluminescence spectroscopy is the determination of composition. Generally this is accomplished near 300°K where the relationship between aluminum content and peak wavelength of band-edge photoluminescence is well known. Recent work has demonstrated that a mole fraction accuracy of 0.003 can be achieved by this technique for aluminum content up to 0.45 molefraction [4]. At higher aluminum content, however, good correlation between photoluminescence and aluminum content has not yet been established. Part of the difficulty is the much reduced tendency of photo-generated electrons and holes to recombine via a band to band process in $Al_xGa_{1-x}As$ with x in the range 0.45 to 1.0. This difficulty can be overcome to some extent by working at higher excitation power density, since the likelihood of producing band to band radiative recombination increases with electron-hole pair density.

A series of $Al_xGa_{1-x}As$ samples were prepared with melt composition controlled to produce epitaxial layers with x ranging up to about 0.60. Photoluminescence spectra were obtained for each of these samples at room temperature, to assess the aluminum content, using either argon-ion 4880Å and He-Ne 6328Å lasers to excite the photoluminescence. Samples with composition $\overline{x} > 0.4$ usually showed weak band-edge photoluminescence at room temperature.

Auger electron spectroscopy showed that the aluminum content was uniform throughout all of the epitaxial layers that were examined in this investigation. A typical AES composition profile is shown in Fig. 8. The profile is plotted in atomic percent, so the mole percentage values are double those indicated. Sputter etching rates for these AlGaAs films were found to be quite uniform ranging from 170Å/min to 190Å/min. The transition from the AlGaAs epi layer to the GaAs substrate was observed to occur over an interval of 500Å for sample #1141, but it is likely that the sputter etching has made the interface appear less sharp than it actually is through surface texturing and ion mixing. The apparent anomaly observed in Fig. 8 for the gallium and arsenic concentrations in the GaAs substrate is the result of slight differences in the matrix which affect the sensitivity factor corrections applied to GaAs and AlGaAs. The gallium and arsenic concentrations shown in Fig. 8 are correct for the AlGaAs only.

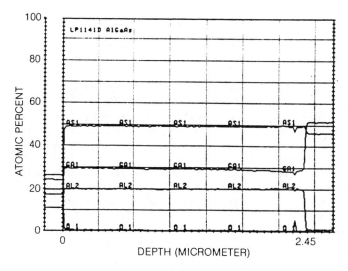

FIG. 8--Profile of AlGaAs sample #1141 determined by Auger electron spectroscopy.

In Table 4 are presented the summaries of photoluminescence data, AES and EPMA results. In all cases, the PL peaks listed are those with the highest energy. The correlation of the room temperature photoluminescence and AlGaAs composition indicates that photoluminescence due to the direct band gap minimum can be used as a measure of aluminum content up to 0.58 mole fraction aluminum arsenide. This is well into the composition range where AlGaAs is an indirect semiconductor. At high and low aluminum concentration, our results agree within 0.012 mole fraction with the photoluminescence vs. composition behavior described in a recent study of AlGaAs [4]. At intermediate concentrations, however (x = 0.40), our measurements indicated composition as much as 0.035 mole fraction higher.

TABLE 4--Photoluminescence of $Al_xGa_{1-x}As$ at 296°K and composition determined by Auger Electron Spectroscopy (AES) and Electron Microprobe Analysis (EMPA).

SAMPLE	PHOTOLUMINESCENCE PEAK (eV) 296°K	COMPOSITION (x)	
		EMPA	AES
1021	1.671	—	0.19
1129	1.764	0.28	0.26
1141	1.916	0.40	0.40
1142	2.029	0.48	0.48
1175	2.101	—	0.53
1144	2.164	0.62	0.58

SUMMARY AND CONCLUSIONS

Room temperature photoluminescence spectroscopy was used to monitor the annealing of Be-implanted GaAs and AlGaAs. The thickness of the as-implanted and electrically activated layers was determined from infrared reflectance spectra. The as-implanted samples showed no photoluminescence at all. Annealing, even at temperatures as low as 600 C produced sharp increases in photoluminescence, attending the recovery of crystallinity and the appearance of electrical activity in the implanted regions. Although more work is required to confirm a definite link between photoluminescence and deep levels in implanted GaAs-based materials, the data suggest that photoluminescence may be useful in choosing annealing conditions that will produce optimum electrical activity and minimum deep level activity.

Band-edge photoluminescence was observed at room temperature in all semi-insulating GaAs materials examined, except those made semi-insulating by ion implantation. Line scans of the room temperature band-edge photoluminescence, with spatial resolution of 40 micrometers or better, revealed strong variations in PL signals on a scale as small as 150 micrometers, indicating corresponding microscopic variations in semi-insulating behavior.

Photoluminescence measurements of semi-insulating GaAs at 10°K showed the carbon content to vary widely from sample to sample. Extremes in average carbon content differed by 20 in samples whose room temperature resistivity differed by less than 2.

Self-quenching of the band-edge photoluminescence was observed in GaAs at room temperature. Some samples displayed 6-fold reduction of the PL signal in 10 minutes. Quenching of the photoluminescence was also observed over a period of hours under normal room illumination, and over a period of days in the dark, though not to the same degree as under laser illumination. Studies of etched wafers and surface analysis by X-ray photoelectron spectroscopy indicate that these quenching effects are related to the formation and thickening of native oxide films less than 35Å thick.

Correlation of the room temperature photoluminescence and composition in AlGaAs, with composition determined by Auger electron spectroscopy, showed that photoluminescence due to the direct band gap minimum could be used as a measure of the aluminum content, well into the composition range where the alloy is an indirect semiconductor.

This work has shown that the application of optical techniques, particularly photoluminescence spectroscopy, to the evaluation of GaAs and AlGaAs, and the monitoring of fabrication processes, such as ion implantation, offers advantages of speed, simplicity and flexibility, However, more work is necessary, including systematic correlation with other methods of materials analysis, to further qualify these quick and easy to use optical methods, and assure their wider acceptance.

ACKNOWLEDGMENTS

The authors wish to acknowledge the following persons: Robert Pastorello and Jeffrey Benoit, for their skill and diligence in executing many of the measurements and sample preparations that were part of this investigation; Claire Clark, for the electron probe microanalysis; D. H. Grantham and T. W. Grudkowski for the support and encouragement that made this work possible.

REFERENCES

[1] Chatterjee, P. K., Vaidyanathan, K. V., McLevige, W. V. and Streetman, B. G., "Photoluminescence from Be-implanted GaAs", Appl. Phys. Lett., Vol. 27, #10, November 1975, pp. 567-569.

[2] Wicks, G. W., "Photoluminescence Studies of III-V Semiconductors", Ph. D. Thesis, Dept. of Applied and Engineering Physiscs, Cornell University, Ithaca, NY, 1981.

[3] Kirillov, D. and Merz J. L., "Laser Beam Heating and High Temperature Band-to-Band Luminescence of GaAs and InP", Journal of Applied Physics, Vol. 54, #7, July 1983, pp. 4104-4109.

[4] Miller, N. C., Zemon, S., Werber, G. P., and Powaznik, W. "Accurate Electron Probe Determination of Aluminum Composition in (Al, Ga)As and Correlation With the Photoluminescence Peak", Journal Appl. Phys., Vol. 57, #2, January 1985, pp. 512-515.

[5] Gibbons, J. F., Johnson, W. S. and Mylroie, S. W., Projected Range Statistics: Semiconductors and Related Materials, 2nd Edition, Halsted Press, J. Wiley and Sons Inc., 1975.

[6] Spitzer, W. G., Liou L., Wang K. W., Waddell C. N., Hubler G. and Kwun S. I., "Infrared Properties of Heavily Implanted Silicon, Germanium and Gallium Arsenide", Advanced Semiconductor Processing and Characterization of Electronic and Optical Materials, SPIE Vol. #463, 1984, pp. 46-55.

[7] Martin, G. M., Verheijke, M. L., Jansen, J. A. G. and Poiblaud, G., "Measurements of the Chromium Concentration in Semi-insulating GaAs Using Optical Absorption", Journal Appl. Phys., Vol. 50, #1, January, 1979, pp. 467-471.

[8] Mircea-Roussel, A., and Makram-Ebeid, S., "A Luminescence Band Associated With the Main Electron Trap in Bulk Gallium Arsenide", Appl. Phys. Lett., Vol. 38, #12, June 1981, pp. 1007-1009.

[9] Yu, P. W. and Walters, D. C., "Deep Photoluminescence Band Related to Oxygen in Gallium Arsenide", Appl. Phys. Lett., Vol. 41, #9, 1982, pp. 863-865.

[10] Blunt, R. T., Clark, S., Stirland, D. J., "Dislocation Density and Sheet Resistance Variations Across Semi-insulating GaAs wafers", IEEE Trans. Elec. Devices, Vol. ED-29, #7, July 1982, pp. 1039-1044.

[11] Dobrilla, P. and Blakemore, J. S., "Experimental Requirements for Quantitative Mapping of Midgap Flaw Concentration in Semi-insulating Wafers by Measurement of Near Infrared Transmittance", J. Appl. Phys., Vol. 58, #1, July 1985, pp. 208-218.

[12] Bournet, M., Visentin, N., Gouteraux, B., Lent, B. and Duchemin, J. P., " Homogeniety of LEC Semi-insulating GaAs Wafers for IC Applications", GaAs IC Symposium, 1982, pp. 54-57.

[13] Tajima, M., "Characterization of Non-uniformity in Semi-insulating LEC GaAs by Photoluminescence Spectroscopy", Japanese J. Appl. Phys., Vol. 21, L227, 1982.

[14] Kitahara, K., Ozeki, M. and Shibatomi, A., "Characterization of Uniformity in Semi-insulating GaAs by Optical and Electrical Methods", Fujitsu Scientific and Technical Journal, Vol. 19, #3, September 1983, pp. 279-302.

[15] Kitahara, K., Nakai, K. and Shibatomi, A., "One-Dimensional Photoluminescence Distribution in Semi-insulating GaAs Grown by CZ and HB Methods", J. Electrochem. Soc., Vol. 129, #4, April 1982, pp. 880-883.

[16] "The Application of Thermal-Wave Microscopy to the Study of GaAs", Technical Report 84.01, Therma Wave, Inc., Fremont, CA.

[17] Chin, A. K., Von Neida, A. R., and Caruso, R., "Spatially Resolved Cathodoluminescence Study of Semi-insulating GaAs Substrates", Journal Electrochem. Soc., Vol. 129, #10, October 1982, pp. 2386-2387.

[18] Suzuki, T. and Ogawa, M., "Degradation of Photoluminescence Intensity Caused by Excitation-Enhanced Oxidation of GaAs Surfaces", Appl. Phy. Lett., Vol. 31, #7, October 1877, pp. 473-475.

[19] Massies, J. and Countor, J. P., "Substrate Chemical Etching Prior to Molecular-beam Epitaxy: An X-Ray Photo-electron Spectroscopy Study of GaAs (001) Surfaces Etched by the $H_2SO_4-H_2O_2-H_2O$ Solution", Journal Appl. Phys., Vol. 58, #2, July 1985, pp. 806-810.

[20] Powell, C. J., "Inelastic Mean Free Paths and Attenuation Lengths of Low Energy Electrons in Solids", Scanning Electron Spectroscopy, Vol. 4, 1984, pp. 1649-1664.

[21] Muilenberg, G. E., Ed., Handbook of Photoelectron Spectroscopy, Perkin Elmer Corp., Physical Electronics Division, Eden Prairie, Minn., 1979.

[22] Carlson, T. A. and McGuire, G. E., "Study of the X-Ray Photoelectron Spectrum of Tungsten-Tungsten Oxide as a Function of Thickness of the Surface Oxide Layer", Journal of Electron Spectroscopy and Related Phenomenae, Vol. 1, 1972/1973, pp. 161-168.

[23] Ashen, D. J., Dean, P. J., Hurle, D. T. J., Mullin, J. B. and White, A. M., "The Incorporation and Characterization of Acceptors in Epitaxial GaAs", Journal Phys. Chem. Solids, Vol. 36, #10-B, 1975, pp. 1041-1053.

[24] Deitch, R. H., "Liquid - Phase Epitaxial Growth of Gallium Arsenide Under Transient Thermal Conditions", J. Crystal Growth, Vol. 7, 1970, pp. 69-73.

Fab Equipment: Automation and Reliability

Carl A. Fiorletta, Roger Lennard, James G. Harper

WAFER FAB AUTOMATION, AN INTEGRAL PART OF THE CAM ENVIRONMENT

REFERENCE: Fiorletta, C.A., Lennard, R., Harper, J.G. "Wafer FAB Automation, An Integral Part of the CAM Environment", Emerging Semiconductor Technology, ASTM STP 960, D.C. Gupta and P.H. Langer, Eds., American Society for Testing and Materials, Philadelphia, 1986

ABSTRACT: A flexible/wafer fab automation system has been developed to perform three basic functions; real time inventory control, material distribution throughout the fab; and automated loading of cassette-to-cassette process machines. This System consists of Automated Guided Vehicles (AGV's) for material transport, intelligent inventory control stations for management of material in queue, and a central computer control for System management and communications to the factory's CAM system.

The wafer fab of the future must have real time control of the manufacturing environment. This includes planning and scheduling of Work In Process, dynamic monitoring of material movement, interactive communications with the process machines and control of the automation system. This interaction permits Just In Time delivery of material to process equipment, and constant monitoring of lots in process. The benefits of an integrated system are lower manufacturing costs, lower inventory levels, shorter cycle time and higher yields.

Dr. James G. Harper is the Vice President and General Manager, Carl A. Fiorletta is Vice President and Product Manager at Veeco Integrated Automation (VIA), 10480 Markison Road, Dallas, Texas 75238. Mr. Roger Lennard is the Marketing Manager for Veeco Instruments Ltd, United Kingdom.

KEYWORDS: automated material handling, fab automation, factory communications, real time shop floor control, just in time material delivery.

INTRODUCTION

An automated material handling system for semiconductor device manufacturing without human intervention has been developed. The system is intended to automatically move material between machines which perform the actual processing of material, to load and unload those machines, and to maintain an accurate real time log of the position of the material which is being processed. The system is particularly of value in semiconductor manufacturing where the manufacturing process is not serial in nature and contains loops of flow between various machines. The system also has the capability of providing movement of material when several different material flows are being run at the same time in the processing machines.

In semiconductor manufacturing, the factory may be composed of a large number of processing machines, more than one hundred, and the material to be made into devices must be processed through most or all of the machines. Different devices require a different sequence of processing machines and many types of devices will be processed in a factory at the same time. This system will provide the automatic movement of the material and process machine loading of all of the device types at the same time.

At present, the movement of material and the loading of material onto machines is accomplished in a variety of ways. Typically it is accomplished by a human. This frequently results in errors in processing sequence and the contamination of the wafers by human particulates. In some cases, conveyors or automatic carts may move material about a factory on a pre-programmed sequence, but human intervention is required for the loading and unloading of the process machines as well as the record keeping of the physical location of the material.

The system described here can provide all of these elements of the manufacturing process automatically and it is the complete system and its benefit to the manufacturer which is the subject of this paper.

FUNCTIONAL RESPONSIBILITY OF THE VEECO SYSTEM

In an automated wafer fab, it is critical that the physical position of all lots in the factory is known in real time to the inventory management system. Lots stored in intelligent WIP Storage Stations, transported on AGV's or in process on various process equipment, all have physical positions described by X, Y, and Z points in space that provide an address in the factory for any lot location. The automation system will therefore assume the responsibility of intelligently storing work in process and moving these materials to specific points in the factory as defined by the MIS host. This responsibility will ultimately require the physical placement of lots on process equipment. The following is a brief description of each major component and how they interract to form an integrated automation system.

SYSTEM ELEMENTS

The System is composed of three major elements; a control and communication computer network, fixed material storage stations and automated guided vehicles with onboard robotics. Each of these system elements provide a portion of the requirements necessary in the manufacturing process. In the most simple form the functions provided by the elements of the system are;

Automated Guided Vehicles

This portion of the system provides the automatic movement of the material to be processed from machine to machine, and between the material storage stations. One form of the vehicle with robotics can provide material movement between the fixed WIP Storage Stations, the fixed stations having or not having robotic elements. Another form of the vehicle with robotics can move material between the fixed stations and the processing machines. At the processing machines it can load and unload the material to be processed. The material itself may be in an open wafer cassette or the cassette may be located in a storage box.

Fixed Material Storage Stations

These stations provide buffers to stage the material between processing steps at a location near the process machines. These stations may vary in size and capacity. They may have robotic elements associated with them to provide the capability to load and unload the material brought to them by the automated guided vehicles. These machines may interact with people in situations where station loading by people is desired.

Control And Communications Computer Network

This portion of the System links all of the system elements together to form the complete system. The network contains a central minicomputer, a wired method of connecting the central computer to each of the microcomputers in the fixed material storage stations, a wired method of connecting the central computer to other computers in the factory which are not a part of this System, and an optical method (infrared) of communicating to the microcomputers housed in the vehicles. The central computer and the microcomputers housed in the fixed stations and in the vehicles contain software which provides the control and communications capability to the System. The software runs the mechanical actions of the System, the wired and the non-wired communications parts of the System, the overall System coordination, scheduling and material routing, record keeping and information transmittal, and the interaction with other computers not a part of this System. A diagram of the control and communications system is shown in Figure 1.

FIGURE 1: SYSTEM OVERVIEW

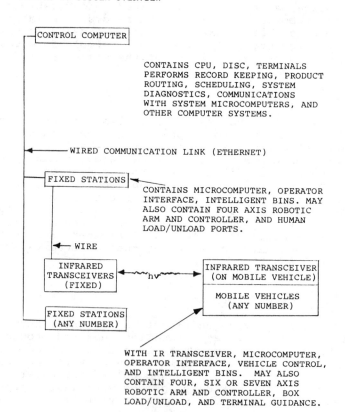

Material is moved by the mobile vehicles from fixed stations to other fixed stations or to process machines, where it is loaded on the machines for processing. Material is stored in the fixed stations between operations on the process machines if the next required process machine is busy or out of order. The distributed computer system, which encompasses both the Central Computer and the microcomputers in the vehicles and the fixed stations, keeps track of all material, the position of the material, and its required action.

SYSTEM SUB-ELEMENTS

The entire system is composed of several subelements, each of these sub-elements may be used once or many times in the total system. Frequently used sub-elements are:

Microcomputer

A 8086 based microcomputer and communications board is housed in all of the fixed stations and in all of the mobile vehicles. They provide local machine control and communications. Communication to other computers which are not a part of this system may be accomplished by an ETHERNET (TM) connection, or an RS-232C or RS-422 connection.

Non-Wired Communication Link

A method of communicating to the system's mobile vehicles which does not use wire is required for mobility. This could be accomplished by the use of a digital radio transmission and receiving system. However a typical semiconductor factory is electrically and magnetically "noisy" and the messages to be communicated are easily interrupted by this interference. This function of transmission and reception is preferably achieved by use of an infrared optical link. The link consists of electronic circuitry and infrared transceivers. The infrared optical transceiver and circuits are mounted on all mobile vehicles to provide communication with the fixed elements of the computing system. They are also mounted on the wall or ceiling of the factory and wired into the fixed stations.

Terminal Guidance

The ability of a robotic system, whether fixed or on a mobile vehicle, to place wafer cassettes on process machines requires that a positioning system control the movements of the robotics. In simple cases, the robot itself can provide enough control to accomplish this task. However, in cases where the distances of motion are large or in cases where the initial position of the robot can vary, such as in a robot mounted on a mobile vehicle, a method of final positioning must be employed. Vision systems can be employed for this control, but they are costly and slow. A method of final positioning of the machine loading robotics has been incorporated in the mobile vehicles with robotics. This system utilizes the fact that the positional accuracy of the vehicle is close to correct. The system then utilizes a small electronic camera to sight a target, calculate the deviation from the correct position and provide corrective control signals to the robotics to cause it to move to the correct position. In all cases the concept of locating a target and adjusting the robotics to the final correct position can be achieved.

The technique employed for vehicle position calibration is infrared optical. The target is a simple decal, with a printed pattern, which is placed near the cassette load platform. A simple camera sights the target decal, and determines the deviation from the correct position. Electronic circuits are used to apply the control signals to the robotics to achieve the final correct position.

Intelligent Bin

In order that the material to be processed is the correct material, it is necessary that it be correctly identified. This can be accomplished by the use of an appropriate coded decal placed on the material or on the container in which the material is contained. The computer system maintains identity between the code and the material. The code can then be read to determine the exact piece or lot of material. This system utilizes a code reader in every position that material is stored, and those positions are called Intelligent Bins. This allows the instant recognition of the exact position of the material when it is in an Intelligent Bin. These bins are employed in the fixed stations and in the mobile vehicles to allow complete knowledge of the position of all material in the factory. The method of reading the code on the material is a static bar code reader, which requires no motion to achieve the reading of the code. The coded decal, placed on the container is a special bar code which allows reading in two directions, so that the direction that the material is placed into the bin is not critical.

Operator Interface

In order that a person can transmit information to the system or can interrogate the system for information, a method of interaction between humans and the entire system must exist. Terminals linked to the system are one method this can be accomplished. An operator interface on each piece of equipment is part of this System. It consists of a display, a keyboard and control circuits which allow the communication with the microcomputers in all parts of the System. Complete diagnostics are available at the operator interface as well as the material position information.

SYSTEM BENEFITS

The automated material handling components described above provide advantages to the user. Productivity is improved. It is estimated that up to one-third of the personnel are involved in material distribution, machine loading, and logging operations. Inventory reductions are achieved by the reduction of material staging. This improves cycle time through manufacturing by as much as 50% and improves product yield. Product yield is also improved by the use of automation to handle the semiconductor material. The robotic approach to material handling reduces the number of operators in the factory. Operators are a source of contamination which causes yield loss. This equipment is far cleaner than people and therefore yield loss from people contamination is reduced.

Clean room studies conducted by Veeco and independent laboratories have concluded that clean robotics is 100 to 1,000 times cleaner than people. Where a clean mobile robot may emit less than thirty (30) particles/minute, an operator will typically emit more than 30,000 particles/minute (0.5 micron particles).

In the study of typical 64K DRAM factories, we found that 34% of the labor tasks are spent performing data logging, material distribution and machine loading functions. If these tasks are performed by the above mentioned "clean" automation, a 34% reduction of people related wafer contamination may be expected.

The economic impact of automated material handling in semiconductor device manufacturing has been discussed in the July, 1984, Solid State Technology and will be summarized here.

The automation of material handling in wafer fabrication has many benefits which reduce the cost of manufacturing semiconductor products. A list of the benefits would include:

1. Probe Yield Improvement by Clean Robotic Material Handling: As noted above, clean automation may reduce wafer contamination (from people generated particulates) by 34%.

2. Eliminate Wafer Breakage Thru Controlled Material Handling: Robotic systems handle material in a smooth, repeatable and consistent manner. This eliminates yield loss due to breakage and yield loss due to wafer damage in rough handling.

3. Short Cycle Time, Small Work-In-Process Inventory and Higher Equipment Utilization: "On Demand" material distribution to work areas and process equipment is effective in maintaining continuous material flow throughout the factory.

4. Match Factory Capacity to Market Demand: Factory automation that is interactive the the factory control system permits on-going changes in process flow priority to meet market demand for one product over another.

5. High Labor Productivity: As stated earlier, automation will perform 34% of labor tasks normally required of operators. When operators are relieved of material distribution, machine loading and data logging functions, they are free to perform other tasks that require skill, dexterity and human judgement.

6. Real Time Inventory Management: An automation system, with intelligent material storage provides real time data concerning lot location. This would permit fab management to interrogate the System for a given lot location or take physical inventory of the factory.

7. Eliminate Misprocessing: An automation system, interactive with the factory control (Host MIS) will eliminate yield loss due to misprocessing.

The benefits listed will reduce chip costs by running wafer fab manufacturing in a more efficient, clearly specified manner (by an algorithm). The benefits of automated material handling will improve the cost of wafer fabrication by the following link between benefits of the material handling system and manufacturing cost problems.

TABLE 1--Manufacturing Cost Problems

Manufacturing Problem	Automated Material Handling
Low Yield	
-Probe Yield	Reduce Contamination Level
	- Reduce total people in fab.
	- Eliminate human handling of cassettes.

Line Yield	Controlled handling to reduce breakage. Eliminate misprocessing with lot input verification.
Labor Productivity	Eliminate operator as material handler. Eliminate operator as process machine loader. Eliminate operator from data logging function.
Cycle Time	Provide "on demand" material mpvement. Move material in priority and sequence set by central factory control.
Equipment Utilization	On demand material movement to process machine. Provide lot location data in real time.

SUMMARY

A material handling system for semiconductor wafer fabrication has been developed which controls all material between process machines and loads and unloads those machines. The elements of the complete system have been discussed as well as the system sub-elements which constitute the key system functions. The benefits of automated material handling are defined and summarized. Cost reductions due to yield improvement, cycle time reduction, labor and inventory reductions can result in a payback on the investment of one-half year. This analysis, published by Veeco in the July issue of Solid State Technology is available from Veeco, 214/349-8482.

Maria S. Ligeti

COMPUTER INTEGRATED MANUFACTURING:
THE REALITIES AND HIDDEN COSTS OF AUTOMATION

REFERENCE: Ligeti, M. S., "Computer Integrated Manufacturing: The Realities and Hidden Costs of Automation," Emerging Semiconductor Technology, ASTM STP 960, D. C. Gupta and P. H. Langer, Eds., American Society for Testing and Materials, Philadelphia, 1986.

ABSTRACT: Automation of semiconductor manufacturing is no longer an option: technological and competitive pressures have made it a necessity. Increasing wafer sizes, the need for reduced particle contamination, increasingly complex process technologies, high capital equipment costs, and continuously declining selling prices are all major challenges for manufacturers. Devices must be produced not only with greater precision and higher quality, but with ongoing unit cost reductions as well. Automation appears to be the top candidate to help meet these challenges.

Automation, however, is not a magic wand. If not approached and managed carefully, it may become a very costly and disruptive exercise, instead of the beneficial one expected.

This paper covers some of the general problems and pitfalls of automation and highlights possible approaches to successful and cost effective solutions. The external and internal factors that determine the level of success also are discussed.

Finally, the recommended approach to automation and the responsibility for implementation will be described.

KEYWORDS: CIM (Computer Integrated Manufacturing), automation, global environment, Distributed system architecture, Dedicated processing, Distributed software functions, system integration.

Maria S. Ligeti is the President and CEO of Qronos Technology, Inc., 19925 Stevens Creek Blvd., Cupertino, CA 95014. The company is a supplier of integrated, distributed systems for manufacturing automation, designed for IBM hardware. Prior to forming QRONOS, she spent twelve years in the communication and semiconductor industries, holding positions in strategic planning, marketing, and information systems development at AT&T, Raychem, Intel and Seeq Technology, which she also co-founded.

WHAT IS CIM?

Computer Integrated Manufacturing, or CIM, is a comprehensive corporate wide network of people, machines and information, interconnected via computerized, electronic links and people to machine interactions. CIM allows for the closing of the gap between information and action that affects the quality of products and services by electronically proceduralizing activities, which have been historically executed by people in various organizations.

Components of CIM

Products and services dedicated to Computer Integrated Manufacturing are projected to create close to a $100 billion market by 1990 [1]. The major components of this market are:

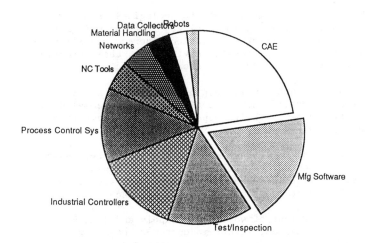

CIM Market Components: 1990

SOURCE: Technology Newsletters

CIM today

Although advances have been made in each area of CIM, it is not yet possible for a company to acquire and implement "off the shelf" automation, or create a true CIM Enterprise.

Among the missing pieces are software designed specifically for the control of manufacturing operations. Other missing links await the technological advances that will make them commercially feasible. These include artificial intelligence, expert systems for diagnostics and analysis, robotics, vision systems, and standardized, low cost data communication links.

It is reasonable to assume that CIM Enterprises will evolve over the next three to five years. Manufacturers serious about automation must plan and execute accordingly.

Planning is most effective when based on "top down", long range corporate strategies, and when implemented "bottom up" with the full participation of the operating entities. Automation must be an organizational success as well as a financial one.

What are the forces behind automation?

Some of the forces behind automation are technological, such as reduced geometries and increasing wafer sizes. But the main forces that drive CIM are improved plant utilization, increased ability to manage complexity in a mixed processing environment, cost reductions across the board, and improved customer satisfaction.

Most merchant IC manufacturers will soon be faced with the challenges described above if they expect to position themselves as strong market participants in their respective areas of expertise, and succeed financially.

To meet these challenges, to improve overall profitability, and to maximize margins for each square inch of silicon processed for a given product mix while at the same time meeting shrinking market windows, changes in operating practices must take place.

First, the dynamic management of world wide supply and demand must be established so that changes, either in customer demand or product availability, can be quickly identified and responded to.

Second, product quality must become predictable and consistent. Audit trails and interactive feedback loops among FAB, Assembly, and Test, must be established. Quality must become the concern of everyone, not just QA.

Third, engineers must bring new products or enhancements to production, and reach and maintain maximum yields, faster than ever before. Increased equipment reliability will be a key factor to success. An unprecedented degree of collaboration between design engineering and sustaining engineering will be required.

Fourth, manufacturing must be able to interact with marketing, corporate planning, and the finance functions, in a real time or close to real time fashion, if production priorities and changes are to be properly managed.

Finally, productivity improvements will have to be experienced in every job, and not just limited to the manufacturing floor. Decision making will need to approach real time. Various parts of the corporation will have to find ways to communicate effectively and quickly. Information must not only be available, but be in a usable form as well.

This will require the establishment of corporate wide, integrated decision networks, comprised of feedback loops among computers, production equipment, and people. Management will be challenged to effectively stage and orchestrate the organizational changes that automation brings about.

The future: The CIM ENTERPRISE

Manufacturing companies of the future will look and function very differently from those of today.

Automation will not necessarily "eliminate" people from the factories; not even from clean rooms. It will, with little doubt, change the skill content required for various job functions as we know them today.

Automation certainly will eliminate some positions, many of them in clerical and direct labor areas. On the other hand, new jobs, with higher skill content, will be created. This will be particularly true in the areas of equipment maintenance, engineering, process control, quality assurance, and business planning and management.

The CIM Enterprise can well be described as a manufacturing environment where the flow of information, as it affects the quality of products and services, is coordinated - from a single piece of equipment on the production floor all the way to the customer, with little or no time delay.

The backbone of such a global environment is a series of
computerized, on-line information management systems, interconnected
in a network that promotes not only the efficient collection, but
virtually instantaneous use, of data.

Such a global communication network must be as easy to use, for all
of the employees of a CIM enterprise, as today's telephone systems.
The burden of making these complex computer systems transparent must
rest with the providers of such systems, not their users.

In order to reach this goal, advancements are required in hardware,
software, communication, and machine control technologies. Most
importantly, a willingness by manufacturers to fundamentally change
operating practices must exist.

This paper focuses on a discussion of approaches to manufacturing
automation, particularly as it relates to global information
systems.

PLANNING AND IMPLEMENTING CIM

Automation is best approached by focusing on expected business
benefits, and by following realistic plans and implementation
priorities.

Approached haphazardly, automation is doomed to fail. If it is
allowed to be driven by the mavericks of an organization, each with
an individual pet project, it is more likely to create resistance,
rather than the support and "buy in" that is necessary to success.

The failure or delay of needed automation programs, caused either by
lack of commitment or cooperation, ultimately may seriously damage a
corporation's long term business viability.

Lacking proper and well communicated plans, effective automation is
easily sidetracked by isolated implementation vehicles, such as a
particular computer to collect factory floor data, or some nice
graphics, or the latest piece of process equipment.

If implemented in a vacuum, such vehicles may never yield the
expected return on investment, simply because they may be totally
incompatible with existing efforts, or the long term needs of the
corporation. CIM uses many different vehicles as discussed earlier.
Individually, they are only pieces of the puzzle. Collectively,
well orchestrated, they become the complete solution.

Computer Integrated Manufacturing is happening for business reasons,
i.e. to maximize margins, to secure a competitive market position,
and to allow long term corporate growth. Guidelines and priorities
must be established accordingly.

By proper planning, management can assure that automation makes sense for everyone, that it can be implemented at a digestible speed for the organization, and that it will yield the expected financial results.

Before automation is initiated in a company, two fundamental questions should be asked: where to start, and what are the expected benefits?

Prerequisites for success of Automation projects

Automation must be planned, and executed properly.

Planning must be global in scope, and be part of the corporation's business strategies. The plan must outline realistic business objectives and expected results. Top management's commitment to automation is a prerequisite to the success of such plans.

Execution should always start with the area of highest impact, according to corporate priorities, be phased for both financial and organizational manageability, and be consistently measured along the way.

Naturally, there are some external factors that will ensure or hinder success as well. They include the availability of needed products and services, the proper selection of vendors and their long term commitment as CIM suppliers, and the wise selection and use of standards, whenever available.

Components of a successful Long Range Automation Plan

1. Corporate business objectives and priorities

2. Information processing strategy

3. Functional areas for automation

4. Dedicated resources and responsibilities

5. Defined decision making process

6. Organizational development

7. ROI expectations and calculated risks

The long range plan must also satisfy the need for consistent future growth, the salvaging of existing investment in capital equipment already dedicated to automation, and to establish consistent and reliable hardware and software environments as much as possible.

Implementation

Based on long range guidelines, short term one or two year stepwise implementation plans can be created for each facility with the desired autonomy of execution. Such plans should be very specific and should establish the following:

1. Particular projects

2. Budgets and deliverables

3. Make versus buy position

4. Vendor selection criteria and procurement processes

5. Customization needs

6. Staffing and skill acquisition

7. Checkpoints and accountability

8. Contingency plans

9. Employee development

10. Measurement and recognition systems.

Who will make CIM HAPPEN in your company?

To implement CIM successfully, the participation and commitment of the following groups are essential both in the planning and implementation phases:

- Top management, with an economic veto power

- MIS personnel, with system technology veto power

- Technical/engineering and line personnel with usability and feasibility veto power

- Support groups such as planning, finance and cost accounting, preferably reporting to the manufacturing organization, either directly or on a dotted line.

Whether an aggressive, or more cautious, "islands of automation" approach is used, groups must work together. It is management's responsibility to ensure that alliances are formed, power struggles and finger pointing is eliminated, and precious resources are wisely spent.

The automation of an average FAB requires between $3 million and $15 million investment, depending on the level desired. If properly handled, ROI may be realized in a matter of months. Incorrectly handled, the return may never materialize.

Extremes in approaches can be observed in several companies today, ranging from MIS only decision making, to technical personnel only decision making. Often, there is already a hodge-podge of ill fitting pieces of personal favorites, with undue effort spent in trying to save them, even at the price of being "penny wise and pound foolish." Often, top management displays a "let's wait and see" attitude, allowing automation to be driven from the bottom up in a fragmented fashion.

Manufacturing automation is computer intensive. But, it is a software and communication intensive, not hardware intensive, problem. Yet, more often than not, selections are excessively biased by hardware preferences. The headaches and costs that will result from attempting to integrate poorly selected automation pieces cannot be underestimated.

WHO WILL MAKE CIM HAPPEN OUTSIDE OF YOUR COMPANY?

While "one stop shopping" of CIM solutions is not yet a reality, it is on the way. Thus, vendor selection criteria should include the considerations listed below.

Reducing the number of vendors a manufacturer has to deal with, exponentially increases the success of integrated implementation. The business goals of individual vendors, their technological competence, and their ability to survive and grow in what will be a highly competitive market, varies widely. The fewer vendors, the fewer of these variables there will be.

Conversely, the more vendors there are, the more they'll need to be managed, and the greater the adverse impact they'll have on such factors as employee training and system service.

SOME KEY CONSIDERATIONS FOR VENDOR SELECTION

Since the information systems selected will become the backbone of the whole automation effort, it is critical that they be selected with care, especially in these early stages of automation. For example:

Distributed system architecture

System architecture should be designed to shadow complex operational networks, be distributed, and be implementable in phases.

The architecture must address the multiple layers of manufacturing functions, from corporate capacity allocation, through factory floor management, to real time process control at each step of production. It also must take into consideration the geographic distribution of manufacturing and support functions, often across international boundaries.

Dedicated processing

Dedicated processing is a must, since different types of information processing requires different types of computers, i.e. mainframes, mini, and microcomputers.

Within these categories, further attention should be paid to the nature of processing required, i.e. batch, on-line, or real time. The need to increase processing capability also must be taken into consideration. Reliability and fault tolerance need to be addressed in critical areas.

Distributed software functions

As with system architecture, software also should be distributed. It should be modular as well. Distributed software can easily be customized to fit local operational needs. Analysis indicates that 80% of data generated within a given function is used locally. However, the other 20% that is "exported - imported" may be critical to the overall efficiency of manufacturing, which in turn drives the ability for the whole corporation to be competitive. The structure, interaction and integration of such global distributed software should be considered as the most critical criteria in system selection.

Detailed functions are important, but should only be considered attractive once the above criteria are satisfied. They are only the icing on the cake.

Local and Wide Area data communication

Vendors that not only follow, but establish, data communication standards are the ideal choices. Long term vendor commitment to such standards is critical, as the trend moves toward more and more decentralized manufacturing operations. As this continues, data communication will play an increasingly important role in factory automation.

System integration with existing and future components:

Newly installed systems will be required to interface with existing systems and networks, including those which may be in use by marketing, finance, engineering, and general office management organizations.

Thus, vendors should be evaluated for their willingness and ability to integrate systems already in place, even when provided by other vendors, and their ability to assume project management responsibilities, if required.

Vendor commitment to both leading edge and proven technologies

The vendor of choice should have a large portfolio of hardware, software and communication products, based on sound, accepted and proven technologies. At the same time, the vendor should demonstrate its recognition of, and commitment to, future technologies, its ability to reduce hardware costs, increase software functionality, employ more and more user friendly tools, and to continually improve data communication efficiencies.

Vendor dedication to customer service

Since vendor-client relationships in manufacturing automation tend to be long term, customer service must be scrutinized just as much as systems capabilities. Vendors should be prepared to offer the following services: pre-installation design and planning assistance; on site installation support; extensive training; system integration; assistance in ROI evaluation; and nation wide and international services.

NEW ALLIANCES AMONG VENDORS AND USERS

The future is likely to see the development of close cooperation between users and vendors of CIM, covering such relationships as joint product development, joint marketing and service arrangements, and joint efforts to standardize.

It is also expected that there will be a finite number of significant CIM suppliers. Large, traditional computer companies will become, more and more, "full service vendors". A new breed of system house will also emerge, with emphasis on dedicated software and systems integration.

Summary of basic guidelines to CIM:

1. Follow a long range strategic plan

2. Choose vendors for long term needs

3. Dedicate resources

4. Take enough time to plan, but know when to start execution

5. Clearly identify who is in charge and accountable

6. Create teamwork among management, MIS, engineering and line personnel

7. Agree to priorities and expectations BEFORE projects start

8. Get the buy in of all employees

9. Establish and communicate a fair measurement system

10. Go overboard on recognition and reward.

   Make every employee a winner.

Conclusion

CIM, correctly implemented, will result in immense financial benefits. These will be measurable by yield improvements, inventory and cycle time reductions, productivity increases in direct, and indirect labor, economies of scale for larger facilities, more effective plant utilization, and finally, significant improvements in customer service.

Those mainstream manufacturers who are not able to improve manufacturing productivity in the next three to five years will not remain competitive.

REFERENCES

[1] Technology Newsletter, Spring, 1985.

JERRY C. GREINER

INDUSTRY CONSIDERATIONS IN DETERMINING EQUIPMENT RELIABILITY

REFERENCE: GREINER, J. C. "INDUSTRY CONSIDERATIONS IN DETERMINING EQUIPMENT RELIABILITY", Emerging Semiconductor Technology, ASTM STP 960, D. C. Gupta and P. H. Langer, Ed., American Society for Testing and Materials, 1986.

ABSTRACT: Automation and Mechanization are essential to the future of our industry. The key to success is understanding not only the raw capability of each component of a system but also the capability of each component with human factors integrated. The SEMI Uptime Document (1) isolates and defines components of real factory time and formulates them in two distinct ways: with and without human interfacing. Part one, Equipment Dependent Uptime, has passed balloting and is nearing implementation as an official standard. The second, which will utilize the same components arranged in a manner to account for human interfacing, is being readied for initial balloting.

KEYWORDS: Equipment dependent uptime, comprehensive uptime, productive time, stand-by time, scheduled time, delays, unscheduled time, logistic time, failure, assist, MTBF, MTBA, MTTR, MTTA.

The ability to separate and understand the elements of time which make-up the total operational (and non-operational) environment of semiconductor processing equipment is essential to effectively planning mechanized and automated facilities. Early research by the SEMI Automation and Equipment Automation & Interfaces Subcommittee Uptime/Downtime Task Force uncovered no cohesive current standard document or definition which was readily understood by the industry. Informal surveys of users and equipment produced common elements of definition but no universal or prevalent mode of formula definition. The terms uptime, downtime, reliability, efficiency, utilization and maintainabilty had as many definitions and variations as the industry had users.

Analysis revealed that in practice there were two major general methods of defining and using uptime and downtime formulas. The first was utilized in an attempt to convey the capability of a given component or piece of equipment to perform the function for which it was designed. The second method attempted to convey the effectiveness of the same component or piece of equipment when

Jerry Greiner is a senior project engineer at Motorola Inc., 5005 E. McDowell, Phoenix, Az 85008, and Co-chairman of the SEMI Equipment Automation and Interface Subcommittee.

working in a factory environment. Analysis also revealed that most of the methods used the same common elements with only the definition of each and final formula arrangement varying. Based on these results the SEMI Automation and Equipment Automation & Interfaces Subcommittee is developing a set of standards for use by the industry. The key elements have been defined and the first section, Equipment-Dependent Uptime, completed and balloted to the SEMI Standards Committee for release as a SEMI Standard. The second section, Comprehensive Uptime, and unifying elements have reached draft form and are expected to be voted to ballot by May of 1986.

A brief overall look at the documents will better demonstrate their content.

MAJOR DEFINITIONS AND GROUND RULES

Definitions

Equipment Reliability: The capability of equipment, with human interfacing, to maintain a pre-determined level of uptime while incurring a minimum level of downtime.

Equipment-Dependent Uptime: An equipment figure of merit which measures equipment performance by extracting human delays from the performance of that equipment.

Comprehensive Uptime: A figure of merit which measures equipment performance by including the human delays to the performance of the equipment.

Rules

Scheduling: Scheduling of equipment is of utmost importance. It is the end user's responsibility to schedule equipment in and out of service in order to fully utilize the equipment effectively.

Uptime Log: In order to utilize the definitions and figures of merit included in this document it behooves the end user to maintain a standard Equipment Reliability Uptime Log.

Definitions: The definitions and figures of merit are not intended to be specifications. It is to be understood that significant differences can and will exist between fabrication areas utilizing different processes and methods of operation. They can, however, be of significant value in helping users and producers work toward a common goal.

Contract Terms: When used as a part of a procurement/sales contract the terms may be further defined with mutually agreed limits; e.g. Equipment Dependent Uptime shall be XX% with no more than Y setups per week.

GREINER ON DETERMINING EQUIPMENT RELIABILITY   675

Scope: The standards are intended as a baseline, not as an absolute final definition. Standards are strictly voluntary and are offered as an aid to common understanding in the industry.

CATEGORIES OF TIME

Specific Time Categories:

All time is to be treated as a part of one of six major categories of time:

1. Production Time
2. Stand-by Time
3. Scheduled Downtime
4. Unscheduled Downtime
5. Delays
6. Unscheduled Time

Each category is defined in general terms and examples cited of time which will fit into that category. It is not intended that the list of examples be all-encompassing or that the list represent the absolute letter of the law in interpreting time categories. Rather, one should look at the examples as a representation of the spirit of the law. One should use the general definition to assign classifications to situations not specifically mentioned.

Definition of Time Categories

Productive time: The period of time during which the equipment performs its intended function. This includes time during which the equipment may exhibit minor defects not judged by the user serious enough to discontinue production mode. It includes:

A. Regular production, including loading and unloading of the machine.
B. Work for third parties.
C. Reworks.
D. Verification runs using production material following scheduled downtime.
C. Verification runs to test new applications.

Standby time: The period of time during which the equipment is in a condition to perform its required function, and is intended to be operated, but is not operated. It includes:

A. Scheduled breaks.
B. No operator or process engineer available.
C. No orders.
D. No products available to be processed (due to delays caused by preceding machines).
E. No facilities available (electrical power, gases, vacuum etc.)
F. No materials.
G. Start up or shut down periods beyond those recommended by the vendor.
F. Initial time during which fault finding checks are performed after suspicion of failure, but no faults are found.

Scheduled downtime: The period of time during which the equipment cannot be used for production due to scheduled maintenance and set-up. It includes:

A. Periodic maintenance checks, servicing and adjustments as required or recommended by the vendor or user.
B. Standstill necessary for the change of consumables.
C. Cleaning.
D. Startup and shutdown periods at the beginning and end of work periods, if necessary for the function of the equipment.
E. Repair time for minor faults found during regular maintenance checks. These can include small repairs that were postponed until a regular check.
F. Test run for verification during scheduled downtime.
G. Set-up time: checks of changes necessary because of change of process when this is a part of a regular schedule.

Unscheduled downtime: The period of time during which the equipment cannot perform its intended function, excluding the logistic delays. Unscheduled downtime commences when user/vendor maintenance personnel are available following notification of equipment failure. It includes:

A. Initial downtime during which maintenance performs fault-finding checks after a suspicion of failure, and finds a fault.
B. Repair time.
C. Test runs for debugging and for verifications during unscheduled downtime.

Delay time: The period of time during which the equipment cannot perform its intended functions, due to delays attributed to either the user or vendor. It includes:

A. User delays:
 1. Waiting for user maintenance personnel after notification of equipment failure.
 2. Waiting for user replacement parts.
 3. User administrative delays.
B. Vendor delays:
 1. Waiting for vendor and/or third party maintenace after notification of equipment failure.
 2. Waiting for vendor and/or third party replacement parts.
 3. Vendor and/or third party administrative delays.

Unscheduled Time: The time during which the equipment is not scheduled for production or preventive maintenance. It includes:

A. Unscheduled time including off shift hours, weekends, or holidays. Any maintenance done to the equipment during unscheduled time will be counted as downtime. Any production or process testing done on the equipment during unscheduled time will be counted as uptime. All other unscheduled time shall be defined as logistic time.
B. Equipment installation, upgrade, overhaul, or retrofit.
C. Waiting for a process/maintenance engineer to release the equipment after an installation, upgrade, overhaul, or retrofit.

EQUIPMENT RELIABILITY

Components

The categories of time can be placed in three components of equipment performance definitions. They are:

<u>Uptime:</u> The time during which the equipment is capable of performing its intended function.

<u>Downtime:</u> The time during which the equipment is incapable of performing its intended function.

<u>Logistic time:</u> The period of time during which delays would affect measured equipment performance, if not substracted from total time.

Uptime

Uptime(%) is defined as uptime divided by the quantity total time less logistic time:

$$\text{UPTIME}(\%) = \frac{\text{UPTIME}}{(\text{TOTAL TIME} - \text{LOGISTIC TIME})} \times 100$$

The difference between Equipment Dependent Uptime and Comprehensive Uptime can be demonstrated as follows:

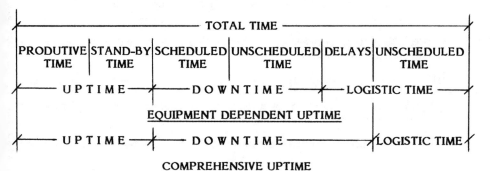

Where Equipment dependent uptime is defined as productive time plus stand-by time divided by the quantity total time less delays and unscheduled time:

$$\text{UPTIME/equip dep}(\%) = \frac{\text{PRODUCTIVE TIME} + \text{STANDBY TIME}}{\text{TOTAL TIME} - \text{DELAYS} - \text{UNSCHEDULED TIME}} \times 100$$

And Comprehensive Uptime is defined as productive time plus stand-by time divided by the quantity total time less unscheduled time:

$$\text{UPTIME/comp}(\%) = \frac{\text{PRODUCTIVE TIME} + \text{STANDBY TIME}}{\text{TOTAL TIME} - \text{UNSCHEDULED TIME}} \times 100$$

Note that for Comprehensive Uptime the delays have been included in the downtime whereas in Equipment Dependent Uptime the delays were included in the logistic time. This allows for the calculation of in-house effeciencies in the use of equipment and the pinpointing of lost equipment time. The effective use of scheduling can also remove many of the avoidable human interface delays which can cause a wide disparity in the two Uptime figures of merit.

Equipment Performance Indicators

Additional common indicators of equipment performance to separately judge each piece of equipment have been proposed in the Comprehensive Uptime Document. The indicators are not meant to be specifications but are proposed as commonly defined and understood terms to be used by vendors and customers in working toward common goals.

MTBF: Mean time between failures; productive time divided by the number of failures during that time.

MTTR: Mean time to repair; total unscheduled repair time divided by the number of failures.

MTBA: Mean time between assist; productive time divided by the number of assists during that time.

MTTA: Mean time to assist; productive time divided by the number of assist.

Failure: Any interruption of operation, due to equipment malfunction, which requires maintenance intervention.

Assist: Any interruption of operation correctable by operator intervention only.

Used in conjunction with the Uptime figures of merit the additional performance indicators can form a foundation of understanding upon which to build effective semiconductor automation and mechanization principals.

ACKNOWLEDGEMENTS

The author wishes to acknowledge with appreciation the help of Paul Davis and Jim King, of SEMI, in supplying background material for use in the preparation of this paper.

REFERENCES

(1) Document No. 1347, Semiconductor Equipment and Materials Institute, 625 Ellis Street, Suite 212, Mountain View, CA 94043, 1985

Appendixes

APPENDIX I

WORKSHOP AND PANEL DISCUSSIONS

The material presented in this appendix has been written by W. Murray Bullis for Precipitation Effects, Oxygen and Carbon in Silicon and by James R. Ehrstein for Dopant Profiling. It is based on their best recollections. The material did not go through the review process and is presented for information only.

1. WORKSHOP ON PRECIPITATION EFFECTS, OXYGEN AND CARBON IN SILICON

   Chairmen: Aslan Baghdadi, National Bureau of Standards, Gaithersburg, Maryland, and

   W. Murray Bullis, Siltec Corporation, Mountain View, California.

The workshop on Precipitation Effects, Oxygen and Carbon in Silicon covered a variety of topics with emphasis on oxygen control and measurement. There were five informal presentations, each followed by a lively discussion.

1. ASTM OXYGEN PRECIPITATION EXPERIMENT - R.B. Swaroop, Fairchild Semiconductor.

The objective of this experiment, being carried out by a task force of ASTM Subcommittee F1.04 on Semiconductor Physical Properties, is to establish a simple procedure to enable prediction of precipitation effects of silicon wafers. Six wafer suppliers provided both n- and p-type wafers which covered a wide range of oxygen content. Two thermal cycles were used in the initial phase of the work: A - 1050°C for 16 h; and B - 750°C for 4 h followed by 1050°C for 16 h. Oxygen content was measured before and after the heat treatments by three laboratories; heat treatments were carried out at two locations, and the same three laboratories measured the oxygen content after the heat treatments.

There was a greater scatter in the change of oxygen content following the A cycle; nearly all wafers, regardless of supplier, followed the same characteristic curve following the B cycle. The oxygen and oxygen change measurements showed significant laboratory-to-laboratory differences, but the characteristic shapes of the observed curves were similar. The greatest variations from one sample to another were observed to occur in the mid range of oxygen content. A more detailed report of the experiment is planned for the October 1986 meeting of The Electrochemical Society [1], and a complete publication is planned for the Spring of 1987 [2].

In the discussion which followed the presentation the following points were brought out:

a. Generally similar results were reported to have been observed at several locations.

b. The desirability of being able to use a shorter test cycle was pointed out.

c. Other, more complex, test cycles might be expected to yield somewhat different results.

d. In earlier work [3], a correlation between the results obtained following the two-step cycle (cycle B) and those following a full CMOS simulation was observed.

e. The 4-h nucleation cycle at 750°C overwhelms the prior thermal history of the wafer; if shorter times at 750°C are used, wafer-to-wafer differences in the precipitation characteristics due to thermal history would be more evident.

2. GRAND ROUND ROBIN IN OXYGEN IN SILICON - R.I. Scace, National Bureau of Standards.

The experimental phase of the world-wide experiment to determine the value and reliability of the calibration factor which relates the oxygen content in silicon (in parts per million atomic or ppma) to the infrared absorption coefficient (in cm^{-1}) at 1107 cm^{-1} was reported to be nearly complete. In the past, values differing by nearly a factor of two have been adopted in various ASTM and other standards. Recent publications report values in a much narrower range: 5.4 to 6.4 ppma/cm^{-1} .

This experiment, begun in the summer of 1985, involves representatives of all organizations which have recently reported values for this calibration factor. Its purposes are to determine the best value for this factor and the reliability with which it can be determined. In addition, a broad base of data regarding the reproducibility of infrared determinations of oxygen content is being developed. A more detailed report of the experimental design may be found elsewhere [4]; at this writing (August 1986), the analysis is not complete so the results cannot be reported. The value, when determined will be included in a revision to ASTM Test Method F 121 for Interstitial Atomic Oxygen Content of Silicon by Infrared Absorption.

The discussion following this presentation revolved around the issue of entering and analyzing all the data obtained without introducing errors in the data; consistency checks are being applied to assure the greatest possible data integrity.

3. ANALYSIS OF INFRARED SPECTRA BY A MULTI-COMPONENT LINEAR REGRESSION TECHNIQUE - R. W. Series, Royal Radar and Signals Establishment.

A curve fitting technique [5] for analyzing the infrared spectra of oxygen- and carbon-containing silicon was described. This technique can be applied to product wafers with a variety of back surface characteristics. It requires calibration by the use of standard wafers with known oxygen and carbon content and two polished surfaces.

Rough back surfaces are assumed to scatter only the tranmitted beam. If appropriately small baseline intervals are taken, interference from most overlapping precipitate bands can be avoided. For the 1107 cm^{-1} oxygen band the appropriate interval is 1130 to 1080 cm^{-1}. The 515 cm^{-1} oxygen band does not suffer from interferences from precipitate bands, but it is not suitable for use on thin samples because of interference fringes from multiple reflections. The resolution used in measuring the carbon band in thin samples must be chosen carefully to avoid too much broadening on the one hand and interference fringes on the other; a resolution of 5 cm^{-1} was found to be optimum.

Questions during the discussion brought out the fact that measurements could be made by this technique on samples with background transmittance as low as 1.0 %; measurements have been made with degraded accuracy on very low resistivity (0.03 ohm.cm) antimony-doped samples which had transmittance of 0.1 %.

4. EVIDENCE ASSOCIATING THE 515 CM^{-1} BAND WITH SUBSTITUTIONAL OXYGEN - W. C. O'Mara, Aeolus Laboratories.

Comparison of the vibration frequencies for various isotopes of oxygen, carbon, and boron were used to support the idea that the 515 cm^{-1} band is associated with substitutional rather than interstitial oxygen. The discussion was an extension of a previously published paper [6].

5. SOME OBSERVATIONS ON OXYGEN PRECIPITATION - N.Inoue Nippon Telegraph and Telephone.

Intrinsic gettering was observed to inhibit carrier lifetime degradation due to reactive ion etching. It was also emphasized that the precipitate density, not simply the change in interstitial oxygen concentration, is an important measure of precipitation. Precipitate measurable change of interstitial oxygen concentration. Precipitate densities were measured by concentration.Precipitate densities were measured by Wright etching and by transmission electron microscopy. It was found that the morphology of the precipitates depends on the ratio of the nucleation temperature to the growth temperature. A two-step low-high, cycle results in platelets, while a single high temperature treatment results in octrahedral precipitates.

REFERENCES

[1] R. Swaroop, W. Lin, N. Kim, M. Bullis, A. Rice, E. Castel, M. Christ and L. Shive, ASTM Oxygen Precipitation Study, to be presented at the Electrochemical Society Meeting, San Diego, California, October, 1986.

[2] -----, to be submitted to Solid State Technology.

[3] H-D.Chiou and L.W.Shive, Test Methods for Oxygen Precipitation in Silicon, VLSI Science and Technology/1985, W.M.Bullis and S.Broydo, Eds., ECS Proceedings, Vol. 85-5, pp. 429-435.

[4] W.M.Bullis, M.Watanabe, A.Baghdadi, Li Yue-zhen, R.I.Scace, R.W.Series, and P.Stallhofer, Calibration of Infrared Absorption Measurements of Interstitial Oxygen Concentration in Silicon, Semiconductor Silicon 1986, H.R.Huff, T.Abe, and B.Kolbesen, eds., ECS Proceedings, Vol. 86-4, pp. 160.

[5] R.W.Series and F.M.Livingston, Measurement of Interstitial Oxygen and Substitutional Carbon in Silicon, Presented at The Electrochemical Society Meeting, Boston, Massachusetts, May 1986.

[6] W.C.O'Mara, Oxygen in Silicon, Defects in Silicon, W.M.Bullis and L.C.Kimberling, eds., ECS Proceedings, Vol. 83-9, pp. 120 - 129.

2. WORKSHOP ON DOPANT PROFILING

Chairmen: James R. Ehrstein, National Bureau of Standards, Gaithersburg, Maryland, and

Marek Pawlik, GEC Research Wembley, Middlesex, U.K.

The intended goals of this workshop were to identify the strong and weak points of the available profiling techniques and their relative suitability for various applications. The workshop was structured around a guided discussion which relied heavily on input from the audience as well as the more traditional exchange between the audience and the speakers from the technical sessions of the symposium. The discussion was guided toward obtaining answers to the stated goals by identifying the profiling applications and needs of the audience as well as some of the difficulties and uncertainties that have been experienced. While reasonable success was obtained in identifying relative strong and weak points of the various techniques, developing a definitive statement of suitability for various applications proved elusive; enough considerations were presented to allow most of the audience to make better informed choices in their future work.

Among the strong and weak points of the various profiling techniques, some of which were widely recognized, others not so well appreciated, are the following:

1. SIMS

This technique has been demonstrated to be amenable to very good precision both within and between laboratories, it does not require electrical activation and can therefore be applied both before and after annealing of ion implants to monitor diffusion, it provides an analysis of the dopant distribution which is of primary interest in process modeling, it has reasonable dynamic range for all common dopants; however, the dynamic range obtainable is a function of dopant species - only for the case of boron is it likely to yield data for a complete profile, the primary limitation on SIMS analysis is likely to be due to sputter crater nonuniformity, sputter redeposition or kinematic knock-on can cause profile shape distortion although this appears to be a negligible effect on boron, quantification of the dopant profile from raw data may be limited by inavailability of suitable reference standards or by inaccuracies of ion implant dose values when such dose values are used to normalize SIMS data, finally, obtaining meaningful values of dopant density at the silicon surface is difficult for some SIMS instruments when there is an oxide cap over the silicon.

2. RUTHERFORD BACKSCATTER

It is a nondestructive measurement and analysis is possible from first principles without use of a reference specimen, like SIMS, it responds to dopant density; however, it is primarily applicable to dopants with a noticeably different atomic mass than silicon and even for these it is limited to fairly high concentrations.

3. SPREADING RESISTANCE AND INCREMENTAL SHEET RESISTANCE

Dynamic range does not depend on dopant species, incremental sheet resistance can profile nearly to a junction while spreading resistance can profile through one or more junctions, both provide information on electrical activation which is an important consideration for novel process development, both can give evidence of electrically active material defects although the interpretation may not be straightforward; both techniques average over a depth below the point of measurement and hence the data must be deconvolved in order to calculate a depth distribution - the deconvolution procedure is relatively straightforward for incremental sheet resistance and relatively complicated for spreading resistance, the profiles directly derived from these measurements are for resistivity or conductivity - as a result the free carrier density profiles cannot be uniquely extracted unless the carrier mobility function is reasonably well known - this may not be the case for novel fabrication procedures, the deconvolution procedures make these techniques highly sensitive to measurement noise, for submicron structures for free carrier concentrations may be displaced from the dopant distribution due to carrier spilling particularly in the tail of the profile - it does not appear possible to work backwards in a unique fashion to calculate the dopant distribution from the carrier distribution.

4. CAPACITANCE - VOLTAGE MEASUREMENTS

This technique is without parallel for profiling lightly doped material, diode-, MOS-, and Schottky-CV measurements allow compatibility with virtually any process being developed; however, it has limited dynamic range due to voltage breakdown, diode and Schottky structures will not allow profiling close to the surface, MOS structures will allow profiling near the surface but results may be degraded by oxide and interface states; for best accuracy, a number of corrections may have to be applied.

DISCUSSION

All these techniques suffer from the limitation of being vertical profilers - none is capable of determining lateral spread of dopants under a mask edge - an area of increasing importance.

An important output of the discussion was the recognition that while precision is the primary requirement for any of the techniques to be acceptable, an understanding of the underlying accuracy of the various techniques as well as the procedures necessary for realizing that accuracy is increasingly important - primarily for process development applications, but also for process control. Discussion of measurement accuracy occupied a significant portion of the workshop. It was noted that, while progress has been made on all techniques, assessing the accuracy of any of them is difficult since each measures a somewhat different characteristic of the dopant profile.

Comparison studies which have attempted to study the relation between different measurements have generally confirmed the theoretical prediction that carrier spilling will cause the dopant profile determined by SIMS to have a deeper tail than the carrier profile determined by spreading resistance. Yet the several reported studies of this type give somewhat different quantitative comparisons. How much of this variability is due to the quality of the measurements and how much is due to differences in the specimens used is not known. One set of boron implant data used to illustrate the results of several profiling techniques resulted in an interesting comment. One researcher in the audience observed that he originally had done most of his original profiling work by CV technique, but now he had become convinced that SIMS was the technique of choice for accuracy; however, for the data being presented, he considered the spreading resistance, not SIMS, profiles to be "correct", because the spreading resistance gave the "right" shape for the profile tail. Another person pointed out that it may be misleading to attempt to verify the accuracy of profile measurements by comparison with the predictions of a processing model such as SUPREM since SUPREM II is heavily tied to spreading resistance measurements as an input, while SUPREM III is heavily tied to SIMS measurements. Evaluation of profiling measurements was thus seen to be somewhat circular. Since there is no fully independent starting point, the physics of the measurements and that of the sample fabrication are intertwined. It is tempting to evaluate the quality of the measurement by what is thought to be known from the models for such processes as implantation, annealing and diffusion; but the models rely on measurements for verification and refinement.

CONCLUSIONS

The conclusions of the workshop were that certain limitations are known for each of the techniques; while some may be minimized by improvements in technique or theory, others, such as carrier-spilling and debye-length limitations, are fundamental. The available techniques were seen to be complementary, rather than competing. Adequate attention to the known limitations may allow a single technique to be used satisfactorily for process control applications. However, the use of two or more techniques will continue to be necessary to improve the understanding of the physics of sample processing steps as well as the physics of the measurements themselves.

APPENDIX II

GRADUATE EDUCATION FOR THE ELECTRONICS INDUSTRY

Pat Hill Hubbard

[The material presented in this appendix did not go through the review process and is presented for information only. It was prepared by Dinesh C. Gupta, Co-Editor from the recordings of the presentation and the excerpts provided by P. Hubbard.]

American Electronics Association, four years ago, launched a program, through its Electronics Education Foundation, to improve the faculty shortages in electrical and computer engineering - viewed as a major bottleneck to a steady and sufficient supply of graduates. Determining that the faculty vacancies were primarily caused by the fact that U.S. citizens with BS degrees prefer industry salaries rather than four to five years of costly doctoral study for an end goal of an academic career with uncompetitive teaching salaries, the Foundation began a national program to make a doctoral study and academic careers more attractive.

A cornerstone to these efforts is the fellowship-loan program. American Electronic Association [AEA] companies are currently supporting 136 U.S. citizens who want to get Ph.D.s and teach - each at a four-year cost to a company of $76,000 at a private and $52,000 at a public University. Nationally, latest data on engineering faculty showed a faculty vacancy rate of 8.5 % [1983] - an alarming figure since enrollments have increased 100 percent over the past decade and new faculty only 10 percent. California is a prime producer of electrical and computer engineers but it produces only one-fourth EE/CE Ph.Ds. annually. This information was the result of a survey of all California public and private colleges which offer EE/CE programs. Other interesting data from this survey are given below.

Pat Hill Hubbard is the Vice President of American Electronics Association, 2670 Hanover Street, Palo Alto, California. She is also the Founder and President of the Association's Electronics Education Foundation.

o In 1984, California ranked first among the states in numbers of graduating BS in EE and CE.

o California's engineering faculty vacancy rate in 1985 was 22 percent or 470 of 2160 full-time equivalent engineering faculty [FTEF] positions were vacant.

o Seventy-seven percent or 362 FTEF vacancies in California engineering schools are filled by 1286 part-time faculty, leaving 23 percent of the remaining authorized FTEF unfilled.

o California engineering schools will lose 178 full-time faculty members or 10.5 percent of the 1690 current full-time faculty over the next five years. About 83 percent of these will retire.

o In the last three years, for every 2 full-time tenure track engineering faculty which have been hired in California, one faculty member has been lost, mostly through retirement. The state is expected to lose, in the next five years, about 11 percent of its full-time engineering faculty members, again, mostly through retirement.

o By 1990, California engineering schools will need to hire 648 new or 30 percent of their current full-time faculty equivalent, just to maintain today's enrollment and the highest level of quality education.

The need to have an adequate supply of quality engineers and sufficient faculty to educate them is considered of paramount importance to the health of the U.S. high tech industry and to the economic health of the nation.

Where is our future faculty to come from?

A major cause of the faculty shortage is that U.S. citizens with B.S. degrees avoid Ph.D. programs and teaching careers in favor of industry salaries. Foreign-born Ph.Ds. remain the major faculty hiring source. Nationally, 39 percent of all engineering Ph.Ds. awarded in the U.S. in 1983-84 went to foreign nationals (all non-U.S. citizens except permanent resident aliens). More than one-fourth (26 percent) of all assistant engineering professors did not receive B.S. degrees from a U.S. university. There are a few other interesting facts. 1. In 1984, one-half to three-quarters of engineering job applicants at Rose-Hulman Institute of Technology were foreigners. 2. Forty-nine percent of the new tenure tract faculty hired in 1983 at the nine California State University engineering departments in the last five years were foreigners and fifty-six percent of electrical and computer engineering faculty new hires were foreigners. 3. Thirty-six percent or 115 of the 322 newly hired full-time faculty over the last three years graduated from a foreign high school or college. While the hiring of foreign nationals may be controversial because foreigners may not communicate well in the classroom due to their English-accent, engineering deans are often faced with a contemporary Hobson's Choice that means either filling a vacancy or leave it as a vacancy.

Nationally, only 693 Ph.Ds. in electrical engineering were awarded by U.S. universities in 1984. This is seven fewer than were awarded in 1974. Sixty fewer computer engineering Ph.Ds. were awarded in 1984 than in 1979, although total CE enrollment doubled during this period. Seventy two percent of U.S. engineering deans indicated the most significant problem in attracting U.S. BS degree graduates to doctoral study was the insufficient funding for graduate student support.

To date, AEA companies have pledged $10 million to support 136 fellowship-loans for U.S. citizens around the country. The program that American Electronics Association launched in order to meet and overcome faculty shortage problem is doing very well.

Subject Index

A

Acceptor compensator, 313
Admittance, metal oxide semiconductor (MOS), 381
Aging of silicon, 313
AlGaAs qualification techniques, 628
Aluminum etching, 250
Amorphization, 108
Amorphous silicon, 241
Arsenic, 521
ASTM Standards, 15
  F 121: 367
  F 123-83: 372
  F 523: 282
  F 723: 474
Atomic
  distributions, 480
  profiling, 535
Auger Electron Spectroscopy, 119, 178, 628
Autodoping, 21, 33
Automated guided vehicles (AGVs), 653
Automated production processes
  automatic resistivity tester (ART), 602
  computer-aided process optimization, 246
  computer-assisted manufacturing (CAM), 653
  computer-integrated manufacturing (CIM), 662
  equipment reliability, 673
  material handling, 653
  particle and material control system, 414
  wafer fabrication automation, 653
Automatic resistivity tester (ART), 602

B

Backscatter, Rutherford, 119, 688
Backsurface damage, 324
Barrel reactor, 79
Beryllium implants, 628
Bonding structure, 173
Boron, 204, 521, 573
BPSG, 615
Buried layers, 21

C

C_2ClF_5, 204
Carbon, in silicon, 365, 683
Carrier distributions, 480, 502, 558
Channel mobility, 21
Charged particle activation analysis, 365
Chemical vapor deposition (CVD)
  bonding structure, 173
  cold wall, 35
  low pressure, 173
  low temperature, 24, 43
  plasma enhanced, 21, 173
  reactor design, 129
  thin epitaxial silicon deposition, 33
  uniformity, 129
Chlorinated gases, 204
Clean room technology, 414, 423
CMOS technology
  epitaxial quality, 51
  silicon epitaxial growth on N+ substrate, 65
Comprehensive uptime, 673
Computer-aided process optimization, 246
Computer-assisted manufacturing (CAM), 653
Computer-integrated manufacturing (CIM), 662
Conductivity, 480
Contamination
  carbon, 365
  control, 336, 414, 423
  hydrogen, 313
  measurement, 119, 365
  oxygen, 365
  process gases, 436
  reactive ion etching damage, 163
Conversion coefficient, 365

696 EMERGING SEMICONDUCTOR TECHNOLOGY

Correction factors, 480, 502
CV, electrochemical, 558

D

Damage
  backsurface, 324
  control, 163
  displacement, 535
  from reactive ion etching, 163
  ingot-to-wafer processing, 297
  plasma, 220
  radiation, 381
Dedicated processing, 662
Defects, 281, 336, 381
Deposition uniformity, 129
Depth profiling, 453, 521
Design rules, 404
Device sensitivity, 393
Diameter scans, 598
Dielectrics, 137, 173, 220, 336
Diffusion, doped oxide spin-on source, 95
Displacement damage, 535
Distributed software functions and system architecture, 662
Doping and dopants
  autodoping, 21
  carrier distributions, 480, 502, 558
  cross-contamination, 119
  implant, 204
  profiling (See Profile control and analysis)
  spreading resistance (See Spreading resistance measurements)

E

Edge profile, 204
Education for the electronics industry, 691
Electrochemical CV, 558
Emission endpoint, plasma etch, 190
Epitaxial films
  growth on N+ substrate, 65, 79
  layers, 586, 598
  quality, 51
  silicon, 79, 281
  surface defects, 281
Epitaxy
  chemical vapor deposition (CVD), 33
  HCL etching, 79
  layers, 586
  low-temperature deposition, 24
  molecular beam, 558
  plasma-enhanced CVD, 21
  quality, 51
  silicon-on-sapphire, 33
  surface defects, 281
Equipment-dependent uptime, 673
Error propagation, 393
ESCA analysis, 313
Etching
  aluminum, 250
  anisotropy, 79
  high-temperature vapor, 79
  in situ HCl, 79
  metal dry, 250
  palladium, 273
  plasma, 220
  plasma etch emission endpoint, 190
  rate, 79
  reactive ion, 220, 250
  uniformity, 190
Exposure ambient, 257

F

Fabrication automation (See Automated production processes)
Factory communications, 653
Films (See also Epitaxial films)
  palladium, 266
  silicon dielectric, 173
  silicon dioxide breakdown, 336
Filtration, 436
Finite element method, 480
Fluorinated gases, 204
Four-point probe, 586
Fourier transform infrared spectroscopy (FT-IR), 353
Fracture, ingot, 297

INDEX **697**

Free carrier absorption, 353
Freon 115, 204

G

Gallium arsenide, 220, 628, 643
Gases, process, 436
Gettering
 effects on epitaxial quality, 51
 impurities, 324, 336
 internal, 8, 65
 intrinsic, 51, 353, 686
Global environment, 662

H

Haze, 79, 313
HCl etching, 79
Hydrogen analysis, 313

I

Implantation, 453
 beryllium, 628
 cross-contamination levels produced, 119
 knock-on, 535
 profile control of plasma-etched polysilicon, 204
Infrared absorption spectroscopy, 353, 628, 684
Ingot processing damage, 297
Insulators, 137
Integration, system, 662
Interface defects, 381
Interstitial oxygen measurements, 353
Ion beam implantation, 453
 cross-contamination levels produced, 119
 damage, 150
 into same-conductivity-type substrates, 586
 into 200-mm wafers, 598
 knock-on, 535
 low-dose, 586
 nitridation, 150

J

Junction
 breakdown voltage, 51
 depth, 573
Junction formation techniques, 95, 266
Just-in-time material delivery, 653

K

Knock-on, 535

L

Laplace's equation, 480
Laser ionization mass analysis (LIMA), 324
Laser scanning, 281
Latch-up, 51
Lithography, X-ray, 257
Loading effect, 190
Logistic time, 673
Low-pressure deposition, 21
Low-temperature deposition, 21, 24

M

Manufacturing (See also Automated production processes)
 computer-assisted manufacturing (CAM), 653
 computer-integrated manufacturing (CIM), 662
 equipment reliability, 673
 mechanical processing, 297
 particle and material control, 414, 423
 standards, 15
 variability, 393
 wafer fabrication automation, 653
Maps, three-dimensional (contour), 598
Materials (See also Gallium arsenide; Silicon)
 contamination measurements, 119

Materials (Continued)
  control, 414
  defects, 54, 65, 281
Mean time between assists (MTBA), 673
Mean time between failures (MTBF), 673
Mean time to assist (MTTA), 673
Mean time to repair (MTTR), 673
Mechanical processing, 297
Metal oxide semiconductor (MOS) devices, 150, 336, 381
Microdefects, 65
Mobility, 150
Modeling
  process, 573
  yield, 404
Molecular beam epitaxy, 558
Monte Carlo method, 190, 535

O

Optical analysis techniques, 628
Optically activated states, 381
Oxide layer analysis, 615
Oxygen, in silicon, 353, 365, 683

P

Palladium silicide-silicon contacts, 266
Parametric variability, 393
Pareto analysis, 393
Particulate control, 414, 423, 436
Phosphorus, 204
Photoconductivity, transient, 241
Photoluminescence quenching, 628
Photoresists, 250
Plasma technology
  deposition of amorphous silicon, 241
  etch emission endpoint, 190
  etching, 190, 220, 423
  plasma damage, 220

plasma-enhanced chemical vapor deposition, 21, 173
  processes, table, 223
  processing of dielectric layers, 220
  profile control of etching, 204
  quality control during plasma deposition, 241
  reactive ion etching (RIE), 163, 220, 250
Poisson's equation, 480
Polysilicon, 204
Precipitation effects, oxygen and carbon in silicon, 683
Process simulation, 404, 573
Processing
  dedicated, 662
  mechanical, 297
Production processes, automated, 241
Productive time, 673
Profile control and analysis
  dopant, 95, 573, 687
  electrochemical CV, 558
  impurity profiles, 521
  of plasma-etched polysilicon, 204
  primary kinematic knock-on in SIMS, 535
  sampling volume correction factors, 502
  spreading resistance, 453, 480, 558
PSG deposition, 204, 615

Q

Quality control
  during plasma deposition, 241
  using laser scanning, 281
Quenching, photoluminescence, 628

R

Radiation
  damage, 381
  deep ultraviolet, 250

INDEX 699

Reactive ion etching
  damage, 163
  on gallium arsenide ICs, 220
  UV radiation, effects of, 250
Reactor design and productivity, 129
Real-time shop floor control, 653
Recoil implantation, 535
Reduced pressure, 33
Reliability of manufacturing equipment, 673
Resistance
  sheet, 95, 598
  spreading (See Spreading resistance measurements)
Resistivity mapping, 586, 598
Rutherford Backscattering Spectroscopy, 119, 688

S

Sampling volume correction factors, 502
  Berkowitz-Lux algorithm, 510
  Local slope technique, 508
  PROF algorithm, 510
  Schumann-Gardner Theory, 505
Scans, diameter, 598
Scheduled time, 673
Secondary ion mass spectrometry (SIMS), 521, 535, 573, 687
Semiconductor Equipment and Materials Institute (SEMI) Standards, 16
Semiconductors
  business forecasts, 7
  equations, 480
  free carrier absorption, 353
  interstitial oxygen measurements, 353
  materials
    gallium arsenide, 220, 628, 643
    silicon (See Silicon)

silicon-on-sapphire, 33
standards (See Standards)
yield enhancement, 404, 423
Sensitivity, device, 393
SF_6, 204
Sheet resistivity, 95, 598, 688
Shot noise, 257
Silicon
  aging, 313
  amorphous, 241
  carbon contamination, 365, 683
  contamination control, 336
  damage, 297
  defects, 281, 336, 381
  dielectrics, 173
  dioxide, 336, 381
  doped oxide source diffusion, 95
  electrochemical CV, 558
  epitaxy
    chemical vapor deposition (CVD), 33
    HCL etching, 79
    layers, 586
    low-temperature deposition, 24
    molecular beam, 558
    plasma-enhanced CVD, 21
    quality, 51
    silicon-on-sapphire, 33
    surface defects, 281
  gettered impurities, 324
  haze generation, 313
  hydrogen contamination, 313
  impurity profiling and measurement, 119, 365, 521
  infrared absorption, 353, 684
  local oxidation of, 150
  mapping, 586
  moisture effects on, 313
  n-type, 353
  oxygen contamination, 365, 683
  oxynitride, 173
  p-type, 353
  process control, 586
  profile control of etching, 204

Silicon (Continued)
  reactive ion etching damage, 163
  spreading resistance (See Spreading resistance measurements)
  X-ray fluorescence analysis, 615
SF_6, 204
Silicon-oxide interface, 381
Software functions, distributed, 662
Solar cells, 297
Source diffusion, doped oxide spin-on, 95
Spreading resistance measurements
  boron profile analysis, 573
  carrier density calculations, 480
  comparison with electrochemical CV, 558
  comparison with four-point-probe measurements, 586
  comparison with secondary ion mass spectrometry, 521
  correction factors, 480, 502
  incremental sheet resistance, 688
  multilayer Laplace equation analysis, 480
  overview, 453
  shallow xenon implantation, 108
Sputtering, 137, 535
Standard mechanical interface (SMIF), 414
Standards
  applications, 15
  ASTM F 121: 367
  ASTM F 123-83: 372
  ASTM F 523: 282
  ASTM F 723: 474
  development organizations, 15
  international cooperation in, 15
Standby time, 673

Statistical modeling, 393
SUPREM computer modeling program, 526, 573, 689
Surface
  analysis techniques, 628
  cleaning, 24, 288
  defects, 281
  roughness, 150
System
  architecture, distributed, 662
  integration, 662
  moments, generation of, 393

T

Thin films, 137
Transient photoconductivity, 241
Transistors, MOS, 150, 336, 381
Transition width, 33
Trend analysis, 598
TRIM Monte Carlo code, 538

U

Ultraviolet radiation, 250
Undercut, 204
Uniformity of plasma etch, 190
Unscheduled time, 673
Uptime
  comprehensive, 673
  equipment dependent, 673

V

Variability, parametric, 393
Vertical reactor, 33
VLSI (very large scale integration) (See also Wafers)
  contacts, 266
  gases, particulate control in, 436
  processing, 8, 266, 414

W

Wafers
  cracks, 297
  inspection, 281
  large diameter, 598
  processing, 598
  thin-layer analysis by X-ray fluorescence, 615
  uniformity, 598
  zone engineering, 8

X

X-ray
  fluorescence analysis, 615
  lithography, 257
  resist, 257
Xenon ionic implantation, 108

Y

Yield analysis and enhancement, 404, 423

Author Index

A

Abernathey, J. R., 173
Abraham, T., 204
Albers, J., 480, 535
Alvarez, T. R., 521
Arai, T., 365, 615
Arst, M. C., 324

B

Baghdadi, A., 353
Banke, G. W., Jr., 573
Bawolek, E. J., 190
Beck, C. H., 404
Beck, G., 241
Berak, J. M., 628
Black, J. F., 628
Borland, J. O., 51
Brain, M. D., 414
Bruel, M., 108
Bullis, W. M., 683

C

Casper, N. D., 423
Chang, H.-R., 24
Chin, B. L., 250
Cockrill, J. R., 220

D

Davidson, J. M., 436
Du Port de Pontcharra, J., 108
Dyer, L. D., 297

E

Ehrstein, J. R., 453, 687
Endo, K., 365

F

Fiorletta, C. A., 653
Fisher, S. M., 33
Fonash, S. J., 163
Fridmann, S. A., 173

G

Gibson, M. L., 173
Gladden, W. K., 353
Greiner, J. C., 673
Groves, R. D., 558
Gupta, D. C., vii, 1

H

Hahn, S., 51
Hammond, M. L., 33
Harper, J. G., 653
Hubbard, P. H., 691

I

Inenaga, S., 313
Inoue, N., 365

J

Johnson, W. H., 598

K

Kar, S., 381
Keenan, W. A., 598
Kirby, B. J., 119
Kubiak, R. A., 558
Kunst, M., 241
Kuppers, U., 241

L

Langer, P. H., vii
Larson, L. A., 119
Lee, H.-S., 150
Lee, S. C., 250
Lennard, R., 653
Leong, W. Y., 558
Liaw, H. M., 281
Ligeti, M. S., 662
Lora-Tamayo, E., 108

M

Maass, E. C., 393
Martin, D. W., 137
Mazur, R. G., 586
Medernach, J. W., 79
Middleman, S., 129
Mizuma, K., 365
Murai, K., 336

N

Nguyen, H. T., 281
Nguyen, S. V., 173
Nozaki, T., 365

O

Ochiai, T., 615

P

Parker, E. H. C., 558
Pawlik, M., 502, 558
Peterson, G. G., 628
Port de Pontcharra, J. Du, 108

R

Ramamurthy, V., 95
Reif, R., 21
Roberts, S., 137
Rohatgi, A., 163
Rosczak, J. S., 24
Rose, J. W., 281
Ruane, T. P., 436
Ryan, J. G., 137

S

Sandler, N. P., 33
Sano, M., 615
Scace, R. I., 15
Shiraiwa, T., 313, 615
Singh, R. N., 266
Slusser, G. J., 573
Smith, A. K., 598
Soren, B. W., 423
Springgate, J. E., 7
Starov, V., 257
Suga, H., 336
Swaroop, R. B., 65
Sweeney, G. G., 521

T

Tada, Y., 615
Tewari, M., 381
Theriault, R., 204
Tributsch, H., 241
Turner, J. A., 220

V

Vanner, K. C., 220
Varahramyan, K., 573

W

Wells, V. A., 79
Werner, A., 241
Wong, C.-C. D., 51